REAL-TIME DATA HANDLING AND PROCESS CONTROL – II

DEUXIEME SYMPOSIUM EUROPEEN SUR

L'INFORMATIQUE EN TEMPS REEL ET LE CONTROLE DE PROCESSUS

Pratiques Communes - Etat de l'Art et Spécifications Futures

Versailles, France, 3-5 Novembre 1982

ZWEITE EUROPÄISCHES SYMPOSIUM ÜBER

ECHTZEIT DATENVERARBEITUNG UND PROZESSSTEUERUNG

Einheitliche Richtlinien - Gegenwärtiger Stand und zükunftige Erfordernisse

Versailles, Frankreich, 3-5 November 1982

SECOND EUROPEAN SYMPOSIUM ON

REAL-TIME DATA HANDLING AND PROCESS CONTROL

Common Practices - Status and Future Requirements

Versailles, France, 3-5 November 1982

Patronné par / Sponsored by / Unterstützt durch

INSTITUT NATIONAL DE RECHERCHE EN INFORMATIQUE ET EN AUTOMATIQUE (INRIA)
BUREAU D'ORIENTATION DE LA NORMALISATION EN INFORMATIQUE (BNI)
COMMISSION OF THE EUROPEAN COMMUNITIES

Comité International d'Organisation / International Organising Committee / Internationaler Veranstaltungs Ausschuss

E. G. KINGHAM, U.K. - ECA (Chairman)

Members:

W. ATTWENGER, Austria	M. LAGASSE, France
G. BULL, U.K.	N. E. MALAGARDIS, France
P. CHRISTENSEN, Denmark	H. MEYER, Belgium - CEC
P. GALLICE, France	K. D. MÜLLER, Fed. Rep. Germany
V. HAASE, Austria	K. THOMPSON, Belgium - CEC
D. JONES, U.K.	K. ZANDER, Fed. Rep. Germany

Comité Local d'Organisation / Local Organising Committee / Lokaler Veranstaltungs Ausschuss

N. E. MALAGARDIS (Chairman)

Members:

J. P. BACONNET	D. LANCIAUX
A. COSTES	G. LE LANN
G. GERMAIN	T. PHAN DUC
B. GIRARD	J. P. VIGNAUD
G. KAHN	

Comité de Programme / Programme Committee / Programm Komitee

G. LE LANN (Chairman)

Members:

G. BULL	K. D. MÜLLER
P. CHRISTENSEN	R. PATZELT
M. GIEN	E. M. RIMMER
H. HALLING	S. SEDILLOT
J. LEWIS	

Referees:

M. ANCEAU, France	H. HUNKE, Fed. Rep. Germany
J. P. BACONNET, France	F. ISELIN, Switzerland
D. BAUM, Fed. Rep. Germany	W. KNEIS, Fed. Rep. Germany
R. BIANCASTELLI, Italy	H. KOPETZ, Fed. Rep. Germany
G. BIANCHI, France	J. LUKACS, Hungary
M. J. CAWTHRAW, U.K.	J. D. NICOUD, Switzerland
M. COSTE, France	G. VERROUST, France
P. DESCHIZEAUX, France	C. A. VISSERS, The Netherlands
S. GOLDSACK, U.K.	B. A. WICHMANN, U.K.

REAL-TIME DATA HANDLING AND PROCESS CONTROL – II

Real-Time Data Processing and Related Standards & Common Practices
- Introduction
- Present Applications and Relevance of Standards
- Future Requirements and Trends – New Technologies and System Structures

Proceedings of the Second European Symposium held in Versailles, France, 3-5 November 1982

edited by

E. G. KINGHAM
Central Electricity Research Laboratories
Leatherhead
U.K.

G. LE LANN
Institut National de Recherche en
Informatique et en Automatique
Domaine de Voluceau
Rocquencourt
France

and

N. E. MÁLAGARDIS
Bureau d'Orientation de la Normalisation
en Informatique
Domaine de Voluceau
Rocquencourt
France

   Institut National de Recherche en
Informatique et en Automatique

1984

NORTH-HOLLAND – AMSTERDAM • NEW YORK • OXFORD

© INRIA, 1984

All rights reserved. No part of this publication may be reproduced, stored in a retrieval system, or transmitted, in any form or by any means, electronic, mechanical, photocopying, recording or otherwise, without the prior permission of the copyright owner.

ISBN: 0 444 86846 1

Published by:
ELSEVIER SCIENCE PUBLISHERS B.V.
P.O. Box 1991
1000 BZ Amsterdam
The Netherlands

Sole distributors for the U.S.A. and Canada:
ELSEVIER SCIENCE PUBLISHING COMPANY, INC.
52 Vanderbilt Avenue
New York, N.Y. 10017
U.S.A.

Library of Congress Cataloging in Publication Data

European Symposium on Real-Time Data Handling and
    Process Control (2nd : 1982 : Versailles, France)
    Real-time data handling and process control.

    In English, French, and German.
    Organized by Institut national de recherche en
informatique et en automatique.
    Includes bibliographical references and index.
    1. Real-time data processing--Congresses. 2. Process
control--Data processing--Congresses. 3. Automatic
control--Data processing--Congresses. I. Kingham, E. G.
(Edward G.), 1928-   . II. Le Lann, G. (Gérard),
1943-   . III. Malagardis, N. E. (Nicholas E.)
IV. Institut national de recherche en informatique et
en automatique (France) V. Commission of the European
Communities. VI. Title.
QA76.54.E87 1982   001.64'404   83-25431
ISBN 0-444-86846-1

PRINTED IN THE NETHERLANDS

FOREWORD

Despite the financial stringency existing at the time of the Second International Symposium on Real-Time Data Handling and Process Control, it nevertheless attracted the support of more than 300 participants from 17 countries.

In a similar way to the First Symposium in 1979, it was designed to present by papers and discussions, the status and future trends of real-time activities in a wide range of applications, in connection with hardware and software aspects of:

- existing standards, common practices and their applications;
- research and development towards new standards;
- the influence of new technology on these standards;
- research on new real-time systems architecture.

It took into account the suggestions made by the participants of Real-Time Data '79 and ensured debate as well as laying greater emphasis on software and system reliability. On this occasion it was also possible to devote more attention to research aspects because of the location of the Symposia in relation to the local organisers.

Industry was more represented at the '82 Symposium and considerable interest was evidenced in the contributions dealing with software - a clear indication of the growing recognition of the relevance of this topic. The analysis of the questionnaires completed by the participants revealed a continuing desire to see the Symposium repeated, preferably at least every 2 years. There was also a frequently expressed wish to see an Exhibition of relevant equipment associated with the Symposium.

The distribution of these Proceedings is anticipated to be of considerable benefit to designers and users, both existing and potential, of real-time systems and their applications in a wide range of activities.

The success of the Symposium was undoubtedly due to:

- the enthusiastic efforts of the speakers and chairmen;
- the local organisational efforts co-ordinated by the Institut National de Recherche en Informatique et en Automatique;
- the services and support of the Commission of the European Communities;
- the French Ministry of Research and Industry;
- the representatives of many other organisations and of course, those participants who contributed so fully to the discussions.

All such support and commitment is gratefully acknowledged.

E.G. Kingham   G. Le Lann   N.E. Malagardis

CONTENTS

Foreword... v

SESSION 1

WELCOMING ADDRESSES AND CONGRESS OPENING

ALLOCUTIONS DE BIENVENUE ET *GRUSSANSPRACHEN UND*
OUVERTURE DU CONGRES *ERRÖFFNUNG*

1.1. RTD 82 - Chairman's Opening Address
E.G. Kingham (United Kingdom).. 3

1.2. RTD 82 - Assistant Secretary General's Opening Address
L.D. Eicher, International Organization for Standardization (ISO)........................ 5

1.3. Community Action in the Field of Standardization for Informatics Technology
K. Thompson (Commission of the European Communities)...................................... 9

SESSION 2

ADVANCED SYSTEM ARCHITECTURES

ARCHITECTURE AVANCEE DE SYSTEMES *FORTSCHRITTLICHE SYSTEM ARCHITEKTUREN*

2.1. Innovative Multicomputer Architectures - The Development of the Eighties
W.K. Giloi (Fed. Rep. Germany).. 15

2.2. Synchronous Parallel Bus as Interconnection Medium for Information Transfer
R. Patzelt and H. Kern (Austria).. 25

2.3. PALME - An Automatic Mixed Packet-Line Exchange
(PALME - Un Autocommutateur Mixte Circuit Paquet)
Y. Corcia, J.L. Dauphin, G. Froger, O. Louvet, and Y. Simon (France)..................... 31

2.4. Verriegelungsmechanismen auf Parallelbussystemen
N. Böck (Austria)... 39

2.5. Database Machines: A Tool for Data Management under Real Time Conditions
H. Schweppe and G. Stiege (Fed. Rep. Germany)... 43

SESSION 3

DISTRIBUTED SYSTEMS

SYSTEMES REPARTIS *VERTEILTE SYSTEME*

3.1. High Level Programming of Distributed Process Control Systems
H. Kopetz, F. Lohnert, W. Merker, and G. Pauthner (Fed. Rep. Germany).................... 49

3.2. Système Logique de Communication pour Applications Réparties: Le LOGIBUS
C. Faulle, B. Gagey, M. Jeandel, and A. Viallevieille (France) 57

3.3. Standardisation Work for Communication among Distributed Industrial Computer Control Systems - A Status Report
G.G. Wood (United Kingdom) 67

3.4. A Process Control Software based on Concurrent PASCAL for Industrial Micro-Computer Systems
J. Grof, G. Révész, and E. Zarándy (Hungary) 71

3.5. A Multiprocessor Network in a Multiuser CAMAC Environment
U. Beyschlag, M.J. Clayton, S. Patel, M. Rabany, R.I. Saban, and A. Thys (CERN) 75

SESSION 4

PROGRAMMING LANGUAGES

LANGAGES DE PROGRAMMATION   *PROGRAMMIERSPRACHEN*

4.1. Programming Equals Specifying Plus Designing
I.C. Pyle (United Kingdom) 81

SESSION 5

METHODS AND TOOLS FOR SOFTWARE DEVELOPMENT

METHODES ET OUTILS DE DEVELOPPEMENT DU LOGICIEL   *METHODEN UND WERKZEUGE ZUR SOFTWARE-ENTWICKLUNG*

5.1. A Survey of Current Trends in Software Tools and Environments
L. Osterweil (U.S.A.) 89

5.2. An Approach to Conceptual Design Methods and Tools for Software Development
(Une Méthode de Conception Conceptuelle. Méthodes et Outils de Développement du Logiciel)
C. Rolland and P. Albin (France) 103

5.3. Quelques Primitives Pour la Programmation Temps Réel et Leur Sémantique Mathématique
G. Berry, J. Camerini, J.P. Marmorat, B. Nguyen Phuoc, and J.P. Rigault (France) 111

5.4. Port-Oriented Synchronisation and Communication in a Local Network
(Synchronisation et Communication par Ports Orientés dans un Réseau Local)
J.P. Elloy and P. Molinaro (France) 121

5.5. Results of a Two-Year Study of a Software Production Method
(Bilan de Deux Ans d'Expérimentation d'une Méthode de Production du Logiciel)
Y. Corcia, J.L. Dauphin, and Y. Simon (France) 131

5.6. Eine Erweiterung von SPECIAL zur Beschreibung asynchroner Aktivitäten
C. Köhler (Fed. Rep. Germany) 139

5.7. PILS, A Portable Interactive Language System
D.O. Williams and R.D. Russell (CERN) 147

5.8. Automatic Generation of PEARL-Programs from an EPOS Specification
V. Scheub (Fed. Rep. Germany) 151

5.9. SECURE - Fortran Implementation of "Subroutines for CAMAC" for NOVA - Compatible Computers
Z. Banasik and J. Zalewski (Poland) 159

SESSION 6

SYSTEM APPLICATIONS IN CONTROL AND AUTOMATION

APPLICATIONS DES SYSTEMES AU DOMAINE DU CONTROLE ET DE LA COMMANDE

SYSTEM-ANWENDUNGEN IN STEUERUNGEN UND BEI DER AUTOMATISIERUNG

6.1. New Computer Control System of 80-in. Hot Strip Mill
  N. Saikawa, J. Nitta, T. Minematsu, T. Mikuriya, T. Tamiya, T. Takechi, K. Yoshie, and H. Ezure (Japan) 165

6.2. The CAB Software Package
  Y. Perrin, P. Scharff-Hansen, L. Tremblet (CERN), E. Barrelet, G. Fouque, and K. De Kerday (France) 175

6.3. Le Système de Contrôle du GANIL
  L. David, E. Lecorche, T.T. Luong, B. Piquet, M. Prome, and M. Ulrich (France) 179

6.4. The CAB System: Applications
  E. Barrelet, R. Marbot, and P. Matricon (France) 189

6.5. A Data Acquisition System for the VAX Computers
  F. Gagliardi, M. Sciré, A. Vascotto, and V. White (CERN, Italy, CERN, U.S.A.) 193

6.6. Système d'Acquisition CAMAC Pour Tomographe à Rayons X Industriel
  J.P. Guerin, J. Huet, and M. Pauton (France) 197

6.7. Emploi des Micro Systèmes dans les Activités d'Essais et de Contrôles du Réseau des Laboratoires des Ponts et Chaussées
  S. Savoysky (France) 201

6.8. CAMAC Serial Highway Implementation at the Joint European Torus (JET)
  V. Schmidt (United Kingdom) 205

SESSION 7

CONFORMANCE TESTING OF HARDWARE AND SOFTWARE

TESTS DE CONFORMITE DU MATERIEL ET DU LOGICIEL

KONFORMITÄTSTESTS VON HARD UND SOFTWARE

7.1. Conformance Testing of Software
  R.S. Scowen (United Kingdom) 211

7.2. An Automatic System for Testing Compilers
  F. Federigi and I. Spadafora (Italy) 225

7.3. Les Procedures de Validation des Compilateurs
  J. Sidi (France) 235

7.4. IDA - Tests des Logiciels Temps Réel Assistés par Ordinateur
  G. Lamarche and P. Taillibert (France) 239

7.5. Moniteur de Tests. Un Outil Pour l'Ecriture Rapide des Tests de Conformité
  Ch. Triolaire (France) 247

7.6. About the Second Generation of Bus Structures
  E.V. Chernykh (U.S.S.R.) 251

SESSION 8

NETWORKING AND DATA HIGHWAYS

RESEAUX ET TRANSMISSION DE DONNEES   *RECHNERNETZE UND DATENÜBERTRAGUNGSLEITUNGEN*

8.1. On Standards, Pre-Standards and De-Facto Standards in Data Communication
    R. Popescu-Zeletin (Fed. Rep. Germany) 257

8.2. Implementation of a Powerful Local Area Network on a Fiber Optic Loop
    (Implementation eines leistungsfähigen Datennetzes mit Ringstruktur auf einem optischen Medium)
    E. Querasser, M. Lindner, H. Preineder, and F. Buschbeck (Austria) 265

8.3. A Q-Bus Interface and an RSX-11/M Driver for the Local Network Danube
    V. Tschammer, K. Emmelmann, and W. Wawer (Fed. Rep. Germany) 273

8.4. Analysis of the Communication and Fault Aspects of Path and Highway for PROWAY Networks
    W. Ansaldi, M. Olobardi, A. Faro, and O. Mirabella (Italy) 277

8.5. FASTBUS, a High Speed Multi-Segment Bus
    H. Müller (CERN) 287

8.6. An Improved Ethernet for Real-Time Applications
    R. Hainich (Fed. Rep. Germany) 293

SESSION 9

SYSTEM AVAILABILITY, RELIABILITY AND MAINTAINABILITY

DISPONIBILITE, FIABILITE MAINTENABILITE DU SYSTEME   *VERFÜGBARKEIT, ZURERLÄSSIGKEIT UND INSTANDHALTUNG VON SYSTEMEN*

9.1. Safety Problems in Advanced Control
    J.R. Taylor (Denmark) 305

9.2. Automation eines Wasserwerkes für unbemannten, vollautomatischen Werksbetrieb mit einem Prozessrechner unter besonderer Beachtung elektrischer und verfahrenstechnischer Fehlertoleranz
    U. Rüdiger (Fed. Rep. Germany) 317

9.3. Evaluating Software Fault Tolerance in a Real-time System
    T. Anderson and M.R. Moulding (United Kingdom) 327

9.4. A Microprogram Production Environment to Support Computer System Reliability
    P. Wilk and M. Norrie (United Kingdom) 329

SESSION 10

"POSTERS"

10.1. Radar Line Watch System
    J. Beyer (Belgium) and J. Heller (Fed. Rep. Germany) 339

10.2. Anforderungen und Entsprechende Lösungsmethoden für Buszugriffssteuerungen
    H. Kern (Austria) 343

10.3. The Equipment Alarm System in the Experimental Areas of the CERN SPS Particle Accelerator
    A. Cojan, M. Rabany, and R.I. Saban (CERN) 347

10.4. A PASCAL-High-Level-Language Interface for an IEC-Bus Multiprocessor Environment
(Ein PASCAL-Sprachinterface für Multiprozessor Betrieb an einem IEC-Bus)
E. Schoitsch (Austria).. 351

10.5. Typenunabhängiges Cross-Entwicklungssystem mit minimaler hardware-Anpassung
H. Schweinzer (Austria).. 355

10.6. Real Time Diagnosis Software with Application to the Diagnosis and the Treatment
of Traditional Chinese Medicine
Wang Zhi-Bao, Lu Gui-Zhang, Wang Xiu-Feng, Wang Shi-Xiang, and
Din Xiu-Wen (People's Republic of China)... 359

10.7. Ordered Bus Access by Low Level Token Passing for the Local Network TOPAS
W. Wawer, K. Emmelmann, and V. Tschammer (Fed. Rep. Germany)........................... 363

CLOSING ADDRESS

Real-Time Data '82 - Closing Speech
P. Christensen (Denmark).. 369

List of Participants.. 371

Author Index.. 385

Session 1

WELCOMING ADDRESSES AND CONGRESS OPENING

ALLOCUTIONS DE BIENVENUE ET OUVERTURE DU CONGRES

GRUSSANSPRACHEN UND ERRÖFFNUNG

RTD 82 - CHAIRMAN'S OPENING ADDRESS

E.G. Kingham
Chairman of the Organising Committee,
Central Electricity Research Laboratories,
Leatherhead, U.K.

Ladies and Gentlemen,

As chairman of the Organising Committee, it is my pleasure to welcome you here in Versailles to the second International Symposium on Real Time Data. The Symposium has been arranged under the combined responsibility of the ESONE Committee, the European CAMAC Association and the European Workshop on Industrial Computer Systems, with the direct support of the Commission of the European Communities and the French Ministry of Research and Industry through the good offices of BNI and INRIA. The organisations are most grateful for the very practical support provided for the Symposium and for the continuing support of the Commission over the years.

It is worth remembering that it is now 21 years since the Commission had the foresight to encourage the various laboratories to join together and form the ESONE Committee. There is no doubt in my mind of the very real value of the support and encouragement the Commission has given both to ESONE and to the Real Time fraternity through the years, leading to the later establishment of ECA and EWICS and I should like to take this opportunity to formally record our thanks to them.

It is therefore with particular pleasure that I welcome Mr. Thompson, representing Directorate General III of the Commission and Mr. Levieux of the French Ministry of Research and Industry. We are also pleased to welcome Dr. Eicher, Assistant Secretary-General of the International Organisation for Standardization, ISO, whose talk will undoubtedly endorse the value and importance of international standardization in our activities.

I must acknowledge the very practical support of the National Institute of Research on Informatics and Automation, INRIA, and its Department of External Relations in facilitating the Symposium in these pleasant surroundings and in undertaking all the necessary secretariate activities. We are most grateful to Mlle. Bricheteau for this and to Dr. Le Lann for his presence with us this morning.

Although these International Conferences commenced as early as 1973 with a CAMAC conference in Luxembourg, this particular Symposium is the second of those dealing with wider issues of Real Time Data arranged by the three organizations. It is therefore a clear indication of the growing collaboration and interdependence of their activities. One of the prime objectives of the Symposium is to foster the interchange and commonality of information and practices within the Real Time community, an aspect made even more important by the increasing complexity of our technology and the atmosphere of economic stringency in which we are now working.

In addition there is the opportunity to renew, or to make fresh, personal contacts and the success of the Symposium will be measured, to some degree, by the extent to which this occurs. I hope it means that you go from here on Friday, with many new ideas and many new friends. In the meanwhile, this is your Symposium; please contribute freely to it and to the discussions taking place.

I must also refer to the telegram of good wishes received from Louis Costrell on behalf of the US NIM committee. This reads as follows: "Best wishes from the NIM Committee for a successful combined Symposium of ESONE, ECA and EWICS. During the past year we have updated and reissued the IEEE/ANSI CAMAC Documents and have published the 1982 issue of the hard cover volume that includes the seven CAMAC standards together with a glossary of CAMAC terms and a tutorial introduction. Our main effort this year has been with the FASTBUS system on which ESONE collaboration has been very useful and very much appreciated. We value the continued rewarding and pleasant working relationship with our European colleagues that we have enjoyed over many years.
Louis Costrell, chairman, NIM Committee National Bureau of Standardisation USA".

Finally, let me welcome our guest speakers and chairmen and let me declare the Second Real Time Data Symposium of 1982 formally open.

RTD 82 - ASSISTANT SECRETARY GENERAL'S OPENING ADDRESS

L.D. Eicher
International Organization for Standardization

Mr. Chairman, Ladies and Gentlemen,

I am very pleased to be here today on behalf of ISO, the International Organisation for Standardization, a co-sponsor of this, the second, Real Time Data Conference.

As many of you already know, ISO together with our sister organisation, the International Electrotechnical Commission (IEC), comprise a complete system for the formulation of International Standards. The scope of ISO is not limited to any particular technology: it covers all standardization fields, except electrical and electronic engineering which is the responsibility of the IEC.

ISO brings together the interests of producers, users (including consumers), governments and the scientific community in the preparation of International Standards, and is the largest international non-governmental organisation for industrial and technical cooperation.

The ISO Central Secretariat in Geneva, comprising about 120 persons, is responsible for the general administration of the organisation, for the day-to-day planning and coordinating of the technical work, for the publication of International Standards and technical reports, for relations with other international organisations, and for public relations and information services on international standardization.

ISO was established in 1947, and in the early years international standardization was concentrated mainly on the mechanical industry. Early technical work of ISO dealt with such matters as the tensile strength of a given type of steel, a test method for rubber, or the dimensions of nuts and bolts. It is of historical interest to note that the first ISO committee - ISO/TC 1 - dealt with screw threads, the second with industrial fasteners, and so on. When ISO was created, 35 years ago, these areas were obviously at the center of the work.

In recent years, however, the scope of ISO has broadened considerably. ISO still deals with screw threads, steel, and rolling bearings, but today's working programme also includes medical equipment, furniture, sizing of clothes, air and water quality, ergonomics, dairy products and information processing systems - to give just a few examples.

Today ISO (and IEC) work covers virtually all fields of human activity. This scope is also reflected in the number of ISO technical bodies, more than half of which have been established in the past 10 years.

The widening scope of ISO goes hand-in-hand with a broader participation in the ISO technical work. Interest in ISO work is no longer restricted to industry, but also extends to an increasing extent to government and intergovernmental agencies, the scientific community, other international organisations and consumer organisations all of which may be granted liaison status with many ISO/TCs and SCs.

Although ISO has several committees dealing with standardization in the area of information technology and interchange, which are illustrated in slide 1 (fig. 1), the ISO work of prime interest to you falls under TC 97 "Information processing systems".

Work in Real Time Data handling and interchange in different environments and between systems obviously requires compatibility between the systems dealing with the data. Technical committee ISO/TC 97, in the centre of figure 1 covers the whole spectrum of application-independent information processing standards, whereas those committees which might, in this context, be regarded as the "application areas" of information processing feed into, and are supplied by, TC 97.

The structure of TC 97 is shown in the second slide (figure 2), and, as you can see comprises some 17 active sub-committees, each of which has below it a more complex structure of working groups, ad hoc groups and panels of experts.

I should however mention that since this diagram was prepared, PLACO, which is the ISO committee responsible for planning the technical work of ISO, has recommended that a new TC be created. This TC will, I am sure, be of great interest to you all as it will be titled "Industrial Processing Systems related to Industrial Automation". It will be formed by the merger, and upgrading, of ISO/TC 97/SC 8 and SC 9, and will absorb the current work programmes of both.

Real time systems, as I said earlier, are dependent on compatibility. By which I do not only mean physical interconnections but also aspects such as the architectures, networking, the portability of languages, and the reliability of systems. All these are aspects of ISO work in Information Systems. One of the perhaps, most far reaching of an up coming family of standards in information processing is that of the "Open Systems Interconnection" series, the first of which is the Basic reference model - ISO/DIS 7498 - which will be published as an ISO International Standard in 1983. Protocols for the various levels described in the Reference Model are being developed by sub-committee 16, and are well on the way to becoming draft International Standards.

EWICS, through its technical committees, has been making valuable contributions to the TC 97 work. The submission of Industrial Real Time FORTRAN, which will shortly become ISO/DIS 7846 (from EWICS/TC 1) and other proposals for new work such as safety and reliability of industrial computer systems (from EWICS/TC 7), are just two examples of the ongoing cooperation between ISO and one of the other organisations supporting this symposium, in these important and rapidly developing areas. The challenge of a timly response to the demands created by stimuli from the data processing community for new standards is now being met by ISO.

We look forward, Mr. Chairman, to an even more productive future and I, on behalf of the International Organisation for Standardization, thank you for inviting us to be a co-sponsor of this 1982 Symposium, and wish you all a most successful and fruitful week here in Versailles.

Thank you.

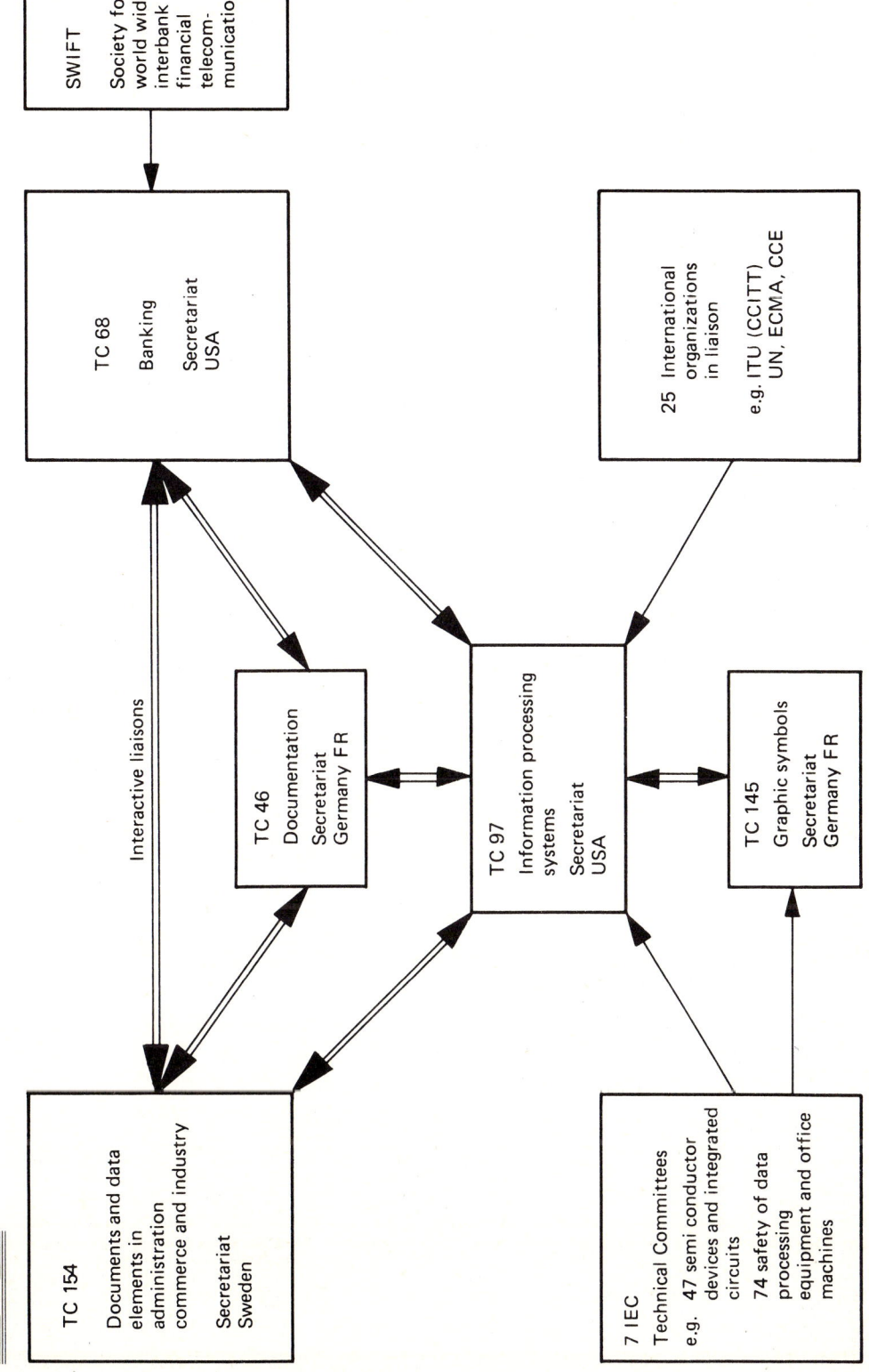

Figure 1.

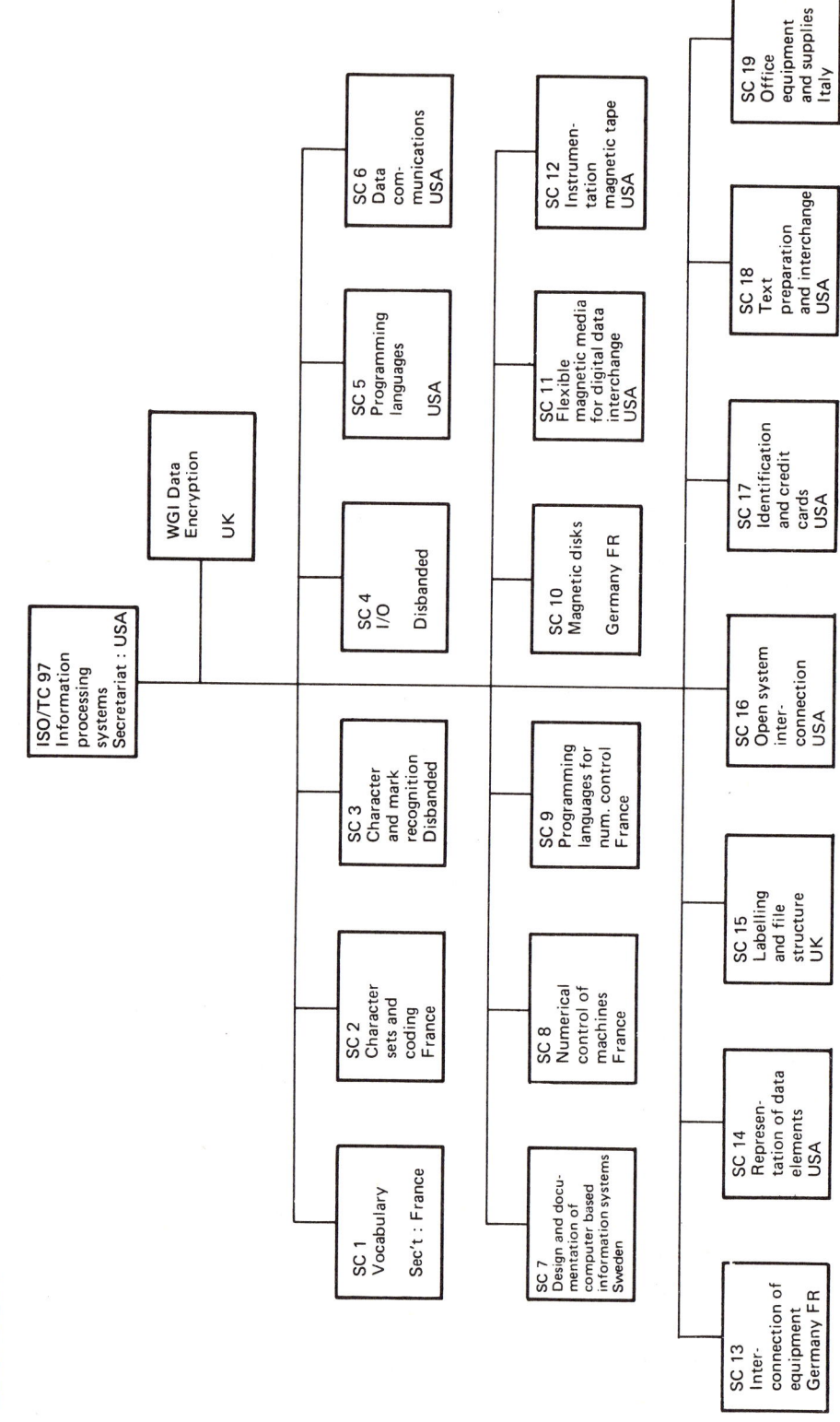

Figure 2.

COMMUNITY ACTION IN THE FIELD OF STANDARDIZATION FOR INFORMATICS TECHNOLOGY

K. Thompson
Commission of the European Communities

It is with great pleasure that I accept on behalf of the Commission of the European Communities your invitation to address this second Real Time Data Conference. The Commission is pleased to have found it possible to sponsor these conferences as focal events. Focal events which are only possible because a considerable number of people from many countries have worked together over several years. This has no doubt been achieved by them despite the daily pressures upon them to solve those urgent daily problems rather than some of the important longer term issues. It is for this reason that we should all recognize the contributions from these people, supported by their organisations, who have given their time and their talents to bring this event to fruition.

The theme of this series of conferences "Real Time Data" is clearly one of great importance to the economic health of the Community, and to the industry within the Community. One important aspect in industry is to plan ahead, to find out what is happening in the rest of the Community and to apply this to normal routine operations. I hope that this conference gives you a short breathing space from those many day to day problems of your own organisations and helps you to raise your heads so that you may see a little further into the future and judge the impact you could make it have on your own operations.

You are all, no doubt, aware that computers have invaded our planet earth. Indeed, if you look carefully you will find some baby invaders in your pockets. Each generation of pocket invaders offers you more facilities and so the invasion extends. It is not necessary to look far ahead to see the time when your pocket invader will combine the functions of a calculator, pocket recorder, diary, office communications bleeper, telephone, and ultimately creates for you your own travelling office, or laboratory. Perhaps then you won't need to go to work but neither will you be able to leave your work in the office or segregate your work from your home or social life.

These pocket invaders will be like offices or laboratories, each different. They will be different because they are, as offices are, different in being designed by different architects and built by different manufacturers, but they will all have the same need to exchange information. They will need to speak the same data information language.

As a slight diversion we can also look further into the future to see a society formed by these little pocket invaders. Their means of transport will be the human beings that carry them in their pockets. The governing authority will be carried out by their larger brethren, the super artificial intelligence computers and they will be made in robot factories by their robot brothers. The rest of this science fiction story I leave to your imagination.

Whether or not this science fiction is welcome or frightening, it is evident that the computer invader population is rather like the brilliant younger generation of our society. They offer great future potential but, as so often, without the necessary self-discipline. Society and parents have a responsibility to assist our children to gain that self-discipline, but who will do it for the computer invaders? The laws of society and the rules and conventions of civilized behaviour provide the parameters and boundaries for human discipline and self-discipline. What are the equivalent laws and rules for our pocket invaders? I put it to you that standards are an important vehicle for that self-discipline. Standards must not only be brought into existence, they must be adopted into equipment and system design and must be demanded by industry and commerce.

The title of this address starts with the words "Community Actions". So you may now sit back in your chairs and relax, whilst I explain all the wonderful actions that we, the Eurocrats in Brussels, are carrying out on your behalf. Unfortunately the words "Community Actions" are intended to refer to actions by the 300 million inhabitants of the European Community and not just by the 0,003 % of the population who are the so called Eurocrats. You, or some of you, are part of that Community. So I am about to address you on the subject of your actions. What are your European Community (EC) Actions? What has been achieved to date at an EC level? I can and will tell you of some of our joint successes and also will invite you to join us, to do even more in the future.

Let me first go back to the beginning of all European Community time. The European Economic Community was born in 1957 with the signing of the Treaty of Rome. The Treaty gave certain responsibilities and duties to be administered. One of these responsibilities (expressed in article 30 of the Treaty) concerned the removal of the barriers to trade between the European Community Menber States. Thus, since standards were referred to in legislation, they became of concern to the Commission. However, in the case of informatics, virtually no legislation or enforceable standards existed. In fact informatic issues did not emerge until the Council of the European Community responded to an initiative of the Commission and passed a Resolution in 1974.

> " Council Resolution of 15 July 1974 on a Community policy on data processing intends to give a Community orientation to policies for encouraging and promoting data processing, and welcomes the Commission"s intention to submit in 1974, after appropriate consultations, priority proposals concerning:
>
> (a) a limited number of joint projects of European interest in the field of data processing applications,
>
> (b) collaboration on standards and applications and in public procurement policy,
>
> (c) the promotion of industrial development projects on areas of common interest involving transnational cooperation. "

This resolution was a major stepping stone for it authorised the Commission to develop an initiative. It took a further 3 years for the efforts of the Commission to bear fruit in the form of the passing of an EC Council Decision in late 1979 authorising a programme of work involving, amongst other things, actions concerning informatics standardization and informatics procurement issues.

Prior to the European Community Council Decision the Commission began to prepare to take up its anticipated responsibilities by giving some initial support to some European cooperative initiatives, namely some of the groups which now sponsor these Real Time Data Conferences. The Commission also founded an advisory informatics standardization working group WGS. The WGS was set up to act as a visible focal point for the discussion of informatics standardization issues, to provide contact between the Commission and Member States and also to form a link between the different Member States. Another group for public procurement, PPWG, was set up as a peer activity for procurement contacts. Unfortunately, staff restrictions imposed on the Commission by the Council were only adjusted with a 2 year delay, so that the PPWG activity collapsed though fortunately WGS survived the crisis and some additional staff were recruited in January of this year.

I have presented this short history to forestall any of those enthusiasts who we know frequently express the view that "The Commission should do" this and that. We have been invited by Member States to help them to cooperate across national frontiers and to prevent the creation of barriers to trade rather than wait until barriers exist before taking action under the terms of the Treaty of Rome. The Commission Services see their prime role, in informatics, as being a supporter and promoter of cooperation between the activities in Member States for the collective benefit of the European Communities. Thus our task is to continually review the European situation within the world context and to take action to ensure that, where appropriate, a mechanism is found to promote the necessary collaborative situation.

This thinking has led us with the cooperation of WGS:

- to encourage, support and initiate regular exchanges between technical experts from European countries, in order to allow them to coordinate their national efforts and to make collective progress.

- to initiate and run specific projects which have a European dimension to complement their national programmes.

This policy has been based on consultations in the EC Member States with their ministry and standardization advisors. WGS has subsequently advised the Commission to adopt a:

" <u>EUROPEAN STRATEGY</u>

A very strong commitment to meeting European needs through world-wide standards has been recognized. There is indeed in view of the international nature of information technology, a very strong resistance to the development of exclusively European standards. This is tempered by the realisation that in some cases it may be impossible to achieve world-wide standards at the right time and in the right form; in such cases, with extreme reluctance, it may be necessary to develop standards within Europe, although the objective will remain of turning these into world-wide standards in due course.

The strategy for the Community should be to support the existing international machinery, particularly ISO and CCITT. Community actions should concentrate on supporting and complementing the ongoing work in these bodies, but the commissioning of technical input is not totally excluded, where this is the most effective means of support.

The Policy advisory document (III/600/82) is in the public domain, like all the WGS papers, and your national representatives can make it available to you.

It is probably an appropriate time to point out again that this is typical of how we work. The above policy was prepared cooperatively, the Commission provided the environment for it to be done, it involved not only an expert external consultant but also the very active participation of the representatives of the national governments and representatives of other organisations working in the field of informatics standardization. The Commission brought them together and they set up the consultation meetings in their own countries.

Collaboration between the activities in the various Member States is mainly organised around bringing technical experts together. This method of working developed partly from the shortage of technical/administrative staff in the Commission and it allows groups to largely organise their own method of cooperation. The Commission's role has not only been to foster such groups but also to guide them on how to integrate their efforts. Many technical experts are, because of their intense, introverted, attention to their own specialization, are naive about how to link their work to official national programmes and standardization institutions. The Commission's role requires a particular mixed expertise; it is necessary to know enough of the technical issues so as not to be fooled and to combine this with knowledge about the national and international official standards institutions.

The Commission now fosters and guides some 30 different technical cooperation groups. The EC not only provides the facilities (i.e. meeting rooms) to allow these groups to meet together but it also pays direct travel expenses (eg. air fares) and in so doing removes the financial inbalance between participants from say Athens and one from Rotterdam. However the contribution from the companies and institutes who make the technical experts available are invaluable. These companies are generally to be applauded for their contributions for the return on their investment is invariably only seen after several years and often benefits the whole EC rather than just the contributing companies.

Some important milestones have been achieved by specially supported projects done in collaboration with these groups:

- The European Workshop of Industrial Computer Systems (EWICS-ICI) has been responsible for drafting a significant part of what will be an international IR Fortran standard taking into account European requirements. Perhaps more importantly though is the fact that they have become part of the ISO machinery. They are now responsible for a working group for the International organisation for Standardization ISO (ISO/TC97/SC5/WG1).

- The Open System Interconnection Group (WGS-OSC) has cooperated in creating a single tutorial document that is available in all the Community languages. This common tutorial was prepared to disseminate easily the complex computing Open System Architecture work going on in ISO to produce a standard. That group has also organised a catalogue of standards related to Open System Interconnection OSI in order to foster better cooperative working.

- The Ada-Europe Group has allowed national programmes to coordinate themselves in Europe without having to go to the USA which is the only other common meeting ground.

Internal to the Commission the advantage of having interworking computers is a goal that has attracted some considerable attention. Most informatics standards, rightly in our view, are restricted to the aspects of information interchange, but unfortunately they do not specify enough detail to ensure real information interchange when applied to computing systems. The Commission services along with many others have recognised the need to have more precise standards for this purpose. In consequence, the Commission has set up an internal Standardization Implementation Committee (SIC) to achieve the information interchange capability between the various Commission computer systems and the Menber States. This requires specifications to be drawn up which select the appropriate ISO, IEC and CCITT (possibly ECMA) standards and specify the options and default value setting.

The Commission has started with the issues of Teletype compatible equipment, information exchange by magnetic tape, multilingual keyboards and teletex compatible equipment. These internal initiatives and others are being linked to the external Subcommittee WGS.

The SIC work has led, very obviously, to the need for conformance to standards testing and independently testing the ability to interchange information. Conformance testing and certification are activities which provide many technical problems, but also bring with them financial and legal issues to compound the problem. A compromise was adopted for the first, relatively trivial but, in practice, important task chosen of the connection of teletype compatiable terminals. A reference testing service was set up at our Joint Research Centre (JRC), which can run a teletype test scenario over a public switched data network (eg. EURONET).

It has become a useful service which has already been accessed from the other side of the world. The scenarios are limited to showing that the terminal can receive, handle and transmit according to the standards and furthermore will work with a computer so operating. This type of service is less expensive and without the legal problems of a full certification service. It also shows that the devices work when playing a data interchange scenario which can be subsequently used for maintenance checking.

What of the future? It seems that we have the current formula for supporting the standards activities in the EC, but clearly the introduction of new standards requires to be integrated with the European manufacturers ability to supply the goods. Despite the fact that they collectively write the first draft of international standards in their joint organisation ECMA - manufacturers still seem to have problems in adopting these standards in their products in a timely manner.

Thus we believe that the synchronisation between demand and supply of standards conformant equipment has to be added to our efforts. This new work area will again only be possible with the full cooperation of industry and commerce with a little help at Community level.

The future is however in our collective hands. The current programme of work authorised by the EC Council (your governments) is due to end in 1983. A new programme is clearly needed and the Commission has to propose it to the Council. The questions to you are: "Do you believe in our method of supporting your work?" Is it worth continuing? If you do please tell us, but more importantly, tell your governments. As I said earlier so many enthusiasts say "The Commission should do this and that", but the Commission can only do what is authorised. Don't forget, someone has to teach these little invaders and their bigger relatives the mainframes, the minis, the micros and the terminals, and word processors how to behave in public.

Thank you for the opportunity to address you and, on behalf of the Commission, I wish you every success in this conference and your future enterprises.

K. Thompson.
1982-10-29

Session 2

ADVANCED SYSTEM ARCHITECTURES

ARCHITECTURE AVANCEE DE SYSTEMES

FORTSCHRITTLICHE SYSTEM ARCHITEKTUREN

INNOVATIVE MULTICOMPUTER ARCHITECTURES
- THE DEVELOPMENT OF THE EIGHTIES

W.K. Giloi
Professor of Computer Science
Technical University of Berlin
and
University of Southwestern Louisiana

ABSTRACT

Multicomputer systems are the only architectural form that allows all major requirements for modern computer architectures to be satisfield such as: software-transparent modular extensibility, fault-tolerance, simplification of system software, high performance obtained not through very costly supercomputer technology but through parallel processing of a number of inexpensive units. For all these reasons, multicomputer architectures are considered to become the most important development of the eighties in computer architecture. Maximum cost-efficiency can be achieved by constructing such systems out of standard "off-the-shelf" microcomputer components that are manufactured in very large volume. In the paper, three cases in point are presented of very innovative multicomputer architectures designed to meet a large spectrum of differing requirements. It is pointed out that the key to the realization of multicomputer systems with complex functionality is the principle of hierarchical function distribution. The paper concludes with a discussion concerning the optimal size of multi microcomputer systems.

RÉSUMÉ

Les systèmes multi-ordinateurs sont les seuls qui presentent une architecture permettant de répondre aux principaux besoins qui doivent être satisfaits par les ordinateurs modernes tels que: extensibilité modulaire, transparence pour les logiciels, tolérance de pannes, simplification du logiciel système, haute performance, obtenues non pas grâce à une technologie très coûteuse de super-ordinateurs, mais grâce au traitement parallèle assuré par un certain nombre d'unités bon marché.

On considère, pour toutes ces raisons, que les architectures multiordinateurs constitueront la majeure partie des développements dans ce domaine pour les années 80. On obtiendra le meilleur rapport efficacité prix en construisant de tels systèmes à partir des micro-ordinateurs standard fabriqués en grand nombre.

Dans ce papier sont présentées trois architectures multi-ordinateurs trés nouvelles conçues pour satisfaire un large éventails de besoins différents. IL est souligné que la clé de la réalisation de systèmes multi-ordinateurs à fonctionnalité complexe réside dans le principe de distribution hiérarchisée des fonctions. La conclusion porte sur la taille optimale des systèmes multiordinateurs.

ZUSAMMENFASSUNG

Viel-Rechner-Systeme sind die einzige Struktur, die es ermoeglicht alle wesentlichem Anforderungen an moderne Rechner-Architektur zufriedenzustellen. Dies sind: Software-transpasente modulare Erweiterbarkeit, Fehlertoleranz, vereinfachte System-Software, hohe Leistungsfaehigkeit nicht durch teure Super-Rechner-Technik, sondern durch Parallel-Verarbeitung auf einer Anzahl preiswerter Einheiten. Aus all diesen Gruenden werden Viel-Rechner-Architekturen als die bedeutendste Entwicklung auf dem Gebiet der Rechner-Architektur der achtziger Johre angesehen. Am kostenguenstigsten kann dies erreicht werden, wenn diese Anlagen aus standardisierten in Gross-Serien hergestellten Mikrorechner-Baueinheiten aufgebaut werden. Drei Beispiele völlig neuartiger Viel-Recher-Architekturen werden beschrieben, die einen weiten Bereich versohiedener Anwendungen abdecken. Es wird darauf hingewiesen, dass der Schluessel fuer die Verwirklichung von Viel-Rechner-Anlagen im Grundsatz der hierarchischen Aufgabenverteilung liegt. Der letzte Teil ist eine Betrachtung ueber die guenstigste Groesse von Viel-Rechner-Anlagen.

1. THE NEED FOR INNOVATIVE COMPUTER ARCHITECTURES

Future, advanced computer architectures are expected to capitalize on the dramatic advances of VLSI technology in order to provide improved cost-effectiveness. Because of the high volume at which they are manufactured, VLSI microcomputer components are highly cost-effective. What they may lack in terms of absolute operating speed can be compensated for to some extent by innovative architectural designs that provide a more efficient way of data access than the conventional "von Neumann machine" and combine it with a certain degree of parallel processing.

Besides a certain minimum performance and improved cost-effectiveness, future, advanced computer architectures are expected to satisfy other major requirements, such as:

Quality software support: The architecture should lead to a strongly modularized and therefore simpler and less costly system software. In addition, the architecture should support the design, implementation, and execution of application software that is comprehensible reliable and maintainable. Software satisfying these attributes shall be called 'quality software'.

Modular extensibility: It should be possible to augment performance and functionality of a computer without any major changes in the system software, just by adding more modules to the system.

Fault tolerance: Critical portions of a computer or even the entire system should be made fault tolerant.

The quality software aspect is more important for the goal of achieving improved cost-effectiveness than any reduction in hardware cost, the reason being the following:

- 80% or more of the cost of the development of a large mainframe computer are software development cost;
- 60 to 70 % of the cost of operating a large mainframe computer are software maintenance cost.

The cost of software development has increased dramatically because of the ever increasing demands on the functionality and convenience of system software and application programs, leading to overly complex systems that require tremendous efforts to design, implement, and debug. For years now this phenomenon is known as the "software crisis". Nevertheless, the present generation of expensive mainframe computers still operate as if there were no software crisis, that is, they provide only little - if any - supporting mechanisms for software reliability and maintainability. We call this phenomenon the "mainframe paradoxon".

A central concern of modern software engineering has been and still is the development of software design methods and tools that lead to quality software. Encouraging progress was made towards the development of adequate specification tools and appropriate programming languages. Program verification tools, however, are still not practically applicable, and it looks as if their large scale use in real world software design is more utopian than it seemed years ago /Bea 82/. Consequently, the postulate is increasingly voiced that it should be the task of the computer architecture to help solving or at least mitigating the software crisis /DEN 81/.

Computers that meet that requirement shall be called quality software machines. A quality software machine supports object addressing, data encapsulation, and information hiding. Quality software machines check program execution consistency at run time. To this end, they may watch over restrictively defined access domain boundaries, object definition status, and consistency of procedure invocation and parameter passing.

In principle, one can put the features mentioned above that characterize a quality software machine into a rather conventional mainframe, by vertically migrating conventional software solutions into the microcode of the CPU. An example for such an approach is the Intel "micromainframe" iAPX 432 /TYN 81/. However, conventional single-processor architectures constitute a rather inhospitable environment for the implementation of object addressing and data encapsulation policies, thus causing problems of program inefficiency /Hea 82/. The inefficiency of program execution is aggravated if the single CPU is to perform the elaborate run time consistency checks mentioned above.

It can be shown that the inefficiency penalty for quality software support and execution can be avoided by distributing the complex functionality of a quality software machine over a hierarcharchy of physical engines in a multicomputer system. We call this the principle of hierarchical function distribution /GGK 82/, /GaB 83/. Hence, the appropriate structure of a quality software machine is the multicomputer system. Since such a structure is a prerequisite for modular extensibility and fault tolerance anyway, it is safe to assume that multicomputer architectures will evolve during the eighties as a major architectural form /Nea 81/.

This leaves the computer architect with the task to devise operational principles for multicomputer systems that reconcile the innovative attributes discussed above with the demand for high performance and improved cost-effectiveness obtained through a high degree of parallel operation and optimal utilization of the multiplicity of resources of such a system. As cases in point, ensuingly we will sketch 3 different multicomputer architectures that were designed to satisfy widely differing sets of requirements. The common denominator of all 3 systems is that each system, though representing a very unconventional, innovative computer architecture, is realized with standard, "off the shelf" microcomputer components, its structural design being based upon the principle of hierarchical function distribution. Another common feature of all 3 systems is the application of the principle of object addressing and typing.

2. CASE IN POINT I: STARLET - A HIGH-PERFORMANCE QUALITY SOFTWARE MACHINE

Access control through capability addressing /LAN 81/ is prohibitively expensive when applied to scalar data objects, i.e., single memory location contents. Capability addressing of memory segments can be exercised somewhat more efficiently. In this kind of "object addressing" /FAB 74/, every single memory location must still be referenced in the machine instructions in order to access it and, consequently, access control is still exercised at the granulation of elementary data items (single memory location contents), with the additional overhead that with every reference an address transformation is required from the referenced logical segment address to the effective physical memory address. By providing a set of associative registers that accomodate the associations of logical segment addresses to physical base addresses for the segments that form the actual working set /CaD 73/, one can bypass to some extent the address transformation as well as the look-up of the capability lists by which the access to the object is controlled. Naturally, the "von Neumann bottleneck" /BAC 78/ is not eliminated that way.

A much more efficient object addressing scheme is obtained if objects are defined in the sense of data structure architectures /GaB 77/. In this case, objects are arbitrarily complex, structured entities. Only the entire object is referenced by a (symbolic) name, whereas the data items of an object are accessed by the execution of access functions. Providing dedicated hardware for the access function execution eliminates totally the von Neumann bottleneck and gives the computer the behavior and performance of an SIMD machine (SIMD: single instruction - multiple data). Data structure architectures generalize the notion of data virtualization at the hardware level to arbitrarily structured data objects.

The STARLET computer, of which a first prototype has become operational since September 1981, is such a data structure architecture. In addition, the arbitrarily structured, user-defined data objects can be encapsulated into user-defined abstract data types, with strict access control exercised by the hardware. In addition, the computer has all the features of a quality software machine as discussed in the preceding section. It turns out that the same descriptor information that is needed in the hardware to build the structured data objects is sufficient also for the hardware-implemented data encapsulation mechanism and for performing all elaborate run time consistency checks.

The STARLET computer is structured according to the principle of hierarchical function distribution /GGK 82/, comprising 7 MC68000 microcomputers, a fast floating-point/fixed-point pipeline processor, and 4 special 'reduced instruction set processors'. The different engines communicate via a fast shared memory. A high degree of parallel operation, the execution of numerical and non-numerical operations on structured data sets in the SIMD mode by a fast pipeline processor, and a most efficient mechanism for object addressing, access control, and data encapsulation result in an excellent performance of the computer. Details are discussed in /GaG 82/.

At present, an improved version of STARLET is under construction which avoids certain bottlenecks found in the first prototype. This led to a major redefintion of object classes, object types, and the reference mechanisms as described in /KAL 82/, as well as to a refinement of the function distribution in the system.

3 CASE IN POINT II: MAC - A HIGH-PERFORMANCE MICRO ARRAY COMPUTER

MAC (micro array computer) has a hardware and software structure similar to that of the STARLET computer; however, it differs from STARLET decisively in terms of its purpose and its machine data types. STARLET is a general-purpose, quality software machine; MAC is a special purpose computer for signal and image processing. STARLET handles user-defined, arbitrarily structured data objects that are encapsulated into user-defined abstract data types. In contrast, MAC handles only the following 3 machine data types:

- type VECTOR with element types integer (16 bit) or REAL (32 bit) and the common vector operations;

- type DIGITAL IMAGE ARRAY (DIA) with element type INTEGER (16 bit) only, in connection with a set of standard image preprocessing and segmentation routines;
- type FFT ARRAY with element type REAL (32 bit) only, with the fast fourier transform and its inverse, respectively, as operations.

In compliance with the primary requirement of performance maximation, MAC does not perform any run time consistency checks, nor does it support user-defined abstract data types. However, the machine ensures that the objects of a machine data type can be subjected only to the operations of the type. The operations of types VECTOR and FFT ARRAY are executed by a floating-point/fixed-point pipeline processor in the SIMD (single instruction - multiple data) mode of operation. The operations of type DIA are performed either by the same pipeline processor, as long as image arrays are transformed in their entirety, or alternatively by a cluster of microprocessors in the MIMD (multiple instructions - multiple data) mode of operation. Data are fetched by special address generators from a fast shared memory. The processing bandwith of a pipeline processor is matched to the bandwidth of the fast memory and amounts in the case of FFT to 10 MFLOPS (million floating point operations per second). Further details can be found in /GaB 82-1/.

4. CASE IN POINT III: UPPER - A HIGH-PERFORMANCE DISTRIBUTED MULTICOMPUTER SYSTEM

UPPER (universal poly processor with enhanced reliability) was designed and is being built to prove the feasibility of a multicomputer system that would combine high performance with fault tolerance and, furthermore, exhibit the following features:

- application software modularization and easy programmability on the basis of cooperating parallel processes;
- system software modularization and simplification through orthogonalization and elaborate hardware support of system functions;
- secured operation through object addressing and rigorous access control.

A fault tolerant system by necessity is a distributed system, consisting of a number of nodes that communicate solely through the exchange of messages. Systems of that type are called losely coupled, and for fault tolerance reasons there must be no centralized system supervision. Rather, control is distributed in the system. In addition, the nodes of the system may be spatially distributed over a certain locality. A system with the latter property is called a local network.

UPPER is a multicomputer system with distributed control that may be operated as a local network as well as a spatially concentrated computer system. In order to give the losely coupled system the performance of a strongly coupled multicomputer system, the underlying communication structure is an ultra high speed packet switching bus that features a gross data transmission rate of 280 Mbit/second. Bus access and message transmission is carried out on the basis of the slotted ring protocol /PaB 79/.

Inter process cooperation takes place on the basis of a built-in high-level policy. The policy is constituted by two unique models, the service request model of inter process cooperation (IPC) and the secured object lending model of data sharing. The mechanisms used in the system to carry out that policy are strongly hardware supported and hidden to the user. All the user sees is an appropriate high-level language construct that reflects the standard IPC policy. Confining the user to a fixed, high-level policy rather than providing him with mechanisms to formulate his own policies on the one hand restricts his freedom (including the freedom of making mistakes) but, on the other hand, renders the system comprehensible and easy and safe to program. Software reliability and data security is furthered by capability-based object addressing. The ultra high-speed packet switching bus and the collision-free, low overhead bus protocol allows entire objects to be used as units of transportation. Single data items of an object therefore are accessed always in the local memory. The UPPER system is in an advanced stage of realization. As result of the project, we will have

- developed a novel form of high-level programming of inter process cooperation;
- designed a node computer that features object addressing, carried out in a very efficient manner (as compared to other examples such as the Intel iAPX 432);
- designed and built the fastest packet switching bus presently existing;
- have proved that the system software implementation effort for a distributed multicomputer system can be kept relatively small through a high degree of functional orthogonalization, system software modularization, and hardware support.

A more detailed discussion of the UPPER system and its underlying notions can be found in /GaB 81/, /GaB 82-2/.

5. THE KEY: HIERARCHICAL FUNCTION DISTRIBUTION

All 3 computer systems discussed above -

STARLET, MAC, and UPPER - have in common that their hardware structure is designed according to the principle of hierarchical function distribution /GGK 82/. By this term we denote the process of classifying the functions of a computer and its operating system into a hierarchy of abstract data types, such that the data types of a given level may be represented by data types of lower levels but never by data types of higher levels, and, subsequently, mapping the thus obtained function hierarchy onto a hierarchy of physical machines.

A data type is defined as a class of data objects and a collection of functions applicable on the data objects. A data type is said to be abstract, if the representations of the data objects and the functions of the type need not be known in order to apply functions to objects. Data types whose functions are represented by microprograms or directly mapped onto hardware functions shall be denoted as machine data types. Table 1 gives an example of a classification of system functions into a hierarchy of data types.

In conventional computers, levels 1 to 3 are executed by the operating system, i.e., software, and levels 4 to 5 by firmware and hardware. In a computer with hierarchical function distribution, one may find a specific 'operating system machine' to perform the functions of levels 1 and 3. The functions of level 2 could be performed by the operating system machine as well; however, it pays to have a separate 'instructions processor' for the following reason: Interrupts to the operating system usually lead to complete process environment switches, whereas interrupts generated in the course of instruction execution usually signal exceptions that may require trap handling only. For the functions of level 4, it pays to have a dedicated 'data access machine', and the data transforming functions should be executed by a powerful 'data processor' tailored for that task. This scheme is roughly the one used in STARLET and MAC.

Table 1 Example of the classification of system functions into a hierarchy

Level	Examples of Abstract Data Types in the Machine
1: Higher-Level Operating System Functions	Communication Objects and Functions for the Interaction With the User; Process Scheduling Objects and Functions
2: Application Process Execution	Run Time Environment Management Objects and Functions; Instructions and Instruction Interpretation Functions;
3: Lower-Level Operating System Functions	Files and File Management Functions; IO Objects and Functions; Inter Process Communication Objects and Functions; Interrupts and Interrupt Handling Functions;
4: Acess Control and Data Acess	Capabilities and Capability Interpretation Functions; Logical Adresses and Address Transformations; Physical Adresses and Memory Access Functions;
5: Data Transformation	Elementary Machine Data Types

In UPPER, there is a (conventional) node computer that takes care of the functions of the local operating system (e.g. resource management), the instruction interpretation, and the data transforming operations. In addition, there exists a cooperation handler machine to perform the functions of the distributed operating system, as well as a data access machine to perform address transformations and access right control.

Hierarchical function distribution offers the following advantages:

- It lends itself toward a "natural" modularization of system software and, thus, decomposition of system software complexity.

- It orthogonalizes computer functions and, thus, allows for parallel execution of such functions.

- It provides "natural fire walls" of protection and, thus, facilitates the protection mechanism.

- It allows the various physical machines to be optimally designed for their function in the system.

- It allows innovative multicomputer architectures with complex functionality to be realized with "off the shelf" microcomputer components.

The last point is the most important one and shall be illustrated by an example.

The "classical" approach of realizing an innovative computer architecture without having to invent an innovative hardware structure is vertical migration. Vertical migration means to take a conventional CPU and move ("migrate") some of the higher-level functions of the function hierarchy (see Table 1) "down" into the CPU's microcode. Since presently available microprocessors are "mask microprogrammed" on the chip, a special processor with special microprograms is necessitated by such an approach. A striking example is the "general data processor' (GDP) of the Intel iAPX 432 computer /TYN 81/. In the GDP, some of the lower-level operating system functions listed in Table 1 as well as the data access functions of a capability-addresses, segmented memory are performed by microprograms and supported by a specific set of registers. This required a very costly design, resulting in a structurally conventional CPU with a rather special instruction set.

It is of interest to compare the iAPX 432 with the STARLET computer, since both systems are quality software machines with similar functionality (except for the processing of structured data objects in the SIMD mode which is unique in STARLET). As can be expected, the STARLET architecture, which is based upon hierarchical function distribution, constitutes the more efficient solution and provides a much better performance than the iAPX 432, which is a conventionally structured CPU with unconventional machine data types realized through vertical migration. STARLET was realized on a relatively small budget in a couple of years by a small research group; the iAPX 432 design reqired a very elaborate, extremely expensive VLSI microprocessor design.

Of course, as soon as we are given the opportunity to design our own chips, we may happily put some special circuits such as the memory management unit of UPPER or the reduced instruction set processors of STARLET on proprietory chips e.g., gate arrays. Such a measure will save us some "real estate" on the printed circuit boards, but it is in no way a prerequisite for the feasibility of our architectures. Rather than being economical (unless a computer is produced in large volume), the incentive primarily is to protect proprietary hardware solutions.

The parallel operation provided by hierarchical function distribution is different from the conventional notion of parallel operation in an array of processing elements or a pipeline processor, which refers solely to the data transforming operations. Of course, both kinds of parallel operation may be combined, e.g., by using a pipeline or an array of processing elements as the data transforming unit in the hierarchy of physical machines. A good example for such an approach is MAC.

In summary, hierarchical function distribution not only is a way to achieve new standards of performance and functionality of innovative computer architectures; it also is the only way to accomplish this with existing micro-computer componentry. This and the additional benefit of system software modularization and simplification is the key to very cost-effective and rather powerful "micromainframes" of the future.

6. WILL WE SEE SYSTEMS EVOLVING WITH 100 OR MORE PROCESSORS?

Arvind /AaK 81/ and other designers of data flow computers envision data flow multiprocessor systems that eventually will consist of several 100 or 1000 processors. The Japanese planners of the "Fifth Generation Computer System" program are talking about giant inference machines that will consist of several 100 or 1000 chip computers, each computer having the power of an IBM 360/168 or more.

Compared to such prospects, the multicomputer systems presented in the preceding sections look rather moderate. STARLET operates with 7 microcomputers and a pipeline processor; MAC contains in maximum 10 microprocessors and 2 pipeline processors; and UPPER can consist of in maximum 16 nodes, each node encompassing a node computer (of arbitrary size and structure), a microcomputer that functions as the "cooperation handler" of the node /GaB 81/, and a special address transformation and memory guard processor. And yet, STARLET with all its quality software supporting features and run time consistency checks is capable of performing in certain cases as many as 1 million floating-point operations per second (MFLOPS), and the fully equipped MAC has a maximum processing power of 20 MFLOPS, one fourth of the performance of the CRAY-1 supercomputer.

The point is that STARLET as well as MAC operate on structured sets of data and, thus, can take advantage of the SIMD gain as well as the pipeline gain. This gives pipeline processing of arrays a competitive edge over processing by an array of non-pipelined processors (e.g., microprocessors). For sufficiently long vectors, the parallel processing gain of a pipeline asymptotically approaches the number of stages of the pipeline. A non-pipelined processor costs as much but does not have that gain. Consequently, if the number of stages is k, one needs at least k processors in the array computer to obtain the same performance as the pipeline computer - provided all k processors can be fully utilized.

Over the past years, the notion of pipeline processing of data streams has been generalized to the systolic array approach. A systolic array is a network of processing elements (PE), into which streams of data are "pumped" from a memory, and from which streams of data flow back to the memory. The difference to a pipeline is that (i) pipelines are usually thought of as one-dimensional arrays of PEs, whereas a systolic array may be multi-dimensional, and (ii) the flow of data back to the memory may in the case of systolic arrays include streams of intermediate results that must be pumped again through the systolic array.

In the case of the processing of data streams by a pipeline processor or a systolic array, there never is any problem of data routing or access collision Since the operating speed of a pipeline or sytolic array can always be made as fast as needed, the performance limiting factor is memory bandwidth. If data are read from the memory in the same order in which they were stored, the effective memory bandwidth can be increased by memory interleaving. In MAC, we might apply such a measure but not in STARLET, in which the logical data structures processed by the machine may differ from the physical storage structure /GaB 77/.

In multicomputer systems with a small to medium number of 'processing nodes', the appropriate communication structure is either the ring or the bus. In a ring, there is a shift register inserted into the bus at each node. In the UPPER system, which features an ultra high speed ring, the propagation time for a bit through the shift register is about 60 ns. With 16 nodes, this adds up to approximately 1 µs as the minimum time it takes a message to travel around the ring - if the propagation delay of the node-connecting cables is neglected. If the nodes are several hundred feet away from each other, the time it takes for a message to travel from a node to its nearest neighbor may already be in the order of magnitude of a microsecond. The following conclusions are to be drawn from our considerations:

- the ring is an interconnection structure adequate only for distributed systems;
- one cannot efficiently connect hundreds or even thousands of nodes through a ring.

The paradigm of a very fast asynchronous computer bus is the shared memory bus of the MAC. This bus allows the conflict free access of the memory at a maximum rate of 50 Mbyte per second. The 80 nanoseconds for a 4-byte memory access in MAC includes memory access cycle, bus arbitration, and bus propagation. Because of the asynchronous organisation, no ECL logic is needed. The MAC bus is the appropriate interconnection structure for multicomputer systems with a small to medium number of single board computers, say, up to 24. Any larger structure would cause difficulties just because of its physical extension.

A fast multicomputer bus, called RAPID-BUS, is under design at Carnegie-Mellon University /ZaS 81/. RAPIDBUS is meant to connect in maximum 25 computers with 25 different memory modules. It is safe to assume that the synchronous bus can operate at a basic frame time of 500 nanoseconds, which is subdivided into 25 time slots of 20 nanoseconds each. Each processor is assigned its individual time slot, during which it can send a request to a memory unit or receive a data word from there. This provides a maximum data rate for each processor-memory communication of 2 Mbyte/second. Theoretically, if there were no access conflicts of any two or more processors to the same memory unit, the total data rate in the case of 25 processors would amount to 50 Mbyte/second, i.e. the same rate as on the asynchronous MAC bus. Of course, such a scenario is unrealistic. In the case of a collision, a processor must at least wait for another 500 nanoseconds before it can repeat its request, and it must be expected that the effective data rate will be considerably below that of the conflict-free MAC bus, which provides for unrestricted communication of each processor with every other processor in the system. In any event, the RAPIDBUS by design could not be used either to connect hundreds or more processors. This leaves the interconnection network /FEN 81/ as the only viable solution to the communication problem in systems with a large number (several hundreds or thousands) of processors.

Interconnection networks connect a number of sources with a number of destinations. They may be

- single-stage or multi-stage networks, designed for
- circuit switching or packet switching, using
- centralized or decentralized control.

Single-stage networks (e.g., the shuffle-exchange network) are also called "re-circulating networks" because a message may have to circulate through the network several times before reaching its final destination. For the best known example, the shuffle-exchange network, it can be shown that as many as $3(\log_2 N)-1$ passes may be required for an arbitrary permutation if N is the network size (number of sources and destinations) /WaF 81/. Multi-stage networks (e.g., the N-cube network) can be designed such that any arbitrary source may connected with any arbitrary destination. A multi-stage network that allows for any arbitrary permutation is the BENES network. A BENES network of size N needs $\frac{N}{2}((2\log_2 N)-1)$ switching elements, each element being capable of connecting at a time one of two inputs with one of two outputs. A simpler network, the N-cube network, requires only $\frac{N}{2}\log_2 N$ switching elements but does not provide the generality of non-blocking interconnections of the BENES network.

In the circuit-switched network, a complete interconnection path between source and destination must be established and maintained for the duration of a communication. Thus, the switching elements in the path remain in their specific states, and the other 3 paths that potentially may lead through each of the switching elements involved cannot be established during that time. Contrastingly, in the case of packet switching, a packet travels its way from stage to stage, releasing links and switching elements immediately after having used them. Therefore, packet switching is advantageous in cases where interconnection paths through the network must be change frequently.

Central control of large interconnection networks, e.g. to connect 1024 sources with 1024 destinations, becomes very complex. Sequential routing algorithms need O(NlogN) steps if N is the network size /FEN 81/. The answer to that problem is decentralized control, in which a message carries a destination tag along with it, and the switching elements perform the routing autonomously on the basis of the destination tag information.

As example, let us consider a circuit switching, multi-stage network of the baseline type /FEN 81/ that interconnects 512 processors with 512 memory units. Such a network would be the core of the Numerical Aerodynamic Simulator proposed by Burroughs /BaL 81/. Each switching element consists in that case of a combinational network and a flip-flop for the switching state. Let us assume that a switching element is a 48-pin LSI circuit that can route 1 byte of information from any of 2 inputs to any of 2 outputs (about 10,000 of such devices will be needed !). Realistically, the delay for a bit to travel from an input terminal to an output terminal of the network is estimated to be 500 nanoseconds. To establish an interconnection path, first a processor sends a strobe signal through the network to a memory unit, to which the memory unit responds by sending back an acknowledge signal (the network must be bidirectional). It is further fair to assume that the rate of data transmission through such a complex network, once a path is established, cannot be made higher than 1 Mbyte/second. Hence, a complete cycle of fetching a 32bit word from memory can be expected to take about 10 microseconds in the best possible case, i.e., if the data path is not blocked and the memory unit is free.

If packet switching is applied, the danger of blocking is reduced. Furthermore, the operations of routing and data transmission can be pipelined to some extent. On the other hand, the overhead of routing and latching packets in each stage is increased. Considerations similar to the ones presented for the circuit switching case have lead us to the conclusion that it is fair to assume that the memory access cycle in the packet switching case will not differ too much from what we found in the circuit switching case.

The conclusion is that a microprocessor of the power of a MC68000 or more would at best run at one tenth of its speed if instructions and data had to be fetched from memory via the network. It needs hardly be emphasized that it is much more efficient and less costly to have 10 processors running at full speed without any danger of collisions than 100 processors running at best at one tenth of their full speed, with a high probability of collisions that will further reduce performance. Therefore, it certainly is much better to have a smaller number of very powerful engines and, consequently, a better manageable communication problem, than a larger number of not so powerful processors and a big communication problem. With microprocessors becoming in the future much more powerful than the presently available ones, we are not so sure whether we shall ever see the "1000-processor-systems" evolving.

REFERENCES

/AaK 81/ Arwind, Kathail V., "A Multiple Processor Data Flow Machine That Supports Generalized Procedures", Proc. 8th Annual Symposium on Computer Architecture (May 1981), IEEE Catalog No. 81CH1593-3, 291-302

/BAC 78/ Backus J., "Can Programming be Liberated from the von Neumann Style? A Functional Style and Its Algebra of Programs", CACM 21,8, 613-641

/Bea 82/ Berg H.K. et al.: Formal Methods of Program Verification and Specification, Prentice-Hall, Englewood Cliffs, N.J. 1982

/BaL 81/ Barn G.H., Lundstrom S.F., "Design and Validation of a Connection Network for Many-Processor Multiprocessor Systems", COMPUTER 14,12 (Dec. 1981), 31-41

/CaD 73/ Coffmann E.G., Denning P.J., *Operating Systems Theory*, Prentice-Hall, Englewood Cliffs N.J. 1973

/DEN 81/ Denning P.J., "Computer Architecture: Some Old Ideas that Haven't Quite Made it", CACM 24,9 (1981), 553-554

/FAB 74/ Fabry R.S., "Capability-Based Addressing", CACM 17,1 (July 1974), 403-412

/FEN 81/ Fen T., "A Survey of Interconnection Networks", COMPUTER 14,12 (Dec. 1981), 12-27

/GaB 77/ Giloi W.K., Berg H.K., "Introducing the Concept of Data Structure Architectures", *Proc. 1977 Internat. Conf. on Parallel Processing*, IEEE Catalog No. 77CH1253-4C, 44-51

/GaB 81/ Giloi W.K., Behr P.M., "An IPC Protocol and its Hardware Realization for a High-Speed Distributed Multicomputer System", *Proc. 8th Internat. Sympos. on Computer Architecture*, IEEE Catalog No. 81CH153-3 (1981), 481-488

/GaB 82-1/ Giloi W.K., Bruening U., "A Configurable Array Computer for Signal and Image Processing", in K.S. Fu/T.L. Kunii(eds.), *Image Engineering*, Lecture Notes in Comp. Sc., Springer, Berlin-Heidelberg-New York 1982

/GaB 82-2/ Giloi W.K., Behr P.M., "Making Distributed Multicomputer Systems Safe and Programmable", *Proc. IEEE Internat. Workshop on High-Level Language Computer Architecture* (Dec. 1982)

/GaB 83/ Giloi W.K., Behr P.M., "Hierarchical Function Distribution - A Design Principle for Advanced Multicomputer Architectures", submitted for presentation on the 10th Internat. Symposium on Computer Architecture, Stockholm, Sweden 1983

/GaG 82/ Giloi W.K., Gueth R., "Concepts and Realization of a High-Performance Data Type Architecture", Internat. J. of Computer and Information Sciences 11,1 (1982), 25-54

/GGK 82/ Giloi W.K., Gueth R., Kallerhoff R., "Hierarchical Function Distribution - A Design Principle for Advanced HLLACs", *Proc. IEEE Internat. Workshop on High-Level Language Computer Architecture* (Dec. 1982)

/Hea 82/ Hansen P.M., Linton M.A., Mayo R.N., Patterson D.A., "A Performance Evaluation of the Intel iAPX 432", Computer Architecture News 10,4 (June 1982), 17-26

/KAL 82/ Kallerhoff R., "Object Classes, Data Types, Instructions, and Referencing in STARLET II", Technical University of Berlin, CAMP Report 2/1982

/LAN 81/ Landwehr C.E., "Formal Methods for Computer Security", ACM Computing Surveys 13,3 (Sept. 1981), 247-278

/Nea 81/ Neumann P.G. et al., "Directions for Future Research and Development in Multicomputer Systems", Final Report SRI Project 1826, SRI International (1981)

/PaB 79/ Penny B.K., Baghdadi A.A., "Survey of Computer Communications Loop Networks", Computer Communication 2 (1979), 165-180 and 224-241

/WaF 81/ Wu C., Feng T., "Universality of the Shuffle-Exchange Network", IEEE Trans. Computers C-30,5 (May 1981), 324-332

/TYN 81/ Tyner P., *iAPX 432 General Data Processor Architecture Reference Manual*, Intel Corp., Santa Clara Ca. 1981

/ZaS 81/ Zoccoli M.P., Sanderson A.C., "Rapid Bus Multicomputer Network", COMPUTER DESIGN (Nov. 1981), 324-332

SYNCHRONOUS PARALLEL BUS AS INTERCONNECTION MEDIUM FOR INFORMATION TRANSFER

Rupert Patzelt, Herbert Kern
Institut für elektrische Meßtechnik
Technische Universität - Wien, Österreich

ABSTRACT:

Many different functional and geometric structures exist for the interchange of information. The state of the microelectronics allow to design devices with high dataprocessing capacity and autonomous control capability. The information to be interchanged ranges from single bits to long data-blocks, responses are needed either immediately or may have a long delay. A parallel bus of specific design allows to satisfy the demands for the interconnection of many processors performing individual tasks in parallel and exchanging data as well as control information. The overhead is low for simple devices and can be adjusted to the actual demands.

ZUSAMMENFASSUNG:

Es gibt viele, nach Funktion und Anordnung verschiedene, Strukturen für den Austausch von Information. Der Entwicklungsstand der Mikroelektronik ermöglicht den Aufbau von Geräten mit hoher Datenverarbeitungskapazität und eigenständigen Steuerfunktionen. Der Umfang der ausgetauschten Information reicht von einigen Bits bis zu umfangreichen Datenblöcken, Reaktionen müssen in machen Fällen sofort, in anderen wieder mit nahezu beliebiger Verzögerung erfolgen.
Ein Parallelbus in der entsprechenden Ausführung kann allen Anforderungen zur Verbindung von vielen Prozessoren genügen, er erlaubt, einzelne Teilaufgaben gleichzeitig auszuführen und sowohl Daten als auch Steuerinformation zu übertragen. Auch für einfache Einheiten kann der notwendige Aufwand gering gehalten und an die gegebenen Anforderungen angepaßt werden.

RÉSUMÉ

Il existe de nombreuses structures fonctionnelles et géométriques différentes qui permettent l'échange des informations. L'état de l'art en microélectronique permet de concevoir des appareils pourvus d'une grande capacité de traitement de données avec commande autonome.

La taille des informations à échanger va du simple bit au long bloc de données; d'autre part, on peut exiger des réponses immédiates ou accepter de grands retards. Un bus parallèle de conception particulière permet de satisfaire les demandes d'interconnexion de nombreux processeurs exécutant des tâches individuelles en parallèle et échangeant aussi bien des données des informations de commande.

Le temps système est faible pour les systèmes simples et peut être ajusté aux demandes effectives.

1. INTRODUCTION

A parallel bus with a synchronous timing for the transfer of the information and control signals can be used as an extremely efficient medium for the fast exchange of data between many source-devices and destination-devices. The connections can be established and used practically independet from each other. A functional handshake-mechanism maintains a degree of safety, equal to that of asynchronous timing systems, by the use of status-response signals, that indicate the availability and acceptance of the information. Established connections can be protected against interfering operations to obtain functionally uninterrupted operation sequences.

It is investigated which features have an essential influence on the performance of the parallel-bus-system. The limits for the data-throughput, the rate of independent operations and alternating accesses and their dependence on the method of controlling the transfers and on the type of the timing are considered. An abstracted ideal system is defined, that allows to separate the different influences and to establish features that give an optimal performance for a system with many processors that execute in parallel individual tasks belonging to a common overall program.

The devices have a layered structure: the drivers and receivers connected to the bus-lines, the bus-interface (BIF) and the application-logic (APL). The bus-lines have groupwise separate functions: they transfer information (INF), transfer-control (TCS), transfer-status (TSS) and timing (TIM) signals. The timing signals define the instants of actions of the bus-interfaces, i.e. when signals are to be transmitted to the buslines, when they may change, when they must be stable and when they are to be accepted from the buslines. The control command (TCS) defines the transfer-path by selecting those BIFs, that take part in an operation, and defining the type of the actions. The status signals (TSS) indicate if the BIFs perform the action defined by the control command. The information signals contain all kind of data to be transferred from one application-logic to another application-logic, like pure data, internal addresses, interrupt information, error messages etc. They are not processed inside the BIF, but passed from one APL to the other.

By this concept the occupation of the bus for every data transfer can be kept at a minimal time, defined by the reaction time of the logic in the BIF and the propagation delay from the line-drivers via the bus line to the line-receivers. During all the delays inside the APL and between APL and BIF the bus is available for operations of other devices. The data-throughput can be high and the number of tasks, that can utilize the bus in a time-sharing manner without essential delays, can be large. Simple reservation mechanisms in the BIFs and appropriate signals in the control-command allow to establish virtual connecting links between devices, that provide for functionally uninterruptable operation sequences (test and set, block-transfers, etc.) as in conventional bus-systems, but without blocking the bus for operations of other tasks.

2. DEFINITION

For the explanation of the parallel-bus-system the following definitions are used and the following assumptions are made:
- A BUS-OPERATION is controlled by a COMMANDER, one or more RESPONDERs take part in the operation. In a bus-operation information is passed from the TRANSMITTER via the bus to one or more RECEIVERs.
- Every DEVICE connected to the bus may be any combination of commander, responder, transmitter and receiver.
- The BUSINTERFACE (BIF) of a device is separate of and operates independently from the APPLICATIONLOGIC (APL) of the same device. Specificly it must accept and analyse every control-command immediately and indicate by status-signals (TSS) without essential delay, if it is able to perform the operation according to the command.
- The BIF of every device in a system reacts on a specific address-code, the BUS-ADDRESS. The internal address-code for memory-cells and registers within a device is called the LOCAL ADDRESS. It should be noted, that the global address used in modular systems is always devided into these two part according to the distribution of address-ranges in the memory-map.
- The INFORMATION-TRANSFER from the APL of one device to the APLs of other devices is the main purpose of the bus. Bus operations for this purpose are called data or information transfers.
- If specific bus operations are used to control the BIF of a device they are called CONTROL-OPERATIONS.
- The CONTROL-COMMAND is the part of a bus-operation, that is relevant for the BIF of a responder. Typically it prepares the succeeding transfer of data or, more general, information. The bus-address is considered to be part of the control-command. A control-operation may consist only of a control-command.
- The selection of a commander for a bus operation out of a group of competing commanders is called ARBITRATION. The same expression is used if a (requesting) commander takes the control from another (active) commander. The (program-controlled) transfer of the access-right by the active commander to another one is called MASTERSHIP-TRANSFER or TOKEN-PASSING.
- The TRANSFER-CONTROL-SIGNALs (TCS) on the bus define the transfer-path on the bus and the actions inside the BIFs. They represent the control-command.
- The TRANSFER-STATUS-SIGNALs (TSS) are generated by the BIF as reaction to the command received and according to the internal status of the BIF. They do not depend on any feedback from the APL, that would cause an additional delay.
- All different types of INFORMATION transferred from one APL to another APL (e.g. data, local address, interrupt vectors, error messages) are handled similarily. The bus and the BIFs are transparent for this (application) information.
- The information to be transmitted by a BIF must be present in the BIF when the command is received, information to be received must be accepted in the BIF independent from the time necessary to transfer it to the APL of the same device. The TSS are used to indicate if these conditions are fulfilled. If they are not fulfilled, the commander is responsible to take care of the situation. It can either repeat or extend the operation.
- The SIGNAL-DELAY between two BIFs is the propagation delay from the input of the line-driver in the transmitting BIF to the output of the line-receiver in the receiving BIF. It includes the transition-time on the bus-line and the travelling-time along the bus-line, that is on short busses in

most applications negligible.
- The COMMAND-REACTION-TIME of the BIF for the control-command consists of the decoding of the received TCS, generating the internal control signals for line-transceivers and the latches, and preparing the TSS, typically a number of levels of logical decisions in sequence.
- The SIGNAL-REACTION-TIME of the BIF on status-signals consists on the reaction on one signal or a simple signal-pattern, typically only one level of logical decision.
The (INFORMATION-)ACQUISITION-TIME of the BIF for the information consists of the time necessary to latch the information-signals into the input-register inside the BIF.
- The COMMAND-TRANSFER-TIME consists of signal-delay plus reaction-time, the INFORMATION-TRANSFER-TIME of signal-delay plus acquisition-time.
- The BUS-OPERATION-CYCLE for an information-transfer consists of the sum of the command and information transfer-times. In write-transfers these times can overlap partially, for read-transfers the full sum is necessary. It is essential to recognize, that the receiving BIF must receive and decode the control-command before it can accept the information. This is true even for a write-operation in a parallel-bus with separate lines for control, address and data. If the command-phase of one operation overlaps with the data-phase of the preceeding, the data-throughput of the bus can be higher, because the duration of every transfer-operation is defined only by the longer of these both phases. If a block of data is transferred via an established connection, the control-command needs in principle not to be repeated for every data-transfer. This can be considered either as one multiple bus-operation or as one full bus-operation followed up by abbreviated ones.

The performance of a bus system is defined by different characteristic values, depending on the configuration and the kind of operations demanded by the application:
- The DATA-(THROUGHPUT-)RATE refers to the transfer of a long block of data via an established connection. It is basically defined by the reciprocal of the data-transfer-time, if a specific mode of operation for block-transfers is provided and only one control-command is used for the whole data-block. If every data-transfer must be prepared by an individual control-command the data-rate is not higher than the rate for independent operations.
- The execution of independent operations of the same commander depends on the full length of the bus-operation-cycle and includes the transfer of all response signals to the commander. It is called (INDEPENDENT-)BUS-OPERATION-RATE.
- If different commanders use the bus alternately by independent accesses an arbitration between every two accesses is necessary. The characteristic value for this mode of operation of the bus-system is called in this article ACCESS-RATE. It is the reciprocal of the sum of arbitration-time, command-transfer-time and information-transfer-time.
- If operations are initiated by responder-devices by a request, the time for recognizing, identifying and processing the request is to be taken into account additionally. If the request can be handled immediately, this is only the REQUEST-PROCESSING-TIME. If the active commander may perform some operations before accepting the request a REQUEST-WAITING-TIME is to be added.

If there may be more than one requesting responder, the REQUEST-LATENCY-TIME is to be added for the responders with lower priority. These times depend on the configuration and the design of a specific system and the momentary situation. In all these cases no fixed values can be stated, but only average values and sometimes also maximum values.

- DATA-ORIENTED SYSTEMS deal with information that is independent of the time and that is not changed by time-delays. A delay or the repetition of operations has no effect on the results.
- PROCESS-CONTROL SYSTEMS deal with information about external irreversibly proceeding operations. The time is an essential parameter of the information, every delay in the system has an effect on the results.

3. GEOMETRICAL STRUCTURE

Star of serial full-duplex links with central exchange, meshed network of serial full-duplex links, serial (simplex) ring, extended medium (reflections absorbing), parallel (half-duplex) bus.

Serial full-duplex links provide very high data-rates, especially for the transfer of data-blocks. The response is fed back after every data-block. The necessary framing of a message (leader and trailer) consist usually of some bytes, the overhead is high for short messages. Every device is connected with the central exchange in a star or with every other device in a mesh. The mesh provides the optimal accessibility, but the amount of circuitry increases with the square of the number of devices and becomes prohibitive above about 8 devices. The star needs a central interconnection. In all serial connections the protocol demands an appreciable overhead for the framing of every message (synchronisation, control information, data protection etc.), that is for short messages a multiple of the data-content.

A serial simplex ring has a very low amount of circuitry. It allows also a very high data-rate, but only one interconnection can be activated at a time. The time for the arbitration or the mastership-transfer is relatively long and increases with the geometrical length of the ring. Every response is delayed by one roundtrip signal-delay.

An extended medium (with absorbing terminations at every end) behaves very similar to the ring. It has still less circuitry than an unidirectional ring, but some disadvantages arise from the uncontrolled propagation of the signals.

A parallel half duplex bus allows a very fast arbitration and a high rate of independent bus-operations. The data-rate on every single line of a parallel-bus is essentially lower than on a serial link because of the necessary deskewing, synchronizing and controlling of the parallel signals. The response is possible within or immediately after every operation. The overhead for synchronization and framing of short messages is very low. All other devices are directly accessible from every device. The circuitry is independent from the number of devices connected to the bus. Different mechanisms exist, that allow very fast and flexible arbitration by exploiting the specific features of a parallel bus. A maximal latency and waiting time can be guaranteed.

The parallel bus is the structure that allows the highest rate of independent transfers of short messages, while the meshed network or the star of serial links allow the highest capacity for transferring messages of long data-blocks. The ring and the medium demand much lower effort of circuitry, but provide lower capacity, specificly for short independent messages.

4. FUNCTIONAL STRUCTURE

Centralized control with initiation of data-transfers by requests and transmission between devices by store-and-forward, partially decentralized control with an essential part of the devices purely passive (with respect to the control of operations), fully decentralized control without passive devices.

The low cost of µPs allows to equip every device with its own processor, primarily for all local operations, but also for the direct transmission of results to other devices. This fully decentralized control results in many short data-transfers via permanently changing connections and controlled by different commanders. A fast arbitration becomes the most essential feature. Interrupts (or service-requests) become less important or unnecessary, since the devices perform all operations under their own control. Messages are deposited in mailboxes, either directly in the receiving devices or in a central device.

Only with decentralized control of the bus-operations it is possible to make real use of the advantages of distributing parts of an overall program to many processors. Centralized control would create an excessive overhead for the management of the tasks running in parallel.

5. MESSAGE STRUCTURE

Message-types: Data-blocks or long messages: time independent, delayed logic response sufficient; control-messages, real-time information (measurement results) and service requests: short, time dependent, immediate functional response necessary.

In data-transfer oriented systems the relevant values are the data-rate and the integrity of the data, the information does not depend on time-values. In systems for the control of external "processes" the time and time-differences are values with the same relevance as the data. The messages are short, the transmission delay and the reaction time are essential values. A mixture of all the message types mentioned above represents the most general case of demands for a system. Since the operation of any kind of system has at least some components similar to the process-control these demands should be included in the considerations for the design of generally applicable systems.

6. TIMING OF BUS-OPERATIONS, PERFORMANCE OF THE BUS

The conventional method for fast and safe execution of data-transfers is the so-called asynchronous or handshake timing. The transfer-protocols of many bus-systems are examples for that.

Protocol of an asynchronous handshake-timing.
An operation consists generally of the following steps in sequence (it is assumed that the commander is already active, possesses the access-right to the bus):

1. Transfer of the control-command: An operation-indicator (BUSY) is set and all those signals (CONTROL and BUS-ADDRESS) are asserted by the commander that define the transfer-path (the addresses of the transmitting and the receiving device) and the type of operation. The commander sets the timing start-signal (MASTER-SYNC or STROBE), representing the first clock-transition of the operation t1. Also WRITE-DATA and LOCAL-ADDRESS are asserted.
   * LOCAL-ADDRESS and WRITE-DATA are asserted before the control-command is decoded, although they can not yet be accepted by the responder.

2. Decoding of the control-command in the BIFs of all existing devices: When STROBE is received, all BIFs analyse BUS-ADDRESS and CONTROL. The BIF, that is addressed, enters the addressed state, accepts the control-command and prepares for the execution of the operation. All other BIFs, that are not addressed, enter an inactive state until the operation is finished.
   * The completion of the decoding is not indicated by a signal.

3. Preparation of the data-transfer: As soon as the addressed state is entered, LOCAL-ADDRESS is transmitted to the APL of the responder and READ-DATA are fetched internally or the acceptance of WRITE-DATA is prepared. The responder indicates the completion of this actions by the timing-response-signal (SLAVE-SYNC or ACKNOWLEDGE), that represents the second clock-transition t2 and adapts the duration of the clock-period to the performance of the device.
   * CONTROL and ADDRESS remain active, although they are not used any more by the responder. The usual ACKNOWLEDGE is a purely timing signal without real status-information about the acceptance of the received data. In operations addressing more than one responder it contains no information, how many of them have really performed the operation, i.e. if the combined READ-DATA or status-response-signals are valid and if all responders have accepted the information of a broadcast-transfer.

4. Data-transfer: When ACKNOWLEDGE indicates the preparation of the data-transfer, WRITE-DATA are accepted by the responder or READ-DATA and possibly status-response signals are present on the the bus-lines to be accepted by the commander.
   * WRITE-DATA remain active, although they are already accepted by the responder.

5. Termination of the data-transfer by the commander: When ACKNOWLEDGE is received READ-DATA are accepted and all signals (CONTROL, ADDRESS, WRITE-DATA and STROBE) except BUSY are withdrawn from the bus-lines by the commander. The reset of the timing start-signal represents the third clock-transition t3 of the bus-operation, that is necessary as preparation of t1 of the next handshake-timing cycle.
   * READ-DATA remain active, although they are already accepted by the commander. The remainder of the bus-operation could in principle overlap with the beginning of the next one, except for the preparation of the timing signals.

6. Signal-deactivation by the responder: When the responder recognizes, that STROBE is reset, all signals (READ-DATA, status-response and

ACKNOWLEDGE) are withdrawn. The reset of the timing response-signal respresents the forth clock-transition t4, that is necessary for the preparation of t2 of the next handshake-timing cycle. The BIFs of all devices enter a waiting state expecting the next control-command.

7. Termination of the operation: When the commander recognizes, that ACKNOWLEDGE is reset, it starts a next operation or withdraws BUSY to indicate that the bus is not any more used and another commander may access the bus.

The handshake timing-signals have multiple functions:
- Clock: The edges of STROBE and ACKNOWLEDGE define the instants when actions in the BIFs are to be executed. They constitute by that the clock necessary for every digital system.
- Adaptation of clock-period: The generation of the two complementary signals by commander and responder respectively provide an adaptive adjustment of the period of the clock to the appropriate speed for the devices taking part in the operation and the signal-delay between them. It is to be noted, that typically inside the BIFs additional timing circuitry is necessary (e.g. for deskewing) with tight tolerances.
- Transfer-status: The states indicate that information is valid or accepted.
- Functional interlock: The mutual generation and acceptance of the both signals provide the interlock between the termination of one step of the transfer-operation and the initiation of the next step.

As a result of the combination of these different functions an essential part of the time is lost, during which the BIFs wait for the arrival of response-signals and the bus-lines are occupied by signals, that are not any more used.

7. POSSIBLE IMPROVEMENTS BY A SYNCHRONIZED TIMING WITH FUNCTIONALLY HANDSHAKING STATUS-SIGNALS

The most essential delay is caused in read-transfers by the step 3., preparation of the transfer, when data are fetched from the APL. The duration of this action depends on the internal features of the APL of one device, that influences by that the performance of the bus as a whole.

By a strictly layered structure of the devices, in which the functions of BIF and APL are separate and independent, any influence of the processing delay inside the APL on the performance of the bus can be avoided. Read-operations are split into a preparatory transfer of the command with the local-address and the corresponding data-transfer (controlled either by the initial commander after a proper delay or by the initial responder as soon as the data are presented to the BIF).

The inherent delays of the handshake-protocol can be avoided by a protocol similar to that used by most µP-busses. A clock (with constant period) and interlocking control-signals and status-response-signals (like wait or ready) provide the adjustment of the data-transfer to the response-time of the individual devices in steps of clock-cycles without mutual signal-exchanges, when they are not necessary.

The signal-delay on the buslines together with the reaction-time of the BIF and the acquisition-time in the BIF constitute the limiting factors for the data-rate and the bus-operation-rate that can be achieved in a bus with a synchronous, or better, synchronized transfer-protocol.

For an optimal performance of a set of devices all the BIFs should have similar values of these limiting factors, but every system is fully operable as long as the clock period is adjusted to the values of the slowest BIF in the system.

A full protection of the operation-integrity as in the handshake-timing can be provided by proper transfer-status-signals (TSS, wait and ready), generated by the transmitter and the receiver (irrespective which of them is commander or responder). Ready-signals are sufficient if only one responder takes part in an operation, wait-signals are needed additionally for operations with multiple responders. Individual signals for the transmitter (=information valid) and the receiver (=information accepted) are needed, if a commander should be enabled to control a transfer from one responder to another. Four TSS, wait and ready for transmitter and receiver make the comprehensive set of status-signals for all kinds of transfer-operations, with single or multiple source and destination.

The integrity of a sequence of operations (like address and data, test and set, read-modify-write) can be guaranteed by a reservation-semaphor that is set in the responder by every initial operation of a commander and reset by the terminating operation.

If every commander transmits together with the command an identification-code and this is stored in the responder and compared with the identifications received in subsequent operations, this identification can be used to lock a responder to a specific commander or task for subsequent operations. This feature allows a kind of decentralized resource allocation and management. Also virtual links can be established, e.g. for block transfers. With a control-command addressed to a responder the commander sets the reservation by depositing its identification. All subsequent operations under the same identification are performed by the reserved responder, until the reservation is reset. Attempts to access the responder under an other identification are rejected. Many such links can be used by interleaved bus-operations between different pairs of devices. A device that is partner of such a link as a responder can, at the same time, use links as a commander to the same device or others.

This feature is an example for the inherent upward compatibility (with restricted performance for the simpler devices): Devices with this more complex type of BIF-logic can be used in a system together with simpler devices without this reservation mechanism. The links can be established only between the more complex devices, for the other devices a full sequence of command and transfer operation is always necessary and the same level of integrity of operation-sequences can be guaranteed only by additional software and the corresponding operations referring to the resource-management (locally or centrally located).

8. PROTOCOL OF A SYNCHRONIZED TIMING WITH FUNCTIONAL HANDSHAKE

1. Transmission of the control-command: After the first transition of the clock t1 the commander asserts all the signals necessary for the responder-BIFs to prepare the information transfer (the TCS, i.e. CONTROL, BUS-ADDRESS, possibly commander or task identification). The INF and TSS lines are not used and may carry overlapping signals of the preceeding bus-operation.

2. Decoding of the control-command: After t2 all responder-Bifs analyse the control-command, the addressed ones enter the addressed state and prepare the actions according to the command.

3. Transmission of the information and the transfer-status-signals: After t3 the line-drivers of the transmitter and the line-receivers of the receiver are enabled routing the INF and TSS according to the command. The state of the TSS indicates if the activated BIFs are performing the commanded actions properly, i.e. the INF is valid and will be accepted in the next step. The TCS lines are not any more used and may be used therefore for the preparation of the next bus-operation.

4. Acquisition of the information and decoding of the transfer-status: After t4 the INF is latched into the register in the receiving BIFs and the TSS are decoded, so that all devices taking part are informed about the progress of the bus-operation and can take appropriate actions, like accepting or rejecting INF, repeating the bus-operation, transmitting status or error information, entering a status-interrogation or an error-recovery procedure etc.

The assertion of signals to the bus-lines takes always place during the same clock-interval as the withdrawal of the signals of the preceeding operation.

9. CONCLUSIONS

A bus-system with the features described above allows many processors to use the bus for the exchange of information with all devices of the system. The access-delay is negligible for low to medium load of the bus. The time consumption for normal operations is minimal. All kinds of exemptions from normal operation and error-conditions are indicated, so that appropriate counter-actions can be performed as necessary. Momentary or permanent overload results in a graceful degradation of the performance, but any destruction of operations or blocking of devices or tasks can be avoided. The same applies for passive failures of devices. The essential features are:

- The layered structure of the design of the devices, separating the specific features of the bus itself together with the bus-interfaces and the application oriented part of the devices, avoiding delays by the APL.
- Fast arbitration, that provides a short access-delay for the highest priority.
- The synchronized timing for the fast transfer of the signals, providing short bus-operations.
- The short duration of every bus-operation results in a low probability that the bus is occupied by another commander and by that in low average access-delay.
- The functional interlock of all operations by the handshake-status-signals, insuring the integrity of every single operation.
- The reservation mechanisms, providing for the integrity of operation-sequences without excluding inserted operations of other commanders.
- The reservation by the commander-identification, providing virtual links.
- The coded control-commands, that provide a high degree of flexibility and adaptability to specific and future demands.

Comparison of the time-consumption of different procotocols.

Times: SD = signal-delay
CRT = command-reaction-time
IAT = Information-acquisition-time
SRT = signal-reaction-time
" indicates overlapping actions.

Abbreviations:

ACK = acknowledge CMDR = commander
INF = information RPDR = responder
STR = strobe TCS = command
TSS = status-response
W-INF = write-information
XX* = XX inactive.

Asynchronous handshake-timing:
(Write operation with address and data parallel)

Actions:	Time:
1. TCS and STR asserted by CMDR W-INF transmitted by CMDR	1 SD
2. received, decoded by RPDR	1 CRT
3. Transfer-preparation in RPDR-APL	undef.
4. W-INF accepted by RPDR ACK asserted, (status-response transmitted by RPDR)	1 IAT" 1 SD"
5. ACK received, detected by CMDR	1 SRT"
6. STR reset, TCS and W-INF withdrawn by CMDR	1 SD
7. STR* received, detected by RPDR	1 SRT
8. ACK reset, (status-response withdrawn by RPDR)	1 SD
9. ACK* received, detected by CMDR	1 SRT

Synchronized timing with functional handshake:

Actions:	Time:
1. TCS asserted by CMDR	1 SD
2. received, decoded by RPDR	1 CRT
3. INF and TSS transmitted	1 SD
4. accepted	1 IAT

For operations transferring local-address and data the same sequence is needed twice, the address-transfer may overlap with the command for the data-transfer:

1. CMD1 asserted by CMDR	1 SD
2. received, decoded by RPDR	1 CRT
3. INF1 + TSS1 transmitted, CMD2 asserted	1 SD
4. accepted, decoded resp.	1 IAT" 1 CRT"
5. INF2 + TSS2 transmitted	1 SD
6. accepted	1 IAT

PALME - AN AUTOMATIC MIXED PACKET/LINE EXCHANGE

Y. Corcia, J.L. Dauphin, G. Froger, O. Louvet, Y. Simon

Centre National d'Etudes des Télécommunications, Lannion, Belgium

ABSTRACT

PALME is an automatic star-network exchange which simultaneously handles traffic in packet mode and in line mode. After giving our reasons for carrying out this study, we present details of the functional architecture.

Part two explains the physical structure and attempts to demonstrate the original features of the solution adopted. After that the detailed software organisation into basic software and application software will be dealt with. The basic software enables the application software to use a multiprocessor machine in a manner which is transparent to the number of processors and to any conflict situations. In addition, it continuously monitors the consistency of operations by the applications software with respect to the tasks managed.

The last section deals with implementation and performance problems.

RÉSUMÉ

PALME est un autocommutateur en etoile traitant de façon simultanee les echanges en mode paquet et en mode circuit. Apres avoir presente les raisons qui nous ont amenes a effectuer une telle etude, nous presenterons de façon detaillee l'architecture fonctionnelle.

Une deuxieme partie expliciters la structure physique et tentera de mettre en lumiere l'originalite de la solution retenue. L'organisation du logiciel detaillee entre logiciel de base et logiciel d'application sere ensuite abordee. Le logiciel de base offre a l'application l'usage d'une machine multi-processeur de façon transparente au nombre de processeurs et aux eventuels conflits. De plus il surveille continuement la coherence des actions de l'application vis a vis des objets manipules.

Dans une dermiere partie nous aborderons les problemes de mises en oeuvre et de performances.

ZUSAMMENFASSUNG

PALME ist eine automatische Sternnetz-Vermittlungsstelle für die gleichzeitige Vermittlung von Datenpaket- und Fernsprechleitungsverkehr. Der Darlegung unserer Gründe für diese Untersuchung folgen Einzelheiten der Funktions-architektur.

Im zweiten Teil wird die physikalische Struktur erläutert und versucht, die Originalität der angenommenen Lösung hervorzustellen. Danach wird die detaillierte Softwareeinteilung in Grundsoftware und Anwendungssoftware behandelt. Mit der Grundsoftware wird der Anwendungssoftware ermöglicht, ein Multiprozessorsystem auf eine für die Anzahl von Prozessoren und für mögliche Konfliktsituationen transparente Weise zu benutzen. Darüberhinaus überwacht es kontinuierlich die Kohärenz der Abläufe der Anwendungs-software hinsichtlich der Aufgabenausfuhrüng.

Im letzten Teil werden die Probleme der Inbetriebnahme und Leistungen angesprochen.

PALME - UN AUTOCOMMUTATEUR MIXTE CIRCUIT PAQUET

CORCIA yvon, DAUPHIN jean louis, FROGER gerard, LOUVET olivier, SIMON yves

Centre National d'Etudes des Telecommunications
BP 40- 22301 LANNION
Tel (96) 383176 ou 381111

PALME est un autocommutateur en etoile traitant de facon simultanee les echanges en mode paquet et en mode circuit. Apres avoir presente les raisons qui nous ont amenes a effectuer une telle etude, nous presenterons de facon detaillee l'architecture fonctionnelle.

Une deuxieme partie explicitera la structure physique et tentera de mettre en lumiere l'originalite de la solution retenue. L'organisation du logiciel detaillee entre logiciel de base et logiciel d'application sera ensuite abordee. Le logiciel de base offre a l'application l'usage d'une machine multi-processeur de facon transparente au nombre de processeurs et aux eventuels conflits. De plus il surveille continuement la coherence des actions de l'application vis a vis des objets manipules.

Dans une derniere partie nous aborderons les problemes de mises en oeuvre et de performances.

1 INTRODUCTION

La diversification des services offerts aux abonnes dans les reseaux des telecommunications conduit a poser le probleme de l'INTEGRATION selon le double point de vue de l'abonne et de l'exploitant. Parmi les activites dans le domaine des reseaux integrant la telephonie et les donnees(RITD), le CNET explore les solutions techniques succeptibles d'etre appliquees, en particulier en matiere de commutation mixte circuit-paquet. Ceci a amene le lancement des 1979 d'un projet de commutateur multiservices, le projet PALME, dont l'objectif est d'evaluer les solutions techniques et d'apprehender les problemes de mise en oeuvre.

Les choix de mode de transport entre circuit et paquet sont les suivants;
    les voies de parole du service téléphonique sont traitées en mode circuit
    les informations de données, ainsi que celles de signalisation, sont traitées en mode paquet selon le protocole standard X25.

PALME offre un service de transport mixte circuit-paquet capable de supporter l'ensemble des applications entrant dans ces catégories de service.

LRC : liaison mode circuit
LRX : liaison mode paquet du XPC
LRP : liaison mode paquet de la périphérie
LRP : liason mode paquet du reseau Transpac
ETTD-P : équipement de transmission de données mode paquet X 25

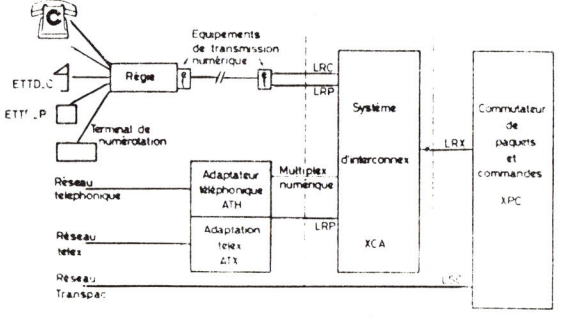

MAQUETTE PALME - ARCHITECTURE GENERALE

2 - ARCHITECTURE GENERALE (figure 2)

PALME est un autocommutateur mixte de liaisons en mode circuit et de liaisons en mode paquet.

Les liaisons desservent:
- soit des abonnés et leurs divers périphériques,
- soit des jonctions appartenant aux grands réseaux publics de télécommunications.

Les fonctions de cet autocommutateur sont regroupées dans deux ensembles principaux:
- un frontal: commutateur de messages, le XCA, qui assure la connexion des canaux en mode circuit LRC et l'aiguillage des trames échangées sur les canaux en mode paquet entre les liaisons LRP et les liaisons LRX
- un calculateur spécialisé (XPC) offrant par les LRX un lot de ressources de traitement. Il assure:
  - la commutation des paquets transitant sur les LRP,
  - le traitement des signalisations d'appel et de libération,
  - les fonctions d'exploitation de l'ensemble: autocommutateur, adaptations, périphériques d'abonné et jonctions des réseaux publics.

2.1. CONNEXIONS DE CIRCUITS ET ALLOCATION DE RESSOURCES (XCA)

L'ensemble XCA est organisé autour d'un commutateur temporel de voies numériques, chargé d'interconnecter, à la demande, les voies entrantes et sortantes de multiplex à 2 Mégabits/s (MUX).

Le commutateur est environné d'un groupe d'interfaces de raccordement au réseau (IRR). Les IRR ont pour mission:
- d'effectuer le multiplexage des liaisons de raccordement LRR, sens entrant et le démultiplexage des voies du MUX, sens sortant;
- d'émettre le code de repos adapté à la nature de la liaison LRR: LRC, LRP ou LRX,
- de détecter le début des trames reçues sur les LRP et de solliciter une connexion immédiate de la LRP sur une LRX disponible.

L'interface de marquage (IMQ) collecte les demandes de connexion en provenance des IRR et du calculateur XPC, mais n'exécute la requête que si la ressource souhaitée peut être allouée.

L'ensemble XCA assure le transit des trames du niveau 2 des liaisons en mode paquet entre la périphérie et l'organe central de traitement. Il aiguille la trame reçue des adaptations vers une ressource traitement, accédée par une LRX. Dans l'autre sens il aiguille la trame émise sur la LRX vers la LRP desservant l'adaptation destinataire, à condition cependant que cette LRP soit libre. La connexion, qui réalise l'aiguillage, ne dure que le temps de transit de la trame échangée.

La commutation de messages "trame" s'accompagne d'un effet de concentration-expansion entre un nombre maximal de 43 liaisons LRP et un nombre maximal de 7 liaisons LRX.

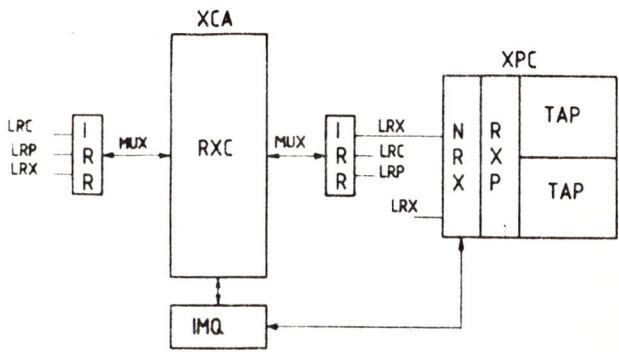

EXP : Exploitation
IMQ : Interface de marquage
IRR : Interface de raccordement au réseau
NRX : Niveau trame X25
RXC : Commutation temporelle numérique
RXP : Commutation de paquets
TAP : Traitement des appels
XCA : Commutation de circuits et allocation de ressource

Figure 2: Architecture fonctionelle de la machine "PALME"

2.2. CONNEXION DE PAQUETS ET COMMANDE XPC

Cet ensemble regroupe:
- le mécanisme de réception et d'émission des trames sur les LRX (NRX),
- le commutateur de paquets (RXP),
- le noyau de traitement des appels (TAP),
- l'ensemble des fonctions relatives à l'exploitation du système (EXP).

Les quatre sous-ensembles sont reliés entre eux par des liaisons fonctionnelles internes au XPC. Ils n'accèdent à la périphérie que par l'intermédiaire des liaisons en mode paquet LRX.

2.2.1 LE NIVEAU TRAME (NRX)

Placé en frontière du XPC, le niveau trame assure la fonction d'entrée-sortie des messages-trame échangés avec la périphérie par l'intermédiaire des LRX, à travers le XCA.

Il obéit aux spécifications du niveau 2 de la norme X25. La réception d'une trame est subordonnée à la recherche d'une liaison LRX libre et à l'allocation par connexions réalisées dans l'IMQ.

2.2.2 LA COMMUTATION DES PAQUETS (RXP)

Le RXP effectue la commutation des paquets échangés sur les diverses voies logiques desservies par l'autocommutateur.

Les paquets de données, reçus sur les voies logiques entrantes des circuits virtuels commutés sont, après transit, émis sur les voies sortantes destinataires.

Les paquets de signalisation sont échangés entre les voies logiques et le noyau logique de traitement des appels TAP.
Des paquets de données sont échangés entre les noyaux fonctionnels TAP et EXP et la périphérie sur des circuits virtuels permanents spécialisés.

La fonction de commutation de paquets inclut les actions
- de contrôle de flux
- de réinitialisation des circuits virtuels,
- d'initialisation et d'allocation des ressources de transit,
- de traitement des paquets d'interruption.

2.2.3. LE TRAITEMENT DES APPELS (TAP)

Le sous-ensemble TAP supervise tous les protocoles d'établissement et de libération des divers types de communications: locale, entrante, sortante, pour tous les types de services assurés: téléphone, données, télex et télécopie.

Il effectue principalement:
- l'interprétation des messages d'appel,
- la recherche du terminal demandé le mieux adapté,
- l'enchaînement des phases de la communication et des signalisations échangées,
- la libération forcée des circuits virtuels commmutés, en cas d'évènement fâcheux.

2.2.4 L'EXPLOITATION (EXP)

Le sous-ensemble EXP coordonne les actions qui concourent à l'exploitation et à la maintenance de l'ensemble: autocommutateur, adaptations et périphériques. Il assure principalement:
- l'initialisation de l'ensemble de la maquette,
- la reprise sur les liaisons en mode paquet,
- la gestion des abonnés et de leurs divers périphériques,
- la surveillance des conditions d'exploitation du système: trafic, charge, observations,
- la maintenance des équipements: traitement des signalisations
   d'anomalie et d'alarme, traitement des positionnements,
- le dialogue homme-machine avec les opérateurs chargés de
   l'exploitation.

3 STRUCTURE PHYSIQUE

3.1 ORGANISATION GENERALE (Figure 3)

La structure physique du système PALME est déduite de l'architecture fonctionnelle. Elle comprend 3 parties distinctes:

1- La PERIPHERIE qui recouvre l'ensemble suivant:

. les régies d'abonnés
. l'adaptation aux réseaux téléphonique et télex (ATR)

Les régies d'abonnés sont pourvues d'équipements de transmission (ETD) permettant leur implantation distante du coeur du système.

2- Le FRONTAL d'interconnexion (XCA) assurant les fonctions de commutation de circuits et d'allocation de ressources "processeurs".

3- Le système MULTIPROCESSEUR (XPC) assurant les fonctions de commande et de commutation de paquets.

Ce système multiprocesseur est relié au Réseau TRANSPAC par une liaison synchrone à 9600 bits/s conforme à la norme X25.

L'ATR assure les fonctions d'adaptation au réseau téléphonique (ATH) et d'adaptation au réseau télex (ATX).
L'interface entre l'ATR et le frontal XCA comprend:
- un multiplex à 2 Mégabits/s transportant 30 canaux de parole provenant du réseau téléphonique,
- une liaison en mode paquet (LRP) sur laquelle sont transmis les paquets de données et de signalisation.

RGI Régie d'abonnés
ATR Adaptation Réseau Téléphonique et télex
ETD, ETL Equipement de Transmission Distant, Local
IRR Interface de raccordement au RXC
RXC Réseau de connexion de circuits
PXC Processeur de commutation de paquets et de commande
MC Mémoire commune

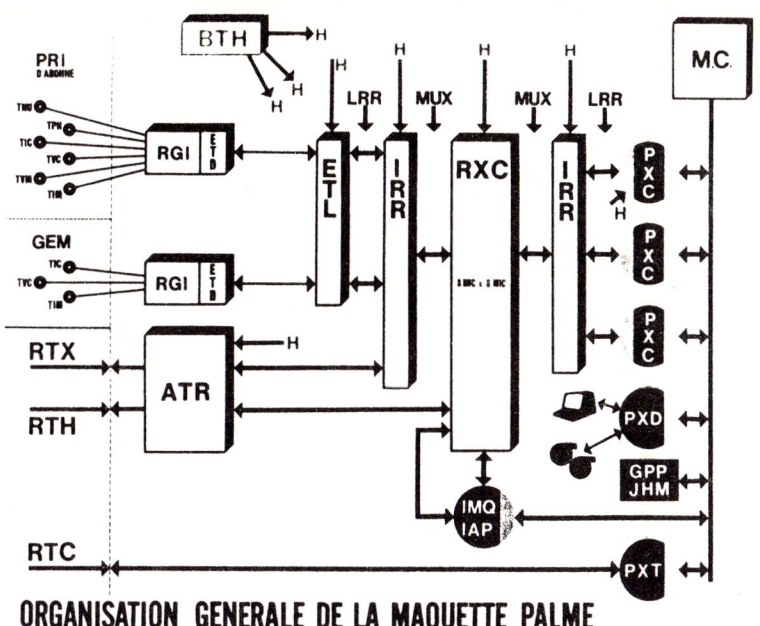

ORGANISATION GENERALE DE LA MAQUETTE PALME

3.2 LE FRONTAL D'INTERCONNEXION (XCA)

Le système d'interconnexion comprend:
- une base de temps générale (BTH) assurant la génération des signaux d'horloge nécessaires aux différents organes du système,
- un réseau de connexion numérique (RXC) qui assure la commutation temporelle de canaux à 64 Kilobits/s en mode "circuit" ou "paquet".
Le réseau de connexion numérique est réalisé essentiellement par un circuit intégré LSI appelé Matrice Temporelle Symétrique (MTS) (7) assurant la commutation de 256 voies entrantes réparties sur 8 liaisons multiplex entrantes vers 256 voies sortantes réparties sur 8 liaisons multiplex sortantes à 2 Mégabits/s.
- des organes de raccordement au réseau (IRR) qui exploitent les informations contenues dans le canal d'exploitation de chaque liaison LRR pour reconnaître le mode d'utilisation du canal principal et qui assurent la détection de la présence de trame sur les liaisons LRP.
- un générateur de tonalités.

3.3 LE SYSTEME MICROPROCESSEUR (XPC)

Dans sa réalisation, ce système est composé d'un ensemble de processeurs 16bits interconnectés par un bus à une mémoire commune.

Parmi les différents processeurs, on distingue:
- un processeur (PXT) relié au réseau Transpac par la liaison LSC,
- les processeurs de commande (PXC) reliés au système d'interconnexion par les liaisons (LRX) en mode paquet. Leur nombre est fonction du trafic de paquets à écouler.
- un processeur de service (PXD), spécialisé dans la gestion du disque et du pupitre opérateur.
- le processeur (PXM), assurant les fonctions d'allocation et de marquage dans le commutateur temporel du système d'interconnexion et le regroupement des alarmes et positionnements.

La structure physique de chaque processeur est identique:
- un microcalculateur 16 bits autorisant le fonctionnement en multiprocesseur (ordre exclusion mutuel câblé),
- une mémoire locale pouvant atteindre une taille de 512 Kilooctets, répartie en mémoires vives et mortes,
- un circuit spécialisé d'entrée/sortie:
  . interface HDLC-X25 pour le processeur raccordé au réseau Transpac (PXT) et les processeurs de commande (PXC)
  . accès au réseau de connexion pour le processeur (PXM) de marquage
  . accès au disque pour le processeur de service (PXD).

Le seul moyen de dialoguer entre les processeurs est la mémoire commune, accessible à tous les processeurs par un bus global à travers leur logique d'accès. Elle contient les données communes aux processeurs et en particulier les paquets en attente. Sa taille est de 512 Kilooctets de mémoire vive, protégée par un code de Hamming (correction d'une erreur, détection de 2 erreurs et plus).

Les niveaux fonctionnels décrits précédemment - traitement du niveau trame (NRX), commutation de paquets (RXP), traitement des appels (TAP), exploitation maintenance (EXP), s'exécutent, de façon banalisée, sur les processeurs de commande (PXC) et sur le processeur relié au réseau TRANSPAC (PXT).

4 ORGANISATION LOGICELLE

4.1. LOGICIEL DE BASE

4.1.1. BUT

La machine de traitement dispose d'un logiciel de base spécifique. Il permet, grâce à la création d'une machine virtuelle, d'affranchir le logiciel d'application implanté sur les processeurs de commutation et de commande (PXC), des contraintes dues à la structure matérielle, à savoir:
- accès simplifié aux données communes (mémoire, disque) et gestion des conflits d'accès,
- dialogue entre tâches d'application,
- émission et réception des trames échangées avec la périphérie,
- commande du réseau de connexion de circuits,
- utilisation d'espaces mémoire dynamique,
- gestion du temps.

La plupart de ces fonctions sont prises entièrement en charge par le processeur sur lequel l'application effectue la requête.

Certaines d'entre elles sont centralisées sur un processeur de service spécialisé, le PXD (Gestion du Temps, Sauvegarde disque).

Le marquage du commutateur de circuit est réalisé par le processeur spécialisé PXM.

4.1.2 NOTION DE TACHE

Une tâche est une entité exécutable par un processeur et caractérisée par un point d'entrée, connu du moniteur, et par le contenu d'un message qui lui est associé.

Une tâche peut être activée par une autre tâche du processeur sur lequel elle s'exécute (tâche locale), ou d'un autre processeur (tâche commune). Chaque tâche doit réserver, par l'intermédiaire des primitives du logiciel de base, les ressources qui lui sont nécessaires. Si l'une d'entre elle est refusée, l'exécution de la tâche est suspendue et les ressources sont restituées dans leur état initial. La tâche sera représentée plus tard par le moniteur. Le nombre de tentatives d'exécution est limité.

En fin d'exécution la restitution des ressources, avec les modifications effectuées, peut être demandée directement par la tâche ou être laissée à la charge du logiciel de base.

En cas d'erreur grave, toute trace de l'activité de la tâche est effacée.

4.1.3 DECOUPE DU LOGICIEL DE BASE

Le logiciel de base est divisé en cinq grandes parties:
- le moniteur local,
- la gestion des ressources locales,
- la gestion des ressources communes,
- les accès aux données communes,
- l'interface d'entrée-sortie niveau trame.

L'ensemble de ces modules est résident et implanté en mémoire REPROM.

a) Le moniteur Local

Il assure l'enchaînement:
- de l'exécution des tâches d'application locales et communes selon leur niveau de priorité.
- des tâches d'émission de trames vers la périphérie.
- d'une boucle d'attente pendant laquelle, le processeur est libre en réception. Il est disponible pour être connecté, par le frontal "commutateur de message", à une liaison périphérique sur laquelle une trame serait reçue. Chaque trame reçue provoque l'activation directe ou indirecte de plusieurs tâches locales et/ou communes.
Le fait que le processeur ne redevienne libre que lorsque toutes ces tâches seront exécutées assure une excellente régulation du trafic.

b) Gestion des ressources locales

Ce module regroupe les primitives nécessaires à l'application, pour la gestion des ressources locales à un processeur, à savoir:
- Activation d'une tâche locale,
- Déroutement d'une tâche,
- Suspension temporaire d'une tâche,
- Armement et désarmement d'une temporisation locale,
- Demande et restitution d'espace local.

c) Gestion des ressources communes

Ce module regroupe les primitives nécessaires à l'application pour la gestion des ressources communes à plusieurs processeurs; à savoir:
- Activation d'une tâche commune,
- Mise en file d'attente d'une trame à émettre vers la périphérie,
- Armement et désarmement d'une échéance commune,
- Demande et restitution d'espace commun,
- Horodatage,
- Dialogue avec le commutateur de circuits.

d) Accès aux données communes

Ce module est constitué d'un ensemble de primitives nécessaires à l'application pour réserver, libérer, accéder, modifier, créer et détruire des objets, sans connaître leur localisation physique en mémoire commune ou sur mémoire de masse.

Ces objets se présentent comme des articles de fichiers. Ils sont totalement définis par le nom du fichier et l'index de l'article dans le fichier.

Les règles suivantes sont appliquées à cette gestion de fichiers:
- Tous les articles de fichiers sont accessibles en lecture,
- Les accès en modification ou en création/destruction d'article sont soumis à un contrôle de permissions,
- Les accès en création/destruction ne sont permis que sur les fichiers déclarés en gestion dynamique.

e) <u>Interface d'entrée-sortie niveau trame</u>

- L'émission de trame vers la périphérie consiste à demander au commutateur de messages XCA, une connexion de circuits entre le processeur et l'organe périphérique.

Le processeur émet alors la trame conformément aux spécifications du niveau 2 de la norme X25.

- La réception d'une trame de la périphérie consiste:
. A analyser la trame reçue. Elle doit être conforme aux spécifications du niveau 2 de la norme X25
. A extraire le paquet qu'elle contient et à le transmettre au bloc fonctionnel "Réseau de commutation de paquet RXP", par création d'une tâche locale.

4.2 LOGICIEL D'APPLICATION

Le logiciel d'application est découpé en blocs fonctionnels reliés entre eux par des liaisons fonctionnelles. Il réalise les différentes fonctions d'un commutateur de paquets et de circuits. (voir chapitre II architecture fonctionnelle).

Ces différents blocs fonctionnels sont découpés en tâches permettant de mettre en oeuvre les différentes fonctions qui leur sont attribuées.

Il existe une tâche par liaison fonctionnelle.

Ces différentes tâches communiquent entre elles. Un échange sur une liaison fonctionnelle se concrétise par l'activation d'une tâche. Les informations échangées sont placées dans le message associé à la tâche.

5-DEVELOPPEMENT, MISE EN OEUVRE ET PERFORMANCES

Le projet a été conduit de manière méthodique et structurée dans une démarche uniforme, les différentes phases sont:
- Définition du projet,
- Spécifications externes,
- Spécifications internes,
- Réalisation,
- Intégration de la maquette.

La méthode de production du logiciel a été définie par l'équipe projet en collaboration avec le département "Génie logiciel" du CNET et la société CAP-SOGETI-LOGICIEL (6). L'utilisation d'une méthode a permis:
- de disposer d'une documentation claire permettant en particulier d'induire la politique de développement et de suivre l'avancement du projet,
- d'assurer la coordination et la communication entre les équipes de réalisation,
- d'aboutir à un produit fiable, documenté et maintenable, dont la gestion des différentes versions soit aisée.

Le travail a été réalisé par une équipe d'environ 35 personnes durant 36 mois.

Actuellement (nov. 82) les diverses parties du système sont opérationnelles, seuls certains modules du sous-ensemble exploitation" sont en phase d'intégration (module de surveillance, anomalie, etc...)

<u>Mise en Oeuvre</u>

Le coeur de chaîne incluant: le commutateur de circuit, la commutation de paquets et commande et les interfaces avec les réseaux existants, occupe 3 bâtis. La surface au sol est d'environ 1,2 m2 pour 2 m de hauteur. Ceci correspond à une réalisation à l'aide de cartes standard disponibles dans le commerce, pour une configuration de 100 abonnés et 578 périphériques divers. (téléphoniques et télématiques).

Le commutateur de circuit est réalisé à l'aide d'un boîtier assurant la commutation temporelle de 8 MIC x 8 MIC à 2 Mbits/s.

Les interfaces de raccordement au réseau (IRR) sont réalisées en éléments standards et occupent chacun 2 cartes format 8U.

Il paraît aisé d'intégrer un IRR dans un boîtier monolithique.

La taille du logiciel global est de l'ordre de 120 000 instructions d'un langage évolué (PLM 86) dont 10 000 instructions pour le logiciel de base.

Le commutateur de paquets et commande occupe environ un bâti et demi en éléments standards, il pourrait facilement être ramené à moins d'un bati en développant du matériel et des alimentations spécifiques.

Performances

Le commutateur de circuits et allocateur de ressources (XCA) à pleine charge est capable de détecter et d'allouer environ 250 trames par seconde. Chaque processeur de commutation et commande peut traiter simultanément 25 paquets par seconde (niveau trame et paquet) grâce à l'utilisation d'un moniteur temps réel et d'accès directs mémoire pour les émissions et les réceptions de trame.

Les performances globales du système ne sont pas validées en vraie grandeur. Toutefois, une évaluation analytique basée sur des hypothèses de trafic moyen permet d'évaluer que 100 abonnés peuvent être traités par 7 processeurs dans la commande avec une probabilité de répétition des trames de l'ordre de 1%. Une trame répétée n'est pas perdue, son temps d'acheminement est simplement augmenté.

6-CONCLUSION

Nous avone ete surpris par les differences des couts de mise au point des logiciels d'application et des logiciels de base. La mise au point des logiciels de base (logiciels temps reel) a necessite des delais bien plus importants que prevus et ceci a provoque des glissements dans les plannings.

Ceci est du a notre avis;
- a la complexite des occurences d'evenements qui ne sont pas prevues au moment de l'analyse.
- a l'utilisation de composants sophistiques (boitiers HDLC et de gestion d'interruption notamment) qui se veulent multiusage et dont l'utilisation dans un cas precis n'est pas simple.
- au fait que les erreurs n'apparaissent que dans certaines configurations.

Un autre enseignement est que la realisation de passerelles entre protocoles necessite, en general, un cout en materiel et en logiciel important. Les études autour de la machine PALME vont maintenant s'orienter vers la validation fonctionnelle et l'expérimentation de nouveaux services offerts aux abonnés.

Une réflexion sera menée sur les répercussions de la prise en compte de contraintes de disponibilité.

Pour envisager l'évolution des réseaux des télécommunications actuels vers le réseau intégrant téléphonie et données, il est nécessaire de disposer de noeuds assurant la commutation mixte circuit-paquet. La machine PALME fait, dans ce domaine, figure de précurseur.

BIBLIOGRAPHY

(1) Brevet d'invention:
"Système de commutation numérique à division du temps de canaux en mode circuit et et mode paquet"
Déposé le 23. 01. 81 sous le no 81. 01. 327

(2) M. J. ROSS, C. M. SIDLO
"Approaches to the integration of voice and data telecommunications"
International Conference on Communications - 1980.

(3) A. ROCHE, J. M. CHADUC, P. LUCAS, G. PAYS, H. SEGUIN
"Development strategies for a service integrated digital network"
3rd World Telecommunication Forum - Geneve 1979.

(4) M. CLOST, A. ROCHE, A. VOMSCHEID
"Perspectives of evolution towards the integrated services digital network"
International Conference on Communications - 1982.

(5) D. HARDY, J. L. DAUPHIN, O. LOUVET, G. PAYS
"Le système PALME: un autocommmutateur intégrant la téléphonie et les données"
International Switching Symposium -Montreal 1981

(6) Y. CORCIA, J. L. DAUPHIN, Y. SIMON
"La méthode de production de logiciel adoptée dans le projet d'autocommutateur PALME"
AFCET - Colloque de génie logiciel - Paris juin 1982

(7) P. CHARRANSOL, C. ATHENES, J. C. AUDRIX, J. P. MOREAU
"M. T. S. circuit standard pour équipements modulaires de commutation temporelle"
Commutation et Transmission - Mars 1982 - pages 63 à 68.

Verriegelungsmechanismen auf Parallelbussystemen

Dipl. Ing. Norbert Böck
Österreichisches Forchungszentrum Seibersdorf
Elektronik-Institut

Zusammenfassung

Dieser Beitrag behandelt Datentransfermechanismen auf einem timemultiplexed Parallelbussystem, analysiert die atomaren Elemente eines Transfers und untersucht ihre Anwendbarkeit auf Semaphoroperationen. Es wird eine Methode benützt, die es erlaubt Semaphoroperationen ohne Blockierung des globalen Bussystems zu implementieren.

Aus dieser Methode wird ein Parallelbuskonzept entwickelt, das die Einschränkungen marktüblicher Mehrprozessorsysteme durch die Reduktion der Zahl der möglichen Elementaroperationen, durch eine Vereinfachung des Synchronisationsalgorithmus und durch die Implemantation eines semaphorähnlichen Verriegelungamechanismus auf Modulebene umgeht.

Abstract

This contribution deals with datatransfermechanisms on a timemultiplexed parallel bus system. It analyses the primary elements of a transfer and investigates their applicability to semaphoreoperations. A method allowing semaphoreoperations to be implemented without a lock of the global bus system is used.

From this method a parallel bus concept is derived. It circumvents restrictions of commercially available multiprocessorsystems by reducing the number of possible elementary operations by simplifying the synchronisation algorithm and by implementing a locking algorithm similar to the algorithm used to implement semaphores described above.

Résumé

Cette contribution traite des méchanismes de transfert de données sur un bus parallèle multiplexé. On y analyse les principaux éléments d'un transfert et y étudie leur possibilité d'application aux sémaphores. On utilise une méthode permettant d'implémenter les sémaphores sans verrouillage de tout le bus.

La concept de bus parallèle est dérivé de cette méthode. Il permet de pallier les restrictions des systèmes multi-processeurs disponibles dans le commerce en réduisant le nombre d'opérations élémentaires possibles, en simplifiant l'algorithme de synchronisation et en implémentant un algorithme de verrouillage similaire à celui utilisé pour l'implémentation des sémaphores.

1. Einleitung

Prozesse, die gemeinsame Aufgaben erfüllen, müssen in der Lage sein, Daten auszutauschen. Ist eine lose Kopplung auf Grund der geforderten Reaktionszeiten eines Systems nicht ausreichend, so können eng gekoppelte Prozessorsysteme angewendet werden, die unter Verwendung eines parallelen Bussystems eine effizientere Kommunikation der Prozesse erlauben.

Die Zahl der auf einem Bussystem arbeitenden Prozessoren kann jedoch nicht beliebig erhöht werden, da durch die auf dem Prinzip der Busverriegelung beruhenden Konzepte des Busprotokolls nur beschränkt hohe Datenübertragungsraten erreicht werden können. /1/

In derartigen Systemen müssen in kritischen Bereichen verwendete Variable verschiedenen, an der Prozesskommunikation beteiligten Prozessoren gemeinsam zugeordnet werden. Daher dürfen Elementarprozeduren, die auf diese Variablen angewendet werden, einander nicht überlappen. Als kritische Bereiche werden Programmabschnitte bezeichnet, in denen der Programmablauf des Prozessors nicht unterbrochen werden darf. /2/

Es wird der Vorschlag aufgegriffen, die Synchronisation von Prozessen über Spezialregister (Synchronisationsregister) zu implementieren, die nach dem erstmaligen Ansprechen des Registers dieses für die Dauer der Elementarprozedur verriegeln. Damit erscheint es für Operationen, die in dieser Zeit die Variable anzusprechen versuchen, gesperrt. Unter Synchronisation ist dabei jede Einschränkung der zeitlichen Anordnung von Operationen zu verstehen.

Sollen sehr hohe Datenraten erreicht werden, so kann dieses Konzept auch auf den Adreß/Datentransfermechanismus angewendet werden. Adreßtransfer und Datentransfer werden dann als separate Informationsübertragungen aufgefaßt, wobei der den Transfer initiierende und der auf den Transfer reagierende Modul durch das Setzen des Synchronisationsregisters bei der Initiierung des Transfers logisch gegen den Zugriff dritter Module gesperrt werden.

Damit kann ein sehr leistungsfähiges Buskonzept, das eine weitgehend technologieunabhängige Implementation erlaubt, realisiert werden. Zusammenhängende Sequenzen von Transfers werden dann durch eine strenge zeitliche Ordnung in der Abfolge der Informationstransfers und nicht mehr durch die Ununterbrechbarkeit des Buszugriffs geschützt.

2.1 Datentransfermechanismen in Parallelbussystemen

In Mikroprozessorsystemen wird jeder Transfer durch entsprechende Synchronisationssignale geschützt und durch Steuersignale gekennzeichnet.

Im Fall eines Adreß/Datentransfers können <u>zwei</u> Informationsübertragungen unterschieden werden:

Kennzeichnung als Adresstransfer
- <u>Senden der Adressinformation</u> - Synchronisationssignal für die Gültigkeit der Adreßinformation - Rückmeldung des addressierten Moduls, Unterscheidung der Datenrichtung
- <u>Senden der Dateninformation</u> - Synchronisationssignal der Datenquelle - Rückmeldung der Datensenke.

Allgemein kann eine Informationsübertragung in Teilschritte, wie Steuerung der Informationsquelle, Synchronisationssignal von der Informationsquelle, Steuersignal der Informationssenke und Synchronisatonssignal von der Informationssenke zerlegt werden.

2.2 Datentransfermechanismen in der Prozeßkommunikation

Bei komplexen Problemen wird die Verteilung der Aufgaben durch eine Entkopplung der Prozesse mittels Nachrichtenaustausch gelöst. Die Daten werden zwischen den Prozessen in Form von Nachrichten übermittelt. Da die Prozesse voneinander unabhängig sind, ist es möglich, daß ein Empfänger noch nicht bereit ist, eine vom Sender produzierte Nachricht zu empfangen. Um eine Blockierung des Senders zu vermeiden, wird ein Puffer eingeführt, der die Geschwindigkeitsvariation zwischen den Prozessen ausgleichen soll.

Die Verwendung von Semaphoren ist ein Spezialfall einer Prozeßkommunikation ohne Datenübergabe. Der Prozeß ist zum Beispiel nur interessiert, ein Zeitsignal von einem anderen Prozeß zu erhalten. Der Puffer wird dadurch auf einen einzelnen Zähler reduziert, der die Anzahl der gesendeten, aber noch nicht empfangenen Nachrichten enthält. /3/

In Bezug auf den Semaphor (v) sind das Senden ('signal(v)') und das Empfangen ('wait(v)') einer (leeren) Nachricht Operationen in kritischen Bereichen und müssen einander zeitlich ausschließen. In der Mikroprozessortechnik sind für diese Operationen heute die Verwendung von 'read modify write' oder die Abwicklung anderer nicht unterbrechbarer Transfers üblich.

3.1 Implementation von Semaphoren mittels Blockierung des globalen Bussystems

Semaphoroperationen können dadurch implementiert werden, daß ein Modul nach Abschluß eines Transfers die Synchronisationsleitungen und die Steuerung des Busprotokolls nicht freigibt, sondern den Bus für die Dauer der Abwicklung einer kritischen Operation blockiert. Damit wird die Datenübertragungsrate eingeschränkt.

In neueren Systemen sind zum Teil diese speziellen Mechanismen für die Kommunikation zwischen Prozessen, die von verschiedenen Prozessoren bearbeitet werden, vorgesehen.

Zum Beispiel werden im Fastbuskonzept verschiedene Adress- und Datenmodes unterschieden. Im mode 'mixed' können 'read modify write' Operationen durchgeführt werden, oder im Mode 'block' ununterbrechbare Blockoperationen.

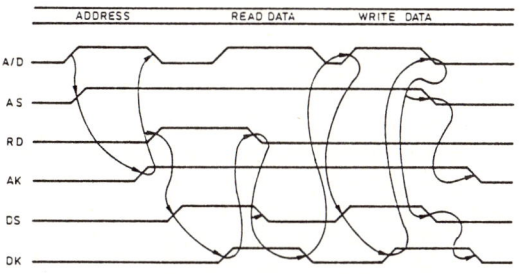

Abb.1: Fastbus "read modify write" /4/

Im Eurobus unterscheidet man zwischen einer 'basic' und 'variant' Form des Informationstransfers. Der 'variant' Mode eines Transfers erlaubt dem Modul, das Bussystem für nachfolgende Transfers zu blockieren.

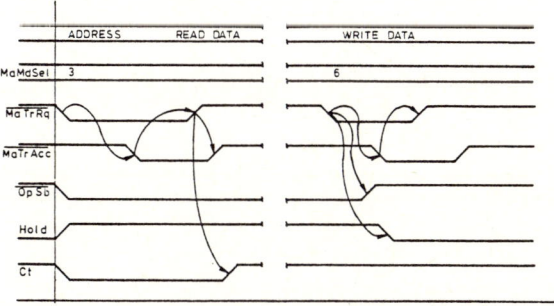

Abb.2: Eurobus "read modify write" /5/

Es ist in keinem dieser Systeme möglich, gleichartige, aber nicht konkurrierende Transfers parallel abzuwickeln.

Die Lösung der Zugriffe konkurrierender Prozesse in ereignisgesteuerten Systemen mit verteilter Intelligenz wird daher nur durch die Serialisierung der Buszugriffe gelöst.

3.2 Implementation eines Semaphores mittels Synchronisationsregister

In Systemen, in denen eine Blockierung des Bussystems nicht erwünscht oder nicht möglich ist, hat das Synchronisationsregister die Aufgabe, die zeitlich wechselseitige Ausschließlichkeit von Elementarprozeduren zu garantieren.

Im einfachsten Fall der Elementarprozeduren für Semaphorvariable 'signal(v)' und 'wait(v)' wird dieses Synchronisationsregister vor dem Zugriff zur eigentlichen Semaphorvariablen gelesen und gesetzt.

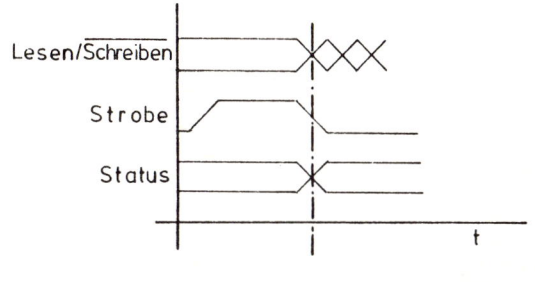

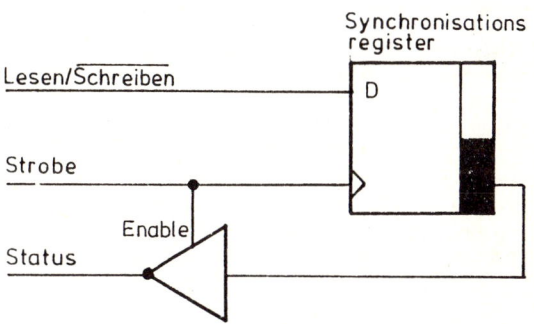

Abb.3: Synchronisationsregister

Wird das Synchronisationsregister nicht gesetzt vorgefunden, so darf zur Semaphorvariablen zugegriffen werden, andernfalls muß die Freigabe des Synchronisierungsregisters abgewartet werden.

Die Freigabe des Synchronisationsregisters erfolgt durch den den Transfer initiierenden Modul nach Abschluß der Elementarprozedur durch einen schreibenden Zugriff auf das Synchronisationsregister.

Die Entscheidung über den Zugriff konkurrierender Prozessoren zu einer gemeinsamen Variablen wird durch die zeitliche Reihenfolge der Buszugriffe der einzelnen Prozessoren auf dieses Synchronisationsregister gelöst. Dadurch wird erreicht, daß im Protokoll des Bussystems keine speziellen Vorkehrungen für die Prozeßkommunikation getroffen werden müssen.

In gleicher Weise muß bei der Verwendung von Variablen, die zu kritischen Bereichen gehören, das Synchronisationsregister behandelt werden.

4. Integration eines Synchronisationselements in das Busprotokoll

In der Zeit, in der der empfangende Modul die Adresse decodiert und die Speicherzugriffszeit abwartet, müßten Sende- und Empfangsmodul den globalen Bus nicht blockieren. Diese Zeit könnte für die Abwicklung gleichzeitiger Datentransfers genützt werden.

Darunter sind Informationsübertragungen zu verstehen, die auf verschiedene Module oder auf verschiedene Register in den Interfaces der Module zugreifen, welche von einem aktuellen, bereits initiierten Transfer nicht benötigt werden.

Dies kann durch die Integration eines Semaphoralgorithmus in das Busprotokoll erreicht werden. Der zusätzliche Hardwareaufwand beschränkt sich dabei auf die Implementation der Abfrage eines Synchronisationsregisters.

Die abgebildete Implementation geht von einer dezentralen Arbitration und Bussteuerung aus. Das Synchronisationselement wird für jedes Interface implementiert.

Mit der Übertragung der Moduladresse wird das Synchronisationsregister selektiert und dabei eine Information übertragen, ob das Register gesetzt oder gelöscht werden soll. Auf der Statusleitung liegt dabei der ursprüngliche Zustand des Registers an.

Stellt der sendende Modul bei der Übertragung der Adreßinformation auf Grund der Statusantwort des Synchronisationsregisters fest, daß der Empfängermodul ausgewählt und gegen nachfolgende Übertragung von Adressoperationen anderer Module gesperrt wurde, so bricht er den von ihm initiierten Transfer ab.

Stellt der Sendemodul keine Verriegelung des Empfangsmoduls fest, so initiiert er nach Ablauf der Speicherzugriffszeit die Datenoperation. Dabei wird das Synchronisationsregister wieder gelöscht.

Durch diese Maßnahmen wird das Interfacesystem von der modulinternen Geschwindigkeit oder Technologie eines Systems unabhängig.

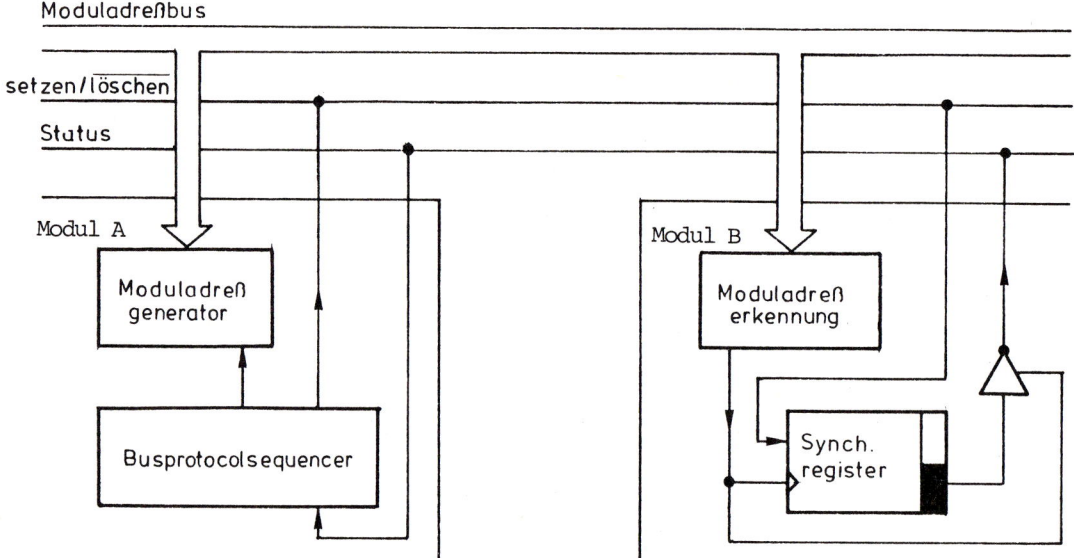

Abb.4: Hardwareimplementation eines Semaphores

5. Literatur

/1/ Hoener S., Roehder W., Efficiency of a Multi Microprocessor System with Time Shared Busses, in Euromicro 1977 (North Holland Publishing company)

/2/ Brinch Hansen P., Betriebssysteme, (Carl Hansen Verlag 1977)

/3/ Schoitsch E., Bedingte kritische Regionen als Werkzeug zur Prozesssynchronisation bei Echtzeitanwendungen, in Real Time Data handling and Process control 79, H. Meyer (ed.) (North Holland Brussels and Luxembourg, 1980)

/4/ U.S. NIM Committee, Fastbus, Modular High Speed Data Acquisition System for High Energy Physics and other Applications, Working Group Document - Tentative Specification, April 1981

/5/ ESONE Committee, Eurobus Modular Data Acquisition, Processing Control System Standard for industrial Applications, Tentative Specification, December 1980

DATABASE MACHINES: A TOOL FOR DATA MANAGEMENT UNDER REAL TIME CONDITIONS

H. Schweppe, G. Stiege
Technical University Braunschweig (FRG)*

Abstract

Management of Databases is one of the most important areas in data processing. While early Database Management Systems have been operated in batch mode, more and more applications need a real-time behaviour of the data manager. In order to meet these requirements concerning response time and throughput, new computer architectures dedicated to the specific task, have been investigated.

One purpose of this paper is to give an overview of research and development activities in this area. Commercial products which are emerging on the market, are described. A discussion of a specific system - the Relational Database Machine, developed at Tech. University Braunschweig - will be given. RDBM is a multiprocessor system with dedicated processors, a common main memory with a record-oriented interface and a secondary memory with data filtering capabilities.

Zusammenfassung

Die Verwaltung großer Datenmengen ist eine der wichtigsten Aufgaben der elektronischen Datenverarbeitung. Während Datenbanksysteme ursprünglich nahezu ausschließlich im Stapelbetrieb arbeiteten, benötigen mehr und mehr Anwendungen ein Realzeitverhalten des Datenverwaltungssystems. Um den wachsenden Anforderungen an Antwortzeitverhalten und Durchsatz gerecht zu werden, wurden in den letzten Jahren neuartige, der spezifischen Datenverwaltungsaufgabe angepaßte Rechnerarchitekturen untersucht.
Der Artikel verfolgt einerseits den Zweck, eine kurze Übersicht der Forschungs- und Entwicklungsarbeiten auf diesem Gebiet zu geben, auch was kommerziell verfügbare Systeme angeht. Die Relationale Datenbankmaschine, ein Forschungsprototyp, der derzeit an der TU Braunschweig entwickelt wird, wird näher vorgestellt. RDBM ist ein Mehrprozessorsystem mit spezialisierten Prozessoren, einem gemeinsamen, objektadressierten Hauptspeicher und einem Sekundarspeichersubsystem mit Datenfiltereinrichtung.

Résumé

La gestion des bases de données est un des plus importants domaines du traitement des données. Alores que les premiers systèmes de gestion de bases de données fonctionnaient en traitement par lots, de plus en plus d'applications nécessitent un traitement en temps réel.

Afin de satisfaire ces besoins relatifs au temps de réponse et au débit, on a étudié de nouvelles architectures spécialisées.

Un des buts de ce papier est de donner une vue d'ensemble des activités de recherche et de développement dans ce domaine. On y décrit les produits en train d'apparaître sur le marché. Y figure une analyse d'un système spécifique - le système de base de données relationnelle (RDBM) développé à l'Université Technologique de Braunschweig -. RDBM est un système multi-processeur comportant des processeurs spécialisés, une mémoire centrale commune avec interface travaillant au niveau de l'enregistrement et une mémoire auxiliaire capable de filtrer les données.

---

* Institut für Theoretische und Praktische Informatik
  Gaußstr. 11, D 3300 Braunschweig

1. Introduction

Management of large amounts of data has become one of the dominating tasks of computer systems. This is true not only for business data processing but also for process control systems and even scientific computing tasks. Requirements as far as response time and throughput is concerned, are continuously increasing. This is due to time critical control applications on the one hand and rapidly changing user behaviour on the other. Since data processing technology changed dramatically in the past 10 years, new computer architectures have been investigated in order to meet the increasing requirements. As far as database management is concerned, these architectures are called database machines.

2. Associative Mass Storage Devices

In the mid seventies, several research projects have been conducted, which investigated the problem of processing data not only in the central processing unit but already while being transferred from the disk. Although access is sequential, data of one track can be processed in one rotation. This is useful, when only a very small fraction of the data is actually needed for further processing, as is the case when searching in a large database. Figure 1 shows the architectural characteristics of these

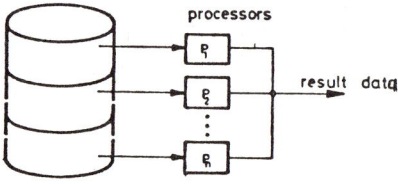

Fig. 1: 'Associative' Mass Storage

systems. RAP /1/, SURE /2/ and CAFS /3/ are well-known systems of this type. Although this architecture works extremely well in situations, where a scan of large parts of the database is needed, there are some drawbacks of this configuration. This is the inevitable rotation time which is a lower bound for system response time, and on the other hand, lack of flexibility in performing different data management operations with reasonable efficiency. Therefore integrated database machines have been designed as an alternative.

3. Architectures for integrated Database Machines

One of the key ideas of DBM is to relieve the main computer from database processing. Therefore the typical configuration is a separate computer (backend) which handles the mass storage management and which is connected to one or more host computers. Several hosts have access to the DBM when it serves as a specialized node in a computer network.

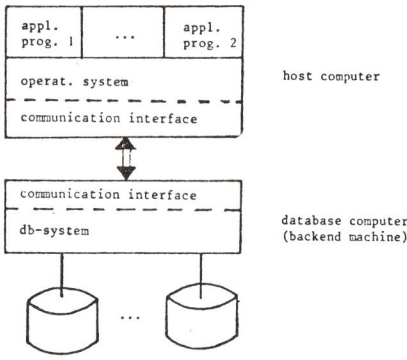

Fig. 2: Host-Backend Configuration

Concurrent operation of host and backend system, as shown in fig. 2 are only one kind of parallelism. Most DBM architectures employ internal parallelism of operations in order to enhance performance. Typically, a database machine is a multiprocessor system with a common main memory. At least three different modes of operation are reasonable: (1) a database task is scheduled to all processors which concurrently process the task on different partitions of the data set, (2) subtasks are created and scheduled to one or more functionally specialized processors, (3) each processor works on a different task in order to maximize overall throughput.

A problem encountered in all multiprocessor configurations is the synchronization of processor activity. Depending on the scheduling policy, processors which cooperatively work on one task, have to synchronize their actions at certain points in time. Strict master control on the one hand and message based systems on the other are employed in most published DBM architectures. A more advanced method - data driven synchronization by object addressing - will be shortly discussed in the next section.

The language interface between programs and DBM usually is a set of highlevel data access and definition primitives, as provided by conventional Database management systems as well. These include set manipulation primitives where data sets are dynamically defined based on complex data qualification predicates, simple record manipulation operations such as updates, and transaction scheduling commands. Furthermore, operations for processing unstructured data are becoming more and more important. Database machines available on the market are minicomputer-oriented. Nevertheless, performance characteristics are not bad. The Intel database processor (iDBP) /4/ is said to perform 3-4 transactions with 10 I/Os each. iDBP is a system of microprocessors connected by a bus. Database processing is concentrated in one processor, while protocol conversion of the host links and message handling is done in a communication subsystem. The open architecture of this database machine allows for integration of functionally specialized components, such as hardware support for text processing or voice understanding. On the other hand, a low-end implementation on a single board (available in 1983) will be a powerful tool for upgrading small systems by database processing capabilities.

Another database machine available on the market is IDM /5/ which is more or less mainframe-oriented. This machine has explicitly been designed as a high performance DBM. Besides conventional microprocessors, a dedicated subsystem for sorting and searching is utilized. Until now, the performance gains of this 'accelerator' are not clear.

IDM has been the first DBM commercially available and it has thus already gained a certain degree of maturity.

4. The Relational Database Machine

RDBM is a research DBM which is developed at Tech. University Braunschweig. One of the primary goals of this research project is to design and implement a backend system which offers a functionally complete Data Base Management System with considerable increase in performance. Another important goal was the investigation of architectural issues which are particularly important in a data processing environment, such as parallelism and object-oriented addressing.

The overall architecture is shown in fig. 3.

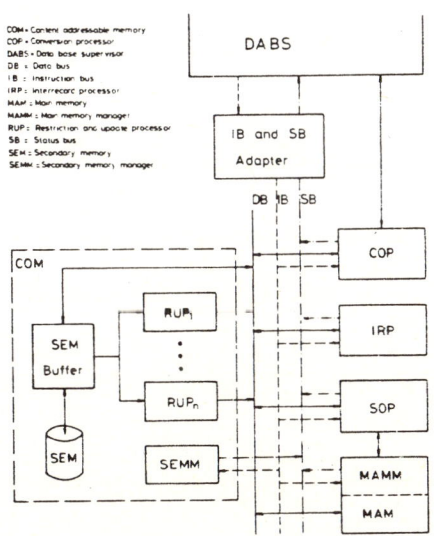

Fig. 3: The RDBM Multiprocessor System
(from /6/)

The system is controlled by the database supervisor, which establishes the communication with the host, schedules concurrent user transactions and controls the functionally specialized processors. The main components are: the secondary memory system comprising a RAM buffer and specialized processors for data filtering and updating of simple records. A sort processor, an interrecord operations processor and a conversion processor have access to a RAM memory, which stores intermediate data. Since sorting is a basic but time consuming function, a hardware implementation of the merge-sort algorithm has been built. The interrecord processor performs all operations with more than one record as operand, such as the relational join or aggregation of data. The conversion processor transforms code and data formats. The components are connected by a bus system (data, instruction, control, 8 MHz/sec).

Data segments transferred from the disk are cached and filtered record by record by the restriction and update processors. The data qualified are stored as a temporary relation in main memory for further processing. The synchronization of subsequent actions of different processors is achieved by the object addressing feature of the main memory. Instead of byte-addresses, requests for records in a temporary relation are send to the memory subsystem by the processors. If P1 requests a record, e.g. for formatting and output, it will be available as soon it has been processed by the predecessor P2, e.g. the data filtering processors. The translation of logical addresses is realized by an address table which holds the byte addresses of each record. For each segment, the address table is set up without any delay when records are stored. The main memory system implements the data flow between the different processing subsystems.

A detailed discussion is given in /7/.

5. References

/1/ Ozkarahan et al.: RAP - An Associative Relational Processor, AFIPS Conf.Proc. 44, 1975.

/2/ Leilich, H.O. et al.: A Search Processor for Database Management Systems, Proc. VLDB 1978.

/3/ Babb, E.: Implementing a Relational Database by Means of Specialized Hardware, ACM TODS 4(1).

/4/ INTEL: Guide to iDBP, order no. 22 Intel Corp., Austin, 1982.

/5/ BRITTON-LEE: IDM-500, Product Description, Britton-Lee Inc., Los Gatos, 1982.

/6/ Auer, H. et al.: RDBM - A Relational Database Machine, Information Syst. 6(2), 1981.

/7/ Schweppe, H. et al.: A Dedicated Multiprocessor System for Database Management, Proc. Int. Workshop on DB Mach., San Diego, 1982.

Session 3

DISTRIBUTED SYSTEMS

SYSTEMES REPARTIS **VERTEILTE SYSTEME**

HIGH LEVEL PROGRAMMING OF
DISTRIBUTED PROCESS CONTROL SYSTEMS

Hermann Kopetz, Frieder Lohnert,
Wolfgang Merker, Georg Pauthner
Institut für Technische Informatik
Technische Universität Berlin

ABSTRACT

This paper introduces some new concepts for programming distributed computer control systems. After a discussion of the requirements for an intertask communication mechanism the basic ideas of MARS are introduced. MARS (MAintainable Realtime System) is a distributed system for process control applications which has been developed from the point of view of maintainability. The consideration of maintenance aspects requires that a system should be easily expandable and support the autonomy of subsystems.

The newly developed concepts include selective message receipion, limited validity time of a message, distinction between state and event messages and group addressing of typed messages. Language constructs for distributed programming in a high level language are introduced.

ZUSAMMENFASSUNG

Es werden neu entwickelte Konzepte zur Programmierung von verteilten Systemen zur Prozeßdatenverarbeitung vorgestellt. Nach einer Diskussion der Anforderungen an einen Mechanismus zur Intertaskkommunikation werden die MARS (MAintainable Realtime System) zugrundeliegenden Ideen dargestellt. MARS ist ein verteiltes System für Anwendungen in der Prozeßdatenverarbeitung. Es wurde unter dem Gesichtspunkt der Wartbarkeit entwickelt. Aus der Betrachtung der Wartungsaspekte resultieren die Forderungen nach leichter Erweiterbarkeit und der Unterstützung der Autonomie von Subsystemen. Die neu entwickelten Konzepte beinhalten selektiven Empfang, begrenzte Gültigkeitszeit und Gruppenadressierung von "getypten" Nachrichten sowie Unterscheidung von Zustands- und Ereignisnachrichten. Für die Programmierung in einer höheren Programmiersprache werden Sprachkonstrukte vorgestellt.

RÉSUMÉ

Le papier décrit des concepts nouveaux pour la programmation des sytèmes distribués des procédés en temps réel. Après la discussion des constraintes imposées à un mécanisme de communication entre tâches paralleles, il présente les idées qui sont la base de MARS (MAintainable Realtime Systems). MARS est un système distribué pour contrôler des procédes en temps réel. Il a été développé sous l'aspect de la maintainabilité et en vue de l'autonomie de sous-systèmes. Les concepts nouveaux concernent la réception sélective, la durée de vie limitée des messages, l'adressage par groupe et la distinction entre messages d'état et messages d'événement. Pour la programmation en langage de haut niveau, le papier décrit des constructions speciales.

1. INTRODUCTION

The widespread use of microcomputers and local area networks in the area of computer process control has placed in the center of interest the problem of high level programming in such a distributed environment.

Especially the question for a suitable mechanism for intertask communication (ITC) has to be answered. Though many proposals for ITC have been made in the last few years (e.g. /Hoar 78/, /Brin 78/, /ADA 80/, /Feld 79/, /Lisk 79/, /Cook 80/, /Mao 80/, /Stro 80/, /Kram 81/), only few proposed mechanisms meet - in our opinion - the requirements for distributed computer control systems (DCCS).

The selection of a proper ITC mechanism is greatly influenced by the general requirements on the overall architecture of a DCCS, e.g.:

- Subsystems in a distributed system should behave in a fairly autonomous manner, i.e. they should be able to make their own decisions on the basis of locally available information. In particular, it has to be an autonomous decision of an individual subsystem whether to accept incoming data (e.g. a request) from other subsystems and how long to wait for them. Autonomy of subsystems is necessary to provide fault isolation between subsystems.

- A system, especially a DCCS, should be easily expandable in order to accomodate the often occurring functional enhancements. For a high degree of maintainability, only few modifications should be necessary to integrate a newly added subsystem.

After a discussion of the requirements on the semantics of the programming constructs for ITC in distributed computer control systems this paper introduces the basic concepts and solutions of the MARS (Maintainable Realtime System) architecture. Some examples show the expressive power of the newly developed constructs for programming distributed computer control systems.

2. REQUIREMENTS FOR ITC IN DCCS

These requirements can be derived from the analysis of information exchange between subsystems in typical process control applications.

The following types of information and information exchanges can be found:

1) Measured Values have to be transmitted from sensors to processing units, display units and recording units. Typically, one measured value has to be delivered periodically to several units, i.e. there is a demand for a multidestination communication pattern. These diverse receiving units may be interested in different aspects of the measured value, e.g. a display unit requires only the most actual value while a recording unit has to record the complete history of the measured values.

2) Control Variables normally are generated by a processing unit; in case of exceptional or emergency conditions they are produced by an operator or a safety unit. As a consequence, actuators should be able to receive control variables from several controlling units.

3) Event Information (which reports the happening of an event such as an actuation of a limit switch) has to be transmitted from sensors to processing units, display units and recording units - similar to measured values. While with measured values the value of a physical quantity is relevant, with event information the time of occurence of the event is of interest. Processing units should be able to wait for events and to detect the correct or incorrent occurrence of a particular event in the domain of time.

4) Alarm Information may be produced by processing units and/or intelligent sensors. In particular alarm information may be directed to other processing units, display units and recording units - like measured values and event information. Alarm information should report about the cause of the alarm situation and the occurence in time of the alarm situation.

Beside the types of information exchanges mentioned above which are unidirectional, some other types can be found:

5) Request/Response Transactions in which service is requested by a master and a response is expected from the server. Indeed, most of the proposals for ITC exactly cover this case of transaction (sometimes called remote invocation). There is a requirement to request service from more than one server at a point in time, i.e. these transactions should not be unnecessarily serialized. Late or missing responses must not lead to arbitrary delays of processing.

6) Exclusive Communication Transactions, i.e. a server performs service for just one master for a period of time (e.g. reservation and exclusive use of a line printer). This requires that a server is permitted to accept data only from one particu-

lar master during this period (and vice versa).

This analysis of types of information, basic transactions and the goal of supporting autonomy and expandability lead to the requirements for ITC as listed below:

1) Provide a multidestination communication pattern

2) Support the different aspects of delivered information at diverse receivers:
    - only the newest one is relevant
    - all information is relevant

3) Support a high degree of parallelism, i.e. producing data should not be tightly coupled with accepting data

4) The demand for autonomy leads to the requirement for a mechanism to provide for selective acceptance of incoming data and to provide for the delimitation of the delay time.

   (Note that ADA's rendezvous technique does only offer a selection according to the name of the entry call (and not, for example, according to the name of the caller). Further, in ADA time supervision does not take place when a rendezvous has already startet. For these reasons, we consider the ADA rendezvous to be insufficient as the only mechanism for ITC in distributed computer control systems which support autonomy of subsystems.
   /Bos 80/)

5) The demand for expandability requires a task should be able to communicate with tasks which are not known to it at development (programming) time. This requirement influences the naming (addressing) scheme of the ITC, i.e. accepting data should not be unnecessarily coupled to the name of the partner task.

6) Data exchanged with ITC should contain additional information (at least for maintenance purposes) such as

   - a key which denotes its semantics
   - the time of creation (note that producing data is an event)
   - an identification of the producer.

3. DESCRIPTION OF MARS

Considering the requirements listed above and general concepts of software engineering and networking, we developed the MARS architecture including language constructs for a new ITC mechnism.

3.1 The architecture of MARS

The hardware architecture consists of components (self-contained microcomputers) which are interconnected by a local communication medium (e.g. a bus or redundant busses). The application software which runs on a component we call a module. The software architecture consists of communicating modules. A module may consist of several tasks. All modules have access to the global physical time (For a detailed discussion of this time base see /Kope 82a/.)

There exists a special kind of components and modules which interact with the environment (e.g. the instrumentation). We call them interface component and interface module, respectively. A task within a module can be viewed as an abstraction of an I/O device. The means for I/O programming is thus ITC.

We assume that the best mechanism for ITC in a distributed process control system is message passing. Messages can be sent from one module to another module or (to satisfy the requirement for a multidestination communication pattern) to a group of modules. Tasks within a module communicate with another in the same way.

A MARS message is a data structure consisting of a predefined header and an user defined record. The header contains the following information:
- name, the message name,
- receiver, the name/groupname of the module/task to which the message is directed,
- sender, the name of the sending module/task,
- send time, the time at which the message has been produced,
- validity time, the time at which the message becomes invalid.

The message name designates the semantics of the message and, as a consequence, the data type of the user defined record; thus messages are strongly typed. Sender and receiver must contain consistent declarations of messages.

In real time systems, data which are not processed in time become invalid. Using invalid data (e.g. an out-of-date measured value or a request) may lead to an error. We therefore introduced the concept of the limited validity time of a message. After the expiration of the validity time the message is discarded by the communication system (see example 1).

Another important concept is the distinction between event and state information. Event information deals with the occurence of events while state information reports about the state of an object. In order to facilitate the exchange of event and state information we introduced the concept of event and state message. The difference between event and state message concerns the handling of these two message classes at the receivers end. The handling of event messages conforms with the "classical" message passing semantics. An event message is queued at the receiver when it arrives and dequeued (i.e. consumed) when it is read. State messages on the other side are not queued. A new state message of a given name overwrites (or updates) the previous one. On reading, state messages are not consumed, i.e. the most actual valid version of a state message can be read several times. Depending on the interest in the delivered information, a given message can be handled differently by different receivers:

- as a state message if the receiver is interested only in the most actual information,

- as an event message if another receiver is interested in the complete history of information.

3.2 Programming

For comfortable programming of the concepts described above we developed some language constructs for module, task and message declarations and for ITC. For the sequential programming of tasks we have chosen PASCAL.

3.2.1 Declarations

Module declarations:

As mentioned above, the application software which runs on a component is called a module. The MARS system supports a one-to-many communication pattern. Modules can be members of one or more groups.

```
MODULE module_name MEMBER_OF group;
    .
    .
    END module_name.
```

Each message addressed to a particular group will be delivered to all modules in this group. The sender of a message does not know the names nor the number of receivers.

External message declaration:

In the header of a module all messages which can be received and generated by this module have to be declared. We call the messages, which are declared in the module header, external messages. The declaration of this external messages forms the interface specification of a module.

```
MODULE module_name;
    IMPORT inmsg = RECORD ... END;
    EXPORT outmsg;
    .
    .
    END module_name.
```

The module 'module_name' receives messages with the name 'inmsg' with a user defined record and it generates messages named 'outmsg'. This 'outmsg' contains only the MARS header, i.e. it is a pure signal message.

Task declaration:

A module consists of one or several tasks. These tasks are not visible from the outside of a module. In the header of each task, all messages which can be received and generated by this task have to be declared. If the messages remain within the module, we call them internal messages. Internal messages are only declared in the task header, but not in the module header.

There exists a special task within each module which we call the priority task. Whenever a message for this task arrives all other tasks are halted and only the priority task may proceed until it waits for further messages. The priority task has also the authority to reset any or all other tasks, i.e. to force them into their initialization state. This priority task is necessary for exception handling purposes.

Task local message declaration:

Each task local message declaration declares a task local variable of the same type as the message. In addition to the message attributes IMPORT, EXPORT and type the task header must contain a declaration determining the handling of the message as an event or a state message.

```
MODULE module_name;
    IMPORT push_button;
    IMPORT position;
    .
    .

    TASK task_name;
        STATE position;
        EVENT push_button;
        EXPORT action;
        .
        .

    END task_name;
    .
    .

END module_name.
```

This module receives one or more messages 'push_button' and 'position'. The task 'task_name' declares the 'position' as a state message and with an input operation the 'position' is assigned to a local variable 'position'. The message 'push-button' is treated as an event message. In addition the task sends a message 'action' to another task(s) (Note that 'action' is not declared in the module header).

3.2.2 INPUT and OUTPUT Statement

Because the semantics of the inter-process communication statements in MARS differs from the semantics normally associated with send and receive statements, two new keywords, INPUT and OUTPUT, are used in MARS.

The OUTPUT-Statement

The OUTPUT statement is used to output a message to the communication system. The issuing task then proceeds (i.e. the issuing task performs a rendezvous with the component local communication system). This no-wait-send property of the OUTPUT statement satisfies the requirement for a high degree of parallelism. The OUTPUT statement requires the specification of a validity time for the message.

```
OUTPUT msg TO receiver VALID time
```

- 'msg' is a declared message that will be transmitted to the

- 'receiver', that is a module name or a module group name or a task name or a task group name.

- 'time' is a time expression (relative or absolute). After this time the message will be discarded by the communication system.

At execution of the OUTPUT statement the message header is constructed from this information. The send time and the sender name are inserted into the header by the communication system.

The communication system delivers a message to the addressed module(s)/task(s) with a high degree of reliability. Communication problems such as queue overflow (obviously, this cannot happen with state messages) are reported by an error message generated by the communication system. In terms of networking, the MARS communication system performs a datagram service with error reporting. It is a good programming discipline to handle error messages by the priority task.

The INPUT Statement

With the INPUT statement a message can be read from the communication system into a task local message variable. It is a 'selective receive' - from all stored messages the denoted one will be selected.

The simplest form is:

```
INPUT message => stmt END
```

This statement has the following semantics:

- if a message named 'message' is already stored in the communication system the message will be
    - for an event message - consumed and assigned
    - for a state message - only assigned

  to the local message variable specified in the declaration.

- if at execution time no 'message' is present the task waits until such a message arrives.

An INPUT statement is called successful, if a message is assigned. The statement 'stmt' will then be executed.

In order to increase the expressive power of the INPUT statement the following extentions are available in MARS.

Filtering:

In the above described form of the INPUT statement the selection of messages is realized by the message name. This selection can be refined by a predicate on the full message contents, i.e. message header and/or user defined record. This predicate is a logical expression on the message contents and/or local data of the issuing task. This predicate is called a filter and denoted by the keyword FILTER.

```
INPUT msg FILTER filter_expression => stmt END
```

Only if the evalution of 'filter' delivers true, the message will be read. Thus the filter forms a 'peep hole' to outstanding messages.

With the aid of the filter mechanism it is possible

- to select messages which are generated by a specific sender (e.g. for performing an exclusive communication transaction),
- to select messages which are generated before or after a specific point in time, i.e. a selection in relation to the send time,
- to select messages which contain specific application data, i.e. a selection depending of the user defined record,

- to combine any of these basic selection criteria.

The filter mechanism simplifies programming:
- It protects a task against messages which are unwanted at a specific state of computation.
- While the selection via message names is static (message names have to be declared) the filter mechanism allows a dynamic selection of messages.

Thus it is never necessary to read any message which is only relevant in a computation at a further point in time.

Thus the filter is a powerful mechanism to support the autonomy of modules.

Conjunction on message reception:

Often an operation can only be done if two or more messages are available at the same point in time

```
INPUT msg1 MSGAND msg2 => stmt END
```

means only if both messages 'msg1' and 'msg2' can be read from the communication system the INPUT statement is successful.

Message reception alternatives:

In many applications a task may expect one message out of a set of different messages and then perform different operations.

```
INPUT
        msg1 => stmt1
    MSGOR
        msg2 => stmt2
END
```

The INPUT statement will complete successfully if either 'msg1' or 'msg2' (exclusive or) is read. If both alternatives are possible the choice is unspecified.

INPUT with time out:

For real time applications an infinite delay in any statement must be avoided. Therefore the delay time in the INPUT statement can be delimited by a time out part.

```
INPUT message => normal
    AFTER time => timeout
END
```

'time' denotes a duration or an absolute time. At execution time of the INPUT statement the time expression 'time' will be evaluated. If no message arrives until the specified time has elapsed, the 'timeout' alternative is executed.

Clearly, all combinations of the possible extensions of the INPUT statement are useful and permitted, e.g.

```
INPUT
        msg1 => stmt1
    MSGOR
        msg2 FILTER f2 => stmt2
    MSGOR
        msg3a FILTER f3a MSGAND msg3b => stmt3
    AFTER
        timeout => stmt4
END
```

4. EXAMPLES

Example 1: Programming a Request/Reply Transaction

Consider a master requesting the service of a server. The master may send a request to the server at any time. The request contains a maximum response time 'limit' for the server. The master expects a reply (i.e. the result of the requested service) before this 'limit' elapses. Using the MARS constructs for specifying the validity time and a time out interval the problem is solved as follows:

```
MODULE master;
    IMPORT reply   = RECORD ... END;
    EXPORT request = RECORD ... END;
    .
    TASK mastertask;
        EVENT reply   = RECORD ... END;
        EXPORT request = RECORD ... END;
        .
        OUTPUT request TO server VALID limit;
        .
        INPUT reply => reply_in_time
            AFTER limit => no_reply_in_time END;
        .
    END mastertask;
    .
END master.
MODULE server;
    IMPORT request = RECORD ... END;
    EXPORT reply   = RECORD ... END;
    .
    TASK servertask;
        EVENT request = RECORD ... END;
        EXPORT reply  = RECORD ... END;
        .
        INPUT request => perform_service END;
        OUTPUT reply TO request.sender
                        VALID request.valid;
        .
    END servertask;
    .
END server.
```

This example shows the synchronization of the start of execution of the time-out alternative with the automatic (remote) destruction of the late reply. In future this message is not only irrelevant but it can also interfere with the further course of the computation, i.e. without the concept of the limited validity time a late reply could be falsely used as a reply to a further request.

Example 2: Processing of Measured Values

Consider an electric motor driving a changeable load. The goal of the control system is to keep the speed of the motor constant and to protect it against dangerous load changes.

The MARS control system is connected to the plant via two interface components. One interface component produces periodically a MARS message 'speed' containing the actual rpm (revolutions per minute) in its user defined record and addresses this message to the module group 'speed_user'. Another interface component expects messages 'adjust' with adjust information in its user defined record. Internally these messages are declared as state messages.

The control component can be programmed as follows:

```
MODULE control MEMBER OF speed_user;
    IMPORT speed  = RECORD ... END;
    EXPORT adjust = RECORD ... END;
    .
    TASK speed_control;
        STATE speed  = RECORD ... END;
        EXPORT adjust = RECORD ... END;
        .
        INPUT speed => control_algorithm END;
        OUTPUT adjust TO ...
        .
    END speed_control;
    .
END control.
```

Note that the period of the 'speed' message and the period of execution of the control algorithm can differ, provided a valid speed message is always available in this module. No detailed synchronization of message production and message consumption is needed because 'speed' has been declared as a state message. The semantics of state messages simplifies the programming in all those cases where a loose coupling of modules/tasks is sufficient. This is common in many process control situations.

The alarm detector has to detect dangerous positive or negative angular accelerations. It accepts the same 'speed' message as the component 'control' but declares it internally as an event message.

```
MODULE alarm_detector MEMBER OF speed_user;
    IMPORT speed = RECORD ... END;
    .
    TASK acceleration;
        EVENT speed = RECORD ... END;
        .
        INPUT speed END;
        compute_acceleration(oldspeed,speed);
        oldspeed := speed;
        .
    END acceleration;
    .
END alarm_detector.
```

To compute an acceleration at least two speed values and the time between the measurements are required. The semantics of the event message guarantees the delivery of all produced speed messages, including their time of origin (message send time), to the task acceleration. The alarm detector should not only detect dangerous accelerations, but also dangerous speeds. For this purpose it is sufficient to incorporate the following additional task in the alarm detector module:

```
    .
    .
    TASK limit_speed;
    STATE speed = RECORD ... END;
    .
    INPUT speed FILTER speed.rpm > limit => alarm END;
    .
    END limit_speed;
    .
    .
```

This example demonstrates the usefulness of filtering on application data. The task 'limit_speed' will be delayed in the input statement until a speed value exceeding the limit arrives. In this case the programmer is not concerned with the administration of 'irrelevant' messages i.e. messages with 'speed.rpm' ≤ 'limit.

5. CONCLUSION

We introduced some concepts and language constructs for high level programming of distributed computer control systems developed in the MARS project. The overall architecture of MARS is described in /Kope 82a/; fault tolerant aspects are published in /Kope 82b/.

At present the implementation of MARS on a distributed system with 5 LSI-11/23 computers is in the works.

This work has been supported by the German Ministry of Research and Technology under Research Contract IT 1018.

LITERATURE

/ADA 80/ Reference Manual for the ADA Programming Language, United States Department of Defense, July 1980

/Bos 80/ Bos, J., Comments on ADA Process Communication, Sigplan Notices, Vol. 15, No. 6, June 1980, p. 77-81

/Brin 78/ Brinch, Hansen, P., "Distributed Processes: A Concurrent Programming Concept", Comm. ACM, Vol. 21, November 1978, p. 934-941

/Cook 80/ *MOD - A Language for Distributed Programming, IEEE Transactions on Software Engineering, 6, 1980, p. 563-571

/Feld 79/ Feldmann, J.A., High Level Programming for Distributed Computing, Comm. ACM, Vol. 22, No 6, June 1979, p. 353-368

/Hoar 78/ Hoare, C.A.R., "Communicating Sequential Processes", Comm. ACM, Vol. 21, No. 18, August 1978, p. 666-677

/Kope 82a/ Kopetz, H., Lohnert, F., Merker, W., Pauthner, G., MARS - An Architecture for a Maintainable Real Time System, Technical University of Berlin, Report MA2/82, April 1982

/Kope 82b/ Kopetz, H., Lohnert, F., Merker, W., Pauthner, G., Fault Tolerance in MARS, Informatik Fachberichte 54, "Fehlertolerierende Rechnersysteme", Springer-Verlag, Berlin, Heidelberg, New York 1982, p. 205-219

/Kram 81/ Kramer, J., Magee, J., Slowman, M., Intertask Communication Primitives for Distributed Computer Control Systems, Proc. Distributed Computing Systems, April 81, p. 404-411

/Lisk 79/ Liskov, B., Primitives for Distributed Computing, Proc. of 7th ACM SIGOPS Symp. on Operating Systems Principles, Dec. 1979, p. 33-42

/Mao 80/ Mao, T.W., Yeh, R.T., Communication Port: A Language Concept for Concurrent Programming, IEEE Transactions on Software Engineering, 2, 1980, p. 194-203

/Stro 80/ Stroet J., An Alternative to the Communication Primitives in ADA, Sigplan Notices, Vol. 15, No. 12, December 1980, p. 62-74

SYSTEME LOGIQUE DE COMMUNICATION POUR
APPLICATIONS REPARTIES : LE LOGIBUS

Claude FAULLE, Bernard GAGEY, Martine JEANDEL, Arlette VIALLEVIEILLE
CIMSA - 10, 12 Avenue de l'Europe,
78140 - VELIZY, FRANCE

RESUME

Cet article montre une approche de structuration d'applications réparties, définie par analogie avec les systèmes matériels, basée sur le concept de fonctions qui communiquent à travers une interface généralisée. Ce concept favorise la construction d'applications modulaires et extensibles, et permet de rendre les applications indépendantes de la structure physique ; de plus, il offre la possibilité de concevoir des systèmes disponibles.

La structure d'applications réparties définie introduit un système logique généralisé de communication appelé le LOGIBUS. Lorsqu'une fonction veut communiquer avec une autre fonction, elle invoque une opération gérée par le LOGIBUS.

ABSTRACT

This paper sets forth an approach to distributed application structuration, defined by analogy with the hardware systems, and based on the concept of functions which communicate with each other through a general-purpose interface. This concept makes it easier to build modular and extensible applications, and the applications do not depend on the physical structure any more. In addition, available systems can be designed thanks to this concept.

The distributed application structure thus defined introduces a general-purpose logical communication system named LOGIBUS. Whenever a function wants to communicate with another function, it calls upon an operation managed by the LOGIBUS.

ZUSAMMENFASSUNG

In diesem Artikel wird der Versuch einer Strukturierung von verteilten Applikationen gezeigt, die mit dem Hardwaresystemen analog bestimmt ist und auf Funktionen beruht, die durch eine generelle Schnittstelle verbunden werden. Dadurch wird die Konstruktion modularen und vergrößerungsfähigen Applikationen möglich sowie deren Unabhängigkeit von der physischen Struktur; außerdem wird es dadurch ermöglicht, verfügbare Systeme zu entwerfern.

Die bestimmte Struktur der verteilten Applikationen führt ein generelles logisches Übertragungssystem ein, das LOGIBUS genannt wird. Wenn sich eine Funktion mit einer anderen in Verbindung setzen will, dan muß sie sich auf eine durch den LOGIBUS gelenkte Operation beziehen.

1. INTRODUCTION

Les architectures classiques de système sont centralisées, même lorsqu'elles supposent la coopération de plusieurs processeurs : ceux-ci partagent des ressources communes, ce qui peut entraîner des difficultés en ce qui concerne la fiabilité d'un système et sa disponibilité.

De plus, comme la spécification d'un système est rarement figée, il est nécessaire de pouvoir modifier les fonctionnalités existantes, d'en retirer ou d'en ajouter de nouvelles ; ces évolutions doivent bien sûr pouvoir se faire au moindre coût ; ici encore les architectures conventionnelles n'offrent pas cette souplesse.

Mais l'apparition des microprocesseurs a bouleversé la vision classique, en permettant d'utiliser des concepts nouveaux de communication entre processeurs, de décentralisation des traitements et du contrôle : c'est la notion de système réparti, caractérisé, entre autres, par l'absence de ressources communes à plusieurs processeurs ; ceci a des conséquences sur la réalisation des applications et conduit à étudier de nouvelles structures pour celles-ci.

Cet article présente une étude en cours de réalisation à la CIMSA dans le cadre d'un contrat avec l'Agence de l'Informatique, visant à définir une structure et une méthode de réalisation de systèmes répartis pour des applications présentant des besoins divers, notamment dans les aspects temps réel et transactionnel.

Dans la première partie, on expose la définition de concepts de structuration d'applications réparties.

La deuxième partie présente une proposition de système logique généralisé de communication : le LOGIBUS.

2. LA STRUCTURATION D'APPLICATIONS REPARTIES

2.1 Le problème

Les microprocesseurs actuellement disponibles permettent de réaliser des unités de traitements ayant une puissance sensiblement équivalente à celle d'un mini-calculateur. Les progrès prévisibles de la technologie laissent espérer que cette puissance augmentera encore à brève échéance. Si toutefois une unité de traitement ne suffit pas à répondre aux besoins d'une application, il est intéressant de pouvoir interconnecter plusieurs processeurs pour constituer un système unique. De plus, utiliser un ensemble de processeurs interconnectés plutôt qu'un seul processeur plus puissant et de performance équivalente semble offrir des avantages économiques : les microprocesseurs sont aujourd'hui fabriqués à des coûts très compétitifs.

Dans des architectures multiprocesseurs classiques, tous les processeurs partagent entre eux un espace mémoire et des unités d'entrées-sorties ; mais la mémoire commune constitue un goulot d'étranglement et l'accroissement des performances d'un tel système est de plus en plus faible au fur et à mesure que l'on ajoute des processeurs ; de plus, la présence d'éléments centralisés ne favorise pas les techniques de détection, de diagnostic et de reprise sur erreurs ou pannes permettant d'assurer une bonne sûreté de fonctionnement.

Dans un système réparti, chaque unité de traitement possède ses ressources propres : mémoire, unités d'entrées-sorties... Si un processus s'exécutant sur un processeur veut accéder à une ressource contenue dans un autre processeur, il doit alors communiquer avec une entité logique qui gère cette ressource.

Une application est généralement représentée par un ensemble de processus dépendant les uns des autres et communiquant notamment par données communes. Or, ce type de structuration ne permet pas de dérouler l'application sur des architectures réparties qui ne possèdent pas de mémoire commune, mais seulement une voie de communication entre processeurs. Aussi, l'environnement induit par les systèmes répartis pose le problème de définition de la structure logique des applications.

2.2 Critères

Les critères retenus dans cette étude pour le choix d'un modèle de structuration d'applications réparties, sont les suivants :

- <u>Extensibilité</u> :

  Une application doit pouvoir évoluer et être modifiée sans reprise de l'ensemble de l'étude. Il doit être possible d'ajouter ou d'enlever une entité logique, afin de prendre en compte facilement les évolutions de l'application.

  Les systèmes répartis répondent bien aux critères d'extensibilité : l'ajout d'un processeur se traduit par la connexion d'un élément matériel sur une voie de communication.

- <u>Modularité</u> :

  La modularité d'une application est la facilité avec laquelle l'application peut être construite à partir d'un ensemble d'entités logiques autonomes avec une interface généralisée.

  Dans les systèmes répartis, l'utilisation de processeurs interconnectés au moyen d'une voie de communication amène une modularité du matériel. La modularité doit de même exister dans le logiciel d'application en profitant au mieux des possibilités offertes par le matériel.

- <u>Disponibilité</u> :

  Un système disponible est un système qui permet d'isoler des modules matériels ou logiciels défaillants sans provoquer de perturbations sensibles dans l'ensemble du système.

Les systèmes répartis offrent un haut degré de disponibilité : ceci est dû à la répartition de la logique de contrôle, mais nécessite une structuration de l'application permettant l'implémentation de mécanismes adéquats (par exemples : redondance, reprise, reconfiguration).

- Indépendance de l'application vis-à-vis de la structure physique

Pour le réalisateur de l'application, il est important de ne pas avoir à connaître l'architecture physique du système ; l'adjonction de processeurs ne doit pas entraîner la reprise de l'étude.

Pour des raisons d'extensibilité, de disponibilité ou d'efficacité, certaines parties de l'application peuvent être regroupées sur un même processeur, tandis que d'autres peuvent être déportées sur d'autres processeurs. Ces opérations doivent être réalisables sans aucune modification du code de l'application. En particulier une application conçue pour un système ne contenant qu'un seul processeur doit pouvoir s'exécuter sur un système comprenant plusieurs processeurs.

2.3 Choix d'une structure d'application

La structure logique d'application répartie retenue dans cette étude a été définie par analogie avec la structure des systèmes matériels [1]. En effet, l'architecture physique d'un système réparti est constituée d'un ensemble d'unités de traitement connectées à une voie de communication ; ceci offre les avantages suivants :

- Extensibilité :

La puissance du système peut évoluer en ajoutant (ou en retirant) une unité de traitement, tout en maintenant l'interface logique et physique avec la voie de communication.

- Modularité :

Le système peut être configuré à partir d'un ensemble d'unités de traitement autonomes.

Les unités sont définies par leur interface logique et physique avec la voie de communication.

- Disponibilité :

La structure d'un système multiprocesseur réparti offre la possibilité de faire exécuter les travaux d'un processeur défaillant par un autre processeur.

Le logiciel d'application répartie devant offrir ces mêmes avantages d'extensibilité, de modularité et de disponibilité, il semble naturel de le construire par analogie avec le matériel. On considère alors une application comme un ensemble d'unités de traitement logiques appelées fonctions, connectées à un système logique de communication. Cette structuration doit correspondre à un découpage logique ; elle aide, en outre, l'utilisateur à spécifier son application.

Une fonction représente un composant logiciel élémentaire ayant une interface logique bien définie permettant de communiquer avec les autres fonctions. Une fonction n'est connue d'une autre fonction que par son interface.

On peut concevoir une interface se situant au niveau d'un système de transport par messages au travers de ports de communication [2], [3]. Dans ce cas, l'utilisateur appelle directement les services du système de transport en fournissant un message et en identifiant un port. Or, une fonction peut être vue comme une séquence de traitement pouvant être invoquée de façon procédurale. Aussi l'interface proposée est de niveau supérieur à celle d'un système de transport, faisant abstraction des notions de messages et de ports. Le mécanisme d'invocation d'une fonction est de type procédural et est associé à une transmission de paramètres (paramètres de données et paramètres résultats).

En conclusion, il est proposé de découper une application répartie en un ensemble de fonctions autonomes qui communiquent et se synchronisent au travers d'une interface généralisée.

La figure 1 illustre l'organisation logique générale d'une application.

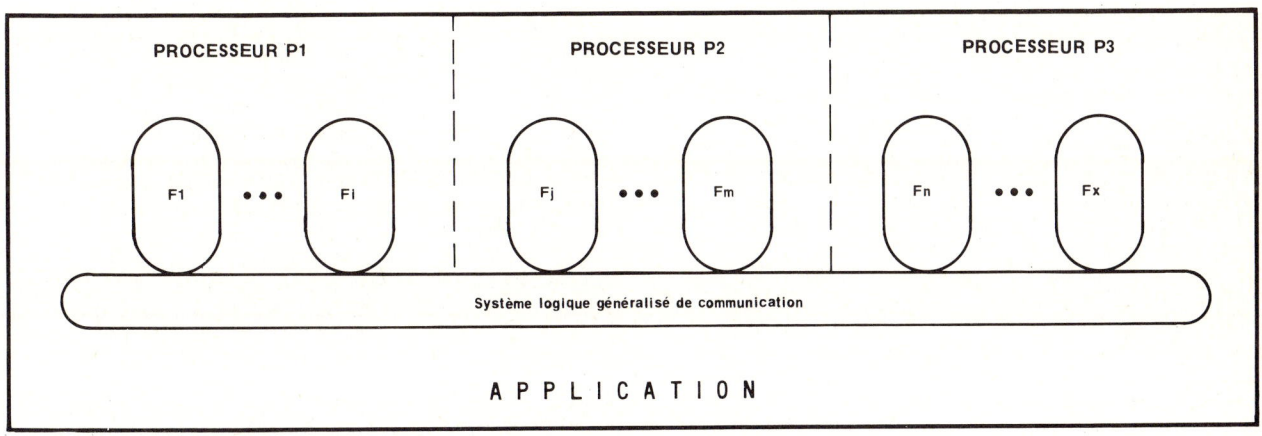

F = Fonction

FIGURE 1 - ORGANISATION GENERALE D'UNE APPLICATION

Ce découpage présente plusieurs avantages :

- décomposition en modules à interface définie,

- possibilité de réutiliser ces modules,

- possibilité de modifier ou changer complètement, un module sans modifier les autres.

2.4 Concepts et définitions

2.4.1 Organisation Logique

Une application répartie est découpée en fonctions qui communiquent entre elles.

Les fonctions ne peuvent pas partager de ressources. Cependant, elles communiquent et se synchronisent via une interface généralisée. Cette interface est définie d'un point de vue logique. Elle est indépendante de la répartition des fonctions sur les processeurs et de la structure d'interconnexion des processeurs.

Une relation de coopération assurant la communication entre deux fonctions est représentée par une session. La session est le support logique de communication entre deux fonctions.

Le système de communication déduit de l'interface des fonctions les actions à réaliser pour la communication ; il prend en charge les problèmes de transport des données (découpage, routage, contrôle).

Le système logique généralisé de communication assure :

- une relation logique entre fonctions par session,

- une communication entre fonctions réalisée par échange de messages.

Une fonction contient ses composants. Elle possède ses ressources logicielles et matérielles propres (par exemple une fonction d'E/S contient le périphérique qu'elle gère).

Une fonction est constituée d'un ensemble de processus coopérants. La gestion de ces objets est réalisée par l'intermédiaire d'un exécutif (système d'exploitation) local à chacun des processeurs. Ceci permet la construction de fonctions englobant des traitements plus ou moins complexes pouvant nécessiter des mécanismes internes de synchronisation adéquats (évènements, sémaphores, etc...), et assurant une bonne efficacité notamment dans les applications temps réel.

En résumé, sur chaque processeur d'un système réparti, s'exécutent des fonctions. La communication et la synchronisation entre fonctions se déroulant sur un même processeur ou sur des processeurs différents sont assurées par un système logique généralisé de communication. Les processus contenus dans une fonction sont contrôlés par l'exécutif local.

La figure 2 illustre l'organisation logique d'un tel système réparti.

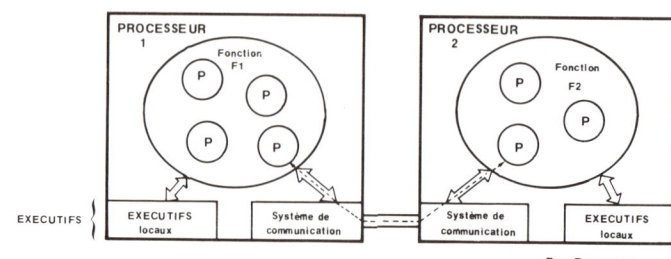

FIGURE 2 - ORGANISATION LOGIQUE D'UN SYSTEME REPARTI

2.4.2 Fonctions

Une fonction est une entité logique de base d'une application. Une fonction est entièrement contrôlée par un processeur physique ; cependant, plusieurs processeurs peuvent réaliser la même fonction. Une fonction peut être entièrement déportée d'un processeur à un autre selon les besoins d'application. Une fonction assure des services.

L'entité exécutant les services d'une fonction doit réaliser les actions suivantes :

- Recevoir la demande d'un service ;

- Interpréter cette demande ;

- Acquérir les données et réaliser le service demandé ;

- Renvoyer des résultats à l'entité qui a émis la demande.

Dans ce mode de coopération, l'entité qui a émis la demande est appelée "Client" et l'entité qui exécute le service est appelée "Serveur". Les échanges d'informations associés à la réalisation d'un service constituent une transaction.

Un serveur est constitué d'un ensemble de processus pouvant se réduire à un seul.

Pour la réalisation d'un service un serveur peut solliciter les services d'une autre fonction. Dans ce cas il devient alors client pour le service réclamé à cette fonction.

Dans une même fonction, les processus communiquent entre eux par couplage serré tel que données communes.

La communication entre deux fonctions, qu'elles soient ou non sur le même processeur, est assurée par le système généralisé de communication, le support de la communication étant la session.

La figure 3 montre le découpage d'une fonction en un ensemble de processus ainsi que les moyens de communication, interne et externe à la fonction.

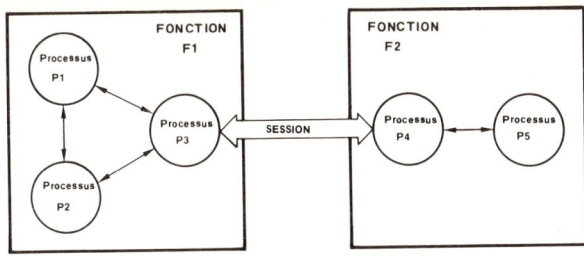

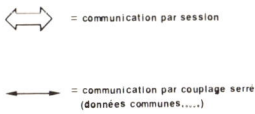
= communication par session

→ = communication par couplage serré
(données communes.....)

FIGURE 3 - DECOMPOSITION DE FONCTIONS ET COMMUNICATIONS

FIGURE 4 - PHASES D'UNE SESSION

2.4.3 Session

Le support logique de communication entre deux fonctions est la session.

Afin que deux fonctions puissent communiquer pour la réalisation de services successifs, une session doit être établie entre ces deux fonctions.

Une fonction demandant l'établissement d'une session pour réclamer des services à une autre fonction, prend le rôle de client.

Une fonction offrant des services à une autre fonction et, par suite, acceptant l'établissement d'une session, prend le rôle de serveur.

Une fonction peut avoir le rôle de "serveur" pour une première session et le rôle de "client" pour une autre session si, par exemple, elle sous-traite à une autre fonction la réalisation d'une partie du service.

Pour la réalisation d'un service, des informations sont transmises entre le client et le serveur. L'ensemble des échanges inhérents à la réalisation d'un service constitue une transaction.

Lors d'une demande d'un service par un client, l'identification du service à effectuer et les paramètres de données du service sont transmis au serveur.

Lorsqu'une session est établie entre un client et un serveur, plusieurs transactions peuvent être réalisées dans cette session. Cependant une transaction ne peut être demandée que lorsque la précédente est terminée c'est-à-dire que le compte rendu de cette dernière a été reçu.

L'ensemble des échanges inhérents à toutes les transactions d'une session constitue une conversation.

La figure 4 illustre les différentes phases d'une session.

2.4.4 Communication par messages

La gestion des échanges entre fonctions est assurée par un mécanisme de communication par messages.

Pour le transfert des données entre deux fonctions, un système de transport interne au système généralisé de communication est utilisé.

2.5 Méthode de réalisation d'une application

En résumé, des concepts proposés pour la structuration d'applications réparties se déduit une méthode de conception d'application.

Les différentes étapes de conception et de réalisation d'une application répartie sont les suivantes :

- Définition du découpage logique de l'application qui permet de déterminer les différentes fonctions ;

- Spécification de l'interface de chaque fonction. Une déclaration d'interface de fonction définit le nom de la fonction, ainsi que les services qu'elle assure. Chaque service est défini par ses paramètres de données et ses paramètres résultats ;

- Définition du découpage logique de chaque fonction en processus ;

- Selon la structure matérielle du système et les différentes fonctions à intégrer, production de l'application en rassemblant les fonctions et en désignant pour chacune d'elles le processeur sur lequel elle doit se dérouler.

3. UN SYSTEME LOGIQUE GENERALISE DE COMMUNICATION : LE LOGIBUS

Ce chapitre présente une proposition de système logique généralisé de communication permettant de réaliser des applications réparties répondant aux principes définis précédemment.

3.1 Présentation générale

Une application répartie est composée d'un ensemble de fonctions qui communiquent et se synchronisent au travers d'un système logique généralisé de communication : le LOGIBUS.

Lorsqu'une fonction veut communiquer avec une autre fonction, elle invoque une opération gérée par le LOGIBUS et la communication est réalisée selon un protocole bien défini.

Le support logique de communication entre deux fonctions est la session.

Dans ce type de relation, une fonction demandant un service à une autre fonction a un rôle de client, la fonction réalisant le service demandé a un rôle de serveur. Les échanges inhérents à la réalisation d'un service constituent une transaction.

Le LOGIBUS assure :

- La relation logique entre fonctions par session : gestion de session ;

- Le dialogue entre fonctions : gestion de transaction.

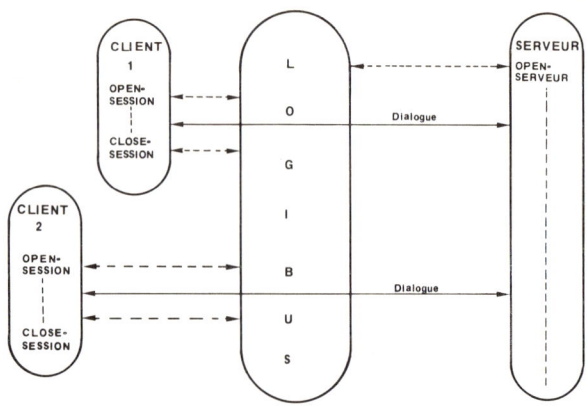

FIGURE 5 - ETABLISSEMENT ET FIN DE SESSION

3.2 Les services fournis

Le LOGIBUS fournit aux fonctions les services suivants :

3.2.1 Etablissement et fin de session

L'établissement d'une session permet de connecter deux fonctions dont l'une est client et l'autre serveur.

Une session est établie par une connexion puis utilisée pour supporter une conversation d'une certaine durée et enfin libérée par une déconnexion.

L'établissement d'une session nécessite l'existence d'un serveur pouvant exécuter la fonction désirée et ne participant pas à une autre session. Les serveurs doivent par conséquent être connus du LOGIBUS. Pour cela, lors de son initialisation, un serveur invoque une opération gérée par le LOGIBUS d'ouverture de serveur (OPEN-SERVER). Par cette opération, le serveur déclare la fonction qu'il exécute ainsi que tous les services qu'il peut offrir, en précisant pour chaque service, une liste des paramètres de données.

Dans une application, il peut y avoir plusieurs serveurs associés à une même fonction.

Lorsqu'un client désire faire réaliser des services à une fonction, il doit au préalable demander l'établissement d'une session avec un serveur pouvant exécuter cette fonction. Pour celà, il invoque une opération gérée par le LOGIBUS d'ouverture de session (OPEN-SESSION) en précisant le nom de la fonction à connecter.

Lorsqu'un client n'a plus de service à demander à un serveur, il peut alors libérer la session. Pour celà, il dispose d'une opération du LOGIBUS permettant de clore la session (CLOSE-SESSION). A partir de ce moment, le serveur n'est plus connecté à un client et peut alors être utilisé pour participer à une session avec un autre client.

3.2.2 Gestion des dialogues

Chaque transaction revient à un échange alterné d'informations, entre le client et le serveur, transportées par le LOGIBUS au moyen de messages :

- Le client et le LOGIBUS initialisent la transaction par envoi de données identifiant le service demandé et représentant les paramètres nécessaires à la réalisation du service.

- A cet instant, le serveur et le LOGIBUS pilotent la transaction en effectuant une suite de demandes de paramètres et de renvois de résultats.

Afin de permettre à l'utilisateur de pouvoir moduler à son gré le contrôle du flot des données, plusieurs modes de transmission de paramètres sont disponibles :

- Mode immédiat - Les paramètres sont définis comme étant immédiats par le client à l'appel d'un service. Ils sont transmis au serveur, par le LOGIBUS, en début de transaction. Le flot des données est contrôlé par le LOGIBUS.

- Mode différé - Les paramètres sont définis comme étant différés par le client à l'appel d'un service. Ils ne sont transmis, par le LOGIBUS, que sur sollicitation explicite du serveur. Le flot des données est contrôlé par le serveur.

A son initialisation un serveur spécifie tous les paramètres de données immédiats inhérents aux services qu'il gère. Un service ne peut alors être exécuté que lorsque tous ses paramètres de données immédiats sont reçus.

Les paramètres résultats, envoyés par le serveur sont transmis :

- au cours de la réalisation du service : contrôle du flot par le serveur,

- en fin de réalisation du service : contrôle du flot par le LOGIBUS.

Les figures 6 et 7 montrent les différents échanges possibles au cours d'une transaction avec contrôle du flot des paramètres, soit par le LOGIBUS (figure 6), soit par le serveur (figure 7).

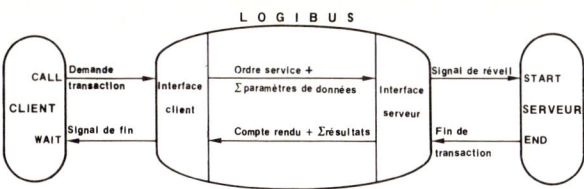

FIGURE 6 - DIALOGUE AU COURS D'UNE TRANSACTION CONTROLE DU FLOT DES PARAMETRES PAR LE LOGIBUS

3.3 Opérations "client"

Les opérations offertes aux clients sont de deux types :

- les opérations de gestion de session,
- les opérations de gestion de transaction.

3.3.1 Opération de gestion de session

OPEN-SESSION

Cette opération assure l'établissement d'une liaison entre le client et un serveur exécutant une fonction donnée.

Pour celà, un serveur est alloué et attribué à la session, ainsi que les ressources nécessaires au dialogue client-serveur.

CLOSE-SESSION

Cette opération assure la suppression de la liaison entre un client et un serveur. C'est la fin normale d'une session.

Les ressources allouées au client par le LOGIBUS sont libérées. De plus, le serveur est libéré, pouvant alors participer à une nouvelle session.

KILL-SESSION

Cette opération provoque la fin exceptionnelle d'une session. Elle permet l'arrêt d'une conversation entre un client et un serveur.

Les ressources allouées au client par le LOGIBUS sont libérés. De plus, le serveur est libéré, pouvant alors participer à une nouvelle session.

TEST-SESSION

Cette opération permet à un client de connaître, à tout moment, l'état de la session.

3.3.2 Opérations de gestion de transaction

CALL-TRANSACTION

Cette opération permet à un client de demander la réalisation d'un service au serveur auquel il est associé par une session.

Le client fournit les paramètres de données du service et précise les paramètres résultats attendus. Au cours de la transaction les données et les résultats transitent entre les deux fonctions.

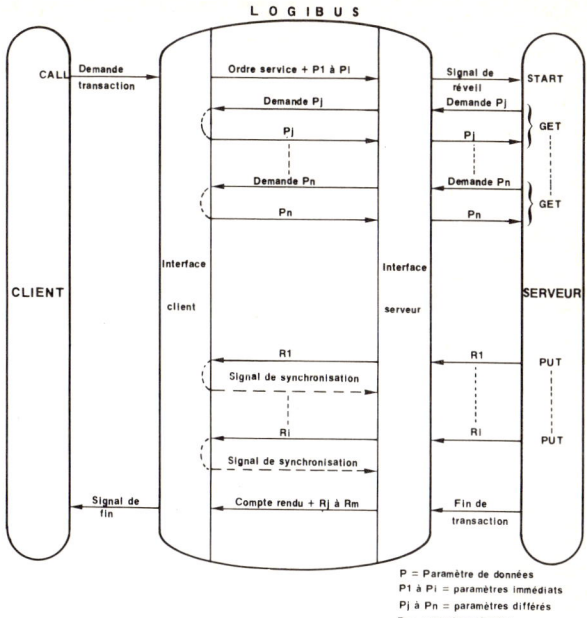

P = Paramètre de données
P1 à Pi = paramètres immédiats
Pj à Pn = paramètres différés
R = paramètre résultat

FIGURE 7 - DIALOGUE AU COURS D'UNE TRANSACTION CONTROLE DU FLOT DES PARAMETRES PAR LE SERVEUR

Le client a la possibilité d'attendre la fin de réalisation du service requis. Dans ce cas, il est suspendu jusqu'à réception du signal de fin de transaction. Sinon, le contrôle lui est rendu et il pourra se synchroniser sur la fin de transaction en invoquant l'opération WAIT-TRANSACTION.

WAIT-TRANSACTION

Cette opération permet à un client d'attendre la première occurence de fin parmi plusieurs transactions, chacune de ces transactions étant supportée par une session différente.

3.4 Opérations "Serveur"

Les opérations offertes au serveur se décomposent comme suit :

- Opérations de gestion de serveur,
- Opérations de gestion de transaction.

3.4.1 Opérations de gestion de serveur

OPEN-SERVER

Un serveur doit être initialisé avant qu'un client ne l'invoque. C'est son existence qui permet au LOGIBUS d'accepter une demande d'établissement de session.

Lors de l'invocation de l'opération d'ouverture serveur, ce dernier déclare la fonction qu'il exécute ainsi que chaque service qu'il peut réaliser. Pour chaque service le nombre total de paramètres de données nécessaires à la réalisation du service est transmis, et les données immédiates attendues sont décrites sous formes de paramètres.

Le LOGIBUS alloue toutes les ressources nécessaires au dialogue du serveur avec tout client possible. Aussi, une session ne pourra pas être refusée, ni même mise en attente, du côté serveur, par manque de ressource disponible.

Le LOGIBUS range les informations communiquées lors de l'invocation de l'opération, dans des tables descriptives du serveur. Ces informations ne peuvent pas être modifiées.

Le serveur pourra alors être choisi pour participer à une session.

3.4.2 Opérations de gestion de transaction

START-TRANSACTION

Cette opération permet à un serveur de gérer une transaction.

Si une transaction est présente, ses paramètres sont dans les zones mémoire réservées à cet effet, le traitement peut donc être lancé.

Si aucune transaction n'est présente, le serveur est mis en attente de l'arrivée d'une transaction. Il sera réveillé dès sa réception.

Lorsqu'une transaction est reçue, tous les paramètres immédiats de données attendus par le serveur sont mis dans les zones mémoire réservées par le serveur et le code du service demandé lui est transmis.

END-TRANSACTION

Par cette opération, un serveur signale au LOGIBUS la fin d'exécution du service et lui transmet en même temps des résultats et un compte rendu d'exécution.

Le LOGIBUS transfère les résultats et le compte rendu dans des zones mémoire réservées par le client et lui signale la fin de la transaction. Le client peut de nouveau invoquer le LOGIBUS pour demander la réalisation d'une autre transaction supportée par cette session.

Le serveur doit invoquer l'opération START-TRANSACTION pour acquérir ou attendre une nouvelle transaction.

GET-PARAMETER

Cette opération permet à un serveur de demander l'acquisition d'un paramètre de données.

Le LOGIBUS transfère le paramètre du client vers le serveur. Le client n'est pas affecté par ce transfert.

Le paramètres est rangé dans la zone mémoire fournie par le serveur.

PUT-PARAMETER

Cette opération permet à un serveur d'envoyer un paramètre résultat.

Le LOGIBUS transfère le paramètre résultat du serveur vers le client. Le client n'est pas affecté par cette opération, ni même averti.

Le résultat est rangé dans la zone mémoire réservée à cet effet par le client. L'adresse de la zone est donnée par le client lors du lancement de la transaction.

Remarque :

Les opérations GET-PARAMETER et PUT-PARAMETER permettent au serveur de contrôler le flot des paramètres.

4. MODELISATION

Une évaluation préalable des performances du LOGIBUS a été réalisée. Son objectif était de valider les concepts du LOGIBUS avant d'entreprendre la réalisation d'une maquette logicielle.

Cette évaluation a été réalisée à partir de modèles à base de réseaux de files d'attente. Pour la description des modèles, il a été considéré une architecture à base de processeurs MC 68000 connectés à un bus à haut débit, le contrôle du bus étant réparti sur l'ensemble des processeurs.

Cette modélisation a permis une estimation du ralentissement apparent d'un système provoqué par la présence de transactions entre fonctions.

Les résultats obtenus ont montré que les concepts du LOGIBUS s'avèrent réalistes en ce qui concerne l'efficacité des services qu'il offre.

5. IMPLEMENTATION

Dans le cadre d'un contrat avec l'Agence de l'Informatique, une réalisation expérimentale du LOGIBUS est actuellement en cours. Cette réalisation est effectuée sur des architectures à base de processeurs MC 68000 reliés par un bus. Le logiciel du LOGIBUS constitue une agence ajoutée à un exécutif développé sur ce matériel. Ce logiciel est réalisé en PASCAL-SOL et est conçu, comme le reste de l'Exécutif auquel il est adjoint, autour d'un noyau SCEPTRE [4].

6. CONCLUSION

Bien que l'étude du LOGIBUS ne soit pas encore terminée, quelques points positifs sur le développement d'application répartie peuvent être retenus :

- L'architecture logique offre une grande souplesse d'utilisation et amène une bonne modularité dans le logiciel d'application ;

- Le découpage logique en fonctions permet :
    . d'intégrer les mêmes modules dans différentes applications,
    . d'étendre ou modifier facilement une application,
    . de ne pas remettre en cause le codage d'une application lors d'une modification de la répartition.

- Les traitements logiques peuvent être rendus largement indépendants, ce qui offre la possibilité d'implémenter des outils de protection.

La généralisation au niveau du logiciel de la structure matérielle des systèmes répartis contribue à la conception de systèmes modulaires, extensibles et ouvre une voie vers les systèmes disponibles.

BIBLIOGRAPHIE :

1 - P. FISCHER, B. GAGEY
    "Bus Logiciel"
    THOMSON-CSF - LCR - Réf. 1338/80 CHM.

2 - J.S. BANINO, A. CARISTAN, M. GUILLEMONT, G. MORISSET, H. ZIMMERMANN
    "Chorus : an architecture for distributed systems"
    INRIA - Rapports de Recherche n° 42.

3 - A. LISTER, J. MAGEE, M. SLOMAN, J. KRAMER
    "Distributed Process Control Systems : Programming and configuration"
    Researd Report n° 80/12
    Département of Computing and Control,
    IMPERIAL COLLEGE, LONDON SW7 2B.

4 - M. DERRIENNICK, P. DESCLAUD, H. FALLOUR, C. FAULLE, J. FEBVRE, M. HANNE, M. KRONENTAL, JJ. SIMON, D. VOJNOVIC
    "Proposition de Standard de Noyau d'Exécutif Temps Réel"
    Projet SCEPTRE - BNI
    Convention n° 79.2.36.0055

5 - E. DOUGLAS JENSEN
    "The Honeywell Experimental Distributed Processor an Overview"
    Computer January 1978.

6 - G. LE LANN
    "Le Contrôle dans les Systèmes Informatiques répartis : nature du problème et quelques solutions".
    INRIA/SIRIUS - CTR 0.00J.

STANDARDISATION WORK FOR COMMUNICATION AMONG DISTRIBUTED
INDUSTRIAL COMPUTER CONTROL SYSTEMS - A STATUS REPORT

Graeme G. Wood,
Manager, European Market Research
and Product Planning,
The Foxboro Company, Redhill, Surrey, RH1 2HL, England

ABSTRACT

The current position of some standardisation groups working in local area communications are discussed and compared with the needs of Distributed Computer Control Systems in industry. The standards groups are International Electrotechnical Commission (IEC), International Standards Organisation (ISO) and Institute of Electrical and Electronic Engineers (IEEE). However, it should be noted that this presentation is not an official statement by any standardisation body and should not be relied upon for decisions about product design or commercial activity. The opinions are those of the author and do not necessarily represent his employer, any other committee member, or their sponsoring organisation.

*The author has been a member of the IEC Working Group on PROWAY since its inception and Chairman since April 1982. He is a liaison member of the IEC Working Group on Interface Systems for Programmable Measuring Instruments.

RÉSUMÉ

L'état actuel des travaux menés par les groupes de standardisation dans le domaine des réseaux locaux est présenté et comparé aux besoins des Systemes de Contide Répartis dans l'industrie. Les groupes concernés sout l'IEC, l'ISO et l'IEEE. On notera que cette présentation n'a rien d'officiel et ne doit pas servir de base à des décisions concernant la conception de produits ou une activité commerciale. Les points de vue exprimes n'engagent que l'auteur et aucunement son employeur, les participants aux groupes de standardisation on les groupes eux-mêmes.

*L'auteur fait partie du Groupe de Travail PROWAY de l'IEC depuis sa création et en est le Président depuis avril 1982. Il est aussi chargé des liens avec le Groupe de Travail "Interfaces pour Instruments de Mesures Programmables" de l'IEC.

ZUSAMMENFASSUNG

Der gegenwaertige Stand in der Arbeit einiger Normierungs-Arbeitsgruppen auf dem Gebiet lokaler Verbundnetze wird betrachtet im Hinblick auf die Anforderungen verteilter rechnergesteuerter systeme im industriellen Bereich. Die Normierungsgruppen sind die Internationale Elektrotechnische Kommission (IEC), die Internationale Normen Organisation (ISO) und das Institute of Electrical and Electronic Engineers (IEEE). Es muss jedoch daraufhingewiesen werden, dass die vorliegende Abhandlung keine offizielle Mitteilung eines Normierungsgremiums ist und dementsprechend nicht fuer entwicklungstechnische oder Kommerzielle Entscheidungen herangezogen werden sollte. Der Autor traegt ausschliesslich seine persoenlichen Ansichten vor, die durchaus nicht mit denen seines Arbeitgebers, der anderen Komittee-Mitglieder oder der beteiligten Organisation uebereinstimmen muessen.

*Der Autor ist seit Beginn Mitglied der PROWAY - Arbeitsgruppe des IEC, seit April 1982 ihr Vorsitzender Er ist liaison member der IEC - Arbeitsgruppe "Schnittstellen systeme fuer programmierbare Messgeraete".

INTRODUCTION

Several communication standards are emerging with potential for use in Distributed Industrial Computer Control Systems. These systems are characterised by up to 100 microprocessor based stations separated by distances up to 2 Km. Such systems are the industrial equivalent of a Local Area Network (LAN).

This paper covers the author's view of the present status among four standards which could support future communication for Distributed Industrial Control:-

- Process Plant Communications (PROWAY[1] and a related standard for Process Application Formats)

- Office Communications: (IEEE 802[3] and Ethernet)[4]

- Post and Telegraph Communications (CCITT X 25)[5]

- Laboratory Communications: (Serial Extension of the IEC 625-1[2] standard).

Present Status of Standards

The IEC PROWAY Committee has defined functional requirements for an industrial data highway. Table 1 lists some of the requirements and reasons for their choice.

1. Multiple master stations sharing the highway with passive connections from each station to the data highway.

2. Single wire, serial, multi-drop connection among stations. Options include redundant paths.

3. Variable length message frame with no restriction on the bits and bytes contained in a user's application message.

4. Operation of the communication system must be deterministic.

5. Automatic acknowledge, within the communication sub-system, for each message sent.

TABLE 1
GUIDELINES FOR COMMUNICATION IN INDUSTRIAL CONTROL SYSTEMS

The PROWAY standards group is working to meet the Table 1 criteria by defining a standard for Layers 1 and 2 of the OSI model[6].

Two other communication standards may be important for future industrial communications. These are the IEEE 802[3] with its near relative, the Ethernet Highway[4]: and the CCITT X25[5] which is a long distance telecommnication standard being used for large networks and packet switching. Both these standards are intended for commercial use rather than industrial environments.

The IEC 621-1 standards group are developing a serial bus for Laboratories and light industrial applications. They are presently cosidering a subset of IEEE 802 which meets only points 1, 2 and 3 in Table 1.

Availability and Scope of Standards

Presently the only widely available standard is the telecommunications standard X25. Its scope includes networking message formats and hardware for message transfer.

Early production samples of IEEE 802/Ethernet units are offered by a limited group of suppliers. Products using PROWAY or IEC 625-Serial are not expected to be available for applications until at least 1985.

The scope of work for PROWAY, IEEE 802/Ethernet and 625-Serial is restricted to providing transfer of digital data messages among stations with complete freedom in the content of a message. If several supplier companies are connected in a plant-wide system, an agreed message format and translation programmes are necessary. This ensures that programmes by one manufacturer can understand a plant-wide message transmitted by equipment from another manufacturer. Standards for networking are evolving under X25 and IEEE 802 and it is expected these will influence the future industrial uses of networking.

Standards for industrial application messages are being studied by IEC TC 65C WG1. However, this group has just begun the task and some years will pass before a standard will be adopted.

References and Notes

1. PROWAY is under development by the International Electrotechnical Commission, Technical Committee 65C, Working Group 6. This group was previously named TC65A, WG6 and some drafts for national comment have been circulated by TC65A. See Table 2.

2. IEC 625-1 and 2 are defined by the International Electrotechnical Commission, Technical Committee 65C, Working Group 3. This group were previously named TC 66 WG3 and a short distance, bit serial, byte parallel data bus has been published as an IEC Standard numbered 625. The group is now investigating a long distance, fully serial bus which will have

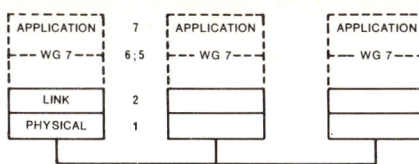

DISTANCE	2 km
DEVICES	100
LINK RATE	10^6 Hz (10^5 Hz FOR APPLICATION DATA)
CONNECTION	PASSIVE MULTIDROP
SERVICES	FRAME TRANSFER WITH IMMEDIATE ACKNOWLEDGE
LINK MANAGEMENT	MULTIMASTER
MASTER TRANSFER	YES. TOKEN PASSING, GUARANTEED ACCESS TIME
SCOPE	PHYSICAL & LINK LAYERS NETWORK OPTIONAL TRANSPORT SESSION PROBABLY NULL
FRAME	TWO VERSIONS. ONE HDLC DERIVED.
ENVIRONMENT	HEAVY INDUSTRIAL HAZARDOUS CHEMICAL EXPLOSIVE RISK
STATUS 1Q82	50%. PART 1 PUBLISHED
RELATED WORK	IEC TC 65C WG 1 "APPLICATION DATA FORMATS" (WAS TC 65A WG 7)

TABLE 2. PROWAY FEATURES

maximum compatibility with 625. This paper refers to the proposed serial extension as "625-Serial". See Table 3.

3. Institute of Electrical and Electronic Engineers, Project 802. Draft documents were published for comment at the end of 1981. The drafts include many sub-options some of which are very close to the Ethernet standard. See Table 4.

4. "The Ethernet, a Local Area Network: Data Link Layer and Physical Layer Specification". Jointly published by DEC, Intel and Xerox Corporations. Also referred to as the DIX Data Highway. See Table 4.

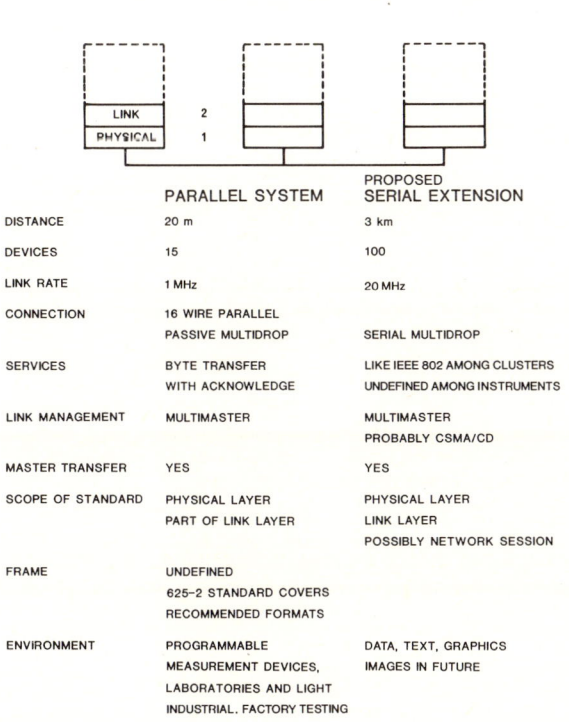

	PARALLEL SYSTEM	PROPOSED SERIAL EXTENSION
DISTANCE	20 m	3 km
DEVICES	15	100
LINK RATE	1 MHz	20 MHz
CONNECTION	16 WIRE PARALLEL PASSIVE MULTIDROP	SERIAL MULTIDROP
SERVICES	BYTE TRANSFER WITH ACKNOWLEDGE	LIKE IEEE 802 AMONG CLUSTERS UNDEFINED AMONG INSTRUMENTS
LINK MANAGEMENT	MULTIMASTER	MULTIMASTER PROBABLY CSMA/CD
MASTER TRANSFER	YES	YES
SCOPE OF STANDARD	PHYSICAL LAYER PART OF LINK LAYER	PHYSICAL LAYER LINK LAYER POSSIBLY NETWORK SESSION
FRAME	UNDEFINED 625-2 STANDARD COVERS RECOMMENDED FORMATS	
ENVIRONMENT	PROGRAMMABLE MEASUREMENT DEVICES, LABORATORIES AND LIGHT INDUSTRIAL. FACTORY TESTING	DATA, TEXT, GRAPHICS IMAGES IN FUTURE
STATUS 1Q82	100%	10%

TABLE 3. FEATURES OF IEC 625-1 AND SERIAL EXTENSION

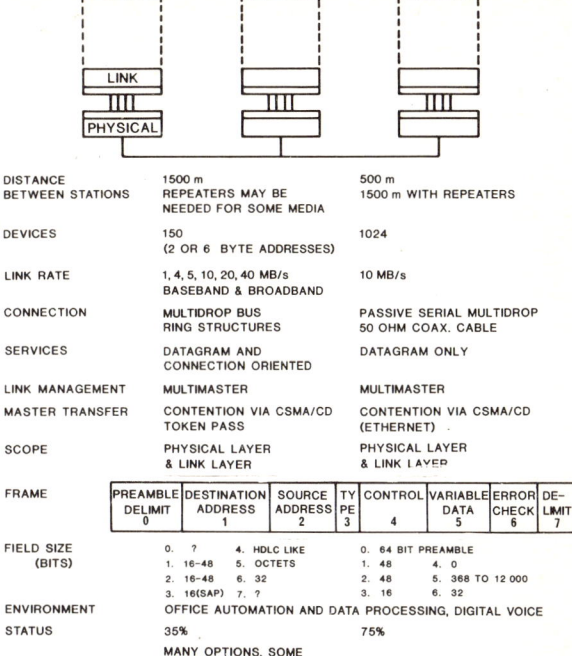

DISTANCE BETWEEN STATIONS	1500 m REPEATERS MAY BE NEEDED FOR SOME MEDIA	500 m 1500 m WITH REPEATERS
DEVICES	150 (2 OR 6 BYTE ADDRESSES)	1024
LINK RATE	1, 4, 5, 10, 20, 40 MB/s BASEBAND & BROADBAND	10 MB/s
CONNECTION	MULTIDROP BUS RING STRUCTURES	PASSIVE SERIAL MULTIDROP 50 OHM COAX. CABLE
SERVICES	DATAGRAM AND CONNECTION ORIENTED	DATAGRAM ONLY
LINK MANAGEMENT	MULTIMASTER	MULTIMASTER
MASTER TRANSFER	CONTENTION VIA CSMA/CD TOKEN PASS	CONTENTION VIA CSMA/CD (ETHERNET)
SCOPE	PHYSICAL LAYER & LINK LAYER	PHYSICAL LAYER & LINK LAYER

FRAME	PREAMBLE DELIMIT 0	DESTINATION ADDRESS 1	SOURCE ADDRESS 2	TYPE 3	CONTROL 4	VARIABLE DATA 5	ERROR CHECK 6	DE-LIMIT 7
FIELD SIZE (BITS)	0. ? 1. 16-48 2. 16-48 3. 16(SAP)	4. HDLC LIKE 5. OCTETS 6. 32 7. ?			0. 64 BIT PREAMBLE 1. 48 2. 48 3. 16	4. 0 5. 368 TO 12 000 6. 32		

ENVIRONMENT	OFFICE AUTOMATION AND DATA PROCESSING, DIGITAL VOICE	
STATUS	35%	75%
	MANY OPTIONS, SOME MUTUALLY EXCLUSIVE	

TABLE 4. FEATURES OF IEEE 802 PROJECT AND THE ETHERNET

5. International Telephone and Telegraphy Consultative Committee. Recommendation X25 covers Physical, Data Link and Network functions. See Table 5.

6. Data Processing - Open Systems Interconnection - Basic Reference Model. Defined in ISO standard DP 7498.

7. High Level Data Link Control. A frame format and control code procedure used by X25 and defined in ISO Standards 3309 on Frame Structure and 4335 on Elements of Procedure.

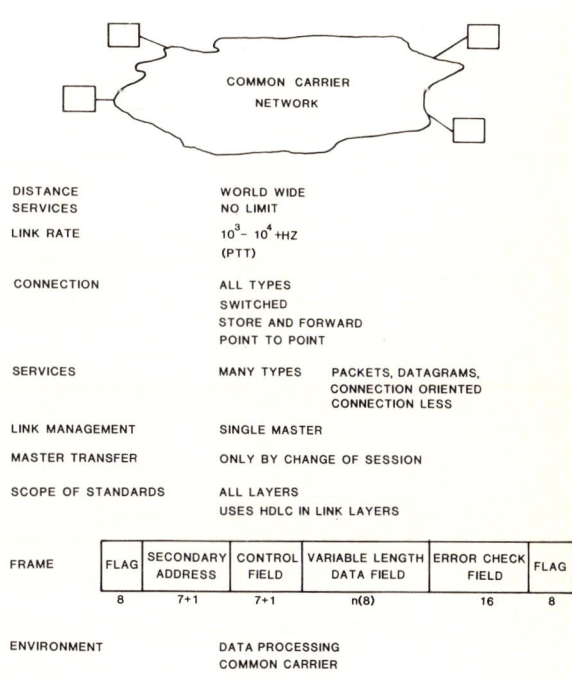

DISTANCE SERVICES	WORLD WIDE NO LIMIT
LINK RATE	$10^3 - 10^4$ +HZ (PTT)
CONNECTION	ALL TYPES SWITCHED STORE AND FORWARD POINT TO POINT
SERVICES	MANY TYPES PACKETS, DATAGRAMS, CONNECTION ORIENTED CONNECTION LESS
LINK MANAGEMENT	SINGLE MASTER
MASTER TRANSFER	ONLY BY CHANGE OF SESSION
SCOPE OF STANDARDS	ALL LAYERS USES HDLC IN LINK LAYERS

FRAME	FLAG	SECONDARY ADDRESS	CONTROL FIELD	VARIABLE LENGTH DATA FIELD	ERROR CHECK FIELD	FLAG
	8	7+1	7+1	n(8)	16	8

ENVIRONMENT	DATA PROCESSING COMMON CARRIER
STATUS 1Q82	≈ 100%

TABLE 5. FEATURES OF CCITT X25

A PROCESS CONTROL SOFTWARE BASED ON CONCURRENT
PASCAL FOR INDUSTRIAL MICROCOMPUTER SYSTEMS

J. GROF, G. RÉVÉSZ, E. ZARÁNDY

Institute for Electric Power Research, Budapest, Hungary

ABSTRACT

This paper describes a process control software called CPPC based on Concurrent Pascal extended by special real-time and control functions making the structure of user programs more simple and clear. Introducing new data types as the structured constants and the BYTE variable the program code and data segments can be reduced. To describe timing functions new real-time elements have been developed, eg. SAMPLE, MARKER, TIMED-RELAY, TIMED-DELAY, etc. An I/O program package has been built in the system for handling the real-time and conventional peripheral units. For the producer-consumer type information exchange between processes a so-called FIFO standard component is implemented. Several control functions are available in the form of classes. The CPPC system is applied in the power industry.

RÉSUMÉ

Cet article décrit un logiciel de contrôle de procédé appelé CPPC basé sur Concurrent Pascal anguel ont été rajoutées des fonctions spéciales temps réel permettant des structures de programmes utilisateur plus simples et plus claires. Le code et les segments de donnés penvent étre réduits grâce à l'emploi de nouveaux types de domées et de la variable Byte. La description des fonctions de temps se fait grâce à de novveaux éléments tels que Sample, Marker, Timed-Relay, Timed-Delay. Un progiciel d'entreé-sortie est utilisé pour gerer les unités périphériques conventionnelles et temps réel. Les echanges d'information de type producteur-consommateur s'effectuent par un composant standard FIFO. Plusicius fonctions de contrôle sont disponibles sous forme de classes. Le logiciel CPPC est utilisé dans une application industrielle énergétique,

ZUSAMMENFASSUNG

Die Abhandlung beschreibt das Prozesssteuerungs-Programm-Paket CPPC, das, auf concurrent Pascal aufbauend, durch Verwendung zusaetzlicher spezieller Echtzeit - und Steuerungs-Funktionen die Struktur der Anwenderprogramme durchsichtiger und einfacher macht. Durch die Einfuehrung neuer Datentypen, wie Strukturierte Konstanten und BYTE-Variable wird die Laenge der Programme und der Datenfelder verringert. Neue Echtzeit-Elemente fuer Zeitsteuerungsfunktionen wurden hinzugefuegt, Z.B. SAMPLE, MARKER, TIMED-RELAY, TIMED-DELAY u.a. Ein Eingabe/Ausgabe-Programmpaket wurde geschrieben, das neben konventiongellen

Auch Echtzeit-E/A-Geraete zu anzusprechen gestattet.

A PROCESS CONTROL SOFTWARE BASED ON CONCURRENT
PASCAL FOR INDUSTRIAL MICROCOMPUTER-SYSTEMS

J. GROF, G. RÉVÉSZ, E. ZARÁNDY

Institute for Electric Power Research, Budapest, Hungary

This paper describes a process control software called CPPC based on Concurrent Pascal extended by special real-time and control functions making the structure of user programs more simple and clear. Introducing new data types as the structured constants and the BYTE variable the program code and data segments can be reduced. To describe timing functions new real-time elements have been developed, eg. SAMPLE, MARKER, TIMED_RELAY, TIMED_DELAY, etc. An I/O program package has been built in the system for handling the real-time and conventional peripheral units. For the producer-consumer type information exchange between processes a so-called FIFO standard component is implemented. Several control funtions are available in form of classes.
The CPPC system is applied in the power industry.

1. INTRODUCTION

The Concurrent Pascal [1] programming language introducing new structures and concepts is an important phase in the history of the real-time and concurrent programming. In spite of its benefits a lot of problems araise in the application of Concurrent Pascal for industrial process control, so on the one hand certain language features are unavailable, on the other hand certain important process control functions can be described only in a very complicated way, with the language facilities of Concurrent Pascal.

A process control software has to provide the following requirements:

(i) the user program should be technology-oriented, its structures should be determined by the tasks to be performed and not by characteristics of the used language;

(ii) high level and application-oriented standard procedures should be avaiable for timing, for describing of control functions and for handling of peripheral units;

(iii) the applied language should be a high level concurrent language suitable for easily and exactly describing different kinds of technical problems.

Considering the above requirements a process control software called CPPC preserving the concepts, structures and elements of Concurrent Pascal has been developed for multi-microcomputer systems that involves the next facilityes and elements.

2. NEW DATA TYPES

In industrial process control systems, it is desirable to separate the data base describing the technology from the programs using them. The data base consists of constant arrays, records and sets besides constant integers, reals and strings. The Concurrent Pascal permits only the latter ones [2]. The Constant types had been extended to the structured constans, moreover the possibility was realized to link data base to the compiled program code. Applying structured constants, the code segment can be reduced with 10-15%, the data segment with 50%.

Nowadays the intelligent peripherals connected with process control equipments are usually based on 8 bit μprocessors. For more efficient communication the BYTE standard type has been introduced. This type differs from the CHAR one in the fact, that the elementary operations on integers are defined on it. In this way, the contradictions of the string access have been eliminated too.

The processes are provided with priorities defined in the parameter list as an integer number following the last parameter. The process scheduling occurs according to the usual priority rules.

3. REAL-TIME ELEMENTS

The control tasks contain a lot of real-time functions [3]. The realization of this functions is very complicated by the original element set of Concurrent Pascal that should use timer monitors and divide logical units to several processes. For describing of this functions the following new facilities and elements are introduced:

(i) On the model of the CYCLE instruction, the SAMPLE instruction has been introduced to ensure the periodical starting of processes:
SAMPLE <time> DO <statement>

The cycle-time of the processes is an integral multiple of time-base of the system, hence in case of 100 msec time-base the longest cycle-time is more then 50 min.

(ii) When several processing of different cycle-time are to be executed on the same data set, it is advisable to describe them in one process omitting the syncronizing and data transmitting monitors. Therefore a new system component the MARKER has been introduced to achieve a longer cycle-time in a periodical process. Two operations are defined on the MARKER type:
PROCEDURE START_MARKER(M:MARKER;T:TIME)
FUNCTION TEST_MARKER(M:MARKER):BOOLEAN
the latter function value signs whether the waited tick has arrived. The time is given in form of hours, minutes, seconds. Using operations on marker the process won't be suspended. The SAMPLE and MARKER elements are realized as the conventional cycle-time handling.

(iii) In control tasks, delays that need restarting and stopping facilities usually occur. It is realized by the RELAY system-type. The above operations are defined on it:
PROCEDURE STOP_RELAY (R:RELAY;T:TIME)
PROCEDURE STOP_RELAY (R:RELAY)
FUNCTION TEST_RELAY (R:RELAY):BOOLEAN
The process is not suspended by the operations on RELAY element too.

(iv) An integer type system time code can be read to determine the exact time differences by the READ_TIME_COUNT procedure.
For the execution of the watch-dog function the following operations are defined on the QUEUE system type:
PROCEDURE TIMED_DELAY(Q:QUEUE;T:TIME)
FUNCTION TIME_OUT (Q:QUEUE):BOOLEAN.
One can examine with the TIME_OUT function whether the suspended process was continued by the CONTINUE operation or by the timing.

4. I/O handling:

For the producer - consumer type information exchange between processes a so-called FIFO standard system component was introduced. This can be considered as a standard monitor. The programmer can access only the procedures of it. The following procedures are defined:reservation for reading/writing, reading/writing of different data types, closing, cencel, test. The memory management subprogram supports the operation of FIFO.
In the Concurrent Pascal the input/output operation are intentionally unsolved. An I/O program package has been built in the system to interface the real-time and conventional units. Process-control requirements have been highlighted for the formulation of the input/output procedures handling the real-time peripheral units. By the help of these procedures, the analog and digital signals can be received and sent; technological events can be recognized in 200 µsec; pulse-width modulated signals can be produced. The information transmission from/to peripherals is realized by the procedures of FIFO.

The RETEX executive program [4] is a part of the system that organizes the running of the process and handles the system clock. Self-testing subprograms have been built in the system to check the hardware moduls.
A configuration consists of several microcomputer subsystems, crates connected on serial lines. The syncronization of system clocks are performed on handler level. In each of crates a common data segment protected by a monitor is defined, the data segments can be mutually accessed by the crates.

The most common control functions are written in form of classes. These are follows: feedforward PI, PID lead/leg, set-point adjuster. Realizing these functions the bumpless manual/auto transfer, automatic prevention of integral wing up, automatic loop start - sequencing etc. have been solved.

It is a general and just demand to ensure segmented compilation in process control systems. Hence further segments can be linked to the already running user program /e.g. to investigate special technological problems/. The segmented compilation has been implemented by modifying the compiler.

5. APPLICATIONS

The CPPC software system is applied in the power industry. The boiler and combustion control on a heating station is one of its applications. Further application is a system that checks and supports the load following, two shifting and start up of 225 MW fossil fired units on a power station.

6. CONCLUSION

A claim for the extension of CPPC system with a modul library has been arisen by certain non computer oriented users. The moduls should solve complete control tasks involving synchronizing procedures too.

7. REFERENCES

[1] Hansen, P.B., The architecture of Concurrent Programs /Prentice-Hall, New Jersey 1977/.
[2] Hartmann, A.C., A Concurrent Pascal Compiler for Minicomputers /Springer-Verlag, New York, 1977/.
[3] Nagy, D., Papp G., Zarandi E., Intelligent Process Control Stations using the CAMAC System. Real-Time Data Handling and Process Control Conference Publication /1980/ 351-354.
[4] Sárközi, E., Grof, J., Videoton Remote Process Terminal, Információ Electronika, No. 4. /1982/ 206-213 /in Hungarian/.

A MULTIPROCESSOR NETWORK IN A MULTIUSER CAMAC ENVIRONMENT

U. Beyschlag, M.J. Clayton, S. Patel, M. Rabany
R.I. Saban and A. Thys

CERN, Geneva, Switzerland

ABSTRACT

The Experimental Areas of the SPS, the large European particle accelerator near Geneva, contain a large quantity of equipment for the monitoring and control of the secondary beam lines. This equipment is computer controlled via CAMAC and Serial CAMAC: the Serial CAMAC loops, which are themselves driven by CAMAC modules allow the connexion of many CAMAC crates distributed over a large area. With the gradual expansion of the zone, the demands for its control now exceed the capacity of a single minicomputer of the type we employ: however, the nature of the zone makes an homogenous system highly desirable. The solution adopted has been to allow several computers to access all the equipment simultaneously, by connecting the computers to the equipment via a network of CAMAC crates which are controlled by local intelligent controllers. The network is self-initialising, and as there are redundant interconnexions, automatically reconfigures itself in case of hardware failure. The computers attached to the network form a pool of computing power which is distributed according to need: computers may be added to the system, or taken down individually for maintenance at any time with little or no disturbance to the users. The initialisation of the CAMAC modules that control equipment both in the Network CAMAC Crates and the dependent Serial CAMAC Crates is performed by programs in the computers which refer to easily changed data files. The entire system is thus designed for maximum flexibility. This paper describes the implementation of the system, and initial experience in use.

RÉSUMÉ

Le zones expérimentales du SPS, l'accelerateur de particules du CERN, contiennent une grande quantité d'équipements assurant le contrôle et la commande du faisceau. Ces equipements sout gérés par un système informatique via des unités CAMAC. Les boucles serie CAMAC, contrôlées elles-mêmes par des modules CAMAC, permettant l'interconnexion de plusieus créneaux CAMAC géographiquement dispersées. Les zones experimentales s'agrandissent régulièrement, la charge de travail liée au contrôle dépasse les capacités du mimordinateur utilisé. La contrainte d'homogeneité est par ailleus très forte. La solution retenue consiste à permattre un access simultané des ces équipements a plusienes ordinateurs, via un réseau de creneaux CAMAC gerées par des contrôleus intelligents. Le réseau peut s'auto-initialiser et se reconfigurer en cas de panue. L'ensemble des ordinateurs constitue une réserve de calcul qui est alloué en fonction des besoins. Des ordinateurs pervent être ajoutés ou retirés du réseau sans que cela cree des perturbations. L'initialisation des modules CAMAC qui contrôlent les equipements daus le reseau des creneaux CAMAC et de modules CAMAC serie est assuree par des programmes lies à des fichiers aisément modifiables. Le système global est danc très flexible. Cet article decrit la réalisation du systemé et les premières experiences.

ZUSAMMENFASSUNG

Die Experimentzonen des grossen europaeischen Teilchen beschleunigers. SPS bei Genf enthalten zahlreiche Geraete zur Ueberwachung und Steuerung der Sekundaeren Strahlstrecken. Diese Geraete sind veber CAMAC und Serielles CAMAC an die Rechner angeschlossen: die seriellen CAMAC-Schleifen, die wiederum von CAMAC-Moduln getrieben Werden, gestatten den Anschluss einergrossen Zahl von CAMAC crates die raeumlich weit verstreut sind. Als Folge der stetigen Ausdehnung der Experimentalzonen veberschreiten die Anforderungen an die Steuerung nun die Leistungsfaehigkeit eines einzelnen Minicomputers des bisher verwendeten Typs; anderseits laesst die Struktur der Experimentalzonen ein homogenes System sehr wuenschenswert erscheinen. Die gewaehlte Loesung besteht darin, mehreren Rechnern gleichzeitig zugriff auf alle Geraete zu gestatten. Dies wind erreicht durch die verwendung eines Netzer von CAMAC Crates, die von intelligenten Controllern gesteuert werden. Das Netz initialisiert sich selbst und verfuegt dank redundantes Verbindungen ueber die Moeglichkeit automatischer Rekonfiguration in Falle eines Hardware-Fehlers.

Die an das Netz angeschlossenen Rechner bilden einen Rechenkapazitaets-Pool, der entsprechend den Anforderungen eingesetzt werden kann: einzelne Rechner Koennen hinzugefuegt oder fuer Wartungszwecke stillgetegt werden, ohne dass nennenswerte Stoerungen fuer die Benutzer auftreten. Die Initialisierung der CAMAC Modul sowohl in den Netzwerk-Crates als auch in den untergeordneten seriellen Crates wird durch Programme in den Rechnern durchgefuehrt, die auf leicht aenderbare Dateien zugreifen. Auf diese Weise Konnte das gesamte system sehr flexibel gehalten werden. Die vorliegende Ab-handling beschreibt die Realisierung des Systems und die ersten Benutzere-fahrungen.

1. INTRODUCTION

The Experimental Areas of the SPS contain the experiments and the secondary beam lines, which are long electromagnetic optical systems for the selection and transport of charged particles. The equipment for the control and monitoring of these beam lines is distributed over large areas, and is connected to the computers by Serial CAMAC Loops{1}. As has been described elsewhere{2}, the whole concept of the control of the experimental areas has been arranged for maximum flexibility of installation and alteration of equipment.

To give an idea of the size of the areas in question, the North Experimental Area is approximately 1000 m by 200 m, and contains 29 Serial CAMAC crates distributed over four loops. Fifty-five terminals are connected to the computers via the serial CAMAC, as well as 269 motors, 302 photomultipliers, 543 scalers and many other types of equipment.

Each Area was originally controlled by a single NORD-10 minicomputer, to which all the equipment was directly connected via CAMAC and Serial CAMAC. This computer provided a terminal service for the users so that they could run programs to control and monitor their beam lines, and also ran various programs at regular intervals to survey the state of the equipment, and to recover (if possible) from any error states that it detected. With the steady growth of the Areas, the computer rapidly became badly overloaded, and a large increase in computing power was urgently needed.

The only acceptable solution to this problem was the connexion of several computers to the same hardware at the same time. This was achieved by building a packet switching network between the computers and the hardware, that enabled each computer to access all the hardware at any time, and resolved any clashes between them.

Each Area is now controlled by four computers. Three of these computers are functionally identical: they provide the terminal service for the users of the Area, who are connected to one or other of them dynamically upon request. If one of the computers fails, the users running on it at the time receive a disconnect message, and may then ask to be reconnected to another computer. This redundancy allows for graceful degradation of the system. The fourth computer holds the database of hardware connexions, and runs the surveillance programs.

2. THE NETWORK HARDWARE

The intelligent controllers{3} on the network CAMAC crates, which are called MACC's, contain a Texas Instruments TMS9900 16-bit microprocessor. These controllers have full access to all the stations, subaddresses and functions of the crate, which are memory mapped onto the upper half of their address space. They also have full access to all the interrupts (LAM's) in the crate: up to 12 of the LAM's may be patched or graded to different interrupt levels of the microprocessor. The lower half of the TMS9900's address space may be divided between ROM and RAM by setting straps in the module. At the moment, the modules are set to have 10k of ROM and 6k of RAM, and the shortage of ROM is now a limitation to further enhancements. A new type of controller has been obtained which contains a paging mechanism, allowing the extension of the ROM to 48k.

The CAMAC Crates that form the nodes of the Network are connected to each other and to the computers by Camac-Camac link modules which allow the transmission of packets of data from one CAMAC crate to another in both directions simultaneously. The modules are single-width CAMAC modules which communicate with a partner module in another crate via a parallel connexion.

Two other modules are necessary for the running of the system. The first is a clock/calendar module, which makes the date and time available to each node. This is used to ensure the correct self-initialisation of the network. The other module is a very trivial one that gives the node number when read. This number is hard-wired into the module. With this arrangement there is nothing in the software in the ROM of any MACC that is dependent upon which node crate it is put in: all the MACCs are identical.

3. THE NETWORK SOFTWARE

The software of the Network is divided into a Monitor and a series of tasks, the Monitor being a simple operating system, and the tasks pieces of reentrant code which perform particular well-defined functions, such as driving a particular type of device. One or more processes are associated with each task and one process, the initialisation process, is created automatically by the Monitor for each task at start-up time. It is the processes that are manipulated by the Monitor, and may be activated, scheduled, deactivated and so on. The initialisation process of a task may, and usually does, create several other processes associated with the task: if the task is a device driver then a process will be created for each device in the crate.

The monitor creates a pool of packets when the system starts up. These are small blocks of memory of fixed size, which are used by the processes to communicate with each other. As the movement of the packets about the system is performed by manipulating links, the movement of data about the system is very fast, and the need for garbage collection is eliminated.

On start-up, the Monitor schedules an initialisation process to run. This process searches the CAMAC crate for link modules, which must respond to the standard CAMAC module identification function. When a link module is found, the initialisation process of the link handling task is called to initialise it. After this, a task called the Handshake task is called to make contact with the MACC or computer at the other end of the link. When the Handshake is successfully completed, the processors at each end of the link will consider that the link is operational, and will know each other's node number. This information is then sent to a task called the Network task, which collects all the information about the nodes which are directly connected to the local node, and broadcasts this information down all the open links. With the aid of this information, and with the information that has been broadcast from all the other nodes, a routing table is constructed in each node which will be used to send packets about the system.

4. EXPERIENCE WITH THE NETWORK

The Network was first installed in the North Experimental Area about a year ago. To ease the transition between the old and the new systems, it was installed in the parallel CAMAC crates of the old North Area computer. This allowed us to move easily from one system to another by stopping the old computer and starting the MACCs, and vice-versa. This was invaluable, as the Area contains a great deal of diverse equipment, and it was impossible to test all the software for it on the test computer in the laboratory. The new system was brought gradually more and more into use, and the old computer has now been removed.

Our experience with the MACCs has been very positive. From the hardware point of view, if a module does not fail within the first 24 hours after installation, then it will run stably for a very long time. From the software point of view, the MACCs are the most stable part of the system. In the last four months they have been restarted manually only once, and that was after a total power failure in the computer room. The bugs that remain in their software are ones that only appear when there are certain rare hardware error conditions in the Area: this is very hard to simulate in a test situation. These bugs normally cause all the packets in the system to be consumed, and under these conditions, the Monitor restarts the node, and the problem is resolved without any outside intervention.

From the point of view of the users, perhaps the most important point of view of all, the new system is very much faster than the old one, which was becoming painfully slow. We have been able to give the users more memory space to run their programs in, and there is space in the systems for new facilities, which was not so before. The new control system in the Experimental Areas has thus not only given an immediate improvement in the service offered to the users, but promises further enhancements to the service in the future.

REFERENCES

{1} M. J. Clayton, M. Rabany and J. C. Wolles, The Serial CAMAC control system in the SPS Experimental Areas, SPS/ELE/Note 82-25.

{2} M. Clayton, M. Rabany, R. Saban and A. Thys, A Multicomputer Architecture in a Serial CAMAC System, <u>Real Time Data Handling and Process Control,</u> North-Holland 1980, 745-749.

{3} C. Guillaume and W. Heinze, The ACC 2421: A versatile CAMAC crate controller and computer, <u>Nuclear Instruments and Methods,</u> 177 (1980), 327-331.

Session 4

PROGRAMMING LANGUAGES

LANGAGES DE PROGRAMMATION**PROGRAMMIERSPRACHEN**

PROGRAMMING EQUALS SPECIFYING PLUS DESIGNING

I.C. Pyle

University of York, U.K.

ABSTRACT

Programming Languages reflect what is considered to be involved in programming. The design of Programming Languages, from Fortran in the late 1950s to Ada 25 years later, demonstrates a substantial shift in emphasis from attention to the mechanics and detailed coding in executable statements to the issues of structure and correctness associated with declarations. It is now possible to identify two distinct activities involved in programming, namely specifying and designing. The implementing is done automatically by compilers. Early Programming Languages said nothing about the specification, concentrating entirely on design. Some of the declarations in modern Programming Languages amount to specifications, but not in any complete way. In Ada for example, there can be partial specification of a data type, but there is no way of specifying the effect of a sub-program other than by stating its design (by giving a body). Recent work on the use of Ada for specifications has shed light on this problem, and suggests the way forward to a new generation of Programming Languages in which semantic specification is the principal subject.

RÉSUMÉ

Les langages de programmation reflétent ce que l'on considère comme important dans la programmation.

La conception de langages de programmation, de FORTRAN, à la fin des années 50, à ADA, 25 ans plus tard, prouve un déplacement de l'intérêt: portant à l'origine sur la programmation détaillée en instructions exécutables, il s'attache maintenant à la structure et à la qualité associées aux déclarations. Il est à l'heure actuelle possible d'identifier deux types différents d'activités en programmation: la spécification et la conception. L'implémentation se fait automatiquement à l'aide de compilateurs. Les premiers langages de programmation ne mentionnaient rien sur la spécification, se concentrant uniquement sur la conception. Certaines des déclarations des langages de programmation modernes remontent aux spécifications, bien qu'incomplétement. En ADA, par exemple, il peut y avoir une spécification partielle du type de donnée, mais il n'y a aucun autre moyen d'indiquer l'effet d'un sous-programme que de préciser sa structure.

Les récents travaux menés sur l'utilisation d'ADA pour les spécifications ont permis de faire la lumiére sur ce probléme et suggérent une orientation à une nouvelle génération de langages de programmation dans lesquels la spécification sémantique est de premier importance.

ZUSSAMMENFASSUNG

Programmiersprachen spiegeln die Vorstellung wider, die man über das Programmieren hat. In der Auslegung von Programmiersprachen von dem FORTRAN vom Ende der fünfziger Jahre bis zu ADA 25 Jahre später zeigt sich eine bedeutende Schwerpunktverschiebung von der Mechanik und detaillierten Codierung in ausführbaren Anweisungen hin zu den mit Vereinbarungen verbundenen Fragen der Struktur und Korrektheit. Heute sind wir in der Lage, in der Programmierung zwei deutlich getrennte Aktivitäten zu identifizieren - Spezifizierung und Entwurf. Implementierung geschieht automatisch mittels Compiler. Die früheren Programmiersprachen befaßten sich einzig und allein mit dem Entwurf, nicht aber mit der Spezifizierung. In modernen Programmiersprachen entsprechen einige der Vereinbarungen Spezifizierungen, jedoch keineswegs vollständig. Beispielsweise kann in ADA zwar eine Teilspezifikation einer Datenart vorkommen, doch besteht keine Möglichkeit, die Wirkung eines Unterprogramms außer durch Angabe seiner Auslegung (mittels des Namens) zu spezifizieren. Durch jungste Arbeiten an der Verwendung von ADA für Spezifikationen wird Licht auf dieses Problem geworfen und der Weg nach vorn zu einer neuen Generation von Programmiersprachen angedeutet, bei denen semantische Spezifizierung das Hauptthema darstellt.

PROGRAMMING EQUALS SPECIFYING PLUS DESIGNING

I C Pyle

Programming Languages reflect what is considered to be involved in programming. The design of Programming Languages, from Fortran in the late 1950's to Ada 25 years later, demonstrates a substantial shift in emphasis from attention to the mechanics and detailed coding in executable statements to the issues of structure and correctness associated with declarations. It is now possible to identify two distinct activities involved in programming, namely specifying and designing. The implementing is done automatically by compilers. Early Programming Languages said nothing about the specification, concentrating entirely on design. Some of the declarations in modern Programming Languages amount to specifications, but not in any complete way. In Ada for example, there can be partial specification of a data type, but there is no way of specifying the effect of a sub-program other than by stating its design (by giving a body). Recent work on the use of Ada for specifications has shed light on this problem, and suggests the way forward to a new generation of Programming Languages in which semantic specification is the principal subject.

Introduction

First I wish to discuss the relationship between design and implementation. Consider the costs or effort involved in making a number of identical systems. The design has to be done once, however many identical systems are made. There is a small incremental amount of work for each of the systems - copying files, writing tapes or disks, distributing and loading. This incremental work per distinct unit (of many identical copies) is very small - usually neglected, particularly since it is unthinkable for sufficient identical copies to be needed to make the aggregate of the unit marginal cost to match the design cost. In other fields of engineering, the unit marginal work is called the implementation. In computing, there appears to be unwillingness to limit the term to the unit marginal work, and it is confusingly applied to the later stages of design: programming and coding.

Design and Programming

The design process covers a gradual development of the program, by a series of transformations. At each stage there is a 'current understanding'; there is no clear separation between designing an information system, and programming it. A design can be viewed as an incomplete program - that is a program in which certain entities or concepts are used which have not yet themselves been designed. There is no single sudden transition from "desired effect" to "means of achieving effect", but many such transitions incrementally (through intermediate specifications) throughout the design process.

During the design stage, the program is necessarily incomplete: some parts which are eventually needed have not yet been considered. At the beginning of the design stage we have only the overall specification; at the end we have the whole program.

A programming language which is capable of stating specifications and progressive detail is also appropriate for expressing a design. A compiler for such a language would reject a design, because it would be an incomplete program. However, the diagnostic messages from the compiler would be important, as they would identify omissions, and draw attention to any inconsistencies.

Design activities

This view of gradual transition from functional specification to eventual program underlies our search for effective information engineering techniques. Our starting point is that there is an essential similarity between the requirements model, the intermediate functional specifications, and the corresponding program. The crucial difference will emerge in the course of the paper.

We can identify two kinds of activity during the design stage which increase the amount of information in the program: creation and discovery. These occur many times over, creation being the recognition and identification of units needed, and discovery the incorporation of facts found to be relevant about the environment of the program. As the new information is obtained, it must be checked for consistency with the design so far, and the relevant specification.

At many times during the design stage, a point is reached when a mistake is made or a prior decision is regretted: some alternative design path is seen to be preferable. No design goes straight from start to finish, and a good design tool is one which permits the designer to locate the relevant mistake or decision point, and recover as much as possible of the work which has been done after the mistake or decision.

Given these points, it is desirable to use a consistent and coherent notation throughout the design process. If there are to be any discontinuities in the design notation, there are sure to be errors (omissions and inconsistencies) which will need further effort to detect and correct.

High level programming languages are obvious tools to use for this purpose. They already include features for separation of concerns and limiting scopes as a means of reducing inter-dependency, and are supported by tools which check for internal consistency.

Requirements and specification

It is essential to have a statement of what an information system is supposed to do, both as a basis for the design work at the beginning of a project, and as an independent input in the verification of the delivered product at the end.

The basic dilemma in negotiations for a software contract is that the client wishes to be assured that he is going to get what he needs, without having to do the system design himself. The specification and the design are usually so intimately related that it is virtually impossible to get an adequate specification without having done the design; so where is the contract to start?

The client wishes to determine as early as possible that a system can be designed to meet his needs subject to his constraints. He will probably not be able to state explicitly what his needs and constraints are, but he will recognise conformance or inadequacy if he is told clearly enough.

How can we give him the information for his decision? The job of the system engineer at the early stages of contract is to provide the client with enough information (but not too much) to permit this check. It is not a formal check: it is 'the satisfaction of the client'.

I have discussed elsewhere [Pyle 81] the issue of formality in requirements specification.

There is a place for formalism in requirements analysis, but it is not at the interface between the client and the system engineer: it is within the system engineer's area of concern, where he can interpret it to the client. The formal representation of the requirement may be considered to be a set of relations involving the inputs and outputs of the system; together with constraints that must be satisfied. The first stage of a contract is therefore to establish these relations and constraints, to the satisfaction of both the client (so that it expresses criteria for an acceptable system) and the engineer (so that it provides the basis for a design and implementation).

Requirements and Design

The requirements are needed to allow design to begin. Requirements state what is to be achieved; design works out how they are to be achieved. Both the requirements and the design have to be specified in some way; as we have shown, the design eventually leads to a program. At many stages during the design, progress consists of identifying subsidiary requirements, and expressing the solution of one part in terms of subsidiaries to be designed subsequently. Thus the above discussion also applies to a considerable extent to the subsequent stages of design, after the initial requirements have been identified.

The significant difference in these later stages is that both 'requirer' and 'designer' are members of the engineer's organisation, and can be expected to have a much closer rapport and common philosophy than between the client and the engineer. In particular the argument rejecting formality no longer holds.

Thus the internal requirements can be expected to be specified formally, forming the interface between parts of the engineer's organisation, concerning one unit that is designed in terms of other units.

Specifying and programming

The specification is of course shorter than the program, so contains less information, much of which is not in the program. In conventional programming teachnology, specifications are written separately, as 'documentation', and the information about overall intention is absent from the program (unless the designer has taken care to incorporate the specification as comments).

Viewed in retrospect, after the program has been designed, it is advantageous (for program maintenance) that the program text should incorporate information about the requirements, namely the functional specification. Present-day programming languages do not permit this.

Inadequacy of Programming Languages

Having noted the relationship between programming and design, we must beware of going too far. Conventional programming languages do not permit the expression of some crucial design information.

We distinguish between declarations and statements, as in ordinary programming languages. For the purposes of specification, declarations are relevant but statements are not. This is because the specification has to define the effect but not the means; the statements express the means but not (explicitly) the effect. To specify a design, we want the functionality implied by a sequence of statements.

Another way of seeing this distinction is to recognise that the functional relationships are a permanent feature of the system, to be described in its specification and satisfied by the designer. Permanence is the crucial fact: invariance of the functionality during execution which is independent of the particular design. Statements are necessarily dynamic, since each is executed at a particular time in the program, and it changes the state of the system. The major deficiency of most programming languages is that they do not allow the effect of a piece of program to be stated, other than by its means - the sequence of statements to achieve that effect.

Statements and declarations

Early high-level languages were considered to be short-hand ways of producing machine code: an important consideration was the expansion factor from a line of source code to lines of machine code.

Note in passing that the normal way of characterising an assembly language is as a "one-to-one" representation of machine code where each line in assembly language generates one word of machine code. The seeds of the present analysis can be seen in this, by recognising that there are important lines in assembly language which are exceptional: equating values to symbols, setting origins, allocating space for variables, giving values for constants etc. These 'pseudo-operations' play a role in assemblers like declarations in higher level languages.

The trend since Fortran has been to put more emphasis on declarations, since it has been discovered that the additional checking that compilers can do as a result pays off in design time. Compilers are more complicated than assemblers because they are (in part) automatically checking the logic of the program, thus taking over some of the design work.

Pascal and similar strongly typed languages (such as Ada) apply this principle to the data in a program, providing automatic checks that data objects are always used in logically consistent ways. But they say nothing about the semantics of the operations on the data: there is nothing in Pascal or Ada concerning the program semantics, that is the functionality of the program units, defining their desired effects.

An Ada subprogram is specified by its heading, which gives the name and parameters. The intended effect of the subprogram is not specified; it is usually expressed implicitly in the name, e.g.

```
function SQRT (X : FLOAT) return FLOAT;
procedure MOVE_VALVE(V : VALVE; TO :
                     VALVE_STATUS);
```

The reader of such a program is simply left to guess that because the name is SQRT, the function takes a square root, and because the procedure is named MOVE_VALVE, it causes the specified valve to be moved to the status given by the second parameter.

Specification and design

Formally, Ada does not deal with the program semantics, above the level of the elementary statements. The body of a procedure expresses the actions to be carried out when the procedure is called, in terms of the individual statements (whose semantics are defined). But this is too mechanistically detailed to be regarded as the semantics of the procedure itself, that is, of a statement calling that subprogram from elsewhere. This is the crucial difference between requirements specification and eventually designed program - not so much level of detail as the difference between intent and achievement: between the effect obtained and the way of obtaining it. Conventionally, the effect is conveyed by the choice of identifier, usually supplemented by comments. We are now moving towards languages that provide a means of expressing the relationship between the states of the system immediately before and immediately after a call of the procedure what in the program proving literature are known as the preconditions and postconditions of the action.

Transitions in State Space

At a sufficiently large scale, and at a sufficiently small scale, the concept of an action is important in information systems. It could be a job in a batch processing system, or a single machine instruction. It carries out a particular effect, in a finite time, by putting the system in a final state which is in some way related to its initial state, taking finite resources to do it. (And it may fail.)

A necessary and sufficient description of the functionality of such an action is the relationship between the initial and final states of the system. Discovering this relationship is the first part of programming: specifying the requirements.

If we imagine the initial and final states as points in a state space, and the functionality specification as a relationship between them, then designing the program consists of discovering a trajectory in the state space which is feasible and compatible with the resources available.

The trajectory is not discovered entirely in one leap (although a breakthrough in realising feasibility usually implies this); we do not locate a continuous trajectory through state space, but a number of intermediate points, with appropriate relationships between them.

Verifying this design consists of showing that the relationships involving the intermediate states can be combined together to give the required relationship between the initial and final states.

Non-procedural Programming Languages

The pattern of development shown here, with the emphasis moving from statements (i.e. executable code) through declarations to specifications, is a steady reduction of the attention on the sequences of actions carried out (the procedural aspect) in order to focus attention on the effects to be achieved (usually said to be non-procedural, although procedures are still involved).

When there are well-established ways of achieving the desired effects, these can be provided by library routines. When the effects can be achieved satisfactorily by a combination of existing operations, the program can be produced automatically. This is the case for rule-based systems, where the language specifies the desired effects and rules of inference for making satisfactory combinations. In such systems, the programming task is entirely one of specifying: all the design is automatic.

However, there are important cases where the automatically generated programs do not satisfy performance constraints, and the programmer has also to make decisions about certain details of the way the effects are to be achieved. This does not detract from the importance of the specification, but provides the escape route to cope with the situation when automatic design is not good enough - until the method of generating the program can take the constraints into account in selecting the algorithm to be used.

Reference

Pyle, I C   Using Ada for specification and design, Informatica 3 (1981) pp 4 .. 10, ISSN YU 0350-5596

Session 5

METHODS AND TOOLS FOR SOFTWARE DEVELOPMENT

METHODES ET OUTILS DE DEVELOPPEMENT DU LOGICIEL

METHODEN UND WERKZEUGE ZUR SOFTWARE-ENTWICKLUNG

A Survey of Current Trends in Software Tools and Environments

Leon Osterweil
University of Colorado at Boulder, USA

ABSTRACT

This paper discusses the important role that software tools must play in increasing software productivity. It proposes that the surprisingly poor acceptance of software tools is caused at least partly by the circular problem of inadequate exposure and experience. Integration of tools into a software production environment is proposed as an approach to breaking this circular loop and providing an experimental framework for the evaluation and improvement of tools. One such project having this as its goal, Toolpack, is described. The architecture of the Toolpack/IST system and the overall experimental nature of the Toolpack project are described in some detail.

RESUME

Cet article considere le role important que devrait jouer les outils logiciel en vue d'accroitre la productivite en programmation. Le fait surprenant de leur faible acceptation y est attribue , au moins en partie , au cercle vicieux cree par un manque de demonstration de leur emploi et d'experience de leur emploi. L'integration de ces outils dans un environnement de production de logiciel pourrait permettre de briser ce cercle vicieux , tout en offrant un contexte experimental pour evaluer et ameliorer ces outils. C'est approche adoptee pour le projet Toolpack decrit dans cet article. L'architecture du systeme Toolpack/IST, et l'approche experimentale adoptee y sont decrits en detail.

ZUSAMMENFASSUNG

Dieser Bericht untersucht die bedeutende Rolle, die Software Werkzeuge spielen muessen um die Produktivitaet im Programmieren zu erhoehen. Wir behaupten, dass die ueberraschend geringe Bereitschaft vorhandene Software Systeme zu verwenden wenigstens teilweise durch einen unglueclichen Kreislauf von ungenuegender Erfahrung und schlechtem Zusammenspiel beider Seiten verursacht wird. Wir versuchen durch Integration von generativen Systemen in die Produktionswelt von Software diese geschlossene Kette zu brechen und hoffen unsere Software Werkzeuge in diesem experimentellen Rahmen beurteilen und verbessern zu koennen. Wir berichten ueber ein solches Projekt, genannt Toolpack. Die Architektur von unserem Toolpack/IST System wird erlaeutert gemeinsam mit der generellen Idee einer Software Werkzeugkiste.

1. Background

There has been a great deal of discussion during the past decade about a "Software Crisis". When examined carefully, most of this discussion centers around the observation that software seems to be very high in cost and unacceptably low in quality. The tone of these discussions seems to be getting more urgent, as software costs have continued to accelerate, while hardware costs, due to a series of breakthroughs in LSI and VLSI fabrication, have steadily gone down.

It seems that the world is awaiting a similar series of breakthroughs in software production, and grows increasingly impatient for them to occur. In truth, it seemed for a while that the only obstacle to universal integration of computing power into virtually every facet of our lives would be the high cost of software. Now it seems that this general assimilation of computing will take place in spite of our apparent inability to produce high quality software inexpensively. Clearly the "software bottleneck" has not been broken, and as computerization of our society continues this is destined to become even more apparent.

Although it does not seem that there is any dramatic breakthrough in software looming on the horizon, it does seem that certain discoveries and initiatives of the past decade or so have created some momentum which is leading to perceptible improvements.

In my opinion, one of the most important of these directions is the increasing use of computer power to assist the process of producing software. In some sense this is a very natural resource for software writers to tap. Who should be expected to be better equipped to understand and harness the power of the computer than the very people who do just this as their professional activity? In another sense, however, software producers are handicapped by this very familiarity with software. Software writers have tended to regard their work as quintessential creativity, while understanding the computer itself to be a very pedestrian device capable only of drudgery. Thus it seems that there has been a built-in bias against thinking of the computer as a source of substantive assistance in software creation. In the past decade, however, it has become increasingly clear that software production, especially large scale software production, entails depressingly little creativity and oppressively much drudgery. Thus it has become increasingly clear that the computer can become a substantive aid in large scale software production by assisting with the considerable amounts of drudgery. Since that early realization, it has become clear that computers can be used not just for drudgery in large projects, but also as a clever (although not inspired) assistant on projects of all sorts. These realizations, and projects exploiting them, do seem to augur well for the future of software production.

Software systems which bring computer power to bear upon software production problems are commonly referred to as software tools. During the past decade there has a been a gratifyingly large number of software tools constructed. These tools have ranged from comparators to testing aids to analyzers to librarians. In fact it seems that a software tool has been written to address the problems of virtually every facet and aspect of the software production process.

It is unfortunate, and surprising to most observers, that these software tools have met with poor acceptance. Thus the promise of computerization of software production has, for the most part, remained just a promise. In order to understand this, I believe it is important to carefully examine the notion of what a tool really is. According to most dictionary definitions a "tool" is a device which is comfortably and conveniently useful in facilitating or multiplying human work. While it seems clear that the multitude of tools which we have produced are capable of facilitating or multiplying the work that users will do, it is also clear that most of the tools which have been built are either uncomfortable or inconvenient or both.

There are some important classes of software tools which are successful and are, in fact, tools in the truest sense of the word. Compilers, loaders, assemblers, and operating systems are certainly tools. They perform the invaluable service of enabling us to write programs in a higher level language, to reuse libraries of already written procedures, to access and store large files of data, and to share access to the computer. All of this is done through the use of software which clearly multiplies and facilitates human effort. Moreover, most users of this software rarely think too much about when and how to use it. Despite occasional nasty surprises when this software fails, or periodic occasions upon which some unfamiliar features of the software must be learned, we generally use these software systems pretty much without a great deal of conscious thought. This software is comfortable to use, and thus these systems deserve the be called software tools.

It is important to reflect upon why these systems have achieved the status of tools in order to understand what must happen in order for the large universe of software-assistance systems to achieve the status of software tools. It is clear that to a large extent compilers and operating systems have become comfortable and familiar simply because of their longevity. At first, these software systems were new and unfamiliar to users. At that time they experienced the same sort of rejection that we see in the case of many software- assistance systems today. Over a period of decades, however, their benefits became recognized, their proper utilization became better understood and the corresponding increases in comfort and convenience led to acceptance. It is important to observe, also, that during this period the quality of the software tools themselves was slowly improved. It is a rare software product indeed which is reliable, robust and well documented right from the start. Early compilers, loaders and operating systems were no exception. Although relatively reliable today, they were not so at first, and their acceptance and transformation into tools took place only after a period of many years during which they were made robust, reliable and well documented.

Thus it seems that our present crop of software-assistance systems is destined to evolve into a set of software tools given the time in which to improve and in which the using public will come to understand the true merits and proper application of this software. Here, unfortunately, we arrive at what I believe is a problem. The user public was willing to tolerate years of poor compilers and operating systems because it understood the role and purpose of these systems, and because it, on balance, believed that, when perfected, these systems would lead to major productivity gains. The same cannot be said for many of our software-assistance systems today. The sheer variety of such systems poses a problem, as does the fact that most are intended to supply aid to the essentially human activities of analysis and synthesis, rather than direct aid in manipulating the computer hardware. Compilers and operating systems are clearly useful to humans because they assist in the necessary but tedious dealings with the actual hardware. As such they are common denominators, as all software writers need this aid at a well-understood, agreed upon time during software production. Our more modern software-assistance aids have been built to address the galaxy of problems which arise either before or after the actual execution of the coded program. As such they form a bewildering array whose proper times and modes of application are neither widely understood, nor widely agreed upon. Many are designed to assist software producers with such human activities as conceptualization, analysis, evaluation and verification. In this, they are designed to help with activities that have previously been the exclusive province of humans. Thus it should be expected that there will have to be a lengthy period during which the bounds of their efficacy are studied and delineated. It appears that this period is still just beginning.

Thus it seems that if our software-assistance systems are to achieve the status of true software tools, it will be necessary to be sure that the proper usage contexts for these systems have been established and agreed to. If this can be done, then I believe it is likely that the using public will have the patience to wait through the laborious and necessary process of incremental improvement which will ultimately lead to systems of quality which is adequate to assure acceptance.

The proper context for the application of these software-assistance systems can be provided by a competent, credible model of the software lifecycle. Thus, here the important current of tool development merges profitably and effectively with another major current of the past decade--namely studies of proper software production methodologies and lifecycle strategies.

2. The Software Production Lifecycle

One of the most far reaching and encouraging advances of the past decade has been the realization and growing acceptance of the fact that software should be thought of as a product and should be developed as most products are, namely by an orderly systematic process. This process has come to be called the software lifecycle. Early attempts to characterize this lifecycle were very naive and simplistic (i.e. see Figure 1), failing to capture the complexity of the processes by which most software is produced. More recently it has been recognized that the software production process is itself an object worthy of considerable study and research. It has become apparent that the processes which are commonly used to produce software are far more complicated than was originally thought. Figure 2 attempts to capture some notion of this considerable complexity. This figure is, however, flawed in two ways. First, it portrays a process which is far too complicated to be grasped without very close inspection, and second it portrays a process which is far more simple than the process which should be used and often is actually used to produce software.

It seems that in order to portray a methodology by which software might be produced in an effective and responsible manner, it is necessary to depict a process which incorporates a great deal of reviewing, verification, iteration, archiving and reporting. Most people who have been engaged in medium and large scale software development have learned this (often the hard way), but are caught on the horns of a dilemma when they attempt to codify and follow such a methodology. While they realize that good practice dictates following a large and complicated procedure, pressures and constraints (both in time and money) inhibit them from doing so.

Clearly a solution approach here is the use of tools. Most of the procedures indicated in Figure 2, and in the more complicated figures which should be drawn to replace or elaborate upon this figure, can and should be supported or replaced outright through the proper application of software tools. The underlying problem in following these complex methodologies is that they entail considerable amounts of drudgery. This problem is exactly the sort that can be profitably addressed by software tools, as it is one which requires multiplication of human work.

Interestingly enough, the need to use effective tool support to implement a software lifecycle also offers a way of approaching the major obstacle to the maturation of software assistance systems into true software tools --namely the lack of understanding necessary in order to make utilization convenient and comfortable. The dictates of the particular software lifecycle model prescribe the circumstances under which the various software assistance systems are to be applied. Thus in understanding and adhering to the software lifecycle methodology the user is guided to the proper appreciation for and application of the tools which support it. Familiarity with the orderly software production process builds familiarity with the software assistance systems which thereby become software tools.

Thus we see that the marriage of software tools and software lifecycle models and methodologies offers the prospect of uniting a whole which will be far greater in power than the sum of its two parts. Methodologies without tool support can be very hard to understand, and very hard to follow. With tool support, the tools can be aids to understanding at the substantive detailed level, and can also serve to enforce the dictates of the methodology. Tools without the guiding influence of a methodological framework must first be thoroughly understood in order to assure their proper application. The process of understanding the tools and adapting ones work habits and procedures to the tools can negate the beneficial effects of the tools themselves. When adapted to support a specific software production methodology or lifecycle model, the proper use of the tools is assured by following the dictates of the methodology. In addition, the tools can be constructed to work optimally in this more limited context. This can relieve the tool writer of the need to produce tools which are unnecessarily general, and consequently unnecessarily hard and inefficient to use.

It is for these reasons that much effort has recently been focussed on the problem of creating large and comprehensive toolsets which are effectively integrated in support of specific software production processes. Such toolsets are commonly referred to as software production environments.

3. Software Production Environments

Although there is general agreement that software production environments have the potential to effect large improvements in software quality and in the productivity of software workers, there is far less understanding and agreement about how such software systems are to be built. Because a software environment is itself a software system, it does not seem unreasonable to begin consideration of how it might be built by establishing a specification of the requirements for such a system.

In an earlier paper [Oste 81] it was proposed that an environment was a collection of software tools which had the following five properties:

(1) Breadth of scope--capabilities spanning the entire range of activities to be performed in order to accomplish a complete specific software job.

(2) User Friendliness--Input language and diagnostic capabilities which would neither intimidate nor harass the user, as well as sufficient adaptability to assure that the tools would remain useful and supportive as the users work procedures and style underwent reasonable changes.

(3) Tight Integration--Tools which are sufficiently aware of each others capabilities to avoid the semblance of overlapping capabilities as well as the possibility of incompatibility.

(4) Internal Reusability--An architecture and design which encourages the reuse of simple modular capabilities in furnishing the various functional capabilities of the environment.

(5) Use of a Central Data Base--an architecture and design in which the various functional tools draw their inputs from, and place their outputs back into, a central repository of information. This repository is to be considered the focus of all knowledge about the software project.

Unfortunately it is unclear how to achieve each of the five characteristics just described with a single software system. There is even considerable question about whether the five are consistent and compatible with each other.

For example, it has been suggested [Oste 81] that tight integration and internal reusability might be inconsistent objectives. It appears, at least on the surface, that tools which are keenly aware of each other might be difficult or impossible to construct from standard self-contained modules.

The need to center the environment around a data base containing all project information also poses a problem. Clearly a given software development project generates an enormous amount of information. If the data base is to contain all information spanning all aspects and phases of the project then it would have to contain all of this information and reflect all of the myriad relations which characterize the project and its status. It is not clear that this can be done in an efficient, cost-effective way. Thus there arises the question of whether a central data base can be created in a manner which is consistent with the need for efficiency and breadth of scope.

In considering how to carry forward this requirements specification to a greater level of detail and how to begin the specification of an architectural design capable of meeting all of these requirements, it is not unreasonable to once again attempt to follow the paradigm of the software development lifecycle. Thus conventional wisdom suggests that one should carry out a detailed requirements analysis to obtain the answers to the questions raised above. For example, careful requirements analysis should make clear the specific actual needs for the various pieces of information and relations which must be in the central data base. Similarly the requirements analysis process should make clear the performance requirements (e.g., access speeds) necessary in order for the environment. to be acceptably fast and inexpensive.

Here, unfortunately the circular nature of this problem starts to become apparent. In order to pinpoint the requirements for an effective software development environment sufficiently to definitively obtain answers to the above questions, it is essential to be able to interview a wide variety of software developers who are knowledgeable and experienced in such matters. Specifically, it is essential to get definitive answers about experiences and judgments concerning specific tool capabilities and the items of information which they utilize and create. It is unfortunately the case that this sort of knowledge and experience is very rare, because of the lack of widespread use of a variety of software development tools. In fact it is the lack of widespread effective utilization of superior tools that has led us to believe that environments must be created. Hence there seems to be a paradox in that the knowledge needed to form the basis for an environment building effort is not available for precisely the same reasons that are prompting the effort in the first place.

Furthermore, there seems to be agreement that access to superior software development environments will rapidly cause developers to change the manner in which they do their work. Thus any guesses about what might be needed in the way of tool capabilities and configurations would probably change as experience with environments grew. Thus it seems that here, as in the case of other software projects which are designed to address a new and evolving problem, it is naive to expect that we will be able to definitively establish a firm baseline set of requirements.

Instead what must be done is to evolve a strategy for studying the requirements for a software development environment which assures that there is steady progress towards the goal of sufficient knowledge and confidence to justify embarking upon a full scale environment development activity. One way of doing this is to embark on a program of constructing a series of increasingly ambitious experimental prototype environments. If each prototype is designed to be the object of study and the sequence is arranged in such a way that the most critical requirements and design issues can be elucidated or resolved by the early prototypes, then there would seem to be good reason to expect that an effective environment would emerge as an end-product of this process.

Thus it seems that what is needed is the creation of an environment which can be used to support the software development work of a significant community, but which is also highly flexible and extensible, so that it can grow and adapt as the need to do so has been established.

In fact, it seems clear that, at least for the present, while people are beginning to explore and determine the proper set of tools and methodologies with which to support their software production activities, it will be necessary to establish a sixth characteristic of software environments--namely flexibility and extensibility. It is true that this characteristic is included in the second characteristic named above--User Friendliness. Nevertheless, it seems that flexibility and extensibility is destined to be such a major factor in assuring the success of early environments, due to user uncertainty about proper application of tools, that it should be enumerated separately as a necessary characteristic of any environment constructed during our early experimental work.

4. The Toolpack Prototype Environment

It is in this spirit that the Toolpack project has embarked upon a plan for producing a sequence of at least three successive releases of an environment for software production.

The main goal of the Toolpack project to is establish a positive feedback loop between environment developers and a broad and diverse base of environment users by supplying those users a sequence of environments that is increasingly responsive to the users needs. The purpose of this feedback process is twofold. One purpose is to create and promulgate a vehicle for the more effective development and maintenance of software. The other is to obtain reliable, detailed, quantitative answers to many of the central questions confronting software development environment builders. Specifically, it is expected that at least the following issues will be elucidated or resolved:

(1) What is an acceptably broad and complete suite of tool capabilities for supporting some specific software development jobs?

(2) How important are various data items and relations and how accessible must they be to various sorts of environment users?

(3) Is there a set of modular "tool fragments" which is sufficiently powerful yet flexible to provide the basis for a broad yet tightly integrated set of tool capabilities?

(4) What are the general characteristics of a user interface language which is sufficiently powerful, yet acceptably "friendly"?

In order to get reliable insights into these questions it is important to construct prototypes in such a way that a large and diverse community of users will become active users of the environment. Thus the Toolpack project has taken great pains to assure that its prototype environments will be of great interest and assistance to a large and significant community of users.

The target community for the Toolpack prototypes is the community of Mathematical Software developers. This community is in many ways a nearly ideal target community. The mathematical software community is among the oldest software production communities, tracing its origins to the small group of scientists who conceived of the stored program computer in the 1940s. Thus it is a coherent group that has well agreed upon goals and procedures. Among the accepted procedures of this community is the utilization of tools (e.g., see [JPL 78], [JPL 81]). In addition, the community has consciously and innovatively striven toward quality for perhaps a longer period of time than any other software community. This has manifested itself for example, in the notable "PACK" projects of the 1970s [Cowc 77].

This community is perhaps unique in that it has long held that portability is a necessary characteristic of high quality software. Thus the community is accustomed to receiving and evaluating new software items as a community, regardless of differences in the hardware/software configurations being used by different members of the community. Of course it is essential that such new software items be presented in portable form.

There are other characteristics of the mathematical software community that cause it to be very desirable for our purposes. It has long ago established a single programming language (Fortran) as its, more or less uniform, standard. Thus a suite of support tools of interest and value to the entire community can be made source language specific. This greatly simplifies the problem of writing generally useful and acceptable tools.

The software produced by this community is ordinarily rather modest in size, usually aggregating less than 10,000 lines of source code. This also simplifies the problem of writing acceptably efficient tools.

In addition, the mathematical software community generally follows a software lifecycle model which is far simpler than the lifecycle models which are widely espoused by and for many other software development communities. Mathematical software development rarely, if ever, begins with _formal_ requirements analysis. Similarly, there is rarely a _formal_ preliminary (or architectural) design phase. This appears at first glance to be paradoxical, especially in view of the high quality and good acceptance of mathematical software over the past decades. The explanation appears to be that mathematical software requirements and preliminary design specifications have been derived over a period of decades (if not centuries) by mathematicians and numerical analysts. These specifications appear as mathematical formulas in books and technical reports. Thus the process of producing mathematical software appears to start on this base and proceed immediately with what other communities would label detailed (or algorithmic) design. These designs are often expressed directly in the form of code for a higher level pseudolanguage but perhaps more routinely they are coded directly in a relatively portable Fortran dialect. There then follows a familiar pattern of testing, documentation and the upgrading and adjustment that is most often referred to as "maintenance".

It is doubtlessly true that mathematical software began encountering, and grappling with, the sorts of data manipulation problems whose solution would benefit from the more formal requirements and preliminary design techniques prevalent in other communities today. It might be interesting to conjecture about why these contemporary techniques have never been adopted by the mathematical software community, but such conjecture would digress from the issue at hand. The issue is that this community currently does not generally perceive the need for these techniques and associated tools. Thus their absence from Toolpack should not endanger community acceptance. This acceptance can be based only on solid support for the lifecycle as it is practiced, rather than as it might or should be practiced.

This prevailing lifecycle model is particularly fortunate for the Toolpack project because is suggests that a tool support set can be relatively modest, addressing the creation, testing, analysis, documentation, and transportation of only code (and perhaps some algorithmic design) and still be considered to be a complete tool set by this community. More fortunately still, most if not all of these tool capabilities have already been produced and evaluated to some extent by some members of the community. Thus a comprehensive tool set will not be totally unfamiliar to the community. Further, the preexistence of such tools means that the environment production activity can focus more on the issues of integration, user interface, and data base contents, and will not need to be preoccupied with more mundane matters such as the recreation of tool capabilities whose reproduction will contribute less to the accumulation of new knowledge about environments.

With these factors in mind, the Toolpack project has set out to build a portable environment capable of extending comprehensive support to the community of people who are engaged in producing, testing, transporting and analyzing mathematical software written in Fortran. Toolpack project environments will be made available through a series of releases, each of which is designed to improve upon its predecessors as a consequence of experience and evaluation obtained through extensive and diverse utilization of those predecessors. In addition, the architecture which has been agreed to for the Toolpack project has been carefully designed so that tool capabilities can rather easily be altered and extended. This should do much to assure that the system will continue to be useful and thus an object of study over a period of years as the needs and desires of the Mathematical Software community undergo expected changes, at least partly in response to this system itself.

The specific approach to the architecture and design of the family of Toolpack environments will now be summarized. A more complete summary can be found in [Oste 82] and [Oste 82a].

5. The Architecture of the Toolpack Integrated System of Tools (IST)

In realization that, despite the obstacles listed above, a set of initial requirements was needed, the Toolpack project was established as a consortium of eight institutions, most of whom are interested more in using the product environment, than in designing and building it. Thus, this group seems to be an appropriate one for collaborating in the creation of at least the initial functional and environmental requirements for the Toolpack support environment. This group has agreed that the following set of tool capabilities is reasonable as a starting point for the series of prototype environments to come:

0. A compiling/loading system
1. A Fortran-intelligent editor
2. A formatter
3. A structurer
4. A dynamic testing and validation aid
5. A dynamic debugging aid
6. A static error detection and validation aid
7. A static portability checking aid
8. A documentation generation aid
9. A program transformer

A compiling/loading capability is generally available on host operating systems. Thus no tool development effort in this area is proposed. Tools are to be developed in all of the nine other areas, however. In fact, significant tool capabilities have already been developed in some of these areas. The interested reader should consult [Oste 82a] for more details on these capabilities and specific tools. The thrust of this paper is to describe the integration strategy for this environment and the associated experimental approach, rather than the tools themselves.

5.1 The Tool Integration Strategy

The tool objectives described above are to be achieved by a software system, currently implemented in prototype form, called the Integrated System of Tools (IST). A primary motivating goal of the architecture and design of the IST is that user support be supplied in as direct and painless a fashion as is feasible. Our approach is to encourage the user to think of IST as a vehicle for establishing and maintaining a file system containing all information of importance to the user and to the tools themselves. Users will consider that the goal of applying a tool is to generate files of information, many of which will be simply diagnostic information. Some of the files, however, will be files of primary interest and use to other tools, to be applied later. Such a file system is potentially quite large and is to contain a diversity of stored entities. Source code modules would certainly reside in the file system, but so would such more arcane entities as token lists, and flowgraph annotations. In order to keep ISTs user image as straightforward as possible this design proposes that most file system management be done automatically and internally to the IST, out of the sight and sphere of responsibility of the user. The user, in addition is to be encouraged to have access to only a relatively small number of files - only those such as source code modules and test data sets which are of direct concern. The user may create, delete, alter and rename these entities. More important, however, the user may manipulate these entities with a set of commands which selectively and automatically configure and actuate the Toolpack tool ensemble. The commands are designed to be easy to understand and use. They borrow heavily on the terminology used by a programmer in creating and testing code, and conceal the sometimes considerable tool mechanisms needed to effect the results desired by the user. In order to best understand this the IST command language will be presented next.

5.2 The IST Command Language

The purpose of the IST command language is to facilitate the use of Toolpack software tool capabilities in the context of software development practices and procedures which are comfortable and convenient for Mathematical Software developers. Thus there has been a conscious attempt to try to create a command language which is a language for manipulating and managing Mathematical Software programs through the coding, testing, verification, transportation and maintenance phases. This should become more apparent through the discussions in this section.

The form of an IST command will always be as follows:

command_name object_list options_specification

where the command_name must be chosen from the list of available tool capabilities, object_list is a list of the source code items to which the tool capability is to be applied, and options_specification is a specification of the options to be used in configuring the selected tool capability for processing of the source code.

The command names have been chosen to, as much as possible, suggest functional activities that Toolpack/IST users would recognize as activities which they already engage in as part of their de facto software lifecycle methodologies. In the current prototype version of the IST, the command names have been defined as two letter sequences because it was believed that users would prefer to avoid verbosity. Thus in order to invoke the formatting tool, the user would input the sequence "fm". In order to invoke the static analysis capability, the user would input the sequence "an", and so forth. Actual user experience with these choices and user reactions to them will dictate whether or not a more verbose form of these command names will eventually be adopted. In addition, actual user experiences, responses and evaluations will suggest whether the right choices have been made in selecting functions to be supported by Toolpack tools, and whether or not the right choices have been made about which functions are to be offered to users as the highest level functional capabilities accessible directly through the command language.

Thus the command names themselves should be viewed as the operations in the Toolpack/IST command language. The operands in this language are primarily source code objects, and the different views and versions of these source code objects which may be of interest and importance to users and Toolpack tools. In Toolpack, source code is considered to be composed of individual compilation objects, such as subroutines, functions subprograms, and main programs. These can and must be individually named (as so-called program units or PUs), but can also be aggregated into arbitrary collections of PUs, called program unit groups, or PUGs. PUGs are best thought of as structures of PUs. The object list which follows the command name is the list of PUs and PUGS to which the specified command is to be applied. Thus if the user wishes to format dozens or even hundreds of PUs, this can be accomplished readily by grouping the PUs into one or more PUGs and then specifying the PUGs after the command name. This ability to group PUs in flexible ways and then have tools process them as conceptual units is seen as one very important feature of the IST.

It should be noted that there is no prohibition against placing a single PU in several different PUGs. The user may wish to group a subroutine library together as a PUG because the library is a conceptual unit to the user. The user may then wish to group the subroutine library with several different test drivers. This can be done by creating several different PUGs which differ from each other only in that they incorporate the different test drivers. This is permitted by the IST. Furthermore, there are no diseconomies in doing so. If, for example, the user directs that one of two overlapping PUGS be formatted and then directs that another overlapping PUG be formatted next, the IST will recognize that the subroutine library shared by both has been formatted after having formatted the first PUG, and will not then repeat the formatting of the subroutine library in formatting the second PUG. The mechanism for effecting this efficiency will be described shortly.

The options specification is optional. IST allows the creation of named files consisting of complete specifications of the options to be used in configuring a specific tool. In this case the named file can be specified as the options specification. The intent here is to offer users the opportunity of attaching an abstract name and representation to different functional tasks that might be carried out by a single multi-purpose tool. Thus, for example, the instrumentor tool might be used to gather information about execution frequencies of program statements or to check executions for violations of array bounds. Different options files would be used to configure the instrumentation tool to do these things. It seems highly desirable that the command invoking the instrumentation tool incorporate some abstract name or denotation to indicate which of the different instrumentation objectives is being pursued.

Another advantage of this is that these options files can be created by experts and used by novices who consequently do not need to become tool configuration experts.

If no options file is specified here, then the IST will access and employ a default options packet which is stored in the file system. Explicit option specifications can also be listed here to augment or override specifications in options files.

It is important to observe that, although the invocation of a tool through the IST may involve a great deal of work, the user is informed of the disposition of the command only by a very terse message. The purpose of this message is merely to advise the user of whether or not the command has been executed successfully, and where further information about the execution can be found. Invariably the further information will be found in a set of files, whose names will be made available to the user. Usually these files will be report files which the user may list out by using file listing commands. However, it is expected that the more sophisticated analytic tools will produce report files which will be best absorbed by the user with the aid of special browsing or perusal tools. These tools will accept report files as their input and digest and format the files in response to user commands for certain types of information.

The various IST tools will create and access various versions and views of the PUs in order to get their work done. The user will be able to access these versions and views, but will be shielded from the necessity to do so. The static data flow analysis capability, for example, will need access to a parse tree, symbol table and flow graph of all PUs of the PUGs it is directed to analyze. The user need not know any of this, however, and need only specify the names of the PUs and PUGs to be analyzed. The IST command language is obliged to understand that these other files are necessary and is empowered to create them by invoking entire complex sequences of lower level tools about which the end user need know nothing. Furthermore, once these lower level tools have created these versions and views, the IST command language interpreter may choose to store them for future reuse. Thus, if the user subsequently asks to have the analyzed program formatted, the IST will recognize that some of the work needed in order to do the formatting has already been done in the process of doing earlier analysis. The IST command interpreter is equipped with sufficient logic to recognize which internal files contain this useful information and to reuse it in formatting. It is expected that these capabilities should enable the IST to effect significant efficiencies in actual use. More details about how this is accomplished will be presented shortly. In order to do so, however, is important to first understand the Virtual File System Concept.

5.3 The Virtual File System

As observed above, IST tools both require and generate a large of number of potentially reusable files. Speed economies can be gained by storing all such files in case they might be found to be useful later. Storage economies can be gained by refusing to store those entities and instead regenerating them as needed. The strategy for retaining or regenerating these entities is the province of the IST file management system, which is embedded in the IST command interpreter, to be described shortly. Clearly this strategy must be adjustable to meet the constraints of a variety of host computer systems.

It is highly desirable that both the end user and the tool ensemble always be safe in assuming that any needed named entities and derived files will always be available. Thus it is necessary that the IST file management system assume the responsibility for either retrieving these items directly or having them created or regenerated (in case storage exigencies precipitated their deletion by IST). A schematic diagram of the virtual file system architecture is shown in Figure 3.

This virtual file system scheme can be stretched even farther. Although the user might expect that the file system will hold in explicit form any formattings and structuring of a given piece of source text, this is not necessary. Such versions could be recreated by the IST file management system only when needed by following a procedure such as just outlined. Even whole static analysis or dynamic testing data bases could be regenerated in this way. This gives the file management system the flexibility to purge large files to regenerate storage while still retaining the ability to recreate these files when necessary.

This feature should prove particularly useful in hosting the IST on smaller storage machines. Here it may be necessary to permanently store only source text. Under these circumstances all derived images and intermediate entities will be routinely purged, requiring that they be recreated whenever needed. This will result in extra computation time to meet the users requests, but seems a very reasonable trade for the lack of storage.

As will be seen shortly, however, it is expected that the greatest benefit of the virtual strategy will be that it will effectively enable the construction of larger tool capabilities out of smaller tool fragments. This should facilitate the alteration of tool capabilities found to be in need of change and should also facilitate the creation of new tool capabilities by enabling them to be built atop a sizeable base of existing tool fragments. The key to understanding the implementation of the virtual strategy, and the overall architecture of Toolpack/IST is an understanding of the IST command interpreter. Hence that will be presented next.

5.4 The IST Command Interpreter

The IST command interpreter processor consists of two phases--compilation and sequential tool invocation. Compilation, in turn, consists of three subphases: command syntactic analysis, semantic analysis, and object code generation (see Figures 4 and 5).

Syntactic analysis will, at least in later releases, be accomplished by a parser generated by a parser generator. The command language is small and uncluttered making it comfortably describable by parser- generator input. Moreover, it is recognized that users reactions to IST may be strongly influenced by the perceived friendliness and ease of use of the command language itself. It thus seems important to enable changes in the language when and if experience indicates they are desirable. This will clearly be facilitated by the parser-generator-created parser.

The second compilation phase, semantic analysis, will be more complex entailing the selection of the standard template of IST files and functions indicated by the command. In particular, the semantics of each IST command will be defined by a standard sequence of IST file system types--namely the file types which contain the information and data objects directly needed to satisfy the users command.

Because IST employs the virtual file strategy just described, it cannot be expected that all of the files of the types which the semantic analysis phase indicates are needed will be physically present. Thus it is the job of the code generation subphase of compilation to infer from the list of needed files and the physical file system status an ordered list of intermediate files to be created and the tool fragments needed to create them.

The end product of this phase is to be a sequential file of IST code describing in detail all steps to be carried out by IST tool fragments in order to effect the specified command in the exact context of the current state of the IST file system. As such this phase might well be viewed as a pseudocompilation into a machine independent intermediate code.

The guidance for this phase of compilation comes from a data structure which contains complete information about which types of files can be generated from which other types of files by means of which particular tool fragments. This structure, called the dependency DAG (Directed Acyclic Graph), is stored in the IST file system. Figure 6 shows a representative portion of this structure. The code generation phase of the IST compiler traverses the dependency DAG, starting at the nodes which the semantic analysis phase indicates are necessary in order to satisfy the users command. The traversal process continues until nodes are reached which indicate file types for which specific files are already physically present in the IST file system. The code generator then uses these files as the bases for execution of the users command. It compiles a list of the tool fragments which are necessary and sufficient to build the needed files out of the files which are known to exist.

In closing this description of the compilation phase, it seems important to observe that the structure of the command compiler makes it amenable to dynamic alteration so as to accept command language extensions. This is due to the fact that each of the three compilation phases is essentially table driven. The parsing phase is driven by parse tables; the semantic phase is driven by the table of templates; and the code generation phase is driven by the dependency DAG. These three tables are to be stored in the IST file system. Thus there appears to be no reason that a user tool might not be written to accept a users specifications of how a new tool fragment is to be invoked by the user, and integrated with other fragments and file types within the file system. Such a tool is currently planned, it is expected that is will make Toolpack/IST the sort of flexible, extensible system that will grow and adapt to the needs of different user communities.

A program chaining approach such as is used by the Software Tools Project [Sche78] will most likely be used, at least in early releases to carry out the second interpreter phase-- sequential tool invocation. With this strategy each tool is made into a separate main program by encasing the tool function (which is a subroutine) in a specially constructed interface main program. The sequence in which the main programs are to be executed is contained in the sequential file produced by the third phase. Effecting the indicated sequencing and error branching is the job of the interface main program. In order to carry out this job the main program must be able to access the command file, identify the next tool or tool fragment to be executed, create a command directing the invocation by the host operating system of the indicated tool or fragments encasing main program, present the command to the host system, access the command file for the arguments needed by encased tools, and manipulate the error flags with which it will communicate with its encased tools.

Experience indicates that it is reasonable to expect all of these tasks to be readily achievable on a wide spectrum of current systems. Thus this chaining strategy seems to offer an ideal combination of flexibility and portability potential.

6. Summary and Future Directions

The foregoing section has described the design and implementation strategy for the early releases of the Toolpack IST. In this section there has also been discussion of the ways in which these releases will be used as experimental vehicles for obtaining answers to the four basic questions outlined in Section 3. Specifically:

(1) The Toolpack tool suite will be evaluated to see whether it is adequate to cover the needs of mathematical software developers.

(2) The Toolpack file system and diagnostic data bases will be instrumented and monitored to help ascertain the precise data needs of mathematical software developers as they perform their jobs.

(3) The specific tool fragments from which the Toolpack tool capabilities are constructed will be evaluated to determine how readily and flexibly they form the basis for these capabilities. Alterations to this set of fragments may be necessitated.

(4) Reactions to the proposed command language (which borrows concepts taken from Lisp, the Unix shell, and the Make processor [Feld79]) will be monitored, and attempts will be made to determine underlying basic requirements for such languages. The adaptability of the Toolpack tool fragments and the ability to readily contrive new tools in response to changing user needs will be carefully observed. The combined efficacy of the flexible command language and the adaptable tool fragments will be studied to determine whether they effect an acceptably friendly user interface. In addition, the choices which we have made concerning the objects and operands which we believe Mathematical Software writers will wish to manipulate with their command language will be evaluated. Our decision to represent Mathematical Software programs as structures of compilation units, each consisting of a diversity of views and versions, is novel and promising, but must be natural and useful to our end users. Only experience will indicate whether this is the case.

It is just this sort of experimentation which seems to me to be most needed now if software assistance systems are to finally achieve the acceptance and perfection which will ultimately lead to widespread use and qualitative improvements in the way in which software is produced. The Toolpack/IST architecture and toolset promise to provide a close enough approximation to the sort of support system that Mathematical Software writers need that this community will use it and gain familiarity with individual tools and support capabilities. This should begin the process of substantive familiarization and evaluation of these tools and aids which should lead to experimentation, modification, codification and eventually the establishment of an agreed upon battery of software assistance systems that will finally deserve the denotation "tools".

Early design and implementation experience with Toolpack/IST seems to indicate that the underlying architectural principles are sound and that these principles should be applicable to the creation of environments to support different user communities and software development contexts. It is important to note, however, that Toolpack is not the only environment project, and IST is by no means the only integrating[architecture which is being implemented and studied. Such other projects as Mentor [Donz 80], the Cornell Program Synthesizer [TeRe 81], and Interlisp[Teit79] all attempt to effectively integrate a variety of tool functions. These projects also provide the sort of palatable and conducive framework that encourages users to gain substantive experience with software assistance, and thereby hastens the process of tool acceptance. Differences in the target user communities for these environments, and differences in the integration architectures of these projects also serve to improve the chance that several effective experimental frameworks for tool evaluation will be developed, increasing the amount of understanding which will be gained towards the goal of being able to construct the large number and diverse varieties of tools which we will need if we are to break the software bottleneck.

7. Acknowledgments

The design of the Toolpack IST is the product of a great deal of collaborative work and discussions with individuals inside and out of the project. In particular the original design concepts of the command language were contributed by Stuart I. Feldman of Bell Telephone Laboratories. The concepts underlying his Make processor were, obviously, also quite influential in the design of IST.

The implementation work has been ably led by Allan L. Shafton. The design and implementation of key command interpreter tool fragments was led by Geoff Clemm, Brent Welch and Mark Maybee.

Finally the financial support of the National Science Foundation, Department of Energy and U.S. Army Research Office is gratefully acknowledged.

8. References

[Cowe 77]   W. R. Cowell, L. D. Fosdick, "Mathematical Software Production," in *Mathematical Software III*, Academic Press, N.Y., pp. 195-224, 1977.

[Donz 80]   V. Donzeau-Gouge, G. Huet, G. Kahn, and B. Lang, "Programming Environments Based on Structured Editors: The Mentor Experience," INRIA Research Report No. 26, INRIA, Rocquencourt, France, 1980.

[Feld 79]   S. I. Feldman, "Make--A Program for Maintaining Computer Programs," Software-Practice and Experience 9 (April 1979), pp. 255-265.

[JPL 78]   Proceedings of Conference on the Programming Environment for Developing Numerical Software, Jet Propulsion Laboratory, Pasadena, Calif., Oct. 18-20, 1978.

[JPL 81]   Proceedings of Conference on the Computing Environment for Mathematical Software, Jet Propulsion Laboratory, Pasadena, Calif., July 15, 1981.

[Oste 81]   L. J. Osterweil, "Software Environment Research Directions for the Next Five Years," *Computer* 14 pp. 35-43, (April 1981).

[Oste 82]   L. J. Osterweil, "Toolpack--An Experimental Software Development Environment Research Project," Proceedings Sixth International Conference on Software Engineering, IEEE Conference Proceedings, Tokyo, September, 1982.

[Oste 82a] L. J. Osterweil, "The Toolpack Mathematical Software Development Environment," Department of Computer Science, University of Colorado at Boulder Technical Report # CU-CS-226-82, July 21, 1982.

[Sche 78]   Scherrer, D., COOKBOOK, instructions for implementing the LBL software tools package. Internal Rep. LBID098, Lawrence Berkeley Laboratory Univ. of California, Berkeley, 1978.

[TeRe 81]   T. Teitelbaum and T. Reps, "The Cornell Program Synthesizer: A Syntax-Directed Programming Environment." Communications of the ACM 24 (September 1981) 563-573.

[Teit 81]   W. Teitelman and L. Masinter, "The Interlisp Programming Environment," *Computer* 14 pp. 25-33 IEEE Computer Society, Los Alamitos, California (April 1981).

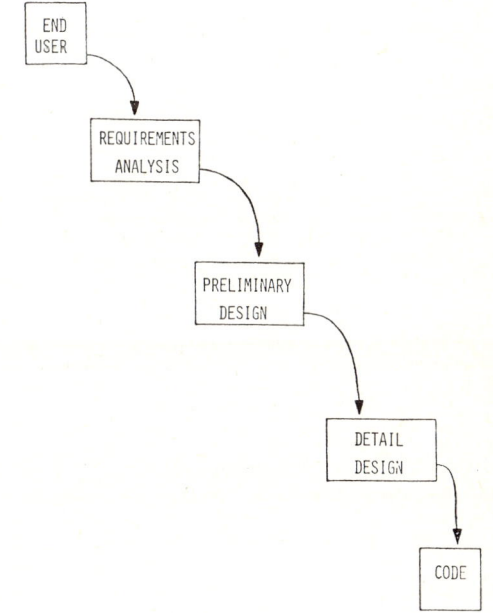

Figure 1: Phased Approach to Software Development

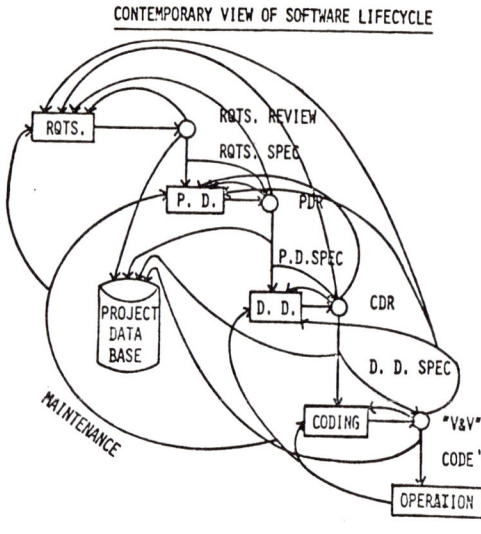

Figure 2

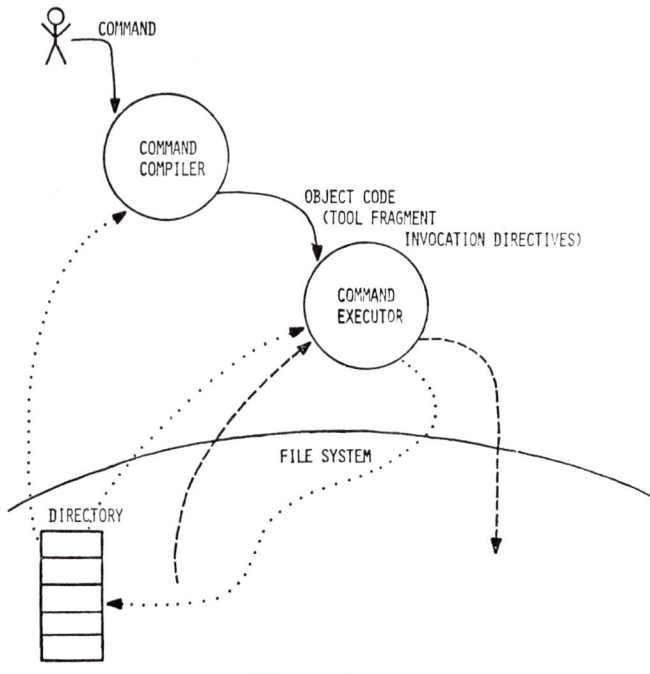

Figure 4:

Overview of the IST Command Interpreter showing two main phases--compilation and execution. Dotted lines show flow of information into and out of the file system directory. Dashed lines indicate data flow through the file system itself.

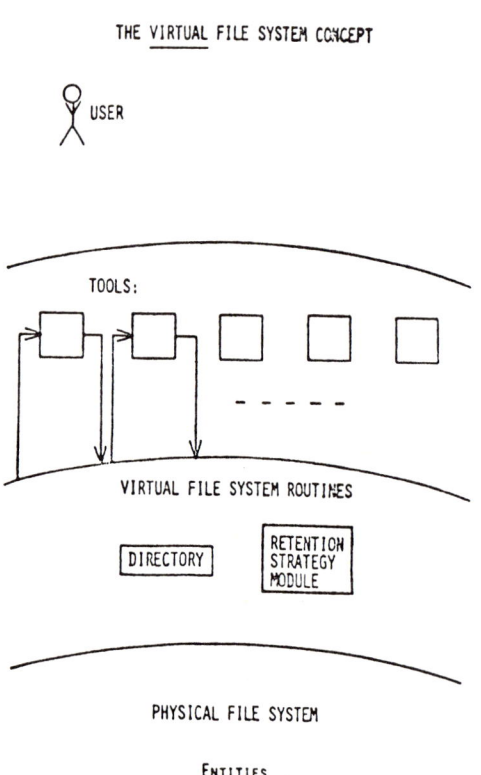

Figure 3

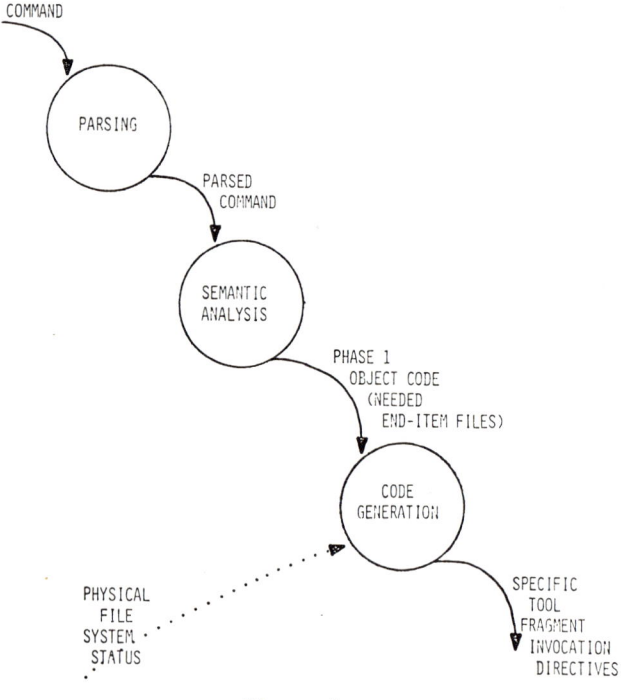

Figure 5:

Breakdown of the Command Compilation process into its three subphases. Note that only code generation requires information about current file system status.

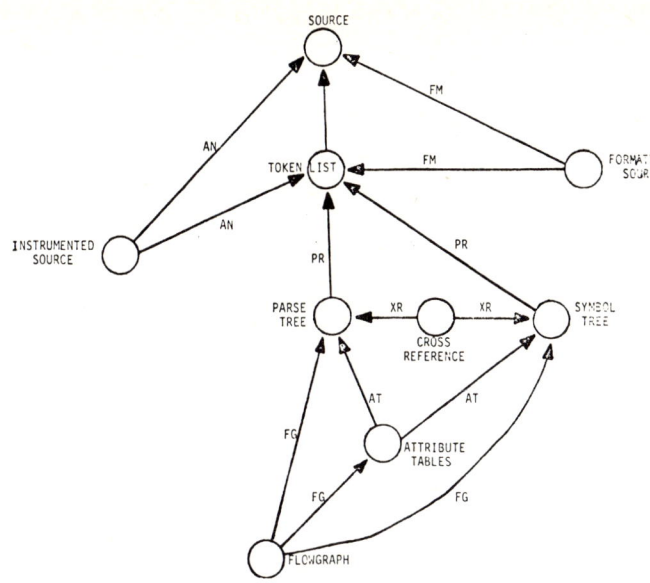

Figure 6:
A sample dependency DAG. Each node represents a type of file to be found in the Toolpack/IST file system. Each edge represents the dependency of this file type on another file type. Specifically, the file type at the head may be required in creating the file at the tail. The label on the edge represents the tool fragment needed to create the file type at the tail. Note that some tool fragments (eg. FM) require more than one file type to produce their output, while others (PR) produce more than one file type.

AN APPROACH TO CONCEPTUAL DESIGN METHODS AND TOOLS FOR SOFTWARE DEVELOPMENT

C. Rolland, P. Albin

Paris University, Paris, France

ABSTRACT

Our concern is with the implementation of real-time software for which we have team developed a specification and implementation method based on solutions obtained in our team for the design of information systems. This can be either real-time control software for connected physical systems or simulated systems of this type.

In the approach adopted, the software is considered as a collection of mechanisms for the control of a highly dynamic system, i.e. one which is subject to a multitude of events to which it must react. Specification consists of defining the behavioural structure or dynamic structure of the system which is being managed. The software is defined as an automaton which controls the operation of the system in accordance with rules determined by the dynamic structure.

The paper presents the specification method and its application to a telephony system.

RÉSUMÉ

Nous sommes concernés par la réalisation de logiciels en temps réel pour lesquels nous avons été amenés à développer une méthode de spécification et de réalisation inspirée des solutions retenues dans notre équipe pour la conception des systémes d'information. Il s'agit soit de logiciels de contrôle en temps réel de systémes physiques auxquels ils sont couplés, soit des simulations de tels systémes.

Dans l'approche retenue, le logiciel est considéré comme un ensemble de mécanismes de contrôle d'un systéme trés fortement dynamique, c'est-à-dire soumis à de multiples événements auxquels il doit réagir. La spécification consiste à définir la structure de comportement ou structure dynamique du systéme géré. Le logiciel est défini comme un automate qui contrôle le fonctionnement du systéme en accord avec les régles décrites par la structure dynamique.

La communication présente la méthode de spécification et son application à un exemple de téléphonie.

ZUSAMMENFASSUNG

Unsere Aufmerksamkeit gilt der Realisierung von Echtzeitsoftware, für die wir eine Spezifizierungs- und Realisierungsmethode entwickelt haben, die auf Lösungen basiert, die in unserem Team bei der Entwicklung von Informationssystemen erzielt wurden. Dabei handelt es sich entweder um Software für die Echtzeitsteuerung von angeschlossenen physikalischen Systemen, oder um Simulierungen solcher Systeme.

Nach der Betrachtungsweise des angenommenen Verfahrens ist die Software eine Ansammlung von Mechanismen zur Steuerung eines höchst dynamischen Systems, d.h. eines Systems, das einer Mehrzahl von Ereignissen unterworfen ist, auf die es reagieren muß. Spezifizierung besteht aus der Bestimmung der Verhaltensstruktur bzw. dynamischen Struktur des verwalteten Systems. Die Software wird als ein Automat definiert, der die Systemfunktion nach den durch die dynamische Struktur bestimmten Regeln steuert.

In dieser Arbeit wird die Spezifizierungsmethode und ihre Anwendung auf ein Beispiel aus der Telfonie erläutert.

UNE METHODE DE CONCEPTION CONCEPTUELLE

METHODES ET OUTILS DE DEVELOPPEMENT DU LOGICIEL

Colette ROLLAND, Pierre ALBIN

Université PARIS I, Panthéon-Sorbonne
12, place du Panthéon
75231 - PARIS CEDEX 05

Nous sommes concernés par la réalisation de logiciels en temps réel pour lesquels nous avons été amenés à développer une méthode de spécification et de réalisation inspirée des solutions retenues dans notre équipe pour la conception des systèmes d'information. Il s'agit soit de logiciels de contrôle en temps réel de systèmes physiques auxquels ils sont couplés, soit des simulations de tels systèmes.

Dans l'approche retenue, le logiciel est considéré comme un ensemble de mécanismes de contrôle d'un système très fortement dynamique, c'est-à-dire soumis à de multiples évènements auxquels il doit réagir. La spécification consiste à définir la structure de comportement ou structure dynamique du système géré. Le logiciel est défini comme un automate qui contrôle le fonctionnement du système en accord avec les règles décrites par la structure dynamique.

La communication présente la méthode de spécification et son application à un exemple de téléphonie.

Nous nous proposons d'organiser le processus de spécification en deux étapes interdépendantes. La solution est obtenue, à chacune des étapes, par un travail de modélisation faisant appel à des concepts et à des règles méthodologiques.

. La première étape ou étape conceptuelle est centrée sur la représentation abstraite du comportement du système réel. Elle aboutit à un schéma conceptuel dynamique extension du schéma conceptuel au sens de l'ANSI/X3/SPARC (1) qui est une expression complète, cohérente, non redondante et structurée de la dynamique du système réel.

. La seconde étape ou étape logique inclut les aspects techniques de la solution ignorés à la première étape. Elle complète la solution en introduisant les paramètres qui n'étaient pas nécessaires auparavant. Elle aboutit à un schéma logique décomposé en deux sous-schémas : le sous-schéma de description de la structure des données du logiciel et le sous-schéma de synchronisation des processus.

I - ETAPE CONCEPTUELLE

Le Schéma Conceptuel aboutissement de cette étape est le résultat d'un travail de modélisation du comportement du système réel au moyen de trois concepts appelés C-OBJET, C-OPERATION et C-EVENEMENT. Ces trois concepts définissent le modèle conceptuel dynamique que nous avons développé et expérimenté dans le domaine des systèmes d'information depuis plusieurs années (4)(5)(6).

I.1. Modèle conceptuel de la dynamique

Les trois concepts ont été définis dans le formalisme relationnel par trois types de relations nommées type-relation c-objet, type-relation c-opération, type-relation c-évènement dans une forme normalisée appelée forme permanente qui assure la minimalité (au sens de couverture minimale (3)) du schéma conceptuel.

C-OBJET : Le concept de c-objet permet la représentation des éléments ou entités du système réel et de leur structure. Plus précisément, un c-objet est la représentation d'un aspect temporel d'une classe d'entités du réel. C'est le plus grand regroupement de propriétés d'une même classe d'entités évoluant de façon homogène dans le temps c'est-à-dire créées, supprimées ou modifiées simultanément. La notion de c-objet est voisine de la notion de relation en troisième forme normale (2). Mais elle tient compte du temps et de sa représentation. Le c-objet représente l'état élémentaire (atomique) du système réel.

Exemple :
Ainsi la relation en troisième forme normale LIGNE TELEPHONIQUE (NLIG, DEBITLIG, CARACTLIG, ETATLIG, NPOSTE) de l'exemple de système de téléphonie locale présenté en I.3 sera décomposé en trois c-objets :

1.- LIGNE (NLIG, DEBITLIGNE, CARACTLIG)
2.- ETATLIGNE (NLIG, DATETATLIG, ETATLIG)
3.- AFFECTATION-LIGNE (NLIG, DATAFF, NPOSTE)

car l'état d'une ligne (relation 2) et son affectation à un poste téléphonique donné (relation 3) n'ont pas le même comportement temporel : une ligne peut être indisponible sans être nécessairement affectée à un poste (dérangement), alors que ses caractéristiques sont permanentes (relation 1).

C-OPERATION : Le concept de c-opération permet la représentation des transformations du système réel. La c-opération est définie par référence à un c-objet. C'est la représentation d'une classe d'opérations du système réel qui modifie les états des entités d'une même classe.

Une opération (une occurrence de c-opération) modifie l'état d'une occurrence unique d'un c-objet donné. Nous dirons par abus de langage qu'une c-opération modifie un seul et unique c-objet.

La c-opération représente la transformation élémentaire (atomique) du système réel.

Exemple :
La c-opération "affectation d'une ligne" de l'exemple de système de téléphonie locale sera représentée par trois relations sous forme normalisée :

1/ AFFECTATION-LIG-PERMANENT (NAL, TYPCHG)
2/ AFFECTATION-LIG-TEXTE (NAL, DATEXT, TEXT)
3/ AFFECTATION-LIG-EXEC (NAL, DATEXEC, NLIG)

parce qu'il y a plusieurs exécutions d'opérations d'affectation d'une ligne (relation 3), plusieurs stratégies d'affectation utilisées à différentes périodes de la vie de la c-opération (relation 2) et un seul type de modification (relation 1) des objets correspondant au c-objet LIGNE (NLIG).

C-EVENEMENT : Le concept de c-évènement permet la représentation des changements d'état particuliers du système réel. Le c-évènement est défini par référence à la c-opération et au c-objet. C'est la représentation d'une classe de changements d'état des entités du réel qui déclenchent des opérations. Plus précisément, un évènement (occurrence d'un c-évènement) est le changement d'état remarquable d'une occurrence de c-objet qui déclenche une ou plusieurs occurrences d'une ou plusieurs c-opérations. Nous dirons par abus de langage qu'un c-évènement correspond à un changement d'état remarquable d'un c-objet qui déclenche une ou plusieurs c-opérations. Le c-évènement représente le changement d'état élémentaire du système qui est un déclencheur d'une ou plusieurs transformations élémentaires. Notons que le déclenchement peut être conditionnel (il n'est assuré que si et seulement si la condition est vraie) et/ou itératif.

Exemple :
Le c-évènement "arrivée d'un chiffre" de l'exemple de téléphonie locale sera représenté par quatre relations :

1/ ARRIVEE-CHIFFRE-PER (NARCHIF, PRED, TYPECHG)
2/ ARRIVEE-CHIFFRE-EFF (NARCHIF, DATARR, NENRLOC)
3/ ARRIVEE-CHIFFRE-DECLENCHT-OP (NARCHIF, NOPTRAD, DATEDEC)
4/ ARRIVEE-CHIFFRE-COND-DECLT (NARCHIF, NOPTRAD, CONDITION)

La relation (relation 1) décrit le changement d'état (via le prédicat PRED) qui définit l'évènement du type "arrivée d'un chiffre" dans l'enregistreur local (NENRLOC) associé à un poste. La relation (relation 2) décrit l'arrivée effective des évènements du type "arrivée d'un chiffre". La relation (relation 3) décrit le déclenchement effectif de l'opération de traduction (NOPTRAD) en adresse physique réseau de ces chiffres. Enfin, la relation (relation 4) décrit la condition de déclenchement associée à la c-opération de traduction.

I.2.- Le schéma conceptuel dynamique

Le schéma conceptuel est une collection de relations appartenant aux trois types c-objet, c-opération, c-évènement. Ce schéma comporte deux sous-schémas relatifs respectivement aux aspects statiques et aux aspects dynamiques du système réel modélisé.

Le sous-schéma des c-objets ou sous-schéma statique est la représentation de la structure des éléments du système physique. Il correspond à une vision statique des phénomènes.

Le sous-schéma des c-opérations et c-évènements ou sous-schéma dynamique modélise la structure de comportement du système physique. Le modèle retenu conduit à une représentation de la dynamique de type causal : les évènements déclenchent des opérations qui modifient les objets; certains changements d'étape d'objets sont à leur tour des évènements... En termes de types sur le schéma, la dynamique est représentée par l'interconnection des trois concepts c-objet, c-évènement et c-opération. Nous proposons une description graphique de ce sous-schéma comme un graphe trialterné de c-objets, c-évènements et c-opérations comme nous le montrons sur la figure 1 dans le cadre de l'exemple de téléphonie retenu.

I.3.- Exemple

Cet exemple concerne une partie d'un système de téléphonie locale. On s'intéresse au processus conduisant du décrochage d'un poste appelant à la sollicitation du poste demandé. Dès qu'un utilisateur décroche son combiné, le système lui affecte une ligne parmi celles disponibles et se prépare à recevoir le numéro que l'utilisateur va composer sur son cadran. S'il n'y a pas de ligne disponible, le poste est dit "mis au noir", c'est-à-dire qu'il n'a aucun moyen de communiquer avec le système téléphonique, et l'usager n'a plus qu'à raccrocher.

Si une ligne a pu être affectée à ce poste, l'usager dispose de quinze secondes pour former le premier chiffre du numéro qu'il souhaite appeler, à partir de l'obtention de la tonalité d'invitation à numéroter. Il en va de même entre chacun des chiffres qu'il compose. Si ce délai s'écoule sans que l'usager agisse, le poste est considéré comme sollicitant un numéro indisponible et reçoit alors une tonalité d'occupation. Si un nouveau délai de quinze secondes s'écoule sans que l'usager raccroche, son poste est "mis au noir" et toutes les ressources qui lui étaient affectées sont libérées.

Dès qu'un chiffre est composé par l'usager, le système vérifie si la suite de chiffres ainsi constituée depuis le début de l'appel est interprétable, et, dans l'affirmative, cette suite est traduite en adresse physique sur le réseau téléphonique local, entraînant ainsi la sollicitation du poste demandé.

Sous-schéma statique

LIGNE (NLIG, DEBITLIG, CARACTLIG)
ETAT-LIGNE (NLIG, DATETALIG, ETATLIG)
AFFECTATION-LIGNE (NLIG, DATAFF, NPOSTE)
POSTE (NPOSTE, CARACTPOSTE, DROITAPPEL)
ETAT-POSTE (NPOSTE, DATETATPOSTE, ETATPOSTE)
TEMPORISATION (NPOSTE, DATETEMPO, DUREE)
ENREGISTREUR-LOCAL (NPOSTE, DATENR, CHIFFRE)
ORDRE-TONALITE (NPOSTE, DATETON, TYPETON)
CODE-ABONNE (NPOSTE, DATECODE, CODABONNE)
ORDRE-EMISSION-SOLLICITATION (NPOSTE, DATESOLL, CODABONNE, TYPEDEM)

Sous-schéma dynamique

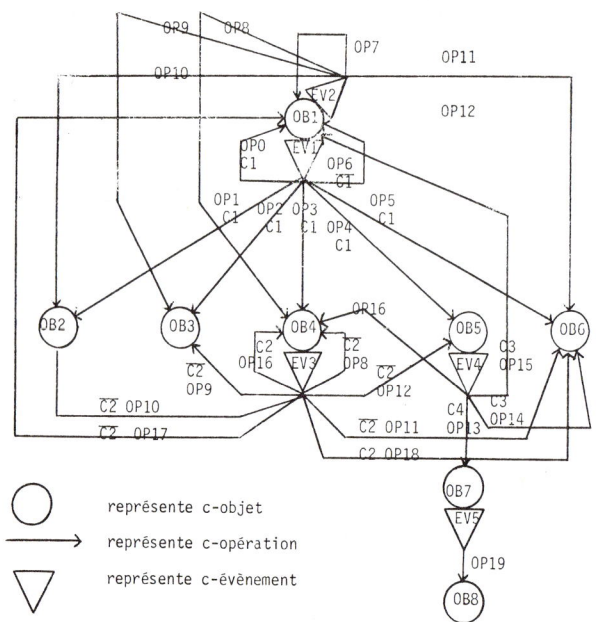

Figure 1 : Sous-schéma dynamique de l'exemple de téléphonie locale

Commentaires

 Le décrochage d'un poste téléphonique est un évènement (EV1) qui conduit, si une ligne est disponible (condition C1), à la prise de cette ligne (OP1) et à son affectation au poste (OP2), à l'affectation d'une temporisation de 15 secondes (OP3) et d'un enregistreur local (OP4) à ce poste, à l'envoi d'une tonalité d'invitation à numéroter (OP5) et au changement d'état du poste (OP0). S'il n'y a pas de ligne disponible, le poste est "mis au noir" (OP6) c'est-à-dire qu'il n'a aucun moyen de communiquer avec le réseau téléphonique local.

 A tout moment, l'usager peut décider d'interrompre le processus d'établissement de communication en raccrochant (EV2). Dans ce cas, le système libère la ligne téléphonique (OP9 et OP10), la temporisation (OP8), l'enregistreur local (OP12) et l'émetteur de tonalité (OP11) associés à ce poste. Le système mémorise également l'état actuel (raccroché) du poste (OP7).

 L'expiration du délai de temporisation est un évènement (EV3) qui conduit si c'est la première fois que le délai expire (C2) à mettre le poste dans l'état d'occupation (OP18) et à réarmer la temporisation pour un nouveau délai de 15 secondes (OP16). Passé ce délai ($\overline{C2}$), si le poste n'a pas raccroché, il est "mis au noir" (OP6) (isolé du réseau) et la ligne téléphonique, la temporisation, l'enregistreur local et l'émetteur de tonalité associés sont libérés (OP8, OP9, OP10, OP11 et OP12).

 Dès qu'un chiffre formé par l'utilisateur sur le cadran du poste apparaît dans l'enregistreur local (évènement du type EV4), le système doit mémoriser l'état actuel du poste (numérotation) via OP15 et supprimer la tonalité d'invitation à numéroter (OP14) s'il s'agit du premier chiffre relatif à cet appel (C3). A chaque chiffre, la temporisation de 15 secondes est réarmée (OP16) et l'opération de traduction du numéro ainsi formé (OP13) est déclenchée si la suite de chiffres de l'enregistreur local est interprétable (C4).

 L'arrivée d'un code physique d'abonné (adresse réseau) consécutif à l'opération de traduction est un évènement du type EV5 qui conduit à la sollicitation du poste appelé (OP19).

II - ETAPE LOGIQUE

L'étape logique a pour but de transformer la spécification conceptuelle en une spécification logique aisément programmable selon les techniques habituelles en tenant compte de l'environnement technique dans lequel le logiciel doit être réalisé. Cette transformation aboutit à la production d'un schéma logique décomposable en deux sous-schémas qui correspondent aux deux sous-schémas statique et dynamique du niveau conceptuel.

. La collection des c-objets devient un ensemble de fichiers, d'enregistrements, de variables dynamiques (piles et/ou listes), de buffers (surtout dans le cas d'interfaces reliant le logiciel à des apareillages physiques émetteurs de signaux) etc... La collection des c-objets définit l'ensemble des informations gérées par le logiciel. Il s'agit au niveau logique de choisir les méthodes de stockage et d'accès de ces informations.

Le sous-schéma conceptuel dynamique devient une structure de synchronisation des modules du logiciel fournissant l'inventaire des modules à programmer et l'inventaire des règles de synchronisation du déclenchement de ces modules qui assureront un fonctionnement cohérent.

Nous développons la méthode de dérivation de ce sous-schéma dans la suite de ce paragraphe.

II.1.- Modèle de synchronisation logique

Le schéma de synchronisation du niveau logique est défini au moyen des deux concepts de module et de déclencheur.

Un module est une séquence d'instructions exécutables dans cet ordre, accessible par son nom, possédant un point d'entrée et un seul. Il est décrit par une procédure (ou un bloc) ayant au moins un paramètres d'entrée et un paramètre de sortie. Il est représenté par un rectangle contenant son nom (la seule voie d'accès aux instructions du module).

$$\boxed{\text{MOD1}}$$

Un déclencheur est une condition portant sur un changement d'état du système permettant de déclencher l'exécution d'un module et d'un seul. Cette condition peut être nommée et être ainsi accessible par son nom. Elle sera décrite par un texte de condition (dans un langage de programmation) ou par une fonction ayant valeur de booléen. Elle peut être satisfaite après l'exécution d'un ou plusieurs modules donnés (déclencheur interne) soit après l'arrivée d'un stimulus extérieur au système (déclencheur externe).

Un déclencheur est représenté par un losange contenant son nom. Un arc orienté (déclencheur → module) traduit le fait que le déclencheur provoque l'exécution du module. Inversement, un arc orienté (module → déclencheur) traduit le fait que l'exécution d'un module contribue à la validation d'un déclencheur.

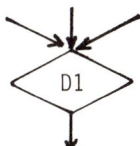

II.2.- Schéma de synchronisation des modules

Le schéma de synchronisation des modules est un graphe bi-alterné sur l'ensemble des modules et l'ensemble des déclencheurs c'est-à-dire un graphe où se succèdent toujours un noeud de type déclencheur et un noeud de type module.

Le schéma de synchronisation des modules est construit à partir de trois structures de base (qui découlent de la définition des concepts et notamment du fait qu'un déclencheur agit sur un module et peut être généré par plusieurs modules).

1/ La structure linéaire

Elle exprime le fait qu'un déclencheur D_i agissant sur un module M_j est soit externe, soit généré par un seul module M_k.

2/ La structure "disjonctive"

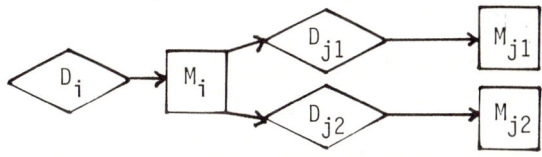

Elle exprime le fait que l'exécution d'un module M_i déclenchée par le déclencheur D_i peut entraîner le déclenchement de l'exécution d'un module M_{j1} par le déclencheur D_{j1} et/ou l'exécution d'un module M_{j2} par le déclencheur D_{j2}.

3/ La structure "conjonctive"

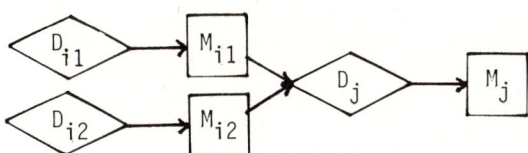

Elle exprime que le déclencheur D_j déclenchant l'exécution du module M_j peut avoir été généré soit par l'exécution d'un module M_{i1} déclenchée par le déclencheur D_{i1}, soit par l'exécution d'un module M_{i2} déclenchée par le déclencheur D_{i2}.

II.3.- Règles de construction d'un schéma de synchronisation des modules

Le schéma de synchronisation des modules se dérive du schéma conceptuel. Nous proposons trois règles de base pour aider le concepteur dans le processus de dérivation.

Règle 1 : Détermination des modules élémentaires

L'ensemble des c-opérations déclenchées par un c-évènement est indissociable : elles définissent la transition élémentaire d'état à état que provoque l'évènement correspondant. Nous associons la notion de module élémentaire à celle de transition conceptuelle.

Un module élémentaire M_j est constitué des instructions réalisant toutes les c-opérations déclenchées par un c-évènement. Le c-objet associé au c-évènement est nécessairement un paramètre d'entrée du module. Les c-objets modifiés par les c-opérations dont les changements d'état sont évènementiels sont nécessairement des paramètres de sortie du module.

Exemple :

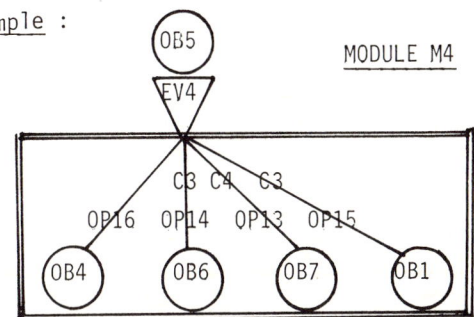

Règle 2 : Concaténation de modules élémentaires

Des raisons économiques évidentes peuvent conduire à regrouper plusieurs modules élémentaires qui sont systématiquement exécutés en séquence. L'application de cette règle s'appuie sur le graphe des dépendances chronologiques entre c-évènements que nous définissons au préalable (7).

A : Graphe de chronologie des évènements

Ce graphe se déduit du schéma conceptuel dynamique et utilise la notion de dépendance fonctionnelle chronologique entre c-évènements.

. Dépendance chronologique permanente directe (DCPD)

Un c-évènement EV_j est en DCPD avec un évènement EV_i si et seulement si :

1/ EV_i déclenche inconditionnellement une c-opération OP_i^k qui provoque le changement d'état d'un c-objet OB_k constaté par EV_j.

2/ Ce changement d'état constitue toujours un évènement de type EV_j (EV_j suit toujours EV_i).

. Dépendance chronologique conditionnelle directe (DCCD)

Un c-évènement EV_j est en DCCD avec un c-évènement EV_i si et seulement si :

1/ EV_i déclenche une c-opération OP_i^k qui provoque le changement d'état d'un c-objet OB_k constaté par EV_j.

2/ Soit EV_i déclenche conditionnellement OP_i^k soit le changement d'état provoqué par OP_i^k n'est pas toujours un évènement du type EV_j (EV_j peut suivre EV_i).

. Graphe de dépendances chronologiques

C'est le graphe qui regroupe toutes les dépendances chronologiques (DPCD ou DCCD) d'un même schéma conceptuel.

En utilisant ⟶ pour la DPCD et ----→ pour la DCCD le graphe des dépendances chronologiques entre c-évènements de l'exemple est le suivant :

EV1   EV2   EV3   EV4
                   ⋮
                   ↓
                  EV5

Remarquons que beaucoup de ces c-évènements sont indépendants. Cela est dû au conditionnement des actions du système téléphonique par des informations "externes" produites directement par l'usager.

B : Règle de concaténation des modules

La règle peut se résumer par le tableau suivant :

$$1/\ EV_i \xrightarrow{P} EV_j \Rightarrow M_i . M_j$$

$$2/\ EV_i \xrightarrow{c} EV_j \Rightarrow \text{choix} \begin{cases} M_i * M_j \\ M_i, M_j \text{ séparés} \end{cases}$$

1/ Si un c-évènement EV_j est en DCPD avec un c-évènement EV_i, EV_j suit toujours EV_i. Donc, le module M_j associé à EV_j suivra toujours le module M_i associé à EV_i et il est inutile de tester le déclencheur associé à M_j. C'est pourquoi nous proposons dans ce cas là la concaténation systématique de M_j à M_i (notée $M_i.M_j$).

2/ Si un c-évènement EV_j est en DCCD avec un c-évènement EV_i, EV_j peut suivre EV_i. Donc le module M_j associé à EV_j peut suivre le module M_i associé à EV_i, et il faudra tester la condition associée au déclencheur de M_j. L'on se trouve alors placé devant l'alternative suivante : soit l'on concatène ces deux modules en insérant le test du déclencheur D_j de M_j entre eux ($M_i * M_j$), soit l'on conserve deux modules séparés M_i et M_j (une des manières de résoudre cette alternative est de prendre la solution soit la moins coûteuse, soit la plus rapide selon les contraintes techniques imposées).

Exemple :

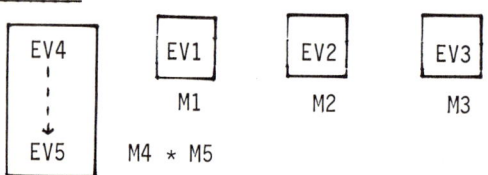

Comme EV1, EV2, EV3 sont indépendants des autres c-évènements, les modules élémentaires associés sont nécessairement indépendants. Comme le caractère conditionnel de EV5 vis-à-vis de EV4 est uniquement dû à l'existence de la condition C4 sur OP13, la solution à la fois la plus rapide et la moins coûteuse est le regroupement des modules élémentaires associés (M4*M5).

Remarque :

Lorsque dans le graphe de dépendances chronologiques un c-évènement EV_k est dépendant de plusieurs c-évènements (par exemple EV_i, EV_j et EV_M), l'application des règles de concaténation des modules peut conduire à la duplication du module M_k associé à EV_k :

Le module M_k est dupliqué dans les modules $M_i.M_k$, $M_j.M_k$ et $M_M.M_k$.

Règle 3 : Détermination des déclencheurs

La condition associée au déclencheur D_j d'un module M_j obtenu après application des règles 1 et 2 est la traduction des prédicats définissant le c-évènement du premier cycle dynamique de M_j.

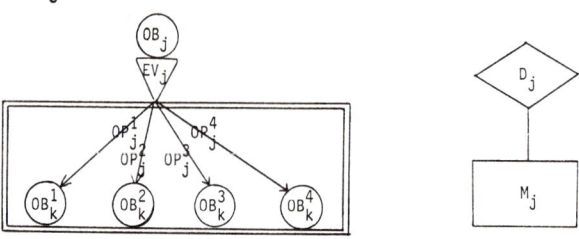

Application des règles à l'exemple de la téléphonie

L'application récursive des règles de dérivation précédentes au schéma conceptuel dynamique permet d'obtenir le schéma logique de synchronisation des modules. Il est représenté sur la figure 2 dans le cas de la téléphonie.

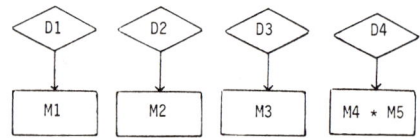

Figure 2 : Schéma de synchronisation des modules du cas de la téléphonie

III - CONCLUSION

L'expérience actuelle de cette méthode nous amène aux conclusions suivantes :

1/ Le modèle conceptuel est simple et puissant. Il est très bien adapté à l'analyse et la modification des systèmes temps réel. La définition sémantique des concepts et leur élémentarité sont deux aspects très positifs du modèle qui facilitent énormément le travail et la rigueur de la conception.

2/ Des outils logiciels sont nécessaires à l'exploitation systématique de la méthode (vérification de la cohérence, simulation de la logique d'enchaînement des évènements, outils de la documentation).

3/ La prise en compte du parallélisme n'a pas été traitée dans cet article. C'est une composante nécessaire du processus de spécification. La solution conceptuelle adoptée s'avère sur ce point intéressante.

IV - REFERENCES

(1) ANSI/X3/SPARC, Study group on data base management systems, Interim Report, FDT bulletin of ACM SIGMOD 7, nb 21, 1975

(2) CODD, E.F., Further normalization of the data base relational model, in : data base systems, Prentice Hall, Englewood Cliffs N.J., 1972

(3) DELOBEL, C., Contribution théorique à la conception et à la définition d'un système d'information appliqué à la gestion. Thèse d'état, Université de Grenoble, France, 1973

(4) ROLLAND, C., A methodology for information systems design, Proceedings of the NCC 81, Chicago, 1981

(5) ROLLAND, C. and FOUCAUT, O., Concepts for the design of an information system conceptual schema and its utilization in the Remora project, Proceedings of the 4th international conference on VLDB, Berlin, 1978

(6) ROLLAND, C., LEIFERT, S. and RICHARD, C., Tools for information system dynamics management, Proceedings of the 5th international conference on VLDB, Rio de Janeiro, 1979

(7) ROLLAND, C. and RICHARD, C., The Remora methodology for information systems design and management, IFIP Conference on "Comparative Review on Information Systems Design Methodologies", Amsterdam, 1982

QUELQUES PRIMITIVES POUR LA PROGRAMMATION TEMPS REEL
ET LEUR SEMANTIQUE MATHEMATIQUE

G. Berry, J. Camerini, J.P. Marmorat, B. Nguyen Phuoc, J.P. Rigault

Centre de Mathématiques Appliquées
Ecole des Mines
Sophia-Antipolis
06560 VALBONNE

RESUME

Nous présentons un ensemble de primitives de programmation temps réel adaptées aux applications de type contrôle de processus, ainsi que leur sémantique mathématique dans le calcul SCCS de Milner. Les primitives utilisent une notion générale d'événement répétitif et pas seulement la notion classique de temps physique. Les événements répétitifs permettent de donner une portée temporelle aux instructions, ce qui autorise une programmation élégante de nombreux algorithmes de contrôle. La sémantique est de type dénotationnel, c'est à dire que nous donnons une traduction syntaxique de nos termes dans ceux de SCCS. Nous indiquons comment on peut utiliser cette sémantique pour valider des équivalences ou des transformations de programmes. Nous pensons que nos primitives sont suffisamment saines pour servir de base à un véritable langage modulaire de programmation en temps réel.

ABSTRACT

We introduce a set of primitives for real time programming, which seem well-suited for process control applications. We give a semantics of these primitives in Milner's SCCS calculus. We adopt the algebraic approach: we propose a small set of primitives in which many high-level constructs have a straightforward translation. Our basic idea is to use a general notion of repetitive event and not just the usual notion of absolute time. This allows us to associate a temporal scope with any instruction. This seems to lead to elegant programs in many applications. Our semantics is denotational: the terms of our language are translated into SCCS in a syntax directed way. We indicate how one can use the semantics for showing program equivalences and the correctness of program transformations. We think that our primitives are sound enough to serve as a basis for a truly modular real time programming language.

ZUSAMMENFASSUNG

Wir beschreiben einen satz von Reqlzeitprogramm-Primitiven, die fuer Prozesssteuerungsanwendungen geeignet sind, und ihre Semantik unter Veiwendung von Milners SCCS Kalkuel. Die Primitive verwenden den allgemeinen Begriff eines Wiederholbaren Ereignisses und beschraenken sich nicht auf den klassischen Begriff der physikalischen Zeit. Die wiederholbaren Ereignisse erlauben es uns, den Instruktion eine zeitliche Reichweite zu geben, was eine elegante Programmierung zahlreicher Steuerung-Algorithmen ermoeglicht. Die Semantik ist denotational, d.h. wir geben eine syntaktische Umsetzung unserer Terme in SCCS-Terme an. Wir beschrelben, wie man diese Semantik benutzen kann, um Equivalenzen oder Programm-Transformationen zu ueberpruefen. Wir glauben, dass unsere Primitive durchaus geeignet sind, um als Basis einer echten, modularen Echtzeit-Programmiersprache zu dienen.

1. Introduction.

Notre objectif est de proposer quelques primitives structurées pour la programmation temps réel et une sémantique mathématique de ces primitives. La classe d'applications que nous envisageons est le contrôle de processus, où les programmes doivent répondre rapidement (si possible "instantanément") à des stimuli externes, en effectuant des actions ayant elles-mêmes un effet sur l'extérieur. L'ordre des événements est une donnée essentielle, contrairement au cas des programmes parallèles asynchrones, mais les constantes de temps des systèmes à contrôler sont généralement assez grandes vis à vis de la complexité des calculs à effectuer. Nous ne visons donc pas la classe des applications pour lesquelles la vitesse est un facteur essentiel, comme le traitement du signal.

Il est bien connu que les programmes de contrôle de processus sont plus difficiles à concevoir, écrire, tester, améliorer et maintenir que les programmes séquentiels classiques ou même que les programmes parallèles de type purement asynchrone (voir par exemple [7]). Ils restent souvent écrits en assembleur, au moins en ce qui concerne les parties critiques du point de vue de la synchronisation. De fait il existe relativement peu de langages prenant en compte les contraintes temporelles. La plupart sont des langages de type "parallèle" augmentés de quelques primitives d'attente d'événements (qui peuvent généralement être des événements externes ou des événements provoqués par programme), ou de délai (le plus souvent par référence à un temps absolu unique). Voir par exemple [12,19,1,4,3]. Ces primitives sont souvent peu structurées et leur sémantique n'est généralement pas définie de façon précise. Or on sait que disposer d'une sémantique exacte est indispensable si l'on veut analyser, transformer ou éventuellement prouver des programmes. Une des difficultés majeures est de créer tout d'abord un formalisme aussi riche que ceux qui ont été développés pour les langages classiques [8,20].

La situation s'est améliorée récemment pour les programmes parallèles asynchrones, en particulier grâce au développement du calcul CCS de Milner [14], dans lequel certains auteurs commencent à exprimer la sémantique de CSP [9], ou encore grâce à l'étude des systèmes de transition [10,16,21]. Mais aucun de ces travaux n'a encore abordé les problèmes de temps réel pour lesquels CCS n'est pas adapté.

Plus récemment est apparu un nouveau calcul qui nous paraît convenir à l'étude du temps réel. C'est le calcul synchrone SCCS de Milner [15], que nous utilisons ici pour donner la sémantique de nos primitives. Nous adoptons la même approche algébrique que Milner: au lieu d'introduire une primitive par besoin pratique, nous cherchons à définir un petit nombre de primitives et à étudier leur articulation. Une fois ces primitives bien comprises, le développement de langages devrait se réduire à l'introduction de constructions syntaxiques rendant la programmation agréable et se traduisant de façon immédiate dans l'algèbre des primitives.

Notre jeu de primitives est fondé sur la notion d'événement répétitif. Les instructions sont construites à partir d'actions élémentaires par des opérations de composition: composition séquentielle, conditionnelle, boucle, composition parallèle, communication, exécution d'une instruction dans un "intervalle de temps" borné - cette dernière construction étant la plus importante car elle permet de donner une portée temporelle à une instruction. Il faut noter que la notion de "temps" que nous manipulons est distincte de la notion de "temps physique", et que les événements répétitifs engendrent en fait plusieurs "temps" asynchrones. Ne pas se limiter au seul "temps physique" permet une programmation très élégante de beaucoup d'algorithmes de contrôle.

Nous définissons tout d'abord les événements répétitifs, puis introduisons nos primitives et leur sémantique intuitive. Nous présentons quelques exemples d'utilisation. Après quelque rappels sur SCCS, nous donnons la sémantique formelle des primitives. Nous avons choisi comme exemple la commande d'une voiture autonome, qui, après un tour de reconnaissance, doit rouler le plus vite possible sur une piste matérialisée par une ligne pointillée: une ligne transversale indique le début (et la fin) des tours, voir [2]. (Cette voiture a été effectivement construite [13], et sa réalisation a permis de dégager certains des principes que nous développons ici.)

2. Evénements répétitifs.

Un des objets de base que nous manipulons est l'événement répétitif. Vu d'un programme un événement répétitif possède un nom <evt> et peut avoir plusieurs occurrences dans le temps. Notre voiture connaît typiquement quatre événements répétitifs:

- la seconde écoulée "s";

- le centimètre parcouru "cm";

- le pointillé parcouru "pointillé";

- le tour de piste parcouru "tour".

Ces événements apparaissent de façon asynchrone mais doivent être perçus de façon simultanée par les différents modules du programme. Aucun d'eux ne joue un rôle privilégié. En particulier la seconde est un événement comme un autre. Ceci n'est pas surprenant en contrôle de processus où le temps est défini souvent de façon relative à un processus, l'utilisation du "temps universel" n'étant pas toujours nécessaire.

Les événements répétitifs peuvent être matériels ou logiciels, et il est essentiel qu'un programme ne puisse pas préjuger de leur cause: autrement il serait impossible de changer la réalisation d'un événement sans changer le programme qui l'utilise - en particulier si un événement réalisé initialement de façon logicielle se trouve ensuite réalisé de façon matérielle, par l'introduction d'un nouveau capteur par exemple. Pour décrire comment les événements sont créés nous devons donc adopter la programmation par "modules" ou "types abstraits", comme en ML [8] ou en ADA [3]. La génération des événements répétitifs doit être décrite dans un module séparé invisible du programme principal. Nous ne nous attarderons pas sur ce point qui fait appel à des techniques tout à fait standard.

Notons qu'il existe une dualité entre la notion d'événement et la notion d'intervalle entre deux événements: les noms que nous avons donnés sont d'ailleurs plutôt des noms d'intervalles. Une fois fixée une origine (mise à l'heure d'une montre), on peut nommer un événement par son déplacement vis à vis de cette origine, en nombre d'intervalles unité. Réciproquement on peut nommer un intervalle en nommant son événement de début et son événement de fin. La structure est celle d'un espace affine (à ceci près qu'on travaille sur l'anneau des entiers et non sur un corps). Voici comment nous nommons les événements:

- "prochain <evt>" désigne la prochaine occurrence de l'événement <evt> lorsqu'on est à un point de contrôle du programme. Si cette occurrence est simultanée avec l'arrivée en ce point, "prochain <evt>" la désigne.

- "prochain <evt> + n" désigne la n-ième occurrence de <evt> après "prochain <evt>".

Nous verrons qu'il n'y a pas besoin de primitive spéciale pour définir une origine à un événement répétitif et nommer les événements relativement à cette origine.

3 Les instructions et leur sémantique intuitive.

Nous introduisons nos primitives et appelons instructions les objets qu'elles permettent de construire. Toutes les instructions s'exécutent dès que le contrôle leur parvient, et peuvent éventuellement se terminer. Nous nommerons "durée" d'une instruction ce qui sépare son début de sa fin. Conformément à ce que nous avons dit précédemment, il ne faut pas comprendre la durée comme relative à un "temps absolu", mais comme relative à la position du début et de la fin de l'instruction vis à vis du flot des événements répétitifs. Une instruction qui ne se termine pas a par convention une durée infinie. Certaines instructions ont une durée nulle: cette hypothèse peut paraître paradoxale ou irréaliste, mais elle nous semble indispensable pour faire des raisonnements simples sur les programmes et montrer la correction de certaines transformations. Dans la même optique les opérations de contrôle nécessaires au déroulement d'un programme prennent un temps nul. Ceci est indispensable si l'on veut avoir des équivalences bien naturelles comme:

délai 2s; délai 3s ~ délai 5s

D'autres justifications seront données plus loin.

3.1. Les actions.

Les actions sont des instructions primitives, en général externes à notre modèle, et que nous ne connaissons que par leurs noms <act>, <act'>,... Ce peuvent être des calculs simples ou complexes, ou encore des communications que l'on ne désire pas présenter sous forme explicite (allumage d'une lampe). La durée d'une action peut être quelconque. Cependant il est essentiel que certaines actions soient de durée nulle. Il existe en particulier une action primitive de durée nulle: l'action rien. (Il est évidemment obligatoire qu'elle soit de durée nulle si on veut pouvoir l'insérer n'importe où dans un programme, et qu'elle n'y fasse réellement rien.)

Certaines actions rendent des valeurs utilisables par d'autres instructions. Nous nous limiterons au cas des valeurs booléennes et entières. Les actions booléennes <bool>, <bool'>, ..., rendent un résultat booléen vrai ou faux. Il existe deux actions booléennes primitives vrai et faux qui sont de durée nulle et rendent instantanément vrai et faux. Les actions entières <ent>, <ent'> rendent un résultat entier, et l'action primitive <n> rend instantanément le résultat n.

3.2. Le séquencement.

Si $<i_1>$ et $<i_2>$ sont des instructions, alors l'instruction

$<i>=<i_1>;<i_2>$

représente la mise en séquence de $<i_1>$ et $<i_2>$. Lorsque le contrôle arrive à <i> il est passé instantanément à $<i_1>$. Si $<i_1>$ se termine le contrôle est passé instantanément à $<i_2>$, c'est à dire que la fin de $<i_1>$ et le début de $<i_2>$ sont synchrones. La terminaison de <i> est alors celle de $<i_2>$. Notons que rien est élément neutre à droite et à gauche pour ";".

3.3. La conditionnelle.

Soient <bool> une action booléenne instantanée et $<i_1>$ une instruction. On construit alors la conditionnelle

$<i>$ = si <bool> alors $<i_1>$ fsi

On évalue instantanément <bool> et on déclenche immédiatement $<i_1>$ si <bool> vaut vrai. Dans ce cas la terminaison de <i> est celle de $<i_1>$. Sinon <i> est de durée nulle. On a donc les équivalences suivantes:

si vrai alors <i> fsi ~ <i>

si faux alors <i> fsi ~ rien

Le fait de supposer <bool> instantanée n'est pas une restriction: si la valeur booléenne désirée est le résultat d'un calcul complexe, on peut exécuter ce calcul avant, <bool> se limitant à la lecture du résultat.

3.4. La boucle.

Soit $<i_1>$ une instruction. On construit la **boucle**

$<i>$ = <u>répéter</u> $<i_1>$ <u>fin</u>

L'instruction $<i_1>$ ne peut pas être de durée nulle. On exécute $<i_1>;<i_1>;<i_1>;...$. L'instruction $<i_1>$ peut contenir des occurrences de l'instruction <u>sortir</u>, dont l'exécution provoque la terminaison instantanée de $<i_1>$ et de $<i>$. Dans le cas de boucles imbriquées, on utilise un mécanisme syntaxique quelconque pour savoir à quel <u>répéter</u> se rapporte un <u>sortir</u>.

3.5. La composition parallèle.

Si $<i_1>$ et $<i_2>$ sont des instructions, alors

$<i>=<i_1>||<i_2>$

est une instruction appelée la composition parallèle de $<i_1>$ et $<i_2>$. Lorsque le contrôle arrive à $<i>$ il est instantanément transmis à $<i_1>$ et $<i_2>$, qui s'exécutent donc en parallèle. La terminaison de $<i>$ est la dernière des terminaisons de $<i_1>$ et $<i_2>$ (qui peuvent être simultanées). La composition parallèle est commutative et associative, et <u>rien</u> en est élément neutre à droite et à gauche.

3.6. La communication.

On peut bien sûr envisager de nombreuses formes de communication synchrone ou asynchrone. Nous avons choisi une communication par <u>ports</u> comme en [11]. On se met en attente d'un signal sur le port P par ?P, et on émet un signal sur P par !P. L'émission d'un signal est instantanée, donc l'action !P est de durée nulle. Elle provoque <u>instantanément</u> la terminaison de toutes les instructions d'attente ?P en activité.

En pratique il faut non seulement émettre des signaux mais aussi passer des valeurs. Si $<ent>$ est une action entière instantanée, alors !P($<ent>$) émet instantanément le résultat de $<ent>$, et ?P devient une action entière instantanée qui se termine en même temps que l'émission !P($<ent>$) et rend le résultat n si $<ent>$ rend n. (on peut émettre des booléens de la même façon.) Notons une difficulté qui n'existe pas dans le cas des signaux purs: il faut interdire à deux instructions d'émettre simultanément deux valeurs différentes sur un même port. On peut par exemple imposer d'avoir exactement une occurrence de !P pour chaque port P dans un programme.

3.7. L'instruction "jusqu'à".

C'est l'instruction la plus importante. Soit $<occ>$ une dénotation d'occurrence d'un événement $<evt>$, c'est à dire une expression de la forme "<u>prochain</u> $<evt>$ + $<ent>$" où $<ent>$ est une action entière <u>instantanée</u>, ce qui ne constitue pas une restriction (cf. la conditionnelle), et soient $<i_1>$, $<i_2>$ deux instructions. On construit l'instruction

$<i>$ = <u>jusqu'à</u> $<occ>$ <u>faire</u> $<i_1>$ <u>anormal</u> $<i_2>$ <u>fin</u>

Le contrôle est passé instantanément à $<ent>$ qui rend instantanément une valeur qui détermine l'occurrence $<occ>$ de $<evt>$. Simultanément le contrôle est passé à $<i_1>$ qui commence à s'exécuter. Si $<i_1>$ se termine avant l'occurrence $<occ>$ de $<evt>$, on attend cette occurrence et $<i>$ se termine instantanément dès qu'elle a lieu. Si $<occ>$ se produit avant la fin de $<i_1>$, l'exécution de $<i_1>$ est <u>instantanément</u> et <u>définitivement</u> arrêtée, le contrôle est instantanément passé à $<i_2>$, et la terminaison de $<i>$ est celle de $<i_2>$. Enfin $<i_1>$ peut contenir une instruction <u>terminer</u>. Lorsque le contrôle parvient à cette instruction, on abandonne instantanément l'exécution de $<i_1>$ et on termine instantanément $<i>$. (comme pour la boucle, on suppose qu'il existe un moyen syntaxique quelconque de dire à quel <u>jusqu'à</u> se rapporte un <u>terminer</u>.)

L'instruction <u>jusqu'à</u> permet donc de donner une "portée temporelle" à l'exécution de $<i_1>$, et d'activer éventuellement une procédure d'exception $<i_2>$ en cas de dépassement par $<i_1>$ de son temps limite (qui peut être donné en secondes, centimètres, tours etc...).

4. Exemples.

Hormis l'instruction <u>jusqu'à</u>, toutes les instructions sont assez classiques. La composition parallèle et la conditionnelle permettent de construire des commandes gardées par expressions booléennes, et on peut ainsi par exemple construire une clause <u>sinon</u> pour la conditionnelle:

<u>si</u> $<bool>$ <u>alors</u> $<i_1>$ <u>sinon</u> $<i_2>$ <u>fsi</u> =
    <u>si</u> $<bool>$ <u>alors</u> $<i_1>$ <u>fsi</u>
|| <u>si non</u> $<bool>$ <u>alors</u> $<i_2>$ <u>fsi</u>

où <u>non</u> est la négation booléenne, de durée nulle. De même on peut poser:

<u>tantque</u> $<bool>$ <u>faire</u> $<i>$ <u>ftq</u> =
  <u>répéter</u>
    <u>si</u> $<bool>$ <u>alors</u> $<i>$ <u>sinon</u> <u>sortir</u> <u>fsi</u>
  <u>fin</u>

L'instruction <u>jusqu'à</u> permet de construire facilement beaucoup d'instructions souvent présentées de manière indépendante [12]. Voici quelques abréviations d'usage courant.

<u>jusqu'à</u> $<occ>$ <u>faire</u> $<i>$ <u>fin</u> =
  <u>jusqu'à</u> $<occ>$ <u>faire</u> $<i>$ <u>anormal rien fin</u>

<u>attendre</u> $<evt>$ =
  <u>jusqu'à prochain</u> $<evt>$ <u>faire rien fin</u>

<u>pendant</u> n*$<evt>$ <u>faire</u> $<i>$ <u>fin</u> =
  <u>jusqu'à prochain</u> $<evt>$ <u>faire rien fin</u>;
  <u>jusqu'à prochain</u> $<evt>$ +n <u>faire</u> $<i>$ <u>fin</u>

```
délai n*<evt> =
    pendant n*<evt> faire rien fin


tousles n*<evt> faire <i> fin  =
    répéter
        pendant n*<evt> faire <i> fin
    fin


entre <occ_1> et <occ_2> faire <i> fin  =
    attendre <occ1> ;
        jusqu'à <occ_2> faire <i> fin
```

Remarquons que dans l'instruction

```
attendre <evt>;
jusqu'à prochain <evt> faire <i> fin
```

l'action <i> est toujours exécutée si elle est de durée nulle, ceci en synchronisme avec la première occurrence de <evt>.

Voici un exemple de "chien de garde":

```
jusqu'à (prochain s + 5)
    faire <i>; terminer
    anormal ! ALARME
fin
```

En contrôle de processus, il existe deux cas de traitements répétitifs: ceux pour lesquels ne pas finir un traitement élémentaire n'est pas important à cause de la stabilité intrinsèque de l'algorithme employé et ceux pour lesquels c'est au contraire inacceptable. Ceci correspondra aux deux formes de programmes suivantes:

```
répéter
    pendant n*<evt> faire <i> fin
fin


répéter
    pendant n*<evt>
        faire <i>
        anormal !ALARME; sortir
    fin
fin
```

Voici l'exemple de la fixation d'une origine pour un événement répétitif. On introduit un module de comptage, qui est activé dès qu'il reçoit "l'heure initiale", envoyée par "! ORIGINE(<ent>)". Il envoie ensuite la nouvelle "heure" en synchronisme avec chaque nouvelle occurrence de <evt>:

```
H := ? ORIGINE;
tousles <evt> faire
    ! HEURE(H);
    INCR(H)
fin
```

L'émission de l'heure H est synchrone avec chaque occurrence de <evt>. Il est inutile de supposer que l'action INCR(H) qui incrémente H est instantanée, mais il est nécessaire de supposer qu'elle prend un temps inférieur à un intervalle de <evt>, ce qui ne doit pas poser de problème en pratique. Pour attendre l'heure H, on peut exécuter en parallèle le programme suivant:

```
tantque (H ≠ ? HEURE) faire
    délai 1*<evt>
ftq
```

Voici un dernier exemple qui fait intervenir deux événements répétitifs: c'est la mesure de vitesse de la voiture. Toutes les M millisecondes on compte combien de centimètres ont été parcourus, et on émet la vitesse mesurée vers les autres processus.

```
répéter
    pendant M*ms faire
        NBCM:=0
        tousles cm faire
            INCR(NBCM)
        fin
    fin;
    ! VITESSE(NBCM/M)
fin
```

En pratique il peut être intéressant de dire que la mesure de vitesse est elle-même un événement répétitif. Dans ce cas le programme précédent pourrait être le corps du module définissant cet événement. Ceci aurait l'avantage d'interdire aux autres modules d'utiliser le synchronisme entre cette mesure et le temps, ce qui permettrait de changer aisément l'algorithme de mesure. On voit ainsi que des communications peuvent servir à produire des événements répétitifs.

5. Rappels sur SCCS. Définition d'opérateurs auxiliaires.

5.1. Rappels sur SCCS.

On se donne un monoïde abélien d'actions $(A,*,1)$ dans lequel $*$ est le produit synchrone et 1 est l'action unité. Si a et b sont des actions, le produit $a*b$ ou simplement ab est l'action obtenue en réalisant simultanément a et b. Lorsqu'il existe, l'inverse de a est noté $\bar{a}$. On construit un ensemble P d'agents et une relation $\to = \bigcup_{a \in A} \xrightarrow{a}$ entre agents. On lit $p \xrightarrow{a} q$ comme "p effectue l'action a et devient q". On définit récursivement P de la façon suivante:

- **0** est un agent qui n'effectue aucune action.

- Si $a \in A$ et $p \in P$ alors $a:p$ est un agent dont la seule action est a, avec $a:p \xrightarrow{a} p$.

- Si $p,q \in P$, alors $p \times q \in P$ avec $p \times q \xrightarrow{ab} p' \times q'$ ssi $p \xrightarrow{a} p'$ et $q \xrightarrow{b} q'$. On nomme $p \times q$ le produit synchrone de p et q.

- Si $p,q \in P$, alors $p+q \in P$ avec $p+q \xrightarrow{a} p'$ si $p \xrightarrow{a} p'$ et $p+q \xrightarrow{a} q'$ si $q \xrightarrow{a} q'$. On nomme $p+q$ la somme de p et q.

- Si p ∈ P, alors δp ∈ P avec δp $\xrightarrow{1}$ δp et δp $\xrightarrow{a}$ p' si p $\xrightarrow{a}$ p'. L'opérateur δ est l'opérateur de <u>délai</u>. Il permet de retarder arbitrairement l'exécution d'un agent.

- Si p ∈ P, alors △p ∈ P avec △p $\xrightarrow{a}$ δ△p' si p$\xrightarrow{a}$p'. L'opérateur △ sert à rendre p <u>asynchrone</u>.

- Si p ∈ P et si B⊂A alors p|B ∈ P avec p|B $\xrightarrow{a}$ p' ssi a ∈ B et p$\xrightarrow{a}$p'. Cette opération essentielle est appelée la <u>restriction</u>. Elle sert à <u>cacher</u> des actions au monde extérieur, et surtout à <u>forcer</u> des actions, voir plus loin.

- Si p ∈ P et si $\bar{\Phi}$ est un endomorphisme de A, alors p[$\bar{\Phi}$] ∈ P avec p[$\bar{\Phi}$] $\xrightarrow{\bar{\Phi}(a)}$ p'[$\bar{\Phi}$] si p$\xrightarrow{a}$p'. Cette opération sert principalement à <u>renommer</u> les actions de p. Si $\bar{\Phi}$ est induit par $\bar{\Phi}(a)=a'$, $\bar{\Phi}(b)=b'$,... on écrit p[a'/a, b'/b,....] = p[$\bar{\Phi}$]. Si $\bar{\Phi}(a)=b$ alors $\bar{\Phi}(\bar{a})=\bar{b}$.

- Nous utiliserons aussi la <u>définition récursive</u> d'agents, sans décrire cette opération de façon précise (voir [15] pour les détails). Par exemple p=a:p définit l'agent infini a:a:a:....

On définit <u>l'équivalence</u> ~ des agents de la façon suivante:

p $\sim_0$ q est toujours vrai

p $\sim_{n+1}$ p ssi

$\quad$ p $\xrightarrow{a}$ p' $\Rightarrow$ $\exists$q'. q $\xrightarrow{a}$ q' & p' $\sim_n$ q'

$\quad$ q $\xrightarrow{a}$ q' $\Rightarrow$ $\exists$p'. p $\xrightarrow{a}$ p' & p' $\sim_n$ q'

p ~ q ssi p $\sim_n$ q pour tout n

Un <u>processus</u> est une classe d'équivalence d'agents modulo ~. On vérifie que la somme et le produit sont associatifs et commutatifs, que **0** et **1**=δ**0** sont neutres pour + et ×, que × distribue sur + et que p×**0**=**0** pour tout p. Voir [15] pour les équivalences concernant la restriction et le renommage.

5.2. Opérateurs auxiliaires.

Nous nous placerons dans le cas où A est librement engendré par un ensemble AS <u>d'actions de synchronisation</u> et par un monoide abélien quelconque AC <u>d'actions de calcul</u>. Donc les éléments de A sont des mots contenant les a, $\bar{a}$, a ∈ AS et les a ∈ AC. On peut toujours considérer des mots <u>réduits</u> sur AS, c'est à dire des mots qui ne contiennent pas simultanément a et $\bar{a}$ pour a ∈ AS.

Les actions de synchronisation nous serviront à modéliser les événements répétitifs et à gérer le contrôle, les actions de calculs modélisant les actions élémentaires. Si a ∈ AS et p ∈ P, on note p\a la restriction de p au sous monoide de A formé par les mots où n'apparaissent ni a ni $\bar{a}$.

Pour introduire un nouvel opérateur dans SCCS, on peut procéder de deux façons: on peut <u>axiomatiser</u> son comportement, ou on peut le <u>définir</u> à partir des opérateurs de base, le mieux étant évidemment de faire les deux. Nous nous attacherons surtout à axiomatiser les opérateurs, ce qui rend leur manipulation aisée dans les preuves.

Etant donné S⊂AS, on notera W_S l'ensemble des mots sans carrés (éventuellement vides) écrits avec les lettres de S, et on notera w, w' les éléments de W. Pour a ∈ A, on dira que S <u>divise</u> a si toute expression de a réduite sur AS contient une lettre de S.

<u>Opérateur</u> => <u>de conditionnement par une action</u>:

Soient a ∈ A, p ∈ P. On axiomatise a=>p ainsi:

a=>p $\xrightarrow{ab}$ p' ssi p $\xrightarrow{b}$ p' ou p $\xrightarrow{ab}$ p'

Donc a sert de signal pour lancer p.

<u>Opérateurs de diffusion</u> $\&_S$ et $\&_S^a$

Etant donnés p et q, on définit le processus p$\&_S$q comme ayant les seules possibilités d'action suivantes:

- Si p $\xrightarrow{wa}$ p', q $\xrightarrow{wb}$ q', w ∈ W_S, a et b non divisibles par S, alors p $\&_S$ q $\xrightarrow{wab}$ p'$\&_S$q'.

Donc les actions de p et q doivent avoir exactement les mêmes composantes dans S. Cet opérateur généralise le synchronisateur de [15]. Il nous permettra d'assurer la diffusion des événements et signaux: tous les processus reliés par $\&_S$ reçoivent simultanément les actions de W_S. Il est commutatif et associatif.

Soit c ∈ S. On définit l'opérateur $\&_S^c$ qui se comporte comme $\&_S$ sauf qu'il ne conserve pas la diffusion de c après la première étape. Il nous servira à définir la composition parallèle. Ses seules possibilités d'action sont les suivantes:

- Si p $\xrightarrow{wa}$ p' et q $\xrightarrow{wb}$ q', w ∈ W_S, ab non divisible par S\{c}, alors
p $\&_S^c$ q $\xrightarrow{wab}$ p' $\&_S^c$ q'.

- Si p $\xrightarrow{cwa}$ p' et q $\xrightarrow{cwb}$ q', w ∈ W_S, ab non divisible par S\{c}, alors
p $\&_S^c$ q $\xrightarrow{cwab}$ p' $\&_S$ q'.

<u>Délai sur S, δ_S.</u>

Etant donné S on définit l'opérateur δ_S de délai sur S de la façon suivante:

δ_Sp $\xrightarrow{w}$ δ_Sp si w ∈ W_S

δ_Sp $\xrightarrow{a}$ p' si p$\xrightarrow{a}$p'

l'agent δ_Sp absorbe tous les mots de W_S ou exécute les actions de p. Noter que 1_S=δ_S**0** est élément neutre de $\&_S$.

Renommage avec portée.

Si p est un agent, alors p[u//a] désigne l'agent $(p[u_1/u])[u/a]$ où u_1 est une variable n'ayant aucune occurrence dans p. Cette opération est essentielle pour éviter les conflits de noms dans les définitions par récurrence structurelle.

6. Sémantique des primitives en SCCS.

6.1. Principes de la sémantique.

A chaque événement répétitif <evt> nous associons une action e ∈ AS. Nous appelons R l'ensemble des actions ainsi obtenues. A chaque port de communication en signal pur nous associons une action notée c_p ∈ AC. A chaque port booléen (resp entier) nous associons deux actions c_p^v et c_p^f (respectivement une action c_p^n pour tout entier n). L'ensemble des c_p est appelé C, et on pose E=R ∪ C.

Nous utilisons deux actions particulières d,f ∈ AC qui nous servent à la transmission du contrôle, et des actions v,f,n ∈ AC correspondant aux booléens et aux entiers.

A tout terme M de notre langage nous associons un agent ⟦M⟧ appelé la _sémantique_ de M et défini par récurrence structurelle sur M. Les composants de M sont reliés par l'opérateur de diffusion sur E, qui assure la transmission synchrone des tops d'horloge et des communications. Les exécutions de M seront par définition les réécritures du terme suivant, qui représente M plongé dans un contexte qui assure la diffusion des communications et donne un signal de départ:

$$((\overline{d}\texttt{=>}H_C) \times \llbracket M \rrbracket)\backslash C$$

avec:

$$H_C = \prod_{c \in C} H_c$$

$$H_c = \delta(\overline{c}:H_c)$$

Les occurrences d'événements répétitifs sont représentées par des facteurs $w \in W_R$ dans les actions du terme précédent.

A cause de l'opérateur de diffusion, on doit prendre certaines précautions pour activer et terminer les agents. Un agent qui n'a pas encore reçu le contrôle doit pouvoir "absorber" toutes les actions de E sans rien faire. Un agent qui a perdu le contrôle (en envoyant son signal de fin) doit se transformer non pas en **0** mais en $\mathbf{1}_E$ qui est l'élément neutre de $\&_E$. Ceci se formalise par la notion d'_agent admissible_ (relativement à E), que nous ne détaillerons pas ici.

On définit les opérateurs suivants (qui préservent l'admissibilité):

p : q = (p[u//f] $\&_E$ q[u//d])\u

p >> q = (p : δ_E(v=>q + f=>rien))\v,f

L'opérateur : sert au séquencement. L'utilisation de u permet de transformer instantanément le signal de fin de p en signal de début de q (remarquer que le renommage [u//f] renomme aussi f en u, puisque c'est un morphisme). L'opérateur >> est la conditionnelle: le contrôle ne parvient à q que si l'action booléenne p rend v.

6.2. Equations sémantiques.

Nous ne donnons ici, en matière d'exemple, que les équations de l'action instantanée et de l'instruction _jusqu'à_. Celles des autres constructions du langage sont disponibles dans [5].

Actions instantanées:

A toute action <act> nous associons une action de calcul $a_{\langle act \rangle}$ ∈ AC, ou a tout court. Nous définissons a_{rien}=1. Si <act> est instantanée, nous posons:

$$\llbracket \langle act \rangle \rrbracket = \delta_E(da\overline{f}\texttt{=>}\mathbf{1}_E)$$

Donc l'action significative $da\overline{f}$ est nécessairement provoquée par l'arrivée d'un d venant du contexte, i.e. par l'arrivée du contrôle. Le signal de terminaison $\overline{f}$ est synchrone avec le début d. Pour une action non instantanée, on introduit un délai supplémentaire:

$$\llbracket \langle act \rangle \rrbracket = \delta_E(d\texttt{=>}\delta_E(a\overline{f}\texttt{=>}\mathbf{1}_E))$$

C'est au moment de leur terminaison que les actions rendent leurs valeurs éventuelles. Voici par exemple l'interprétation de l'action instantanée <n>:

$$\llbracket n \rrbracket = \delta_E(d\underline{n}\overline{f}\texttt{=>}\mathbf{1}_E)$$

Instruction jusqu'à.

Pour simplifier nous nous limitons au cas où l'occurrence est de la forme _prochain_ <evt> + n. Soit e l'action associée à <evt>.

⟦ jusqu'à prochain <evt> + n
  faire $\langle i_1 \rangle$ anormal $\langle i_2 \rangle$ fin ⟧ =

$$(\delta_E \triangle ((d\texttt{=>}A) \&_E^d \llbracket \langle i_1 \rangle \rrbracket [f_N//f]) \times C)\backslash f_N, f_A, f_T, t$$

avec

$$A = B_n \times \delta(\sum_{n=0}^{\infty} t^n \overline{f_T})$$

$$B_{n+1} = \delta(e:B_n)$$

$$B_0 = \delta(e\overline{f_A}:\mathbf{0})$$

$$C = \delta(\llbracket \langle i_2 \rangle \rrbracket [f_A//d]$$

$$+ f_A f_N \overline{f}\texttt{=>}\mathbf{1}_E$$

$$+ \; f_T\bar{f} => \mathbf{1}_E$$
$$+ \; f_T f_A \bar{f} => \mathbf{1}_E$$
$$+ \; f_N : \delta(f_A \bar{f} => \mathbf{1}_E))$$

$[\![\underline{terminer}]\!] = \delta_E(d\bar{t} => \mathbf{1}_E)$

Ici A est un "chien de garde" et C un "collecteur". L'agent A est chargé d'attendre la n-ième occurrence de e, sur laquelle il émet $\bar{f}_A$, ou des signaux t émis par des instructions <u>terminer</u> qui peuvent être exécutées simultanément en nombre quelconque, et qui provoquent l'émission par A de $\bar{f}_T$. Le collecteur C attend trois signaux de terminaison: $\bar{f}_N$, terminaison normale de $\langle i_1 \rangle$; $\bar{f}_T$, envoyé par A en cas d'exécution d'un <u>terminer</u>; $\bar{f}_A$, envoyé par A sur l'occurrence attendue de e. L'arrivée de $\bar{f}_A$ seul provoque le lancement de $\langle i_2 \rangle$. L'arrivée de $\bar{f}_N$ seul provoque l'attente de $\bar{f}_A$. Toute autre combinaison possible de $\bar{f}_A, \bar{f}_N, \bar{f}_T$ provoque la fin du <u>jusqu'à</u>. Dans tous les cas le terme A passe à **0** et le <u>jusqu'à</u> se transforme en le seul collecteur puisque $\delta\backslash \mathbf{0} = \mathbf{1}$.

6.3. Utilisation de la sémantique.

Notre sémantique est évidemment assez complexe, et son utilisation exige une bonne habitude de SCCS. Citons sans détail les deux principaux types de résultats que nous obtenons:

- Les événements répétitifs sont <u>infiniment persistants</u>: toute exécution finie peut se prolonger en une exécution infinie où tous les événements répétitifs apparaissent infiniment souvent. Autrement dit, il ne peut pas y avoir de situation de blocage pour les événements répétitifs.

- Toutes les équivalences de programmes décrites dans la sémantique intuitive sont <u>valides dans la sémantique</u>. Ceci se démontre à l'aide des lois de [15] sur l'équivalence. Plus généralement la sémantique peut être utilisée pour valider des équivalences ou des transformations de programmes.

Conclusion.

La sémantique que nous avons donné est du type "dénotationnel", c'est à dire qu'elle relativise notre calcul à un calcul préexistant, et ceci en opérant une traduction dirigée par la syntaxe. Elle nous permet de vérifier la cohérence interne de nos primitives, et de démontrer des équivalences comme celles données au chapitre 3. Elle suggère de plus une implémentation possible des primitives: la partie la plus délicate dans l'élaboration de la sémantique est l'explicitation du contrôle: de ce point de vue SCCS est un langage d'assez bas niveau, peu éloigné d'un "langage machine". Signalons que d'autres calculs actuellement en cours de développement devraient permettre de donner une sémantique de même nature mais plus simple. Il nous paraît également souhaitable de développer d'autres types de sémantiques plus intrinsèques et plus intuitives, comme les sémantiques opérationnelles de [16]. Ce travail reste entièrement à faire. Il faudrait ensuite montrer que les différentes sémantiques coincident, comme c'est classique dans les langages séquentiels. Le fonctionnement des algorithmes de contrôle nécessite souvent que certaines contraintes entre les événements répétitifs soient satisfaites. Il serait souhaitable de pouvoir exprimer ces contraintes au niveau des preuves. On pourrait par exemple utiliser des logiques temporelles [17,18] ou encore le formalisme des algèbres d'événements de [6].

Revenons sur nos hypothèses très fortes quant à l'instantanéité de certaines actions et de tout le contrôle. Sans elles, il serait impossible d'avoir un quelconque calcul sur les programmes. Mais il y a plus que cette justification interne. D'abord, dans presque tous les cas il nous semble important de spécifier une application sans se préoccuper des contraintes temporelles dues aux contraintes physiques de réalisation, mais seulement de celles dues au problème lui-même. Ensuite il faut aussi remarquer que beaucoup des processus que l'on souhaite contrôler ont des constantes de temps relativement grandes vis à vis des vitesses de calcul actuelles. Dans ce cas on peut raisonnablement supposer en pratique la simultanéité des actions. Pour cette classe de processus, nos primitives pourraient former l'ossature d'un véritable langage de <u>programmation</u> et pas seulement d'un langage de spécification. Lorsque les constantes de temps des processus sont elles mêmes très faibles comme en traitement du signal, les problèmes qui se posent sont beaucoup plus des problèmes d'algorithmique, et il n'est pas sûr que des primitives comme les nôtres soient d'une quelconque utilité.

Sur un plan pratique, il reste bien sûr à concevoir et à implémenter des langages réellement <u>modulaires</u> fondés sur nos primitives. Par "implémentation" nous entendons en particulier la réalisation de systèmes de manipulation de programmes, de systèmes de simulation, de systèmes de vérification dynamique ou statique de contraintes temporelles, et enfin de compilateurs. La conception passe par une étude poussée des relations entre événements répétitifs et événements de communication: certains événements peuvent parfaitement être considérés comme répétitifs à un certain niveau logique, mais être engendrés par des communications à un autre niveau.

D'autre part il est nécessaire de formaliser les problèmes de visibilité de manière plus fine car les approches modulaires classique (de type ML par exemple) ne semblent plus utilisables en univers parallèle

References

1. <u>LTR Manuel Officiel de Référence</u>, Ministère de la Défense (1978).

2. "Premier Championnat International de Voitures Robot," MICRO SYSTEMES (janvier 1980).

3. Reference Manual for the ADA Programming Language, CII Honeywell-Bull (1980).

4. J.G.P. Barnes, RTL/2 Design and Philosophy, Heyden & Sons Ltd. (1976).

5. G. Berry, J. Camerini, J.P. Marmorat, B. Nguyen-Phuoc, and J.P. Rigault, Quelques Primitives pour la Programmation Temps Réel et leur Sémantique Mathématique, Centre de Mathématiques Appliquées Ecole des Mines Sophia-Antipolis (1982).

6. P. Caspi and N. Halbwachs, "Algebra of Events: a Model for Parallel and Real Time Systems," RR.285, IMAG, Grenoble (1982).

7. R. L. Glass, "Real-Time: The 'Lost World' of Software Debugging and Testing," Comm. ACM Vol. 23(5) (1980).

8. M. Gordon, R. Milner, and C. Wadsworth, Edinburgh LCF, Lecture Notes in Computer Science 78, Springer Verlag (1980).

9. M.C.B. Hennessy, W. Li, and G.D. Plotkin, "A First Attempt at Translating CSP into CCS," Research Report, Edinburgh University (1980).

10. M.W. Hennessy and W. Li, "An Operational Semantics of Tasking in ADA," Research Report, University of Edinburgh (1981).

11. C.A.R. Hoare, "Communicating Sequential Processes," Comm. ACM Vol. 21(8), pp.666-678 (1978).

12. F. LeCalvez, "Définition d'un langage de description globale des applications en temps réel," Thèse de 3ème cycle, Université PARIS VI (1979).

13. J.P. Marmorat, "Applications de la Micro-Informatique à la Commande d'une Voiture Robot," Rapport Centre de Mathématiques Appliquées, Ecole des Mines (1981).

14. R. Milner, A Calculus of Communicating Systems, Springer-Verlag, LNCS 92 (1980).

15. R. Milner, "On Relating Synchrony and Asynchrony," Research Report, Edinburgh University (1980).

16. G.D. Plotkin, "An Operational Semantics for CSP," Research Report, Edinburgh University (1981).

17. A. Pnueli, "The Temporal semantics of concurrent Programs," TCS Vol. 13, pp.45-60 (1981).

18. J.P. Queille and J. Sifakis, "Fairness and Related Properties in Transition Systems - A Time Logic to Deal with Fairness," RR.292, IMAG, Grenoble (1982).

19. W. Schild and H. Lienhard, "Real-Time Programming in PORTAL," Sigplan Notices Vol. 15(4) (1980).

20. Ravi Sethi, "Semantics of Computer Programs: Overview of Language Definition Methods," Bell Laboratories Report (septembre 1977).

21. J. Sifakis, "A Unified Approach for Studying the Properties of Transition Systems," Research Report n.179, IMAG, Grenoble (1980).

PORT-ORIENTED SYNCHRONISATION AND COMMUNICATION IN A LOCAL NETWORK

J.P. Elloy, P. Molinaro

E.N.S.M. Laboratoire d'Automatique, Nantes, France

ABSTRACT

In order to implement communication between processes distributed in a local computer network, several languages have been proposed such as C.A.R. Hoare's "Communicating Sequential Processes". One of the disadvantages of the CSP specification is the necessity for the respective partner to be explicitly designated by each process in the input/output commands for setting up such a link. The extension proposed by Silberschatz consists of introducing dedicated ports into these commands - every process then communicates through one or several ports without explicit designation of a partner being required.

This extension has not yet been effectively implemented, i.e. for reducing the number of messages transmitted over the communication paths of the network. In the present paper we are proposing an implementation of these ports which is subdivided into processes running in parallel. The mechanism of data exchanges obtained is divided hierarchically into four levels called application, session, transmission and network protocols which communicate with each other inside the same computer by means of a simple letter box. The implementation described allows any process to be ignorant of the location of its partner in the network, which is established only at the transmission level. And even though this aspect has not been developed here, this data exchange mechanism enables Dijkstra's guarded commands, and thus a CSP extended to dedicated ports, to be easily implemented. This work is intended to provide a tool for managing real-time tasks for controlling large-scale industrial processes.

RÉSUMÉ

Pour exprimer la communication entre processus répartis dans un réseau local de calculateurs, plusieurs langages ont été proposés dont "Communicating Sequential Processes" par C.A.R. Hoare. L'un des inconvénients de la spécification de CSP réside dans la nécessité à tout processus de désigner explicitement son correspondant dans les commandes d'entrées/sorties qui établissent une telle communication. L'extension proposée par Silberschatz consiste en l'introduction des ports orientés dans ces commandes; alors tout processus communique au travers d'un ou de plusieurs ports sans désignation explicite du correspondant. Cette extension n'a pas encore fait l'objet d'une implémentation efficace, c'est à dire résuisant le nombre de messages transmis sur les voies de communication du réseau. Nous proposons dans cet article une implémentation de ces ports décomposee en processus évoluants en parallèle. Le mécanisme d'échanges obtenu se décompose hiérarchiquement en quatre niveaux appelés protocoles d'application, session, transport et réseau qui, dans le même calculateur, communiquent entre eux au moyen d'une simple boite aux lettres. L'implémentation décrite permet à tout processus d'ignorer la localisation dans le réseau de son correspondant qui n'est établi qu'au niveau transport. En outre, même si cet aspect n'est pas développé dans l'article, ce mécanisme d'échanges permet une réalisation facile des commandes gardées de Dijkstra et donc de CSP étendu aux ports orientés. Ce travail est destiné à obtenir un outil de gestion de tâches temps réel destinées à la commande de procédés industriels de grande dimension.

ZUSAMMENFASSUNG

Zur Realisierung des Datenaustauschs zwischen in einem örtlichen Rechnernetz verteilten Prozessoren sind mehrere Sprachen vorgeschlagen worden, unter ihnen "Communicating Sequential Processes" von C.A.R. Hoare. Einer der Nachteile in der CSP-Spezifikation liegt in der Notwendigkeit, daß der entsprechende Teilnehmer für jeden Prozess ausdrücklich in den Eingabe/Ausgabe-Befehlen zur Herstellung dieser Verbindung angegeben werden muß. Die von Silberschatz vorgeschlagene Erweiterung besteht aus der Einführung von reservierten Ports in diese Befehle - jeder Prozess verkehrt dann über einen oder mehrere Ports ohne ausdrückliche Angabe des Teilnehmers.

Diese Erweiterung ist noch nicht wirksam realisiert worden, d.h. zur Verringerung der Anzahl von über die Netzkommunikationswege übermittelten Nachrichten. In diesem Artikel schlagen wir eine Realisierung dieser Ports vor, wonach eine Aufteilung in parallel ablaufende Prozesse stattfindet. Die so erhaltenen Datenaustauschmechanismen werden hierarchisch in vier, Anwendungs-, Sitzungs-, Übertragungs-, Netzprotokolle genannte Ebenen eingeteilt, die im selben Rechner mittels eines einfachen Briefkastens miteinander verkehren. Mit der beschriebenen Realisierung braucht für keinen Prozess der Teilnehmer in Netz geortet werden, der jetzt nur auf der Übertrangungsebene ermittelt wird. Und obwohl dieser Aspekt in diesem Artikel nicht weiter entwickelt worden ist, wird mit diesem Datenaustauschmechanismus eine leichte Realisierung der geschützten Befehle von Dijkstra und somit des auf reservierte Ports erweiterten CSP-Programms ermöglicht. Mit diesser Arbeit soll ein Werkzeug zur Verwaltung von Echtzeitaufgaben zur Steuerung von industriellen Großprozessen bereitgestellt werden.

SYNCHRONISATION ET COMMUNICATION PAR PORTS ORIENTES DANS UN RESEAU LOCAL

J.P. ELLOY - P. MOLINARO

E.N.S.M. Laboratoire d'Automatique - E.R.A. n° 134
1 rue de la Noë - 44072 Nantes Cedex - France

Pour exprimer la communication entre processus répartis dans un réseau local de calculateurs, plusieurs langages ont été proposés dont "Communicating Sequential Processes" par C.A.R. Hoare. L'un des inconvénients de la spécification de CSP réside dans la nécessité à tout processus de désigner explicitement son correspondant dans les commandes d'entrées/sorties qui établissent une telle communication. L'extension proposée par Silberschatz consiste en l'introduction des ports orientés dans ces commandes ; alors tout processus communique au travers d'un ou de plusieurs ports sans désignation explicite du correspondant. Cette extension n'a pas encore fait l'objet d'une implémentation efficace, c'est à dire réduisant le nombre de messages transmis sur les voies de communication du réseau. Nous proposons dans cet article une implémentation de ces ports décomposée en processus évoluants en parallèle. Le mécanisme d'échanges obtenu se décompose hiérarchiquement en quatre niveaux appelés protocoles d'application, session, transport et réseau qui, dans le même calculateur, communiquent entre eux au moyen d'une simple boîte aux lettres. L'implémentation décrite permet à tout processus d'ignorer la localisation dans le réseau de son correspondant qui n'est établi qu'au niveau transport. En outre, même si cet aspect n'est pas développé dans l'article, ce mécanisme d'échanges permet une réalisation facile des commandes gardées de Dijkstra et donc de CSP étendu aux ports orientés. Ce travail est destiné à obtenir un outil de gestion de tâches temps réel destinées à la commande de procédés industriels de grande dimension.

1. INTRODUCTION

Avec CSP, C.A.R. Hoare [7] a introduit un langage d'expression de la synchronisation et de la communication de processus répartis dans un réseau local de calculateurs sans mémoire commune. Les caractéristiques fondamentales de CSP peuvent être résumées ainsi :

(i) tout programme CSP se compose d'un nombre fixe de processus séquentiels évoluants en parallèle. Un programme CSP ne se termine que lorsque tous ses processus ont terminé leur exécution.

(ii) Tout processus contient des variables locales et une liste d'instructions, appelée commande, qui manipule ces variables. Les valeurs des variables déclarées dans un processus sont inaccessibles aux autres processus.

(iii) Les communications inter-processus sont assurées au moyen de commandes d'entrées/sorties Dans ces commandes, les processus correspondants doivent désigner chacun l'autre explicitement.

(iv) La structure séquentielle d'un processus est exprimée par les commandes gardées de Dijkstra [5].

Ces caractéristiques permettent ainsi de décrire tout système composé d'un nombre fini de processus disjoints communiquant entre eux.

Hoare souligne dans son article que sa proposition ne constitue qu'une réponse partielle à la définition d'un langage de programmation concurrente et que plusieurs de ses aspects devront faire l'objet d'études et de développements ultérieurs, notamment sur la façon d'éviter la désignation explicite du correspondant dans les commandes d'entrées/sorties ainsi que sur la façon d'implémenter efficacement son langage.

A. Silberschatz a proposé, dans "Port Directed Communication" [13], une alternative intéressante à la désignation explicite du correspondant : les commandes d'entrées/sorties (E/S) nomment un port à travers lequel les processus établissent entre eux des communications. Un tel port peut être utilisé par plus de deux processus, ce qui permet à une même commande d'E/S d'établir une communication entre plusieurs processus possibles.

Nous nous proposons, dans cet article, de définir la communication par ports orientés et d'en décrire une implémentation respectant les spécifications de Silberschatz.

2. COMMUNICATIONS PAR PORTS ORIENTES

Dans CSP, les commandes d'E/S nomment explicitement le processus correspondant. Jazayeri, Ghezzi et al. [8] ont proposé une première solution destinée à remédier à cet inconvénient : chaque commande d'E/S y nomme un port dont le nom est local au processus ; la correspondance entre deux noms locaux de ports est assurée par une déclaration spécifique. Toutefois, il n'existe en fait aucune véritable différence sémantique entre les commandes d'E/S d'Hoare et celles introduites par Jazayeri et Ghezzi : faire communiquer deux processus par plusieurs ports distincts tels que définis dans [8] revient en fait à utiliser les commandes de CSP avec autant de constructeurs différents.

Dans l'approche proposée par Silberschatz, les commandes d'E/S nomment également un port, mais le concept de port est alors attaché à une communication et non à des processus correspondants. C'est pourquoi une communication peut s'établir au travers d'un même port entre plusieurs processus différents. Ce concept de port comme mécanisme de communication n'est pas nouveau, mais par son introduction dans CSP il prend une signification sémantique différente.

2.1 Déclaration des ports orientés

Un port est défini globalement pour un ensemble de processus communicants entre eux. La façon dont un port peut être utilisé par un processus doit y être explicitement déclarée, soit par l'instruction "port", soit par l'instruction "use". L'instruction "port" :

⟨liste de noms de ports⟩ : port ;

déclare le processus qui l'exécute propriétaire de tous les ports nommés dans cette instruction. Un processus peut être propriétaire d'un nombre quelconque de ports, mais tout port ne doit avoir qu'un et un seul propriétaire. Cette notion de propriété joue un rôle prépondérant dans l'établissement d'une communication, rôle décrit § 2.2.
Tout processus peut utiliser un port dont il n'est pas propriétaire par :

use (⟨liste de noms de ports⟩)

On dira qu'alors le processus est utilisateur de ces ports. Tout port peut accepter un nombre quelconque d'utilisateurs ; aucun utilisateur n'a à connaître l'identité des autres utilisateurs du même port.
Remarque : on supposera par la suite que les noms des ports sont globaux et donc tous distincts. Il serait possible d'étendre la définition des fonctions "use" et "port" de façon à permettre d'attribuer à chaque port un nom local dans chaque processus ; il serait alors nécessaire d'établir par une déclaration particulière la correspondance entre ces divers noms locaux d'un même port. Cette proposition revient sémantiquement à attribuer un nom unique et distinct à chaque port.

2.2 Instructions de communication

A l'image de CSP, la communication à travers un port s'établit au moyen de commandes d'E/S, symbolisées par les caractères "?" et "!". Seule différence syntaxique avec CSP : le nom du port y remplace celui du processus correspondant.

⟨commande d'entrée⟩::=
    ⟨nom du port⟩? ⟨variable-cible⟩
⟨commande de sortie⟩::=
    ⟨nom du port⟩! ⟨variable-source⟩

Bien que syntaxiquement voisines, les commandes d'E/S de Silberschatz présentent deux différences sémantiques fondamentales avec celles d'Hoare. La première porte sur les conditions d'établissement d'une communication : dans CSP, il suffit que les processus correspondants nomment chacun l'autre pour que la communication puisse s'établir. Par contre, une communication par port orienté ne pourra s'établir que si les trois conditions suivantes sont respectées :
(i) les deux processus doivent nommer le même port dans leurs commandes d'E/S
(ii) l'un des deux processus contient une commande d'entrée, l'autre une commande de sortie
(iii) l'un des deux processus est propriétaire du port.
La seconde différence est que dans CSP une commande d'E/S n'établit de communication qu'entre deux processus, tandis qu'une communication par ports permet au processus propriétaire d'un port de communiquer avec l'un quelconque de ses utilisateurs.

Remarque : les commandes d'E/S de CSP demandent la présence d'un constructeur qui identifie le type de l'échange. Il est naturellement possible d'introduire ce même concept dans les communications par port. Toutefois la possibilité à deux processus de pouvoir communiquer entre eux par plusieurs ports distincts ne justifie plus la nécessité d'introduire des constructeurs dans l'expression des commandes d'E/S.

2.3 Commandes gardées

La notion de commande gardée a été introduite par Dijkstra [5] ; Hoare a élargi ce concept en autorisant la présence d'une commande d'entrée dans une garde ; Kieburtz [9], Bernstein [1] ont à leur tour introduit les commandes de sortie dans les gardes.
Dans l'extension proposée par Silberschatz, les commandes d'E/S par port sont autorisées dans les gardes à la seule condition que le processus contenant cette garde soit le propriétaire du port nommé dans cette commande. Cette limitation ne restreint que très peu la puissance du langage et elle en simplifie sensiblement l'implémentation. En effet, chaque couple de processus correspondant (par exemple P_1 et P_2) peut déclarer deux ports A_1 et A_2 de façon à ce que P_1 soit propriétaire de A_1 et P_2 de A_2 ; on peut ainsi établir une communication entre les deux processus dans des commandes gardées soit à travers A_1 (pour P_1) ou A_2 (pour P_2).

Les commandes gardées peuvent être combinées dans une commande alternative ou une commande répétitive [7]. L'exécution d'une commande alternative consiste en le test successif dans un ordre quelconque de toutes ses gardes. Si plusieurs gardes sont vraies, une commande gardée parmi celles dont la garde est vraie est arbitrairement choisie et exécutée. Le contrôle est ensuite donné à la commande qui suit la commande alternative.
La commande répétitive est une commande alternative qui est indéfiniment réexécutée jusqu'à ce que toutes ses gardes soient fausses. L'exécution du processus se poursuit alors à l'instruction qui suit la commande répétitive.

3. STRUCTURE DU MECANISME D'ECHANGES

La structure du mécanisme d'échanges destiné à réaliser la communication par port entre deux processus peut être logiquement décomposée en quatre niveaux principaux : niveau d'application, de session, de transport, de réseau. Ce mécanisme est implanté identiquement dans tous les calculateurs du réseau ; à partir des seules commandes d'E/S il génère la succession des évènements qui conduisent à la synchronisation et au transfert d'informations entre deux processus quelconques qui peuvent appartenir au même calculateur ou à deux calculateurs différents du réseau.

(i) Au niveau supérieur (niveau 0), le protocole d'application est constitué de procédures appelées par les processus qui désirent communiquer au moyen des commandes d'E/S.
(ii) Le protocole de session est le responsable du déroulement de la communication : il synchronise les processus communicants, supervise le

transfert des données et réveille les processus
à la fin de l'échange (niveau 1)
Le protocole de session est sollicité par le
protocole d'application de chaque calculateur.
Les procédures du protocole de session communiquent entre elles par l'intermédiaire des protocoles de niveau inférieur. Ceux-ci rendent
transparente au protocole de session la localisation physique des processus correspondants
dans le réseau.
(iii) Le protocole de transport (niveau 2) est
le seul à connaître la correspondance port/pro-
cessus propriétaire/calculateur. Il transmet
les ordres émis par les procédures du protocole
de session soit au protocole de réseau (échange
entre deux processus résidants dans des calculateurs différents), soit à lui-même (échange
local).
(iv) Le protocole de réseau (niveau 3) est chargé de la constitution des messages et de leur
acheminement sur les voies de communication du
réseau. Ses fonctions incluent la détection de
la perte d'un message, de sa détérioration, de
sa duplication. Le protocole de réseau est constitué de deux processus évoluant en parallèle
respectivement chargés de l'émission et de la
réception des messages.

Un tel mécanisme peut être implémenté de deux
façons différentes :
(i) dans une "implémentation parallèle" les niveaux de session et de transport sont assurés
par deux processus respectivement appelés
SESSION et TRANSPORT évoluant en parallèle avec
les processus d'un programme CSP. Cette structure illustrée par la figure 1 présente deux
avantages : les algorithmes des processus sont
très simples, et l'accès en exclusion mutuelle
des procédures de ces protocoles est implicitement réalisée par leur regroupement dans des
processus. Cette solution par contre introduit
de nombreux échanges inter-processus entre les
différents niveaux des protocoles au cours d'un
échange.
(ii) L'implémentation "série" diminue, aux dé-
pens du parallélisme, le nombre des échanges
inter-processus entre les différents niveaux.
Dans cette solution il devient toutefois nécessaire de protéger l'accès des procédures des
protocoles. Cette structure est illustrée par
la figure 2.

L'objet de cet article est de décrire la première de ces implémentations, cet outil de
synchronisation et communication étant destiné
à la constitution d'un logiciel de commande de
procédés industriels complexes dont la commande
de robots.

Sur les figures 1 et 2 :

▭ symbolise une procédure

⬡ symbolise un processus.

    PA : protocole d'application
    PS : protocole de session
    PT : protocole de transport
    PR : protocole de réseau

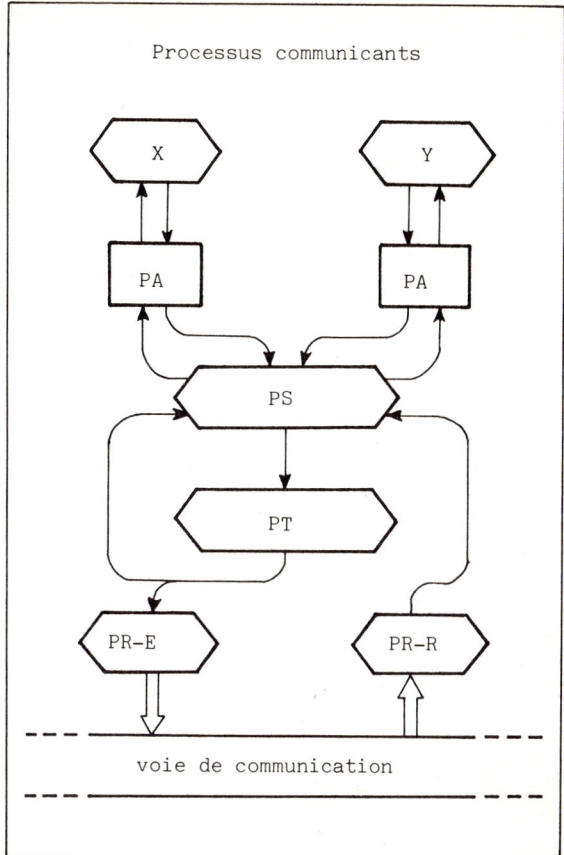

Fig. 1 : Implémentation parallèle

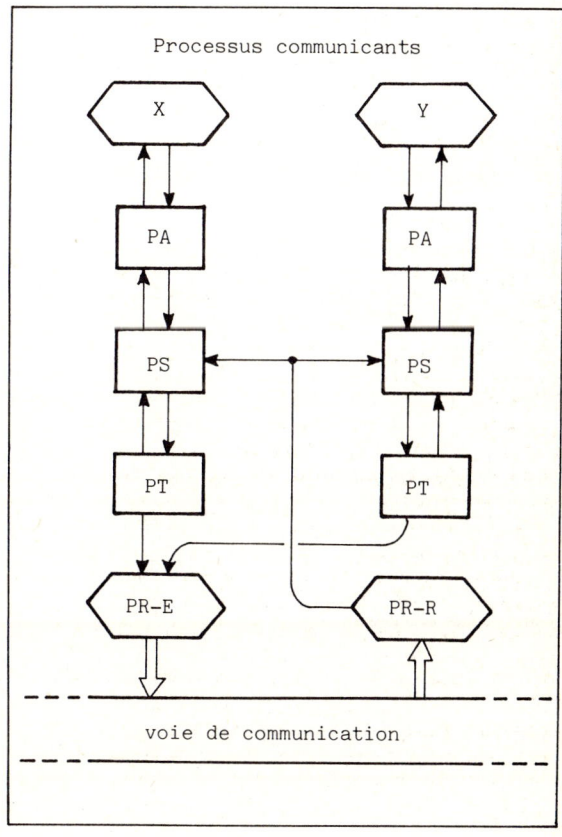

Fig. 2 : Implémentation série

4. IMPLEMENTATION DU PROTOCOLE D'APPLICATION

Dans les algorithmes des procédures présentées dans les paragraphes suivants, la transmission de signaux et la synchronisation entre les différents processus résidants dans un même calculateur sera symbolisée par des commandes d'entrée et de sortie respectivement notées "??" et "!!" :
(i) ?Q?v qui représente l'attente par le processus Q d'une information v émise par un processus quelconque
(ii) X!!w qui représente l'émission de la valeur de w vers le processus X.
Un tel mécanisme, du type "boite aux lettres" peut être aisément implémenté au moyen de sémaphores privés avec message.
En outre, on utilisera dans ces procédures la notation des commandes alternatives et répétitives pour raison de clarté.

Le protocole d'application est constitué de procédures appelées par les processus communicants. Une commande de sortie dans un processus X :
[[X:: ..... ; A!x; ...]]   sera traduite par l'appel de la procédure PORT! si X est propriétaire du port A, ou par l'appel de la procédure USE! si X est utilisateur du port A. De même, une commande d'entrée dans Y :
[[Y:: ... ; A?y; ...]] sera traduite par l'appel de PORT? ou USE?.
L'exécution de ces procédures consiste en la transmission d'un signal au protocole de session l'informant de l'échange requis, signal accompagné des paramètres de l'échange (nom du processus, nom du port, donnée...) ; après transmission du signal, ces procédures restent bloquées en attente de la réponse du protocole de session sous la forme d'un signal de synchronisation.

---

procédure PORT!(P,A,d)
  P : process_id ; A : port_id ; d : data ;
  début
    SESSION!!PORTSEND,A,d ; ?P?
  fin

---

procédure USE!(P,A,d)
  P : process_id ; A : port_id ; d : data ;
  début
    SESSION!!USESEND,P,A,d ; ?P?
  fin

---

procédure PORT?(P,A,d)
  P : process_id ; A : port_id ; d : data ;
  début
    SESSION!!PORTREC,A,d ; ?P?
  fin

---

procédure USE?(P,A,d)
  P : process_id ; A : port_id ; d : data ;
  début
    SESSION!!USEREC,P,A,d ; ?P?
  fin

Vu du protocole d'application, le chronogramme d'un échange entre les processus X et Y à travers un port A dont Y est propriétaire aura l'allure indiquée figure 3.

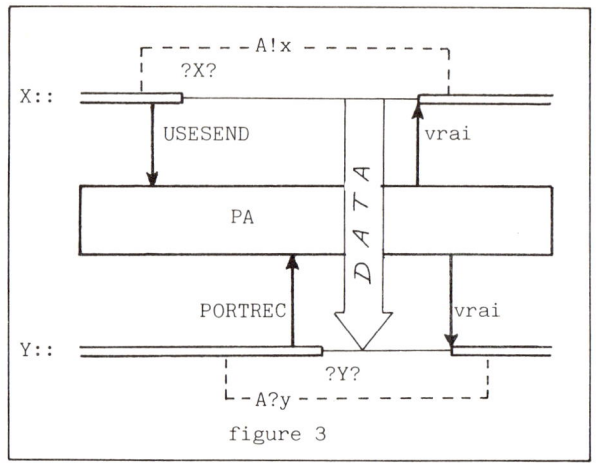

figure 3

5. IMPLEMENTATION DU PROTOCOLE DE SESSION

Le protocole de session d'un calculateur est sollicité d'une part par les processus résidants dans le même calculateur (directement par le protocole d'application) et d'autre part par les protocoles de session des processus éloignés (via leur protocole de transport et de réseau). Le protocole de session gère l'établissement, le déroulement et la fin de toute communication. Il doit donc détenir l'état d'avancement de tout échange, en déduire le séquencement des signaux de synchronisation et procéder au transfert des données. A l'image de l'implémentation de CSP proposée par Chetto [2,3], Elloy [6] nous proposons pour ce faire d'utiliser des blocs de synchronisation et de contrôle (SCB) pour gérer l'échange.
La communication par ports orientés nécessite l'introduction de deux types de SCB : les SCBO (SCB Owner) attachés à un port, et les SCBU (SCB User) attachés à un processus. La déclaration d'un port dans son processus propriétaire provoquera la création d'un SCBO, la déclaration d'un nombre quelconque de ports dans un processus utilisateur provoquera la création d'un seul SCBU. Un SCBU est donc banalisé : tout processus emploiera le même SCBU pour toute communication au travers d'un port quelconque dont il est utilisateur. Par la suite, on désignera par SCBO(A) le SCB attaché au port A, et par SCBU(P) le SCB attaché au processus P.

5.1 Composition des SCBU et SCBO

Un SCBU contient les champs suivants :
(i) champ PROCESSUS : contient le nom du processus attaché au SCBU.
(ii) champ PORT : contient le nom du port concerné par l'échange en cours.
(iii) champ INFO : contient soit l'adresse et la longueur de la donnée à transmettre (commande de sortie), soit celles de la donnée à émettre (commande d'entrée).
Un SCBO contient les champs suivants :
(i) champs PORT et INFO : identiques à ceux du SCBU
(ii) champ PROPRIETAIRE : nom du processus propriétaire du port.
(iii) champ ATSORTIE : contient la liste des processus utilisateurs de ce port suspendus en attente du processus propriétaire après avoir effectué une commande de sortie à travers ce port.
(iv) champ ATENTREE : identique à ATSORTIE, mais pour une commande d'entrée. Cette liste est en

outre accompagnée des adresses et longueurs de la zone mémoire des processus utilisateurs destinée à recevoir le message émis par le processus propriétaire.
(v) champ ETAT : il contient la valeur REPOS hors échange et PRETEMET ou PRETREC si le propriétaire a effectué une commande de sortie ou d'entrée et attend la réponse d'un utilisateur.

5.2 Les différentes phases de l'échange

La communication entre le propriétaire d'un port et l'un de ses utilisateurs s'effectue en trois phases : phase de synchronisation, de communication, de déconnection. La figure 4 représente le chronogramme d'un tel échange.
La première phase permet de synchroniser les protocoles de session du processus propriétaire et de l'un de ses utilisateurs ; elle est effectuée de façon indépendante par ces divers protocoles. Quand un processus effectue une commande d'E/S nommant un port dont il est utilisateur, le protocole d'application en informe le protocole de session qui à son tour en informe le protocole de session du propriétaire. Le processus utilisateur est alors suspendu.

Le protocole de session du propriétaire peut être sollicité soit par son protocole d'application, soit par le protocole de session d'un utilisateur via ceux de niveaux inférieurs. Une demande de communication formulée par un processus utilisateur se traduit par la mise à jour de la liste ATSORTIE ou ATENTREE. Un appel par le protocole d'application entraîne l'examen du contenu de ces listes et établit éventuellement la communication. Si aucun utilisateur n'est en attente dans le sens demandé par le propriétaire alors le champ ETAT du SCBO prend la valeur PRETEMET ou PRETREC ; si l'échange est effectué, ce champ est réinitialisé à la valeur REPOS (sans attendre la fin effective de la communication).

Figure 4a

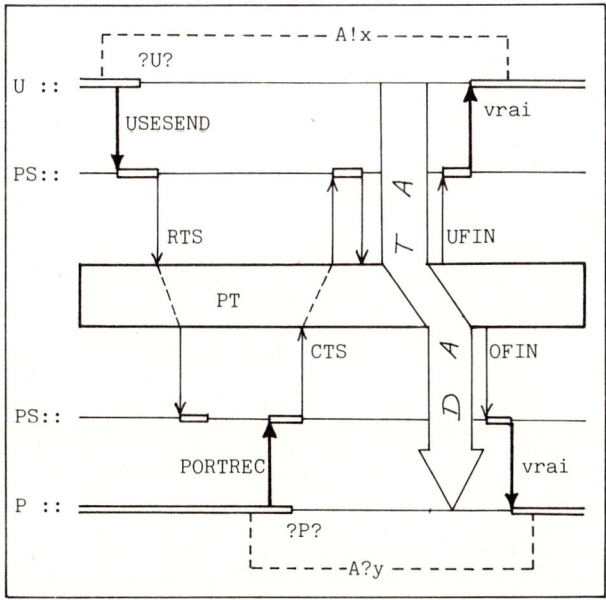

Cas d'une communication utilisateur vers propriétaire où l'utilisateur commence l'échange.

Figure 4b

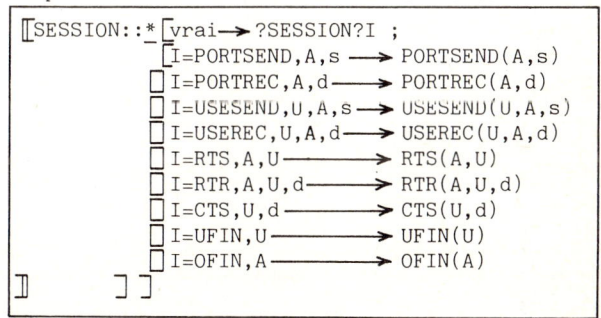

Cas d'une communication propriétaire vers utilisateur où le propriétaire commence l'échange.

La seconde phase (phase de communication) consiste en l'émission d'un signal par le protocole de session du propriétaire vers le protocole de transport du même calculateur. Ce signal est alors utilisé par ce protocole pour effectuer le transfert des données dans le sens requis.
La phase de déconnection est effectuée en parallèle par le protocole de session du propriétaire et celui de l'utilisateur. Le réveil des processus communicants est effectué directement et indépendamment par les protocoles de réseau de leurs calculateurs dès la fin de la transmission des données.

5.3 Procédures du protocole de session

Le processus session a l'allure suivante :

```
[SESSION::*[vrai→?SESSION?I ;
          [I=PORTSEND,A,s ──→ PORTSEND(A,s)
         □I=PORTREC,A,d ───→ PORTREC(A,d)
         □I=USESEND,U,A,s ─→ USESEND(U,A,s)
         □I=USEREC,U,A,d ──→ USEREC(U,A,d)
         □I=RTS,A,U ───────→ RTS(A,U)
         □I=RTR,A,U,d ─────→ RTR(A,U,d)
         □I=CTS,U,d ───────→ CTS(U,d)
         □I=UFIN,U ────────→ UFIN(U)
         □I=OFIN,A ────────→ OFIN(A)
         ]      ]]
```

Les procédures appelées par les fonctions PORT!, PORT?, USE!, USE? du protocole d'application sont les suivantes :

```
procédure PORTSEND(A,s)
  U : process_id ; A : port_id ; s,d :data ;
  début
    [ATENTREE(A)=vide → INFO(A):= s ;
                        ETAT(A) := PRETEMET
    □ATENTREE(A)≠vide → U,d:=dequeue(ATENTREE(A))
                        TRANSPORT!!UDATA,U,A,s,d
    ]
  fin
```

```
procédure PORTREC(A,d)
  U : process_id ; A : port_id ; d : data ;
  début
    [ATSORTIE(A)=vide →INFO(A) := d ;
                       ETAT(A):=PRETREC
    []ATSORTIE(A)≠vide→U:=dequeue(ATSORTIE(A));
                       TRANSPORT!!CTS,U,d
    ]
  fin
```

```
procédure USESEND(U,A,s)
  U : process_id ; A : port_id ; s : data ;
  début
    INFO(U):=s ; PORT(U):=A ;
                 TRANSPORT!!RTS,A,U
  fin
```

```
procédure USEREC(U,A,d)
  U : process_id ; A : port_id ; d : data ;
  début
    INFO(U):=d ; PORT(U):= A ;
                 TRANSPORT!!RTR,A,U
  fin
```

Les procédures activées par un signal du protocole de réseau sont les suivantes :

```
procédure RTS(A,U)
  U : process_id ; A : port_id ;
  début
    [ETAT(A)=PRETREC →ETAT(A):=REPOS ;
                      TRANSPORT!!CTS,U,INFO(A)
    []ETAT(A)=REPOS  ou  ETAT(A)=PRETEMET
                      → enqueue(ATSORTIE(A),U)
    ]
  fin
```

```
procédure RTR(A,U,d)
  U : process_id ; A : port_id ; d : data ;
  début
    [ETAT(A)=PRETEMET→ETAT(A):=REPOS ;
                      TRANSPORT!!UDATA,U,A,INFO(A),d
    []ETAT(A)=PRETREC ou ETAT(A)=REPOS
                      → enqueue(ATENTREE(A),(U,d))
    ]
  fin
```

```
procédure CTS(U,d)
  U : process_id ; d : data ;
  début
    TRANSPORT!!ODATA,PORT(U),U,INFO(U),d
  fin
```

```
procédure UFIN(U)
  U : process_id
  début
    U!!
  fin
```

```
procédure OFIN(A)
  A : port_id ;
  début
    PROPRIETAIRE(A) !!
  fin
```

6. IMPLEMENTATION DU PROTOCOLE DE TRANSPORT

Tout signal reçu par le protocole de transport contient les informations suivantes : type de l'information transmise, nom du processus ou du port destinataire du signal, suivi éventuellement de plusieurs arguments fonctions du type de l'information transmise. Ces signaux sont traités dans un même processus appelé TRANSPORT. Ce processus, grâce aux tables CAL (Correspondance processus/calculateur) et PROP (Correspondance port/propriétaire) obtient l'identité W du calculateur destinataire du signal. Si ce calculateur n'est pas celui du protocole de transport, le signal, auquel l'identité du calculateur destinataire est ajoutée, est transmis au protocole de réseau.
Le protocole de transport aura donc l'allure suivante :

```
[TRANSPORT::*[vrai→?TRANSPORT?TYPE,X,Y,Z,T
             [TYPE=UDATA ou CTS→W:=CAL(X)
             []TYPE=RTS ou RTR
                        ou ODATA→W:=CAL(PROP(X))
             ]
  commentaire : CALC est l'identité
    du calculateur émetteur du signal;
  [CALC≠W → RESEAU-E!!W,TYPE,X,Y,Z,T
  []CALC=W ; TYPE≠ODATA et TYPE≠UDATA
             → SESSION!!TYPE,X,Y,Z,T
  []CALC=W ; TYPE=ODATA → recopie des
                         données de Z vers T ;
                         SESSION!!OFIN,X ;
                         SESSION!!UFIN,Y
  []CALC=W : TYPE=UDATA → recopie des
                         données de Z vers T ;
                         SESSION!!UFIN,X ;
                         SESSION!!OFIN,Y
  ]] ]]
```

7. IMPLEMENTATION DU PROTOCOLE DE RESEAU

Dans chaque calculateur, le protocole de réseau est chargé de l'émission et de la réception des messages en gérant leurs pertes, leurs parasitages et leurs duplications. L'étude complète de ce protocole dépasse le but de cet article ; aussi le décrivons nous succinctement. Il est constitué de deux processus évoluant en parallèle RESEAU-E et RESEAU-R chargés respectivement de l'émission et de la réception des messages. L'allure de ces processus est donc la suivante :

```
[RESEAU-E::*[vrai→?RESEAU-E?W,TYPE,X,Y,Z,T ;
            [TYPE=RTS ou RTR ou CTS → émission
                        vers W du message(TYPE,X,Y,Z,T)
            []TYPE=ODATA → émission vers W du
                           message
                           (ODATA,Y,T,données
                           spécifiées par Z) ;
                           SESSION!!UFIN,X
            []TYPE=UDATA → émission vers W du
                           message
                           (UDATA,Y,T,données
                           spécifiées par Z) ;
                           SESSION!!OFIN,X
            ]]]
```

```
[RESEAU-R::*[vrai→attente d'un message de la
                  forme (TYPE,X,Y,Z,T) en
                  provenance du réseau ;
            [TYPE=RTS ou RTR ou CTS →
                       SESSION!!TYPE,X,Y,Z,T
            []TYPE=ODATA →chargement des données
                          Z dans la zone mémoire
                          Y ; SESSION!!OFIN,X
            []TYPE=UDATA →chargement des données
                          Z dans la zone mémoire
                          Y ; SESSION!!UFIN,X
            ]]]
```

8. CONCLUSION

L'efficacité et la puissance du langage CSP et de son extension par les ports orientés proviennent du fait qu'il a été conçu comme un langage de haut niveau sans considération sur son implémentation. Nous avons proposé dans cet article cette implémentation pour un réseau local de calculateurs faiblement couplés ne se partageant aucune mémoire commune et sans mémorisation des messages sur les voies de communication. Cette implémentation nécessite la présence dans chaque calculateur d'un exécutif élémentaire possédant quelques primitives de synchronisation centralisées ainsi qu'un moyen de transmission des messages dans le réseau. La décomposition du mécanisme d'échanges en différents niveaux hiérarchisés (protocoles d'application, de session, de transport, de réseau) a permis une écriture modulaire et simple des procédures. Le protocole de session a pour mission d'orchestrer le déroulement de toute communication à travers les ports orientés, c'est à dire d'effectuer la synchronisation des processus communicants et la transmission des données. Il a été conçu de façon à minimiser le nombre de messages entre calculateurs, tant pour les commandes d'E/S par ports que pour l'implémentation ultérieure des commandes alternatives et répétitives. La localisation des processus communicants dans le réseau demeure transparente jusqu'au protocole de transport de chaque calculateur qui seul a accès aux tables de correspondance processus/port/calculateur. La topologie du réseau n'intervient qu'au niveau du protocole de réseau.

L'implémentation proposée est actuellement en cours de réalisation dans un réseau local en boucle composé de trois calculateurs destinés à la commande de procédés industriels complexes. La version en cours de réalisation tient compte des cas de destruction dynamique de processus, cas non commentés dans cet article. Elle devrait conduire à une étude comparative des performances de l'implémentation parallèle décrite dans cet article avec l'implémentation série en cours de conception.

9. BIBLIOGRAPHIE

[1] Bernstein (A.J), Output Guards and Nondeterminism in Communicating Sequential Processes ACM trans. on Prog. Lang. and Systems (TOPLAS), Vol.2, n°2, Avril 1980, pp 234-238

[2] Chetto (H), Mécanismes de synchronisation dans les systèmes centralisés et distribués, Thèse 3ème cycle, ENSM, Nantes, Laboratoire d'Automatique, Oct. 1981

[3] Chetto (H), Elloy (J.P), Chaudouard (C) Implementation of CSP in Distributed Systems ENSM, Nantes, Laboratoire d'Automatique, Rapport interne n°04.81, Juillet 1981

[4] Coffman (E.G), Elphick (M.J), Shoshani (A), System deadlocks, Computing Survey, Vol. 3, n°2, Juin 1971, pp 67-68

[5] Dijkstra (E.W), Guarded commands, Nondeterminacy and Formal Derivation of Programs, Comm. Ass. Comput. Mach. Vol.18, n°8, Aôut 1975, pp 453-457

[6] Elloy (J.P), Molinaro (P), Implantation de CSP dans un réseau local par protocole multiniveaux, Mini and Microcomputer Applications, MIMI'82, Davos, Mars 1982

[7] Hoare (C.A.R), Communicating Sequential Processes, Comm. Ass. Comput. Mach. Vol.21, n°8, Aôut 1978, pp 666-677

[8] Jazayeri (M), Ghezzi (C), Hoffman (D), Middleton (D), Smotherman (M), Design and Implementation of a Language for Communicating Sequential Processes, Proc. Intern. Conf. on Parallel Processing, Août 1980, pp 173-180

[9] Kieburtz (R.B), Silberschatz (A), Comments on "Communicating Sequential Processes", ACM Trans. on Prog. Lang. and Systems (TOPLAS), Vol.1, n°2, Oct. 1979, pp218-225

[10] Shrira (L), Francez (N), An Experimental Implementation of CSP, Proc. 2nd Intern. Conf. on Distributed Computing Systems, Paris, Avril 1981, pp 126-136

[11] Silberschatz (A), Communication and Synchronization in Distributed Systems, IEEE Trans. Softw. Engin., Vol.5, n°6, Nov. 1979, pp 542-547

[12] Silberschatz (A), A survey Note on Programming Languages for distributed Computing COMP CON 80, Proc. distrib. com., Washington DC, Sept. 1980, pp 719-722

[13] Silberschatz (A), Port directed communication The Computer Journal, Vol.24, n°1, 1981, pp 78-82.

RESULTS OF A TWO-YEAR STUDY OF A SOFTWARE PRODUCTION METHOD

Y. Corcia, J.L. Dauphin, Y. Simon

Centre National d'Etudes des Telecommunications, Lannion, Belgium

ABSTRACT

After explaining the motives behind using one method and the project in which this method has been employed, we will describe the principles and concepts involved.

In the second part the rules will be described which allow these concepts to be realized in the actual project and the aids which enable the rules to be followed more easily.

In the third part it will be attempted to draw the main lessons to be learned from this study by discussing not only the effectiveness of the organisation but also what experience the users have made with the method. To that end the results of an anonymous inquiry will be discussed which was reduced and analysed by a team outside the project itself and composed of a psychologist and a software production specialist.

Finally, our thoughts on important factors to be considered when wishing to select and introduce a software production method will be presented.

RÉSUMÉ

Apres avoir expose les motivations pour l'utilisation d'une methode et le projet dans le cadre duquel cette methode a ete employee, nous nous attacherons a decrire les principes retenus; les concepts.

Dans une seconde partie nous decrirons les regles qui permettent de faire que ces concepts se retrouvent dans la realite du projet, puis les outils qui guident et facilitent le respect des regles.

Une troisieme partie tentera de tirer les principaux enseignements de cette experience en s'attachant non seulement a l'efficacite de l'organisation mais aussi a la facon dont la methode a ete ressentie par les utilisateurs. A cette fin on discutera les resultats d'une enquete anonyme depouillee et analysee par une equipe externe au projet lui meme et composee d'un psychologue et d'un specialiste des problemes de production du logiciel.

En conclusion nous exposerons ce qui nous semble important a prendre en compte lorsque on desire choisir et introduire une methode de production du logiciel.

ZUSAMMENFASSUNG

Nach der Darstellung der Motive fuer die Anwendung eines methodischen Vorgehens sowie einer Beschreibung des Projektes in dessen Rahmen die Methode benutzt wird, werden die angerwendeten Prinzipien ("Konzepte") beschrieben.

Im zweiten Teil beschreiben wir die Regeln, die es ermoglichen die Konzepte in das reale Projekt zu eubertragen, und die Werkzeuge, die die Einhaltung dieser Regeln bewerkstelligen.

Im dritten Teil werden die wesentlichen Folgerungen aus der Erprobung gezogen, wobei nicht nur die Effiziene der Organisation, sondern auch die Akzeptanz der Methode fuer die Anwender betrachtet wird. Zu diesem Zweck werden die Resultate einer anonymen Untersuchung vorgestellt, die von einer nicht dem Projekt angehoerenden Gruppe erstellt wurde, und die aus einem Psychologen und einen Spezialisten aufdem Gebiet des Programmerstellung bestand.

Am Schluß werden wir darstellen, welche Punkte nach unserer Meinung in Betracht gezogen werden sollten, wenn man eine Methode zur Programmerstellung auswaehlen und anwenden will.

BILAN DE DEUX ANS D'EXPERIMENTATION
D'UNE METHODE DE PRODUCTION DU LOGICIEL

CORCIA yvon, DAUPHIN jean louis, SIMON yves

Centre National d'Etudes des Telecommunications
BP 40 - 22301 LANNION Tel (96) 38 31 76 ou 38 11 11°s12

Apres avoir expose les motivations pour l'utilisation d'une methode et le projet dans le cadre duquel cette methode a ete employee, nous nous attacherons a decrire les principes retenus; les concepts.

Dans une seconde partie nous decrirons les regles qui permettent de faire que ces concepts se retrouvent dans la realite du projet, puis les outils qui guident et facilitent le respect des regles.

Une troisieme partie tentera de tirer les principaux enseignements de cette experience en s'attachant non seulement a l'efficacite de l'organisation mais aussi a la facon dont la methode a ete ressentie par les utilisateurs. A cette fin on discutera les resultats d'une enquete anonyme depouillee et analysee par une equipe externe au projet lui meme et composee d'un psychologue et d'un specialiste des problemes de production du logiciel.

En conclusion nous exposerons ce qui nous semble important a prendre en compte lorsque on desire choisir et introduire une methode de production du logiciel.

1 -LE PROJET PALME

Le Projet d'Autocommutateur Local Multiservice Experimental (PALME) est un système de commutation développé au Centre National d'Etudes des Télécommunications à LANNION dans le cadre des études sur les Réseaux Numériques à Intégration de Services (RNIS). Ce projet prend en compte les services de téléphonie et de données et assure l'établissement des communications locales ou avec des abonnés raccordés sur les réseaux téléphonique, télex et Transpac. Il inclut l'ensemble de la chaine de transmission et de commutation: raccordement des terminaux sur les régies d'abonnés: réalisation du commutateur et des adaptations aux réseaux extérieurs.

Le volume de logiciel à écrire dans un langage de haut niveau d'abstraction s'élève à environ 120000 instructions. Le travail a été réalisé par une équipe de 35 personnes sur 2 ans selon le concept d'une démarche descendante et structurée englobant le suivi du projet et la méthode de production du logiciel.

Le projet a été conduit de manière méthodique et structurée, l'enchainement des différentes phases de son développement est le suivant:

- Définition du projet
- Spécifications externes
- Spécifications internes
- Production du logiciel
- Mise au point et recette.

Dans la phase de production du logiciel, la méthode appliquée a été définie par l'équipe projet en collaboration avec le département "Génie Logiciel du CNET" et la société GAP-SO-GETI. Le choix et la définition des concepts des règles et des moyens retenus dans cette méthode répondent principalement aux motivatons suivantes:

- Nécessité de planifier les délais et les moyens
- Nécessité d'organiser la coordination entre les différentes équipes de production de logiciel.
- Nécessité d'aboutir à un produit fiable documenté et maintenable, dont la gestion des différentes versions de logiciels soit aisée.

2 - LES CONCEPTS ET LES OBJECTIFS DE LA METHODE DE PRODUCTION DE LOGICIEL

2-1. DEMARCHE DESCENDANTE

Nous sommes particulièrement peu originaux en nous fixant comme objectif une démarche descendante uniforme depuis les premières spécifications jusqu'au codage et au test.

Le problème consiste à être assez discipliné et rigoureux pour s'assurer qu'elle est collectivement bien suivie. Au-delà de la prise de conscience, les habitudes et les règles de documentation sont le plus sûr moyen de l'observer.

On peut tenter, une fois de plus, de définir une démarche descendante; cela consiste à:

- diviser un problème complexe en quelques problèmes plus simples dont on ne connaît pas encore la solution mais qu'on estime solubles (s'ils ne sont pas facilement solubles, il y aura des retours en arrière!!!)
- pour chaque problème, se contenter, dans un premier temps, de bien <u>le poser</u>, aussi complètement que possible - spécifier - sans se soucier de la façon dont il sera résolu.
- pour chaque problème, analyser la solution qui paraît la plus adaptée en fonction des moyens de résolution dont on dispose et sans à priori sur le mode de résolution.

On voit bien qu'il s'agit d'une éthique personnelle, presque d'une philosophie (et les maîtres à penser sont nombreux).

2-2. STRUCTURATION EN PHASES CLAIRES

La structuration en phases permet d'atteindre quelques objectifs majeurs:
- s'assurer que la démarche descendante est bien suivie et que la documentation relative à chaque phase est bien approuvée avant de passer à la phase suivante.
- avoir une vision claire de l'état d'avancement du projet et permettre d'induire une politique de développement.
- par l'observation de la durée des phases, se faire une idée de plus en plus précise de la durée totale du projet (à partir de données statistiques sur les durées relatives des phases).

Les phases que nous avons retenues sont:

- analyse générale
- analyse détaillée
- codage et test
- intégration

Au déroulement de chacune des phases est associée la production de la documentation propre à la phase.

La visibilité des phases est liée à la visibilité de la documentation produite.

2-3. FIABILITE DES PROGRAMMES

Le but final du projet est bien d'obtenir des programmes fiables, c'est à dire correspondant à leurs spécifications, faciles à maintenir et prêts à recevoir des extensions.

Un moyen très puissant de tendre vers ce but est de produire des programmes LISIBLES:

a) afin de s'assurer qu'ils sont bien le reflet exact de l'analyse qui a été faite et non quelque chose d'approchant.

b) lors de test unitaire et de l'intégration on découvre des erreurs de programme, il est alors nécessaire de détecter s'il s'agit d'une erreur de codage, de mise en oeuvre, d'analyse ou de spécification.

2-4. CENTRALISATION DES REFERENCES

Par référence, on entend aussi bien la documentation que les objets communs ou encore certains utilitaires appelés lors des compilations ou éditions de liens.

<u>La documentation</u>

Sa centralisation permet d'avoir une référence unique à tout instant et d'être en permanence accessible et à jour.

Nous avons fait le choix de saisir <u>à la composition</u> toute la documentation.

<u>Les autres références</u>

Elles seront centralisées sur un site accessible au moment des compilations et des éditions de liens.

Ce sont les:

- constantes
- modules bibliothèques
- modules communs
- utilitaires.

3 - LES REGLES : DES MOYENS

3-1. LA DOCUMENTATION

Nous avons déterminé une nomenclature de documents, rattachée à chaque phase projet:

- chaque document est composé de rubriques normalisées propres à ce type de document.

3-1-1. NOMMAGE DES DOCUMENTS

Chaque document a UN NOM qui renseigne sur:

- le sous-ensemble auquel il appartient
- le type de document
- la phase de production à laquelle il a été créé
- la hiérarchie fonctionnelle dont il est issu.

    S - MNIL - GT : S
        est la spécification de la fonction "gestion de temporisation"

du module "NIVEAU LOCAL" (NIL)

du sous ensemble " SYSTEME" du projet.

Les documents ont une syntaxe interne orientee vers le traitement specifiques du type du document.

3-2. STRUCTURATION EN PHASES

La phase d'analyse générale constitue une étape entre la conception et la production. Elle a pour but de déterminer l'architecture des sous ensembles et de recenser les grands modules qui la composent elle est conclue par la production:

- du document d'architecture du sous ensemble.
- des documents de modules recensés
- des documents de données associées.

La phase d'analyse détaillée consiste à affiner les modules en fonctions.

Les documents produits sont alors des documents de spécifications de fonction, d'analyse de fonction ou de données.

La phase de codage conduit à la constitution des documents de code qui sont obtenus par adjonction de lignes de code dans le document d'analyse.

La phase de test consiste à opposer les procédures opérationnelles aux programmes de test.

En principe les programmes de test devraient être spécifiés et analysés parallèlement aux programmes opérationnels, dans la pratique, ces programmes sont spécifiés et réalisés dans la phase de test unitaire.

Chaque phase s'achève par la relecture de tous les documents produits dans la phase.

3-3. REGLES DE PROGRAMMATION

Les règles de programmation ont été établies afin de faciliter la relecture des programmes et de guider le programmeur vers une déontologie que certains appelleraient "structurée".

3-3-1. REGLES DE NOMMAGE

Nous travaillons en PLM 86 ou ASM86; les identificateurs peuvent être très longs. (255 car)

Nommage des Variables

Un identificateur de variable est de la forme:

    <XY>$< nom sémantique >

- la première lettre X indique le type:
Pointeur, Mot, octet, entier, réel, structure.
- la deuxième lettre Y est l'attribut de gestion
    E pour exporter (au sens module de compilation)
    A pour fixer à une adresse absolue
    B pour basé
    S pour simple.

EX: SB$CTXVOIE est une structure basée.

3-3-2. NOMMAGE DES PROCEDURES

Pour chaque procédure, ces règles permettent de savoir s'il s'agit d'une procédure typée, son type et son exportation éventuelle.

3-3-3. REGLES DE MISE EN PAGE DES STRUCTURES DE CONTROLES

Ce sont les règles classiques d'indentation des blocs.

3-3-4 PARAMETRISATION DES CONSTANTES

On ne doit pas trouver dans le code (sauf cas exceptionnel) de constantes numériques. Les constantes seront déclarées par valeurs littérales (constantes de compilation); si elles sont partagées elle sont obtenues par inclusion de fichiers de référence.

Dans ce cadre on a considéré comme constantes les structures de données partagées et les messages d'interface.

3-3-5. TRAITEMENT DES ERREURS

Les anomalies détectées à l'exécution font appel à des procédures spécifiques de traitement d'erreurs (1 par sous ensemble, 7 dans tous le projet).

3-3-6. LES REGLES DE CODAGE

Consistent à écrire des algorithmes extensibles qui n'escamotent pas les boucles (temporairement) non significatives. On veille de plus à avoir des programmes "ROBUSTES" c'est à dire toujours méfiants vis à vis des données d'entrée et piègant toutes les assertions possibles sur les variables. Enfin il faut veiller à ce que la taille des procédures reste raisonnable: de 20 à 100 lignes de code.

4-LES OUTILS DE LA METHODE

Les outils regroupent l'ensemble des moyens, de natures très diverses, qui sont mis en oeuvre d'une façon spécifique dans le cadre de la méthode. Ils ont pour effet d'aider l'équipe chargée du développement du logiciel à respecter, dans l'esprit et dans la forme le concept retenu ainsi que les principes et les règles préconisés.

4-1. ATELIER DE PRODUCTION DE LOGICIEL

Les moyens matériels sont intégrés dans un atelier de production de logiciel.

4-1-1. DESCRIPTION DE L'ATELIER

Réparti entre les bureaux et les laboratoires occupés par les membres de l'équipe, cet atelier se présente sous forme d'un réseau de stations reliées à un centre de calcul banalisé constitué d'un IRIS 80 bi-processeur.

Ces stations sont:

- soit des consoles simples équipées d'écran-clavier ou d'écran-imprimante-clavier;
- soit des systèmes de développement, micro-centres de calcul à poste unique, adaptés à la technologie retenue pour la réalisation matérielle de la maquette.

Ces stations sont connectées au temps partagé du centre de calcul principal à travers un commutateur local utilisant une technique de transmission de paquets suivant le protocole standard X25. L'adoption d'un moyen souple d'interconnexion permet d'envisager une éventuelle reconfiguration architecturale très aisée du réseau.

4-1-2 ARTICULATION DES MOYENS

L'utilisation des équipements de l'atelier de production du logiciel est représentée sur la figure 4-1-2.

Les stations SD (système de développement) possèdent des missions spécifiques:

- les stations SD1 et SD2 sont affectées à l'intégration, sur la maquette, des logiciels préalablement validés unitairement sur SD3 et SD4.
- SD3 est particulièrement destinée, par l'utilisation d'une pré-maquette simplifiée, à l'essai et la mise au point des programmes de la couche-système.
- la réalisation des programmes se fait sur SD5 et SD6: compilation, édition de liens, affectation des adresses, gestion temporaire des sources et des objets binaires durant la phase de mise au point.

Les stations-console servent à la saisie des documents de spécifications, d'analyse et de codage sur le centre de calcul principal.

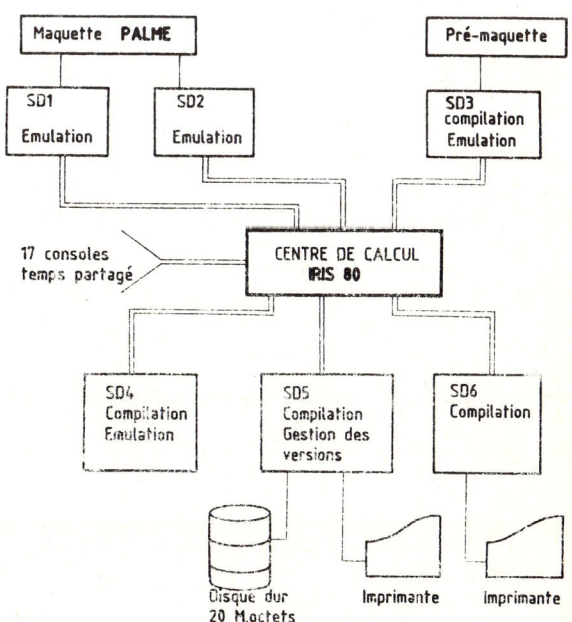

Figure 4.1.2. Organisation de l'atelier de production du logiciel.
(SD: Système de Développement)

4-2 ORGANISATION DE L'EQUIPE

Placé sous la responsabilité d'un chef de projet, secondé d'un expert "matériel" et d'un expert "logiciel", l'ensemble du personnel se divise en groupes de travail, chargés respectivement:

- de la réalisation d'un des sous-ensembles d'application
- de la réalisation du logiciel de la couche-système.
- de l'intégration et de l'exploitation de la maquette.

La coordination s'effectue:

- au sein des groupes, par les réunions d'avancement
- entre les groupes, par les réunions périodiques d'une équipe-projet constituée des animateurs de chaque groupe de travail, du chef de projet et des responsables "matériel" et "logiciel".

Les assemblées de l'ensemble du personnel sont organisées au besoin. Leur but est: soit d'informer, soit d'examiner un problème spécifique d'ordre technique ou organisationnel.

4-3 GUIDES DE LA METHODE

Outre les documents d'information technique fournis par les industriels ou relatifs aux moyens informatiques standard, le personnel dispose de deux guides spécifiques:

4-3-1. LE MANUEL D'UTILISATION DE LA METHODE

Ce document rappelle les principes de la méthode. Il précise les règles retenues:

- organisation et démarche,
- moyens disponibles: équipement, langages,
- normalisation: documentation, programmation,
- gestion de la documentation,
- manuels d'utilisation des outils développés dans le cadre de la méthode.

4-3-2 LE MANUEL DU PROGRAMMEUR D'APPLICATION

Ce document informatisé a la particularité d'être l'objet d'une mise à jour permanente. Il décrit:

- l'ensemble des pseudo-codes permettant d'activer les primitives offertes par la couche-système,
- la bibliothèque des utilitaires disponibles,
- l'organisation, les règles d'interprétation, les déclarations des objets communs, en particulier des données partagées.

4-4. PRODUCTION DE LA DOCUMENTATION

La documentation étant considérée comme le support essentiel de la démarche, un effort particulier a été consenti pour la réalisation des outils relatifs à son élaboration et à sa mise à jour.

Les facilités de production d'un document sont offertes par un outil logiciel: le générateur de pavé (GENPAV) qui permet suivant le cas:

- l'élaboration d'un document
- sa modification
- son édition

4-4-1. CREATION D'UN DOCUMENT

Du fait de la normalisation de nommage, le générateur de pavé guide, d'une façon conversationnelle, l'établissement précis du document. Il oblige une rédaction standardisée, ordonnée, cohérente et exhaustive,

4-4-2. MODIFICATION D'UN DOCUMENT

Ce même outil facilite les modifications des documents. Il garantit la mise à jour du pavé de gestion: état et date, facilite l'insertion en bonne place d'une rubrique donnée et permet la destruction de paragraphes standards complets.

4-4-3. EDITION DE DOCUMENT

Une série d'outils permet l'édition sur divers sites (console de station, imprimante du centre de calcul à chaîne ou à tambour) des documents normalisés, suivant un nombre d'exemplaires choisi.

Il est possible d'obtenir l'édition d'un seul document ou celle de l'ensemble des documents répertoriés dans un document "directives".

4-5. GESTION DE LA DOCUMENTATION

L'objectif de la gestion est:

- d'assurer la maintenance aisée et la validation fondée sur la relecture.
- de permettre l'accès de tous à l'ensemble des documents soit en lecture, soit en modification tout en garantissant la cohérence du document manipulé.

Le principe de la gestion consiste à placer les documents dans un site centralisé et de les étiqueter à l'aide d'états de gestion: écrit, en relecture, approuvé.

Un document écrit est confié par l'auteur en relecture à un autre membre de l'équipe. Le document passe dans l'état "approuvé" lorsqu'un accord est intervenu entre auteur et relecteur. Chacun est suivant le cas auteur ou relecteur.

Le centre de calcul, du fait de la qualité et de la permanence de son exploitation, offre une bonne fiabilité aux fichiers qu'il gère. Un catalogue de travail est affecté à chaque groupe, les membres du groupe disposant d'une clé d'accès à ce catalogue. Les travaux d'élaboration, de modification et d'édition s'y effectuent.

Tout document établi n'existe qu'en un exemplaire de référence, placé sur un catalogue-bibliothèque central. Chaque sous ensemble dispose sur le catalogue central d'un dictionnaire qui contient la liste des documents relatifs à ce sous ensemble et l'état daté de chacun d'eux.

4-5-1. GESTION MANUELLE

Le catalogue central n'est accessible en écriture que par une personne responsable, l'assistante de programmation.

Toute action sur le catalogue central, autre qu'édition, est signalée à l'assistante de programmation qui effectue les transferts et les mises à jour des dictionnaires. Les requêtes lui sont transmises par messagerie, l'acquittement étant intégré à ce service.

4-5-2. GESTION AUTOMATIQUE

Pour des raisons multiples (impossibilité pour l'assistante de programmation d'effacer, sur le catalogue privé d'origine, la copie du document transféré, temps de réponse à la requête, non respect de certaines règles par l'ensemble des utilisateurs) la procédure manuelle ne garantit pas totalement l'exclusion mutuelle.

Une procédure automatique est adoptée. Elle offre trois primitives d'action directe qui assurent, par leurs contrôles, la cohérence des demandes et effectuent la mise à jour des dictionnaires. Les commandes sont: "lecture", "modification" et "envoi".

4-6. LANGAGE D'ANALYSE

Outre le langage de haut niveau adopté pour la programmation, la méthode retient l'utilisation d'un langage d'analyse.

Nous avons, dans un premier temps experimenté PDL qui a été ressenti à la fois trop rigide dans sa presentation et pas assez precis dans les controles effectués. Nous avons donc retenu un langage plus libre avec des structures de controle.

Ce langage offre les avantages suivants:

- il favorise le respect de la démarche descendante
- il propose une organisation algorithmique du raisonnement qui s'approche des structures de contrôle du langage de programmation
- il permet une présentation normalisée et ordonnée du document

Le texte d'analyse sert ensuite de commentaire au programme. Le codage s'effectue par insertion de blocs d'instructions entre les paragraphes d'analyse.

5. EVALUATION DE LA METHODE

Après deux années d'utilisation, une enquête anonyme a été effectuée auprès des utilisateurs de la méthode. Dépouillée par un spécialiste de la méthodologie et un psychologue indépendants de l'équipe projet, elle a permis d'obtenir une appréciation globale sur tout ce qui touche à son emploi: outils, matériels et logiciels, documentation, assimilation de l'intérêt d'une méthode, adhésion à la méthode proposée.

5-1. EVALUATION DES OUTILS MATERIELS ET LOGICIELS

C'est en quelque sorte une gageure que de vouloir créer et expérimenter une méthode de production de logiciel dans le cadre même d'un projet de l'ampleur de PALME.

L'enquête révèle pourtant que l'expérience est positive. Les outils ont été utilisés dans de bonnes conditions bien que certaines limitations ponctuelles aient été mises en lumière.

Le réseau d'interconnexion utilisé dans l'atelier de production s'est avéré sous dimensionné en débit, interdisant le respect pendant la phase de compilation et de mise au point des règles de gestion centralisée des références. Certains outils logiciels spécifiques ont dû être améliorés, leur efficacité relative et l'absence d'une disponibilité permanente les rendant d'un usage inconfortable.

Les systèmes de développement n'offrent pas le potentiel d'accès -compilation suffisant. L'utilisation de stations multi-utilisateur aurait mieux répondu aux besoins de l'équipe.

5-2. EVALUATION GLOBALE DE LA METHODE

Pour le programmeur qui n'a jamais vécu d'expérience d'utilisation d'une méthode de production de logiciel, les contraintes d'une telle méthode paraissent à priori lourdes devant des avantages peu palpables.
L'adhésion à l'esprit de la méthode, nécessaire à son utilisation efficace, implique pour l'analyste et le programmeur, une mutation psychologique qui n'intervient qu'après la mise en évidence d'échecs dus à une mauvaise utilisation de la méthode ou à un refus de l'utiliser.

L'enquête réalisée dans le cadre du projet PALME a fait apparaître divers obstacles à cette mutation: obstacles généralement liés à l'élaboration de la documentation. La documentation informatisée représente en effet un changement des habitudes de travail: le passage du travail avec papier et crayon à celui sur ordinateur exige une excellente disponibilité des moyens informatiques pour être accepté.

Le temps relativement long passé à effectuer la saisie conduit à considérer le travail de documentation comme un but en soi en faisant perdre de vue qu'il n'est que le moyen et le guide d'une démarche aboutissant finalement au programme. En tout état de cause si certains outils de traitement de la documentation avaient été plus confortables, l'adhésion de l'ensemble du personnel de l'équipe à la méthode aurait pu être plus rapide. Si l'aspect planification na pas ete pris en compte, du moins la documentation totalement rédigée dans le cadre décrit ici a-t-elle permis de produire un logiciel fiable, d'assurer dans les meilleures conditions l'échange d'information dans les équipes et l'articulation technique de leurs travaux.

En conclusion, si l'idéal dans l'introduction d'une méthode de production de logiciel est d'obtenir sur cette méthode le concensus et la conviction de tous ses utilisateurs, il nous est apparu que l'élément fondamental pour obtenir ce concensus est de proposer des outils confortables et efficaces. C'est pourquoi, le développement d'atelier logiciel avec postes de programmeurs nous parait être un axe de recherche très important.

Une conclusion importante est que l'approche démarche descendante bien adaptee aux logiciels d'application pose des problemes pour les logiciels de base dans lesquels on doit descendre vers des interfaces qui sont fixes a priori par le materiel.
Il nous a semblé qu'il manque une methode d'analyse qui soit adaptee au temps reel et qui permette de prendre en compte le parallelisme des processus et les occurences d'evenements.

BIBLIOGRAPHIE

°BOEHM 79§
  B W BOEHM
  Software ingineering as it is
  4th INT. CONF. on software engineering, munich sept 79.

°DIJKSTRA 68§
  E W DIJKSTRA
  "GOTO statement considered harmful"
  com ACM, Vol 11, p147-148, mars 1968.

°LOUVET 81§
  O LOUVET, JL DAUPHIN, D HARDY, G PAYS
  PALME a switching system integrating voice and data
  International Switching Symposium -1981- MONTREAL.

°MEYER 78§
  B MEYER, C BAUDOIN
  METHODES DE PROGRAMATION
  Edition EYROLLES PARIS 1978

°MOREAU 81§
  B MOREAU, B ROUGEOT, D SALEMBIER
  GALAAD un systeme de gestion de logiciel
  L'ECHO DES RECHERECHES - oct 81
  Centre National d'Etudes des telecommunications
  92131 ISSY (editeur)

°KRAKOWIAK 80§
  S. KRAKOWIAK
  SYSTEMES INTEGRES DE PRODUCTION DU LOGICIEL
  Journees d'etudes du SESORI, Genie Logiciel-PERROS GUIRREC-janvier 80

EINE ERWEITERUNG VON SPECIAL ZUR BESCHREIBUNG
ASYNCHRONER AKTIVITAETEN

Christian Koehler
Industrieanlagen-Betriebsgesellschaft m.b.H.
Einsteinstrasse, 8012 Ottobrunn, Deutschland

Fuer die Spezifikationssprache SPECIAL, die Teil der Software- Entwicklungsmethode HDM ist, werden drei eng verbundene Aenderungsvorschlaege gemacht: Einfuehrung spontaner O-Funktionen, explizite Darstellung der Ausnahmebedingungen durch V-Funktionen als Teil des Systemzustandes und Einfuehrung globaler Ausnahmebedingungen. Dadurch wird es moeglich, innerhalb einer abstrakten Maschine parallele Ablaeufe zu beschreiben. Das Beheben der Ausnahmebedingungen kann als Teil der Maschine spezifiziert werden, in der die Ausnahmebedingungen auftreten.Und als Folge davon wird es moeglich, Reihenfolgen von O-Funktionen in SPECIAL zu beschreiben. Die Einfuehrung von globalen Ausnahmebedingungen ermoeglicht es, auf einfache Weise die O-Funktionen eines Moduls auf bestimmte Systemzustaende zu beschraenken. Durch die Aenderungen kann auf ILPL innerhalb von HDM verzichtet werden.

Three closely related changes to the specification language SPECIAL, which is part of the software design methodology HDM, are suggested: spontaneous O-functions are added, exceptions are represented explicitly by V-functions as part of the state of the system, and global exceptions are introduced. By these changes it becomes possible to specify concurrency within a single abstract machine. Recovery from exceptions may be specified as part of the abstract machine, where the exception happens. In consequence, it is possible to have O-functions composed of sequences of O-functions. Global exceptions are a simple method to restrict the O-functions of a module to those states of the system to which they are applicable. Given the changes ILPL is no longer needed within HDM.

Trois transformations associées entre elles sont proposées pour le langage de spécification SPECIAL, qui est une partie de la méthode HDM pour le développement du Software. Ce sont: Introduction des O-fonctions spontanées, la description explicite des conditions d'exceptions par les V-fonctions, comme partie de l'état du système, et l'introduction des conditions d'exceptions globales. Ainsi il sera possible, à l'intérieur d'une machine abstraite de définir des traitements parallèles. La correction des exceptions peut être spécifiée comme une partie de la machine, dans laquelle se présentent les exceptions. Comme conséquence de cela, il sera possible d'avoir les O-fonctions dans l'ordre en SPECIAL. L'introduction des exceptions globales facilite la réduction des O-fonctions d'un module aux différents états du système où elles sont appliquées. Grâce à ces changements, on pourra supprimer ILPL à l'intérieur de HDM.

Die vorliegende Arbeit entstand unter dem Auftrag T/R626/A0019/63615 des BMVg, Rue VI 6.

INHALT

1. SPECIAL Bestandsaufnahme
1.1 Das Automatenmodell von SPECIAL
1.2 Die typische Aufgabe der Automaten
1.3 Schwaechen aufgrund des Modells

2. Erweiterungsmoeglichkeiten
2.1 Spontane Transitionen
2.2 Einbindung der Ausnahmebedingungen in das Modell

3. Ausnuetzung der Erweiterungen auf die Beschreibungsmoeglichkeiten
3.1 Nebenlaeufige Automaten
3.2 Behandlung von Ausnahmesituationen
3.3 Beschreibung von Reihenfolgen

4. Zusammenfassende Bewertung und Ausblick

Zusammenfassung

Gegenwaertig stellt SPECIAL zusammen mit HDM (Hierarchical Development Methodology) [9] die fuer industrielle Softwareproduktion wohl am haeufigsten eingesetzte Methode zur formalen Spezifikation von Softwaresystemen dar (z.B. [5], [6]). Ungeachtet dessen hat SPECIAL jedoch einige Schwaechen, die dazu fuehren, dass wichtige Bereiche der entworfenen Systeme nicht oder nur unvollstaendig spezifiziert werden koennen.

So ist in SPECIAL zwar die Beschreibung von Ausnahmebedingungen nicht aber die Spezifikation der Reaktionen auf diese Ausnahmebedingungen moeglich. Ausserdem gibt es in SPECIAL keine Moeglichkeit, Nebenlaeufigkeiten und deren Synchronisation zu beschreiben.

Damit die Behandlung der Ausnahmebedingungen mit den gleichen Mitteln (O- und OV-Funktionen) wie andere Aktionen innerhalb des entworfenen Systems beschrieben werden kann, werden hier diese Ausnahmebedingungen (die bisher nicht zum Systemzustand gehoeren), als Teil des Systemzustandes in Form von booleschen V-Funktionen eingefuehrt. Gleichzeitig wird das SPECIAL-Konzept erweitert, so dass neben den bisher vorhandenen lokalen Ausnahmebedingungen auch globale definiert werden koennen. Das Eintreten der Ausnahmebedingungen wie auch die Synchronisation paralleler Aktivitaeten wird durch Erweiterung der moeglichen Transitionen durch sogenannte spontane Transitionen beschrieben. Die spontanen Transitionen werden wie die anderen Transitionen durch O- und OV-Funktionen spezifiziert. (Da eine Unterscheidung zwischen O- und OV-Funktion in der vorliegenden Arbeit nicht notwendig ist, wird im weiteren Verlauf auf diese Unterscheidung verzichtet.)

Die beschriebenen Erweiterungen ermoeglichen es, die Behandlung sowohl von Ausnahmebedingungen als auch die Darstellung von Nebenlaeufigkeiten durch die bereits vorhandenen Sprachkonstrukte von SPECIAL zu beschreiben.

Auf Grundlagen und Konzepte von SPECIAL wird hier nur kurz eingegangen. Eine ausfuehrliche Beschreibung von SPECIAL und HDM findet der interessierte Leser in dem HDM Handbuch [9].

Die Erweiterung der SPECIAL-Syntax wird nicht ausgefuehrt, da sie zum Verstaendnis der Idee der Erweiterung nicht wesentlich beitraegt.

1. SPECIAL Bestandsaufnahme

SPECIAL nimmt im Rahmen der formalen Spezifikationssprachen zwischen den algebraischen Spezifikationssprachen und denen, die eine Erweiterung von Programmiersprachen darstellen, hinsichtlich des Abstraktionsgrades und der Einordnung in den Entwicklungsprozess eine Mittelstellung ein. So enthaelt eine SPECIAL-Spezifikation einerseits erheblich mehr Information bezueglich der Struktur des spezifizierten Softwareproduktes als algebraische Spezifikationen. Dies ermoeglicht eine weitgehende Festlegung der Systemstruktur und eine Ueberpruefung des Entwurfs vor der eigentlichen Implementierungsphase. Die Umsetzung in ein Programmsystem ist leichter und weniger fehleranfaellig als die einer algebraischen Spezifikation. Andererseits werden Umfang und Komplexitaet vermieden, die aufgrund der exakten Festlegung von Daten-, Speicherstrukturen und Algorithmen in ein Softwaresystem gebracht werden.

SPECIAL ist Teil von HDM. In HDM wird ein System durch hierarchisch geordnete Ebenen abstrakter Maschinen beschrieben. Die abstrakten Maschinen sind in SPECIAL beschriebene Zustandsautomaten.

Eine abstrakte Maschine wird mit Hilfe der Maschine der direkt darunterliegenden Ebene realisiert. Die Kopplung der Maschinen wird durch eine Abbildung der Zustandsraeume beschrieben. Die Zustandsuebergaenge auf der oberen Maschine werden durch ILPL-Programme auf der unteren dargestellt (ILPL ist als "abstrakte" Programmiersprache Teil von HDM).

1.1 Das Automatenmodell von SPECIAL

Die mit SPECIAL beschriebenen abstrakten Maschinen sind Zustandsautomaten mit i.a. nicht endlichem Zustandsraum und nicht endlichem Eingabealphabet.

Der Zustandsraum wird durch Zustandsvariable beschrieben. Die Zustandsvariablen werden als V-Funktionen beschrieben. (Der Zustandsraum ist das Kreuzprodukt ueber den V-Funktionen.) Mengen von Zustandsvariablen koennen zu einer V-Funktion zusammengefasst werden. Z.B. koennen die Zustandsvariablen, die die einzelnen Eintraege eines Stack darstellen, durch eine V-Funktion S beschrieben werden. S(i) wuerde dann den i-ten Eintrag in dem Stack darstellen. Den Wert der Zustandsvariablen liefert die V-Funktion mit dem entsprechenden Argument.

SPECIAL unterscheidet bei den V-Funktionen zwischen primitiven und abgeleiteten. Die Aenderung der Werte der primitiven V-Funktionen wird explizit beschrieben (Zuweisung zu den Zustandsvariablen), die abgeleiteten aendern ihre Werte mit den primitiven, von denen sie abgeleitet sind.

Die Zustandsuebergaenge werden durch die O-Funktionen beschrieben. Die Namen der O-Funktionen koennen als die Zeichen des Eingabealphabets des Zustandsautomaten interpretiert werden (genauer ist der Funktionsname fuer jedes moegliche Parametertupel als ein Zeichen anzusehen), die die Zustandsuebergaenge des Automaten ausloesen.

Die O-Funktionen setzen sich im wesentlichen aus dem Exceptions- und dem Effects-Paragraphen zusammen.

Die Ausnahmebedingungen (Exceptions) sind Praedikate auf dem Zustandsraum, die auf dem Definitionsbereich der Funktion (d.h. den Zustaenden, in denen die Eingabe des Funktionsnamen einen Zustandsuebergang ausloesen kann) nicht wahr sind.

Der Effects-Paragraph beschreibt den Folgezustand. Es werden dabei nur die Veraenderungen von Zustandsvariablen beschrieben.

1.2 Die typische Aufgabe der Automaten

Von einem Zustand aus einer Menge von Zustaenden, die durch eine bestimmte Semantik charakterisiert ist, soll der Automat in einen Zustand aus einer - wiederum durch eine Semantik bestimmte - Menge von Folgezustaenden uebergehen; i.a. werden dazu mehrere Zustandsuebergaenge notwendig sein.

Zum Beispiel: Der erste Eintrag des Stack wird gelesen und somit entfernt. Ausgangspunkt ist die Menge aller Zustaende, in denen der Stack mindestens einen Eintrag enthaelt. Zu jedem Zustand dieser Menge wird der Folgezustand beschrieben, in dem jeder Eintrag eine Position vorgerueckt ist.

O-FUNCTION Popp
   EXCEPTIONS:
      S(1) = null;
   EFFECTS:
      FOR ALL i:'S(i)=S(i+1);

('S:Der Wert von S im Folgezustand. Von einem n an ist fuer alle m>n S(m)=null d.h. undefiniert).

1.3 Schwaechen des Automatenmodells von SPECIAL

Es ist in SPECIAL moeglich, in den Funktionen Ausnahmebedingungen (Exceptions) zu definieren. Die Reaktion des entworfenen Systems auf diese Ausnahmebedingungen kann aber nicht beschrieben werden. (Eine O-Funktion soll angewendet werden, aber eine ihrer Ausnahmebedingungen ist wahr, d.h. das Eingabezeichen kann nicht verarbeitet werden, der Automat ist blockiert.)

Als Folge davon kann die Loesung fuer die typische Aufgabe nicht als Wort ueber dem Eingabealphabet (Reihenfolge von O-Funktionen) beschrieben werden. HDM weicht an dieser Stelle auf die abstrakte Programmiersprache ILPL aus.

In einem ILPL-Programm, das eine O-Funktion einer oberen Maschine auf einer unteren realisiert, muss beschrieben sein, welche Aktionen die untere Maschine durchfuehren soll, wenn bei der Ausfuehrung einer O-Funktion der unteren Maschine, die in dem Programm aufgerufen wird, eine Ausnahmebedingung wahr wird.

Mit der Programmiersprache ILPL wird das Modell des Zustandsautomaten verlassen und es tritt ein erheblicher Verlust an Abstraktion ein. Weiterhin wird eine zweite Sprache als Beschreibungsmittel benoetigt.

In vielen realen Faellen kann die Ausnahmebedingung einer O-Funktion nicht als Praedikat auf dem Zustandsraum der betrachteten Maschine dargestellt werden (z.B. ein benoetigtes Betriebsmittel ist nicht verfuegbar, die Betriebsmittelverwaltung ist aber auf einer tieferen Maschinenebene realisiert und somit nicht mehr sichtbar).

SPECIAL fuehrt hier die Ausnahmebedingung RESOURCE ERROR ein. An der Stelle innerhalb einer Spezifikation, an der diese Ausnahmebedingung verwendet wird, hat sie eine wohldefinierte Bedeutung, die aber auf der betrachteten Maschinenebene nicht erklaert werden kann. In der HDM-

SPECIAL Spezifikation ist eine Beschreibung ihrer Bedeutung daher nicht mehr vorhanden. Ein Beispiel hierfuer ist ein Stack ohne feste obere Grenze. In der Maschine, die diesen Stack beschreibt, gibt es keine Begruendung dafuer, dass der Stack voll werden koennte. Die Speicherverwaltung fuer die Stackmaschine wird aber auf einer unteren Maschine realisiert, die das reale Betriebsmittel Speicher verwaltet. Ein realer Speicher kann aber nur endlich gross sein, d.h. er kann voll werden. Die Ausnahmebedingung Stack_voll der oberen Maschine beruht also auf Eigenschaften der unteren Maschine.

SPECIAL hat keine Sprachmittel, um Wechselwirkungen zwischen Automaten der gleichen Hierarchieebene zu beschreiben.

Nebenlaeufigkeiten koennen daher nur beschrieben werden, soweit keine Wechselwirkungen auf der gleichen Ebene auftreten.

Die nebenlaeufigen Automaten muessen also Aufgaben fuer die naechst hoehere Ebene durchfuehren, die vollstaendig unabhaengig voneinander sind, d.h. aber: nur die trivialen Faelle der Nebenlaeufigkeit sind beschreibbar.

2. Erweiterungsmoeglichkeiten

Bisher wurden einige Probleme aufgezeigt, die das Automatenmodell von SPECIAL aufwirft. Im folgenden soll eine Erweiterung dieses Automatenmodells beschrieben werden, die es gestattet, sowohl die Beschreibung von Nebenlaeufigkeiten als auch die Behandlung von Ausnahmebedingungen in die Sprache zu integrieren.

2.1 Spontane Transitionen

In dem SPECIAL-Automatenmodell werden die O- bzw. OV-Funktion zusammen mit ihren Argumenten als die Eingabezeichen des Automaten aufgefasst.

Die aus diesem Eingabealphabet aufgebaute Halbgruppe hat kein neutrales Element, was aber fuer das theoretische Modell nicht gefordert ist.

Der Automat kann in einen Zustand geraten, der kein Endzustand ist, in dem das Eingabewort nicht vollstaendig abgearbeitet ist und das erste noch nicht verarbeitete Zeichen unseres Eingabewortes nicht verarbeitet werden kann, da eine Ausnahmebedingung der entsprechenden O-Funktion wahr ist. Es soll erreicht werden, dass der Automat entweder in einen (Fehler-) Endzustand uebergeht, wenn die Aufgabe nicht erfolgreich abgeschlossen werden kann, oder durch Recovery-Massnahmen in einen Zustand, in dem die Bearbeitung fortgesetzt werden kann.

Hierzu ein Beispiel von Stacks und Arrays, die den benoetigten Speicherplatz dynamisch aus einem gemeinsamen Pool erhalten. Die Speicherverwaltung ist auf einer unteren Maschinenebene realisiert. In dem Pool kann Speicherverschnitt entstehen.

Es kann der Fall eintreten, dass die Speicherverwaltung ueber keinen freien Speicherplatz mehr verfuegt - die Ausnahmebedingung Stack_voll in der Stackmaschine also wahr ist -, aber Verschnitt im System ist, der wiederverwendet werden koennte.

Wird jetzt die Funktion Push eingegeben, so soll die Speicherverwaltung zunaechst den Verschnitt aufsammeln (garbage collection), ehe eine endgueltige Fehlermeldung gegeben wird.

Bezogen auf unseren Automaten heisst das, dass ohne Eingabe fuer diesen Automaten eine Folge von Zustandsuebergaengen ablaufen soll, bis ein Zustand erreicht ist, in dem Stack_voll falsch ist.

Beobachten wir unseren Automaten von aussen, so stehen diese Uebergaenge (Stack_voll wird wahr, weil ein Array angelegt wurde, Stack_voll wird falsch, weil ein anderer Prozess endet und dadurch Speicher verfuegbar wurde) nicht in direktem funktionalen Zusammenhang mit der Eingabe Push - sie sind also spontan fuer die Stackmaschine.

Ein weiteres mehr Hardware-orientiertes Beispiel:

Ein Programm PG enthaelt einen Puffer P der Laenge L. Waehrend der Ausfuehrung des Programms wird der Puffer zeichenweise von einem Terminal aus gefuellt. Mit jedem uebertragenen Zeichen wird der Pufferzeiger Z erhoeht und wenn der Puffer voll ist, muss der Puffer gerettet werden, da von dem Terminal weitere Zeichen kommen koennen.

Mit jedem ankommenden Zeichen aendert also die V-Funktion P ihren Wert. Bezogen auf die Maschine PG, die unabhaengig von der Eingabe vom Terminal ablaeuft, sind das spontane Transitionen. Nach Ankunft jedes Zeichens muss der Ablauf in PG unterbrochen werden und Z erhoeht werden (analog der garbage collection). Bezogen auf die "normalen" Ablaeufe in PG sind auch das spontane Transitionen.

Wird das Eingabealphabet um ein neutrales Element ¢ erweitert, so koennen wir dieses Eingabezeichen nutzen, um ohne Veraenderung des Eingabewortes die gewuenschten Uebergaenge herbeizufuehren. Trivialerweise ist dann a¢b = ab, wenn a und b Teilworte ueber unserem Eingabealphabet sind. Die zu dem Zeichen ¢ gehoerenden Transitionen werden genau so beschrieben, wie die zu jedem anderen Zeichen gehoerenden, naemlich durch eine O-Funktion.

Die Recovery- und Fehlermassnahmen stellen in einem System haeufig den umfangreichsten Teil dar. Sie in einer einzigen O-Funktion darstellen zu wollen, hiesse hier ohne Strukturierungsmoeglichkeiten zu spezifizieren.

Es bietet sich daher an, in der Eingabehalbgruppe nicht ein einziges neutrales Element, sondern linksneutrale Elemente (und zwar beliebig viele) einzufuehren. Das formale Modell kann problemlos in dieser Richtung erweitert werden. Wir gewinnen dadurch die Moeglichkeit, die Recovery- und Fehlermassnahmen genauso zu strukturieren, wie die anderen Teile des Systems.

2.2 Einbindung der Ausnahmebedingungen in das Modell

Die Ausnahmebedingungen werden als boolesche V-Funktionen dargestellt und damit Teil des Zustandsraums. Sie sind abgeleitete V-Funktionen, wenn sie als Praedikat auf der betrachteten Maschine dargestellt sind und aendern ihren Wert, wenn das Praedikat seinen Wahrheitswert aendert.

Sind sie primitive V-Funktionen, so koennen sie ihren Wert durch eine O-Funktion oder durch spontane Transitionen aendern. Es wird definiert, dass eine Funktion nicht ausfuehrbar ist, wenn eine ihrer Ausnahmebedingungen den Wert wahr hat und dass das Eingabezeichen an dieser Stelle nicht verbraucht wird. (In SPECIAL ist die Reaktion der Maschine und ihre Interaktion mit den ILPL-Programmen bei Ausnahmebedingungen nicht mit letzter Klarheit beschrieben.)

Dies ist in Uebereinstimmung mit den linksneutralen Elementen, da das neutrale Zeichen quasi vor dem aktuellen Eingabezeichen eingefuegt wird.

Die Verwendung rechtsneutraler Zeichen wuerde hier zu einer anderen Definition fuehren. Allerdings waeren bei rechtsneutralen Zeichen die Recovery-Massnahmen nicht mehr unabhaengig vom momentanen Eingabezeichen. Obwohl die Verwendung rechtsneutraler Zeichen andere Vorteile hat, soll hier aus diesem Grund nicht naeher auf sie eingegangen werden. Rechts- und linksneutrale Zeichen sind aber im gleichen Modell nicht moeglich.

3. Auswirkungen der Erweiterungen auf die Beschreibungsmoeglichkeiten

Im folgenden wird naeher ausgefuehrt, welche Probleme durch die Einfuehrung der spontanen Transitionen und der Darstellung der Ausnahmebedingungen durch boolesche V-Funktionen geloest werden.

3.1 Nebenlaeufige Automaten

Um zwei (oder mehr) Automaten zu koppeln, nehmen wir an, dass bestimmte V-Funktionen des einen Automaten mit entsprechenden Funktionen des (der) anderen Automaten identifiziert werden.

Veraendert der erste Automat eine dieser V-Funktionen, so aendert der zweite spontan seinen Zustand und umgekehrt, aendert der zweite eine dieser V-Funktionen, so aendert der erste spontan seinen Zustand.

Das Verhalten jedes dieser Automaten bezueglich der gemeinsamen V-Funktionen, muss innerhalb des anderen durch spontane O-Funktionen beschrieben werden.

Ueber die Ausnahmebedingungen dieser spontanen O-Funktionen kann ein beliebiges Synchronisationsprotokoll, einschliesslich der moeglichen Verstoesse (z.B. durch Uebertragungsfehler), beschrieben werden.

Im folgenden wird aufgezeigt, wie das zu den "normalen" Ablaeufen in PG nebenlaeufige Fuellen und Retten des Puffers in der Zustandsmaschine PG beschrieben werden kann. Alle O-Funktionen von PG erhalten die Ausnahmebedingungen

```
Zeichen_uebernehmen:Ankunft=false;
Puffer_leeren:Puffer_voll;
```

(Hier ist es sinnvoll, Sprachmittel zur Beschreibung von globalen Ausnahmebedingungen zur Verfuegung zu stellen.)

PG wird um drei ¢-O-Funktionen erweitert:

```
¢-O-FUNCTION Ankunft_Zeichen
    EXCEPTIONS:
        Puffer_voll:Z>L;
        Ueberholen: Ankunft=false;
    EFFECTS:
        'Z=Z+1;
        'Ankunft=false;
```

Diese spontane Transition kann nur auftreten und den Pufferzeiger erhoehen, wenn ein Zeichen ankommt, das zuletzt angekommene Zeichen uebertragen ist und der Puffer nicht voll ist.

Durch die globale Ausnahmebedingung Zeichen_uebernehmen kann aber in PG keine andere Transition mehr stattfinden. Das Ueberholen von ankommenden Zeichen ist ausgeschlossen, da Ankunft_Zeichen eine eigene Ausnahmebedingung auf 'wahr' setzt.

¢-O-FUNCTION Zeichen_uebertragen
    EXCEPTIONS:
        Gesperrt:
            Ankunft_ausgeschlossen;
        Noch_nicht_angenommen:Ankunft=
                                true;
        Puffer_voll:Z>L;
    EFFECTS:
        'Ankunft=true;
        'P(Z) in ASCII_Zeichen;

Ein neues Zeichen wird nur uebertragen, wenn Ankunft_Zeichen die Ankunft signalisiert hat. Zeichen_uebertragen setzt seine eigene Ausnahmebedingung auf 'wahr', so dass ein Zeichen nur einmal uebertragen werden kann.

Durch die zusaetzliche Ausnahmebedingung Gesperrt koennte z.B. eine beliebige O-Funktion in PG ausschliessen, dass eine Uebertragung stattfinden kann (Deadlock Gefahr!).

¢-O-FUNCTION Puffer_abliefern
    EXCEPTIONS:
        Warten_auf_voll:NOT Puffer_voll;
    EFFECTS:
        -- P retten, in diesem Beispiel
        -- nicht ausgefuehrt

        'Z=0;

Da fuer die "normalen" O-Funktionen von PG die globale Ausnahme Puffer_leeren gilt, wie auch fuer die anderen ¢-O-Funktionen, ist Puffer_abliefern die einzige Transition, die in der Zustandsmaschine PG stattfinden kann, wenn der Puffer gefuellt ist.

Wenn wir das Terminal als einen unabhaengigen Zustandsautomaten betrachten, so sind dort folgende O-Funktionen von Interesse.

O-FUNCTION Senden (X)
    EXCEPTIONS:
        Puffer_voll: Z>L;
        Gesperrt: Ankunft_ausgeschlossen;
        Ueberholen:Ankunft;
    EFFECTS:
        'P(Z)=X;
        'Ankunft=true;

Diese Funktion hat in PG eine entsprechende ¢-O-Funktion: Zeichen_uebertragen.

¢-O-FUNCTION Zeichen_angenommen
    EXCEPTIONS:
        Kein_Zeichen_senden:
                Ankunft=false;
    EFFECTS:
        'Z=Z+1;
        'Ankunft=false;

In PG entspricht diese ¢-O-Funktion der Funktion Ankunft_Zeichen.

¢-O-FUNCTION Neuer_Puffer
    EXCEPTIONS:
        Nicht_Voll: Z>L;
    EFFECTS:
        'Z=0;

In PG entspricht diese spontane Transition dem Retten des Puffers.

¢-O-FUNCTION
    Sperren_und_Entsperren_durch_PG
    EFFECTS:
        'Ankunft_ausgeschlossen=
            Not Ankunft_ausgeschlossen;

Dieser Funktion werden in PG i.a. mehrere O-Funktionen entsprechen, naemlich alle, die Ankunft_ausgeschlossen veraendern und dadurch die Uebertragung zulassen oder ausschliessen.

Die zwei Zustandsmaschinen PG und Terminal haben die V-Funktionen

    Ankunft
    Ankunft_ausgeschlossen
    P, Z und L

"semantisch" gemeinsam.

Die Ablaeufe in jedem Automaten koennen aber vollstaendig getrennt betrachtet werden. Nur fuer das Verstaendnis der Kommunikation wird die Funktion Ankunft aus PG mit der Funktion Ankunft des Terminals identifiziert und entsprechend fuer die anderen Funktionen.

3.2 Behandlung von Ausnahmesituationen

Die spontanen Transitionen geben uns die Moeglichkeit, das Verhalten des Systems im Fehlerfall zu beschreiben, also insbesondere auch die Recovery-Massnahmen.

Ausnahmebedingungen, die auf der beschriebenen Maschinenebene nicht erklaerbar sind (z.B. Speicher_voll), sind durch die Darstellung als boolesche V-Funktionen in der Maschine beschrieben und damit kann in dieser Maschine der Wahrheitswert der Ausnahmebedingung bestimmt werden (mit der Ausnahmebedingung RESOURCE_ERROR aus dem Original SPECIAL ist das nicht moeglich). Ihr Wahrheitswert aendert sich durch spontane Transitionen. Ihre Bedeutung aber wird durch die Abbildung, die die Maschine auf der darunterliegenden Ebene realisiert, genau beschrieben. Die Ausnahmebedingung RESOURCE_ERROR (deus ex machina) wird damit in SPECIAL nicht mehr benoetigt.

3.3 Beschreibung von Reihenfolgen

Eine vollstaendige Spezifikation eines Systems muss die Beschreibung aller Recovery-Massnahmen wie auch das Verhalten des Systems in nicht aufloesbaren Fehlerfaellen enthalten.

Da das System, wenn es vollstaendig spezifiziert ist, durch die Ausnahmebedingungen nicht mehr in undefinierte Haltsituationen kommen kann, koennen jetzt Worte ueber dem Eingabealphabet (Reihenfolgen von O-Funktionen) beschrieben werden.

Die abstrakte Programmiersprache ILPL ist innerhalb von HDM nicht mehr notwendig, da die Spezifikation vollstaendig in SPECIAL geschrieben werden kann. Dabei wird der Sprachumfang von SPECIAL nur unwesentlich veraendert (Einfuehrung eines Reihenfolgeoperators fuer den Effects-Paragraph).

4. Zusammenfassende Bewertung und Ausblick

Mit der Einfuehrung der spontanen Transitionen wurde die Maechtigkeit der Spezifikationsmethode erheblich erweitert.

Das Konzept der Ausnahmebedingungen wurde auf ein solides Fundament gestellt und auf das unscharfe Konstrukt RESOURCE_ERROR konnte verzichtet werden. Durch die Bewaeltigung der Problematik der Ausnahmebedingungen wurde es moeglich, Reihenfolgen zu beschreiben und gleichzeitig auf die Programmiersprache ILPL zu verzichten. Dadurch tritt eine wesentliche Vereinfachung der Methode ein.

Die spontanen Transitionen ermoeglichen es, praezise und mit geringfuegigen zusaetzlichen Sprachmitteln Nebenlaeufigkeiten zu beschreiben. Damit wurde ein wesentlicher Defekt der Spezifikationsmethode behoben. Wir glauben, dass die Methode damit einer praktischen Anwendbarkeit einen deutlichen Schritt naeher gekommen ist.

Die formale Definition der vorgeschlagenen Erweiterungen durch die Definition der erweiterten Syntax und exakte Beschreibung der Semantik muesste allerdings noch vorgenommen werden. Auch sollten dabei globale Ausnahmebedingungen eingefuehrt werden, die z.B. fuer alle Funktionen eines Moduls gelten, wodurch der Aufschreibungsaufwand erheblich reduziert werden kann.

Der Schwerpunkt weiterfuehrender Arbeiten sollte das Mappingkonzept sein, mit dem eine abstrakte Maschine auf der darunterliegenden Maschinen-Ebene realisiert wird. Auf das Mappingkonzept konnte hier leider nicht naeher eingegangen werden.

Da es in realen Systemen aber schon sehr praktikable, intuitive Loesungen fuer dieses Problem gibt (z.B. mikroprogrammierte Rechner), sollte auch hier mit vertretbarem Aufwand eine Loesung erarbeitet werden koennen.

Ich danke insbesondere Herrn H. vor der Brueck, der sich mit mir in zahllosen Diskussionen um die Loesung der Probleme bemueht hat.

Literaturverzeichnis

[1] Cheheyl, M.H.: Secure System Specification and Verification: Survey of Methodologies; MITRE MTR-3904, 1980

[2] Guttag, J.: Abstract Data Types and the Development of Data Structures CACM Vol. 20, No. 6, June 1977

[3] Guttag, J.; Horowitz, E.; Musser, D.R.: Abstract Data Types and Software Validation, CACM Vol. 21, No. 12, December 1978

[4] Liskov, B.H.; Zilles, S.N.: SPECIFICATION TECHNIQUES FOR DATA ABSTRACTIONS, IEEE Transactions on Software Engineering Vol. SE-1, No. 1, March 1975

[5] McCauley, e.al.: Secure Minicomputer Operation System (KSOS) Development Specification (Type B5), Ford Aerospace & Communications Corporation, Palo Alto (Sept. 1978).

[6] Neumann, P.G.; Boyer, R.S.; Feiertag, R.J.; Levitt, K.N.; Robinson, L.: A Provably Secure Operation System. The System, its Applications, and Proofs, SRI Project 4332, Final Report (Febr. 77)

[7] Nibaldi, G.H.: Specification of a Trusted Computing Base MITRE M79-228, 1979

[8] Robinson, L.; Levitt, K.N.: Proof Techniques for Hierarchically Structured Programs; CACM, Vol. 20, No. 4, April 1977

[9] Robinson, L.; Levitt, K.N.; Silverberg, B.A.: The HDM Handbook, Vol. I-III, June 1979, SRI International

[10] vor der Brueck, H.; Koehler, Chr.; Seiderer, M.: "Spezifikations- und Verifikationswerkzeuge fuer sichere Systemsoftware", IABG-Bericht B-SZ 1228/02 (Mai 1981).

PILS, A Portable Interactive Language System

David O. Williams, Robert D. Russell[*]

Data Handling Division, CERN
1211 Geneva 23, Switzerland

PILS - A Portable Interactive Language System - is designed to provide an identical interactive language, command language and terminal interface on a wide range of 16- and 32-bit computers. PILS is written in 'lowest common denominator' Pascal. The paper gives an overview of the system and discusses the present status.

PILS - un Système de Langage Interactif Portable a eté conçu pour fournir un langage interactif, un jeu de commandes, une 'interface utilisateurs' identiques sur une large gamme d'ordinateurs 16- et 32-bits. La communication donne une vue d'ensemble du système et de son état d'avancement.

PILS - ein portables interaktives Programmiersystem - bietet eine interaktive Sprache verbunden mit einer einheitlichen Benutzerschnittstelle auf verschiedenen 16- und 32-Bit Rechnern. Zur Implementierung wurde ein PASCAL des 'kleinsten gemeinsamen Nenners' benutzt. In einem Überblick werden die Eigenschaften des Systems und der gegenwärtige Stand der Implementierungsarbeiten beschrieben.

(*) Permanent Address: University of New Hampshire, Department of Computer Science, Kingsbury Hall, Durham, NH 03824, United States of America.

BACKGROUND

One of the areas where interactive languages have been particularly successful at CERN is in the testing of new or faulty electronic equipment. Unfortunately the portability between computers from different manufacturers of programs written in these interactive languages has been poor, and this has created problems both for the staff using the programs, who need to learn multiple different environments, and for the support programmers, who cannot find an adequate language in which to develop libraries of transportable interactive code. The development of a portable interactive language system, written in a 'lowest common demoninator' Pascal, and aiming to provide a uniform interface on 'dumb' alphanumeric terminals, is an attempt to solve this problem.

In view of the restricted space available this paper must concentrate on an outline of the features of PILS and its status and there is little discussion of the reasons behind some of the design choices.

AN OVERVIEW OF PILS

The language

ANSI standard Minimal Basic is a proper subset of PILS. There are in fact three known exceptions to this statement but we do not expect that the normal user will ever discover what they are. In addition there are a number of features which can be thought of as 'extensions' to Minimal Basic but which in total create a very different flavour of language. In particular we hope that PILS will be much better adapted than the classical interactive languages for use by a team of collaborating programmers working on a sizeable software project.

° Six data types and their associated literals are supported. These types are 16- and 32-bit integers, 16- and 32-bit words, reals and character strings.

° It is easy to switch between implicit typing for 'quick and dirty' programming and full explicit typing for production quality software. Either Fortran or Basic style implicit typing is available.

° A Module structure is supported to simplify the writing of PILS programs by more than one person.

° Identifiers can be of any length.

° Users may define named subroutines and functions with named parameters. The parameters may be of three types. input, output and input/output and there is full checking that the declarations are consistent with any calls.

° Labels may be symbolic.

° Complex program structures such as subroutines, functions or IF-ENDIF loops may be labelled at both start and end. PILS will check for the consistency of this labelling.

° The full FORTRAN-77 IF-THEN-ELSE-IF...ELSEIF-ELSE-ENDIF control structure is available.

° A LOOP-EXITIF-ENDLOOP control structure is available.

° Access to non-PILS library routines is supported. At present this support assumes that the library routine conforms to FORTRAN parameter passing conventions.

° Character strings are dynamic, of length 0 to 255, and there is a set of string operators.

° Arrays may be dynamic.

° Character input/output to files is supported.

° The user may receive control when certain events occur or upon the occurence of errors.

° In order to preserve runtime efficiency, as much of the checking of the consistency of a PILS program as possible is carried out in a check pass which is separate from the actual running of the program. It is during this pass that, for example, any type conversions that may be required are resolved, and any inconsistent program structures are indicated.

The Command Interface

In reply to the Command? prompt the user can give 1) a true command or 2) a line of PILS program or 3) an 'immediate' statement.

The commands can be divided into three main categories

° General commands are HELP, RUN, CHECK, SET, SHOW, STEP, CLEAR, EXIT and QUIT.

° File manipulation commands are USE, OLD, MERGE, SAVE, EXTEND, SCRATCH RENAME and DIRECTORY.

° Editor commands are LIST, POSITION, FIND, SPOOL, AFTER, BEFORE, COLLECT, DELETE, UNDELETE, REPLACE, CHANGE, COPY, MOVE, NUMBER and EDIT.

Debugging

The job of debugging in an interactive language is already eased with respect to compiled languages. In PILS there is a further improvement because of the major checking that is performed to ensure that, as far as possible, the program is self-consistent with respect to program structures and data types before execution is attempted. Extra assistance is given through commands such as SET TRACE which turns on full tracing of all stores and transfers of control, SET TRACE <id> which provides selective tracing, and SET or SHOW PROFILE which provides information on the frequency with which lines of PILS code have been executed. The STEP command may be issued at any time and forces the PILS program to execute one line at a time. The SET BREAK command allows the user to set breakpoints in a selective manner.

Implementation

PILS is implemented in a 'lowest common denominator' subset of Pascal that has **proved** to be highly transportable. In order to deal with the slightly different conventions required by separate compilation under different compilers we always pass the master copy of PILS through an editor prior to any compilation, removing the few lines that are not required for that machine.

We normally work with PILS source divided into some eleven modules. The division has been made in such a way that on machines where it is necessary PILS can be overlayed.

In order to port PILS onto widely different computers we have defined a PILS-Host Interface consisting of five layers, which we refer to as the PHI. The zero level of the PHI is concerned with bit-handling, the first level with the terminal interface, the second level with the interface to the file system, the third level with handling traps and interrupts and the fourth level with non-PILS libraries.

The Terminal Interface

PILS aims to provide the same terminal interface on all computers. To achieve this goal it requires that PILS should 'own' the terminal and be responsible for processing and echoing ALL characters typed on the terminal. The following terminal control functions are implemented: End of line, End of file, Erase last input character, Erase current input line so far, Interrupt the running program, Re-echo current input line so far, Freeze output, Unfreeze output, Throw away output.

STATUS

The transportability of the Pascal source has been confirmed by successful compilation under nine Pascal compilers, namely those for Vax/VMS, Nord-100/Sintran III, HP 1000 Series, HP 9826 and four different cross compilers for the Motorola 68000; full PHI implementations have been available for some time now for the Vax and the Nord, and partial implementations have been produced for the HP 9826 (which we had on loan for a short period) and for the Motorola 68000 which relies on a remote (Nord) filebase connected through a serial line. The Nord-100 implementation is overlayed. We are waiting for delivery of the newly-announced RSX-Pascal before proceeding further on the PDP-11.

Although a major part of the language has been available on Vax and Nord for several months now, and some significant programs (up to 1000 lines of code) have been written, we have decided not to release for general use any of PILS until a coherent body of the language is complete. For this first release the major work remaining is a tidy-up and thorough test of file input-output and inbuilt functions and the implementation and test of 'immediate mode' commands. We would expect this to be ready for **release** in about November/December 1982.

The documentation in general is in a healthy state. The reference manual[1], in three sections, The System (30 pages), The Language (65 pages) and The PILS Host Interface (24 pages), is complete and the only minor problem will be to keep it in phase with the exact features implemented. There is a 20 page 'Programmers Introduction to PILS'[2], intended for people with general programming experience and there is a fairly extensive HELP file which can be interrogated from inside PILS.

REFERENCES

[1] PILS, A Portable Interactive Language System, R. D. Russell, DD/OC/82-4, CERN, Data Handling Division.

[2] Programmers Introduction to PILS, R. D. Russell, DD/OC/82-5, CERN, Data Handling Division.

AUTOMATIC GENERATION OF PEARL-PROGRAMS

FROM AN EPOS SPECIFICATION

V. Scheub

Instute for Control Engineering and Process Automation
University of Stuttgart, Federal Republic of Germany

ABSTRACT

The software-development tool EPOS (Engineer and Process-Oriented development Support System) which supports the design of realtime systems from the requirements analysis to detailed program design is briefly presented. A new feature of EPOS is the automatic generation of PEARL programs, which is the first step towards a system of generators for the mostly used high level languages for process-automation. It is described how such a generator converts an EPOS design into a PEARL-program. Such a generator is necessary to support the error-prone step from program design to coding.

Keywords: Specification languages, software design, program documentation, PEARL, real-time programming language, program generator.

RÉSUMÉ

Cette contribution présente brièvement l'outil de développement de logical EPOS (Engineer and Process-Oriented development Support system - système d'aide au développement orienté process et destiné aux ingeniéurs) qui permet de concevoir des systémes temps réel, de l'analyse des besoins à la conception détaillée des programmes.

Une nouvelle caractéristique d'EPOS est la génération automatique de programmes PEARL qui constitue la première étape vers un système de générateurs des langages de haut niveau les plus utilisés dans le domaine de l'automatisation des process. On explique également comment un tel générateur convertit un produit EPOS en un programme PEARL.

Un tel générateur est nécessaire pour passer de l'étape de la conception du programme à celle de l'écriture proprement dite, passage souvent générateur d'erreurs.

Mots-cles: Langage de spécification, conception de logiciel, documentation programme, PEARL, langage de programmation temps réel, générateur de programmes.

ZUSAMMENFASSUNG

Es wird das Software-Entwicklungsprogramm EPOS (Engineer- and Process-Oriented development Support system) kurz vorgestellt, das die Erstellung von Echtzeitsystemen von der Bedarfsanalyse bis zur detaillierten Programmauslegung unterstützt. Neu bei EPOS ist die automatische Erstellung von PEARL-Programmen als erster Schritt in Richtung eines Systems von Generatoren für die meistbenutzten höheren Programmiersprachen für die Prozessautomatisierung. Es wird beschrieben, wie durch einen solchen Generator eine EPOS-Konstruktion in ein PEARL-Programm umgesetzt wird. So ein Generator wird zur Unterstützung des fehlerträchtigen Schrittes von der Programmauslengung bis zum Codieren benötigt.

AUTOMATIC GENERATION OF PEARL-PROGRAMS FROM AN EPOS SPECIFICATION

V. Scheub

Institute for Control Engineering and Process Automation
University of Stuttgart, Federal Republic of Germany

Abstract: The software-development tool EPOS (Engineer and Process-Oriented development Support system) which supports the design of realtime systems from the requirements analysis to detailed program design is briefly presented. A new feature of EPOS is the automatic generation of PEARL programs, which is the first step towards a system of generators for the mostly used high level languages for process-automation. It is described how such a generator converts an EPOS design into a PEARL-program. Such a generator is necessary to support the error-prone step from program design to coding.

Keywords: Specification languages, software design, program documentation, PEARL, real-time programming language, program generator.

1. INTRODUCTION

During the last years a number of computer aided specification and design systems have been developed to turn over the creativity of development of software and hardware to man, but however, to make software development available for him. (PDV 80, Lau 79).
These software design systems mostly cover different activities in the life cycle of a software system. They either support the specification of software requirements and the preliminary program design or they support the preliminary and detailed program design. The step from detailed program design to coding is supported only little and therefore entails a number of mistakes.
For the design tool EPOS, which supports the design phase from requirement definition to detailed program design, having been developed at the institute for Control Engineering and Process Automation (University of Stuttgart), a program system has been implemented which generates automatically code for the realtime programming language PEARL from a design written in EPOS.
The advantages of such a computer aided automatic conversion can be found in a computer aided methods available for the user of the EPOS system. Thus he gets a consistant documentation of the program, related to the stage of the design. Especially the various check programs of the EPOS system enable the user to avoid faults in the program early in the stage of design and to generate precise programs (f.e. by checking the scope and visibility of data, by checking the synchronization specification, by checking inconsistency of design).
In EPOS a user optimized and easy to survey modification of the design is possible and thus finally of the program generated from it.
That means changes are done in the design and not in the program, in order to find all relations and effects of the modification to the system automatically.

2. WHAT IS EPOS?

EPOS (Engineer and Process Oriented development Support system) is a computer aided specification and design system (Bie 80).
EPOS is for engineers who project, develop, set into service, maintain and work with realtime systems. EPOS is placed in the category of indirect design methods, which leave the original creating activities to the user, but, however try to take over inconvenient routine operations (f.e. preparation of project data, documents), find mistakes and inconsistencies in the design, and show immediately what possible changes of specification or design effect.

EPOS has following special characteristics:

1. EPOS is a computer aided tool for the whole life-cycle of an automation system, i.e. it can be used in all phases of development and especially in maintenance.

2. EPOS is especially for the design of real-time systems. Therefore, EPOS has description features for interrupts and time-events, for controlflow, for synchronzation of parallel activities and for interfaces to the process. It enables the user to check for certain special characteristics and to operate with static and dynamic simulation.

3. EPOS is meant for automatization engineers. It means, that the formal notation needed for computer aid is added with an informal notation and graphic descriptions, which is for the communication with the people who take part.

4. EPOS is implemented on mini-computers, hence EPOS is also available for small firms.

5. EPOS is built in such a way to easily convert the specified software system in a high level realtime programming language.

The structure of the EPOS-system and the principle way of the design of an automation system is shown in fig. 1.

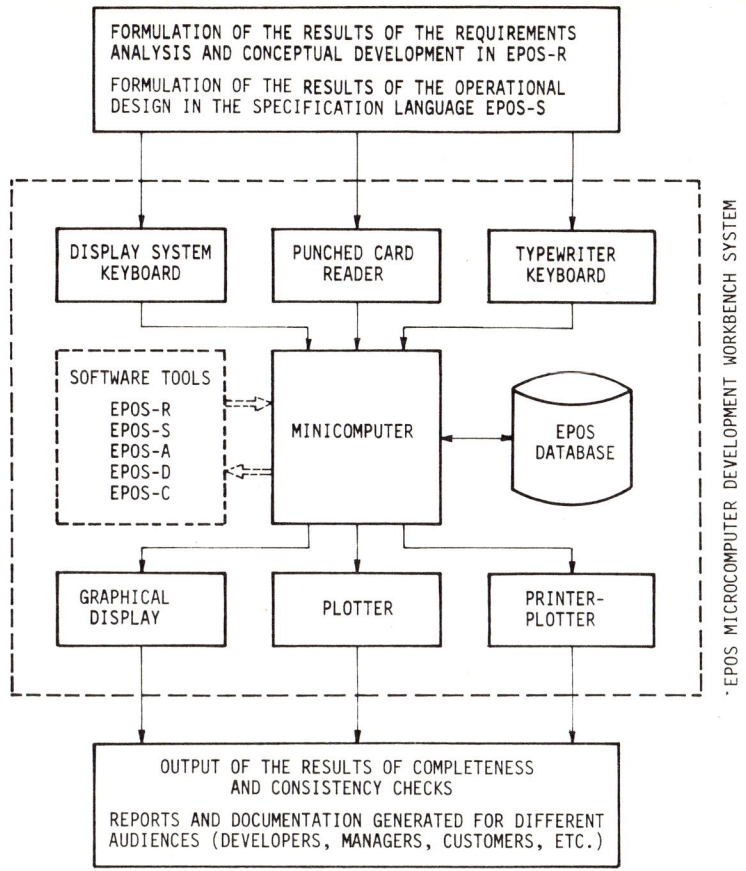

Figure 1:

When designing in EPOS, the requirements for the automation system are described with the requirement definition language EPOS-R and then stored in the database (Lau 81).
EPOS-R is based on a fixed scheme. The requirements entail the functional aim as well as the definition of devices and software packages which shall be used, and build the base for the operational design, which is described with the formal specification language EPOS-S.
The design objects in EPOS-S are described gradually step-by-step at preliminary and detailed design with EPOS so long, until one has reached a level with an adequate detail grade. Therefore EPOS-S has six different design object types:

1. Design Object Action

    For the description of a logic operation, conversion or processing of data (f.e. processing of process data for control or logging).
    The formal description entails remarks among others referring to use data, the triggering event and the refinement with the pertaining control-flow.

2. Design Object Data

    This object serves to describe all data occuring the design process. f.e. controller outputs, measured values, times with type, range, validity and hierarchical refinement with definition of data structure (like Jackson).

3. Design Object Interface

    It describes the interface between the computer system and its environment or several components of an automation system. It entails informations about realization and hierarchical refinement.

4. Design Object Event

    It describes events, which affect the actions.

5. Design Object Condition

    A data-dependent condition for the branch in the control flow of actions with the logic relations of sub-conditions and datas.

6. Design Object Device

    To describe the technical realization of actions and process interfaces.

An EPOS specification consists of many of such design objects, which can be identified by their identifier. (fig. 2)

The formal description of the 6 design objects consists of:

a) The description part in which the function and purpose of the design object as well as relations to the design of EPOS-R and test and management aids are given in verbal form.

b) The information of relations to other design objects, f.e. data which are used or generated in an other action or the specification of synchronisation mechanism.

```
ACTION SWITCH-CONTROL .
DESCRIPTION :
            PURPOSE : "THIS ACTION TESTS IF IT IS NECESSARY TO
                      THROW THE SWITCH FOR THE OBJECT ENTERING THE NODE.
                      IF SWITCHING IS NECESSARY (NEW-POSITION=TRUE),
                      IT CHECKS, WHETHER THERE IS ANY OBJECT WITHIN
                      THE BRANCHING AREA. IF SWITCHING IS POSSIBLE
                      (SWITCHABLE=TRUE) THE SWITCH WILL BE THROWN.
                      OTHERWISE, (SWITCHABLE=FALSE) THIS OBJECT CANNOT
                      BE TRANSPORTED TO ITS DESTINATION, AND THE
                      OPERATING STAFF WILL BE INFORMED BY 'REPORT-
                      ERROR".
            FULFILS : REQUIREMENT 1   , REQUIREMENT 2  ,
                      REQUIREMENT 3   .
DESCRIPTIONEND.

DECOMPOSITION : FIX-SWITCH-POSITION;
                IF NEW-POSITION
                    THEN IF SWITCHABLE
                            THEN SET-SWITCH
                            ELSE REPORT-ERROR
                         FI
                FI;
                ADJUST-NODE-STATE.

TRIGGERED : ENTRY-INTERRUPT.

INPUT : DESTINATION-QUEUE,DESTRIBUTION-NET-CONFIGURATION.

OUTPUT : SWITCH-CONTROL-SIGNAL TO SWITCH-ACTUATOR.

ACTIONEND
```

Figure 2: EPOS-S Specification of an ACTION

c) The description of the decomposition of the design object with the exact definition of the control-flow of the design objects in the refinement.

On the lowest level of the refinement of actions no more design object must be specified. Here a direct program code can be inserted, which entails the realization of this actions in a programming language. For EPOS doesn't entail until now description features for algorithmic, algorithmical specifications must be undertaken in this so-called CODE-part.
To generate a PEARL program from a EPOS-S specification it must be guaranteed that the design is complete and without any contradictions. That means, type, range and scope of data must be specified. Decomposition of conditions must be defined and checked whether they are correct without any type mismatch.

For this cannot be checked manually in large designs, EPOS offers a number of error checks made by the computer. Without a positive result of these checks, it is senseless to start the automatic program generation.

3. IMAGE OF EPOS-DESCRIPTION FEATURES ON PEARL STATEMENTS

If one converts an EPOS specification in a PEARL program a PEARL analogon must be found for each EPOS design object and for the control flow constructions in the decomposition part. Synchronization mechanisms described in the object type ACTION must be imaged on a PEARL-semaphore solution. The design objects ACTION, DATA, EVENT, CONDITION and INTERFACE can be imaged definitely on PEARL-language elements.
The control-flow construction, provided in the decomposition part of ACTIONS can be converted into PEARL-statements for the most part without too much trouble. But the image of event- and data-dependent continuing problems within a refinement is not definitely, because there exist no language features in PEARL.
The description of control-flow in the decomposition part took pattern from language elements used in most high level languages. It is therefore no problem to image the description into control-flow constructions of all high level languages.
The problem is to find the program structure within a large specification and to convert the relations of design objects into a sequence of statements.

3.1 Conversion of Actions and their Refinement into the structure of a PEARL program

PEARL meets the principle of step-by-step refinement used in the design with EPOS-S through its block structure. The design object ACTION therefore is imaged on a correspondent PEARL block, i.e. module, task, procedure or begin-end-block. The structure of the generated PEARL-program must be find in the description of refinement. The used control-flow constructions are similar to those of a higher leveled programming language and can mostly be converted according in PEARL-statements or language elements of other high ordered languages.
In the first version we had no possibility to generate automatically PEARL-modules. Now the EPOS language had been extended by new features dealing with structures like modules as described in MODULA or Ada.

In the following some rules and methods, used by the automatic conversion will be described. Many of the features, mentioned in the following, concern realtime features given in EPOS and in similar way in PEARL, so it was the first step to try automatic generation of PEARL programs.

3.1.1 Conversion in PEARL-Tasks

When an ACTION is converted in a PEARL-task following criticisms are checked by the program generator:

a) If the ACTION has the attribute task, it will be imaged on a PEARL-task.

b) An ACTION which has no predecendant, that means, it stands on the head of a design hierarchy, is, as a rule, converted in a task. This task generally entails the scheduling of other tasks.

c) In case subactions are specified as independent from the rest of the control-flow by the operators (/and/) in the refinement, they are also imaged in PEARL tasks.

The EPOS specification

ACTION TASK DISTRIBUTION-OF-PARCELS
:
ACTIONEND
:
DECOMPOSITION: (/ES-CONTROL, VS-CONTROL/).

is therefore converted in 3 PEARL tasks DISTRIBUTION-OF-PARCELS, ES-CONTROL and VS-CONTROL.

d) If an action is refined in concurrent sub-processes, so these and the action which entails the refinement, are realized as tasks. The latter entails task control-demands for concurrent tasks. A reduction of the number of tasks can be reached by realizing the refined action and one of its concurrent sub-processes as common task.

EPOS

ACTION VS-CONTROL
:
DECOMPOSITION: PARALLEL (VS-CONTROL-1,
                         VS-CONTROL-2).
:
ACTIONEND

PEARL

a) Realization with 3 Tasks

VS-CONTROL: TASK;
:
ACTIVATE VS-CONTROL-1
ACTIVATE VS-CONTROL-2
END; /*of Task VS-CONTROL*/

VS-CONTROL 1: TASK;
:

b) Realization with 2 Tasks

VS-CONTROL: TASK;
:
ACTIVATE VS-CONTROL-1
BEGIN;
      /*VS-CONTROL-2*/
END;
END; /*VS-CONTROL*/

When the control-flow is re-united after the execution of the parallel part, the action must be accordingly synchronized by the introduction of a semaphor variable and adequate synchronization instructions, which is automatically done by the generator.

3.1.2 Scheduling

The tasks specified as independent refinements, mentioned above must be related with the arrival of an event to be able to appear in schedules. This is done in EPOS by the set-part within the refinement of an action.

ACTION PARALLEL DISTRIBUTION
:
DECOMPOSITION: SET (ES-INTERRUPT, VS-INTERRUPT).
:
ACTIONEND.

The specification of events is done with the design object EVENT.

3 types of events have to be differenciated:

1) Events which of the appearance is not predictable (INTERRUPT).
   The adequate conversion in PEARL:
   WHEN interrupt ACTIVATE task;

2) Cyclical Events (CYCLIC)
   PEARL: ALL duration ACTIVATE task

3) Events which appear only one time given by the user (CLOCK): AT time ACTIVATE task;

3.1.3 Conversion of ACTIONS in Procedures and BEGIN-END Blocks

Actions which appear in the refinement of several other ACTIONS are realized in PEARL as a procedure.
In EPOS language features exist for the specification of actual parameters when an ACTION in a decomposition is used as procedure.
The formal parameters, too, which appear within an ACTION used as procedure, can be specified. Herewith the procedure-call with the conversion of actual parameter and the declaration of the procedure itself can be realized in PEARL.
All ACTIONS which cannot be classified as tasks, modules or procedures are imaged in BEGIN-END-blocks.

3.2 Conversion of EPOS-Data Specification

The data formally described by the design object DATA can be classified in EPOS by the attributes TYPE, INITIAL, IDENTICAL.
EPOS-S entails single data as well as data structures. The conversion to PEARL in a single date, a data structure or an array on account of the specification is definitely.
Moreover, EPOS entails data-sets, queues, stacks and files. There are no PEARL-equivalents for these EPOS-data types. If converted automatically standard solutions are used for these EPOS-data types, which of the realization can be interactively supressed.

When converting data, the scope and therefore the place where a data is declared, plays an important role. Is a data specified and a so-called SCOPE given, then the data declaration stands at the beginning of the ACTION which was given in the SCOPE-part of a data specification. Is there no SCOPE given, the declaration of the data is executed automatically in the program block, that includes all sub-ACTIONS in which the data is used.

3.3 Conversion of Synchronization Specification into PEARL-Semaphor-Solutions

EPOS offers several possibilities to specify the synchronization of parallel ACTIONs and their sub-ACTIONs: /Göh 81/

- mutual exclusion of ACTIONs (EXCLUSIVE)
- definition of the sequence of ACTIONs (SEQUENCE)
- cyclical activation of ACTIONs (CYCLIC)
- data or event dependent continuing of ACTIONs (WAIT FOR, WAIT UNTIL)

The realization of this synchronization specification is processed in PEARL by semaphores and adequate semaphore operations.
Herewith the synchronization problem is classified, and an adequate standard solution is used and set into the program.

```
ACTION VS-CONTROL
    :
DECOMPOSITION : PARALLEL (VS-DATA-ACQUISITION,
                          VS-PROCESS)
SYNCHRO: CYCLIC (WRITE-BUFFER   READ-BUFFER).
    :
ACTIONEND.
```

WRITE-BUFFER and READ-BUFFER are Sub-ACTIONs of VS-DATA-ACQUISITION and VS-PROCESS.

The generated PEARL-program has the following structure:

```
DCL SEMA1 SEMA PRESET(1);
DCL SEMA2 SEMA PRESET(0);

VS-DATA-ACQUISITION: TASK;
REQUEST SEMA1;
    BEGIN;
        /*WRITE-BUFFER*/
    END;
RELEASE SEMA2;
    :
END;  /*VS-DATA-ACQUISITION*/
VS-PROCESS: TASK;
    :
REQUEST SEMA2;
    BEGIN;
        /*READ-BUFFER*/
    END;
RELEASE SEMA1;
    :
END;  /*VS-PROCESS*/
```

4. WAY OF PROCESSING AN AUTOMATIC CONVERSION AND ITS LIMITS

The automatic conversion of a specification in the EPOS-database into a PEARL program is done in several steps:

- In the first step the synchronization specification is converted in PEARL-semaphore solution and filed in lists.

- In the second step all tasks and procedures are stated. This is the structure of the PEARL program.

- In the third step the refinement of the tasks and procedures is stated and also stored in lists.

- Now the semaphore solution is joined in the detailed structure having been received in the third step.

- In the next step the data declarations are set into adequate places of the program.

- In the last step the PEARL-code is generated of the lists and issued on printers and/or disks.

This last step is principally independent from the target language, if this is a high level language. At the moment a generator is in preparation, which generates PASCAL-programs. Naturally with this generator only the sequential part of the program design can be generated because synchronization and tasking is not provided in PASCAL.

The program system has been implemented on a AEG 80-20/4 in BCPL. The limits of automatic conversion can be received from the hardware-description, beeing hardly supported in EPOS 80 until now.
For, in EPOS-S user defined devices and process signal names are included, but no system defined device-names, the PEARL-system part cannot be generated automatically. It can, however, be editioned in the automatically generated module. There is another problem when converting the EPOS-INTERFACE-object in a PEARL-DATION. The information being available in EPOS 80 is not sufficient to specify a correct PEARL-Dation. This, too, must be completed after the automatic program generation.

Presently an extension of EPOS 80 is in development which has as goal a description and support of the hardware-design.
As soon as this extension is available, the mentioned restrictions for the automatic program generation doesn't take place anymore.

5. CONCLUSION

It was shown in which way an EPOS specification can be used to generate a PEARL-program especially how the different EPOS-language-elements can be imaged on PEARL-statements.
Naturally the generated code is not every time optimal and must therefore perhaps be optimized before used on the target machine.

Nevertheless the automatic program-generation has two essential advantages:

- When a design must be changed this changement is repeated automatically in the program, and this is at every place of the program related to that changement.
  So coding errors and unrecognized effects can be prohibit in the program. Furthermore, the program remains identical with documentation.

- The other advantage is that the specification of the synchronization in EPOS can be done in a high abstraction level and can easily be checked whether it is complete without any fault.

When the specification of the synchronization is converted automatically it is guaranteed that the solution with the language elements (e.g. the semaphores in PEARL) is correct, without any dead- or lifelock.
The disadvantage of the semaphore-complexity, also given with other synchronization mechanisms, hardly no checking by the compiler and therefore detection of errors at runtime is compensated herewith.

6. LITERATURE

/Lau 79/ Lauber,R.: Modelle zur Beschreibung des Entwurfs von Prozessautomatisierungssystemen.
Regelungstechnik 27(1979) Heft 12

/PDV 80/ Hommel,G.: Vergleich verschiedener Spezifikationsverfahren am Beispiel einer Paketverteilanlage
PDV-Bericht 186, August 1980

/Bie 80/ Biewald,J. Göhner,P. Lauber,R. u. Schelling,H.: Das Software-Werkzeug EPOS zur Unterstützung der Ingenieurtätigkeiten beim Entwurf und bei der Wartung von Prozessautomatisierungssystemen
Regelungstechnik 28(1980) Heft 1

/Bie 80a/ Biewald,J. Schelling,H.: Rechnergestützte statische Analyse des Funktions- und Softwareentwurfs in EPOS.
ACM-Tagung: Software Engineering - Entwurf und Spezifikation
Berlin, 15. Sept. 1980

/Lau 81/ Lauber,R. Jovalekić,S.: Wie formal soll und darf die Beschreibung des Pflichtenhefts für ein Prozessautomatisierungssystem sein?
11. Jahrestagung der GI
München Oktober 1981

/Göh 81/ Göhner,P.: Spezifikation der Synchronisation paralleler Rechenprozesse in EPOS
Fachtagung Prozessrechner 1981
GI, VDI/VDE-GMR, KfK, München März 81

/DIN 81/ DIN 66253 Teil 1: Programmiersprache PEARL, Basic PEARL, Juli 1981

/Joh 81/ Joho,E. Jovalekić,S.: Rechnergestützte Umsetzung von EPOS-Spezifikationen in PEARL-Programme
Fachtagung Prozessrechner 1981
GI, VDI/VDE-GMR, KfK, München März 81.

APPENDIX

Part of an automatically generated PEARL-program.

```
0010    MODULE;
0020                    /* MODUL 2 */
0030
0040
0050    PROBLEM ;
0060
0070
0080
0090    /*------------------------------------------------------------*/
0100
0110    /* PERIPHERIE, ATTRIBUTE EVTL. UNVOLLSTAENDIG ! */
0120
0130    SPC LSTEUERUNG () DATION OUT BASIC GLOBAL ;
0140    SPC FREIGABEORGANSTEUERUNG DATION OUT BASIC GLOBAL ;
0150    SPC RUECKFUEHRBAENDERSTEUERUNG () DATION OUT BASIC GLOBAL ;
0160    SPC STANDARDPERIPHERIEANSCHLUSS DATION OUT ALPHIC CONTROL(ALL) GLOBAL ;
0170
0180
0190    /*------------------------------------------------------------*/
0200
0210    /* INTERRUPTEINGABEN */
0220
0230    SPC BEDIENINTERRUPT INTERRUPT GLOBAL ;
0240
```

```
0250
0260        /*----------------------------------------------------------------*/
0270
0280        /* DATEN */
0290
0300        SPC LESEN BIT(1) GLOBAL ;
0310        SPC FREIGEBEN BIT(1) GLOBAL ;
0320        SPC FREIGABEORGANSTELLSIGNAL ( ) BIT(1) GLOBAL ;
0330        DCL ALTEZIELINFORMATION FIXED GLOBAL INITIAL(0) ;
0340        DCL PAKETZAEHLER FIXED GLOBAL INITIAL(0) ;
0350        DCL MAXPAKETANZAHL FIXED GLOBAL INITIAL(10) ;
0360        DCL UMRECHNUNGSTABELLE (2,8) FIXED GLOBAL ;
0370        DCL MAXAUSLAUFZEIT DUR GLOBAL INITIAL(10 SEC) ;
0380        DCL UNBELEGT INV BIT(1) GLOBAL INITIAL('0'B1) ;
0390        DCL BELEGT INV BIT(1) GLOBAL INITIAL('1'B1) ;
0400        DCL ESBELEGUNG BIT(1) GLOBAL INITIAL(UNBELEGT) ;
0410        DCL EIN INV BIT(1) GLOBAL INITIAL('0'B1) ;
0420        DCL AUS INV BIT(1) GLOBAL INITIAL('1'B1) ;
0430        DCL RUECKFUEHRBAENDERFLAG BIT(1) GLOBAL ;
0440        DCL BAENDERSTELLSIGNAL1 BIT(1) GLOBAL ;
0450        DCL BAENDERSTELLSIGNAL2 BIT(1) GLOBAL ;
0460        DCL UNTERBRECHUNGSFLAG BIT(1) GLOBAL INITIAL('0'B1) ;
0470        DCL ANGEHALTENMELDUNG INV CHAR(14) GLOBAL INITIAL('PVA ANGEHALTEN') ;
0480        DCL MAXNACHFUEHRZEIT DUR INITIAL(20 SEC) ;
0490
0500
0510        /*----------------------------------------------------------------*/
0520
0530        BEDIENUNG : TASK GLOBAL ;
0540
0550
0560        DCL BEDIENFUNKTION FIXED INITIAL(3) ;
0570        DCL LINKS INV BIT(1) INITIAL('0'B1) ;
0580        DCL RECHTS INV BIT(1) INITIAL('1'B1) ;
0590        DCL LENKORGANSTELLSIGNAL BIT(1) INITIAL(RECHTS) ;
0600
0610
0620              /* BEGIN ACTION BEDIENUNG */
0630        WHILE BEDIENFUNKTION GT 2  REPEAT
0640              /* BEGIN ACTION BEDIENARTBESTIMMEN */
0650              /* CODE-PART FEHLT ! */
0660              /* END ACTION BEDIENARTBESTIMMEN */
0670          IF BEDIENFUNKTION EQ 1
0680          THEN
0690                  /* BEGIN ACTION BETRIEBSSTART */
0700                  /* CODE-PART FEHLT ! */
0710                  /* END ACTION BETRIEBSSTART */
0720          ELSE
0730            CASE BEDIENFUNKTION
0740              ALT :                   •
                                          •
                                          •
0880                      /* BEGIN ACTION UTABEINGEBEN */
0890                      /* CODE-PART FEHLT ! */
0900                      /* END ACTION UTABEINGEBEN */
0910            ALT
0920                      /* BEGIN ACTION BETRIEBSUNTERBRECHUNG */
0930                      /* BEGIN ACTION UNTERBRECHUNGSFLAGSETZEN */
0940                      /* CODE-PART FEHLT ! */
0950                      /* END ACTION UNTERBRECHUNGSFLAGSETZEN */
0960                      /* BEGIN ACTION PAKETVERTEILUNGSTOPPEN2 */
0970              IF ESBELEGUNG EQ UNBELEGT
0980              THEN
0990                      /* BEGIN ACTION RUECKFUEHRUNGAUSSCHALTEN1 */
1000                      /* CODE-PART FEHLT ! */
1010                      /* END ACTION RUECKFUEHRUNGAUSSCHALTEN1 */
1020                TERMINATE NACHFUEHRUEBERWACHUNG ;
1030                AFTER 2 SEC RESUME ;
1040                IF ESBELEGUNG EQ UNBELEGT
1050                THEN
1060                      /* BEGIN ACTION FREIGABESTELLUNG */
1070                      /* CODE-PART FEHLT ! */
1080                      /* END ACTION FREIGABESTELLUNG */
1090                  AFTER 10 SEC RESUME ;
1100                      /* BEGIN ACTION PAKETABFUHRSTOPPEN */
1110                      /* BEGIN ACTION RUECKFUEHRUNGAUSSCHALTEN2 */
1120                      /* CODE-PART FEHLT ! */
1130                      /* END ACTION RUECKFUEHRUNGAUSSCHALTEN2 */
1140                      /* BEGIN ACTION UNTERBRECHUNGSFLAGRUECKSETZEN */
1150                      /* CODE-PART FEHLT ! */
1160                      /* END ACTION UNTERBRECHUNGSFLAGRUECKSETZEN */
1170                      /* BEGIN ACTION PVAANGEHALTENMELDEN */
1180                      /* CODE-PART FEHLT ! */
1190                      /* END ACTION PVAANGEHALTENMELDEN */
1200                      /* END ACTION PAKETABFUHRSTOPPEN */
1210                FIN ;
1220              FIN ;
1230                      /* END ACTION PAKETVERTEILUNGSTOPPEN2 */
1240                      /* END ACTION BETRIEBSUNTERBRECHUNG */
1250            FIN ;
1260          FIN ;
```

SECURE - FORTRAN IMPLEMENTATION OF "SUBROUTINES FOR CAMAC"
FOR NOVA - COMPATIBLE COMPUTERS

Zbigniew BANASIK and Janusz ZALEWSKI
Institute of Nuclear Research, Dorodna 16
03-195 Warszawa, Poland

ABSTRACT

SECURE is a set of subroutines compatible with ESONE/SR/01 "Subroutines for CAMAC" for SEN Electronique CC2023 crate controller and NOVA-compatible computers running the RDOS operating system. It was designed for experimenters which apply CAMAC but have programming experience only with Fortran. Its main remarkable feature and the real benefit of use is to make possible all CAMAC programming, including all interrupt service and message transfer between tasks, in Fortran. The straightforward consequence of this fact was that programming of quite sophisticated control systems was possible to be made by technicians. When combined with other experimenters' experience, this may prove the extraordinary efficiency of software standardization efferts.

RESUME

Le système SECURE est un ensemble de sous-programmes étant en accord avec la recommendation ESONE/SR/01 "Subroutines for CAMAC", ecrits pour le contrôlleur du chassis CAMAC du type CC2023 /produit par SEN Electronique, Geneve, Suisse/ pour les ordinateure NOVA operants sous le moniteur RDOS. Le système est destine pour les experimentateurs, qui utilisent CAMAC, mais ils ne connaisent que Fortran. L'avantage le plus important de SECURE c'est la possibilité de programmer en Fortran des fonctions de tous le blocks CAMAC, le service des interruptions et la communication parmi les taches y compris. Ça donne une possibilité de programmer des dispositifs complexes de CAMAC meme par des techniciens. Ce constantation et les opinions des autres specialistes font le preuve du l'efficacité extrême des travails sur le normalisation du logiciel.

ZUSAMMENFASSUNG

Das SECURE System ist eine Unterprogrammenbibliothek, die gemäss der ESONE/SR/01 Empfehlung "Subroutines for CAMAC" für die CAMAC-Kassettensteuereinheit von Typ CC2023 /der Firma SEN Electronique/ sowie für die NOVA Kleinrechner mit dem Betriebssystem RDOS, erbereitet wurde. Das System ist für diejenigen Experimentatoren vorgesehen, die CAMAC, anwenden, aber Erfahrung nur bei der Anwendung von Fortran haben. Das wichtigste Merkmal des Systems, das die Vorteile seiner Anwendung bezeichnet, ist die Möglichkeit der kompletten Programmierung der CAMAC Module, einschliesslich mit Unterbrechungenbedienung und Nachrichtenüberweisung zwischen einzelnen Aufgaben, in Fortran. Die unmittelbare Konsequenz dieser Tatsache ist die Möglichkeit der Programmierung von zusammengesetzen Steuermodulen auch für Techniker. Die obenerwähnte Feststellung, zussamen mit Meinungen von anderen Experimentatoren, beweist einer enormen Wirksamkeit der Normierungsarbeiten im Software-Bereich.

1. INTRODUCTION

SECURE stands for "SEN Electronique Controller Under RDOS Executive" and is a set of subroutines designed for CAMAC systems controlled by CC2023 crate controller-interface /SEN Electronique, Geneva, Switzerland/ to NOVA minicomputer. It is written in compliance with "Subroutines for CAMAC" document /ESONE/SR/01/ and runs under the RDOS operating system /Data General, Westboro, MA, USA/.

Although there existed CAMAC software for this configuration, based on Basic /1/ or Fortran /2/ calls, these seemed to be more complicated, home-made solutions, not compatible to standards. The rationale behind using standard Fortran calls to program all CAMAC I/O, including interrupt service, was to enable the experimenters to do all required programming by themselves. This is extremely important when a user non-familiar with computer software details can request and program by himself all computer functions he wants. His savings in money then, are evident. On the other hand, this directly leads to high increase in efficiency of professional programmers.

2. DESCRIPTION

2.1 "Subroutines for CAMAC" reminder

The document titled "Subroutines for CAMAC" defines a unique set of callings, preserving the distinction between declaration statements, which are used to name and specify computer and CAMAC entities, and action statements to implement various data movements and condition tests in CAMAC. The subroutines are grouped for convenience in three subsets /levels of implementation/.

2.2 Primary subroutines /level A/

There are two subroutines, which can handle all of the CAMAC programming even for a moderately sophisticated experiments: CDREG - defining CAMAC address on a basis of station number and subaddress values, according to the controller command register format, CFSA - performing the desired CAMAC function /upon the CAMAC address defined previously/ and transferring 24-bit data and Q-response. Some subroutines from level B are implemented with the use of level A code, so in fact they are also included into the primary set. These are: Crate Clear, setting, clearing and testing Dataway Inhibit.

2.3 LAM handling /level B/

The most important of the LAM handling routines allow declaring LAMs /CDLAM/ and linking LAMs to service procedure /CCLNK/. The CCLNK subroutine performs an association between LAM-interrupt specified by LAM-variable /defined previously by CDLAM/, and a procedure identified by LABEL. As a result of this association the procedure is executed whenever the LAM-interrupt is recognized by the system.

The actual service priority is determined by the sequence of CCLNK calls: the earlier a definite LAM is assigned a routine, the higher is its service priority. The service procedure must be parameterless and can be written totally in Fortran, with use of the further calls to the standard subroutines /e.g. from the scope of this paper/. Then, though the speed of service is sacrificed, the user has an opportunity to write the CAMAC interrupt service routines by himself, without knowledge of assembly language.

Three other LAM-related subroutines - enabling/desabling, clearing and testing LAMs - make use of CFSA calls and fall within the scope of primary subroutines rather than level B.

2.4 Multiple action routines /level C/

It is obvious from the definition of a single action subroutine, CFSA, that in fact there is a possibility to take advantage of the multiple use of certain arguments /function, address or data/ in one call, rather than perform several similar calls. This is done in multiple action subroutines: General Multiple Action, Address Scan, and Block Transfers. These subroutines however, include some peculiarities, and for complete description the reader is referenced to the user's manual /available from the authors/. Suspending execution until a specified LAM occurs, is not supported in current version.

2.5 Remaining subroutines

SECURE comprises 17 of the total 21 recommended subroutines. Those not implemented are: Crate Demand subroutines /important in multicrate systems - the authors have had a possibility to fully test SECURE with single crate only/, and Crate Initialize subroutine /unuseful for this particular minicomputer/. The only system dependent subroutines are: FIXMT - set up to enable sending messages from Fortran interrupt service routine to another task, and FLINC - performing floating-point to 24-bit integer conversions and vice versa.

3. EXPERIENCE

3.1 Program example

To illustrate the real simplicity of programming we present a piece of code for servicing touch panel interrupts: the service routine should identify the button pressed, and send a message to activate another task. The main program defines the association between a specific LAM and a routine to service it, called PANEL:

CALL CCLNK (LAM∅∅, PANEL)

When a specific button is pressed, e.g. to read and display data from a selected channel, the following routine is executed:

```
SUBROUTINE PANEL
...
;READ AND CLEAR LAMREG
CALL CFSA(2,LAMREG,DATA,Q)
;IDENTIFY BUTTON
IF (cond) MSG=source
...
CALL FIXMT(KEY,MSG)
RETURN
```

The proper service, dependent upon function of the interrupting button, is performed by a separate task, which accepts a message by executing a statement:

CALL REC (KEY,MSG)

3.2 Multitasking and interrupts

When using multitask access to CAMAC crate, it is the responsibility of the user to provide some means for scheduling tasks in order to protect CAMAC operations /e.g. block transfers/ against distortion, and to ensure data integrity. Although such tools were not envisaged in "Subroutines for CAMAC", there are various possibilities under RDOS, e.g. by message transfer with XMT and REC calls, or by .SINGL/.MULTI task calls. Crucial for the safe operation of LAMs is the reentrancy of Fortran routines. At the time the interrupt occurs the total environment is being frozen, and the need arises to adjust some stacks /3/.

3.3 Implementation remarks

Since the crate controller /CC2023/ station number is ∅ -- not allowed in CAMAC -- there is a danger that CFSA and other operations on undefined CAMAC variables will be executed, due to high probability that undefined address is taking the value of zero. This may cause severe troubles when debugging programs. Though there is some room in "Subroutines for CAMAC" for non-standard or system-dependent subroutines, some of the CC2023 intristic features are not exploited in SECURE. This concerns, in particular, Data Channel transfers /NIOP computer instruction/, and in part the possible subaddress and station scans /when bits ∅ or 1 of controller command register are set/.

4. CONCLUSIONS

The most important conclusion has become obvious after first months of the work with SECURE. This is that the real benefit of implementing the "Subroutines for CAMAC" was the possibility to program all CAMAC interrupt service in Fortran. The straightforward consequence of this fact was that programming of quite sophisticated neutron spectrometer and diffractometer control systems was possible to be made by technicians. When combined with other experimenters' statements that such modular software could increase support personnel efficiency by several times, this may prove the extraordinary efficiency of software standarization efforts. /5/

Although SECURE plays its role quite well, "Subroutines for CAMAC" might have been utilized in more effective manner in a language like Ada. This concerns, in particular, CAMAC definitions package. Language properties can significantly complete the constructs which are missing in "Subroutines for CAMAC", e.g. guarding mechanism /4/.

The authors are aware that when simplicity of use is of primary concern, the speed and response time are sacrificed. However, the interrupt service time is of the order of milliseconds, thus not slower than in other similar systems /1,2/. Further technical details and a full listing of SECURE are available upon request.

5. REFERENCES

/1/ Uotila I., CAMAC Extension for Basic on the NOVA 1200 minicomputer, Report STL-A23, Institute of Radiation Protection, Helsinki, September 1976

/2/ Feenstra R., Johnson R.R., Winter C., Rapid CAMAC Data Handling in Fortran Real Time Programs, Nuclear Instruments and Methods, Vol.160, pp. 511-518 /1979/

/3/ Banasik Z., Zalewski J., Report on Data General Fortran IV Run Time Reentrance at Interrupt Time, Draft, Institute of Nuclear Research, Warsaw, January 1982

/4/ Zalewski J., Example of CAMAC Programming in Ada, submitted to 3rd IFAC/IFIP Symp. on Software for Computer Control, Madrid, 5-8 October 1982

/5/ Rimmer E.M., Requirements for Standard Practices in Future Large High Energy Physics Experiments, IEEE Trans. NS, Vol. 27, No. 1, p. 631 /1980/

Session 6

SYSTEM APPLICATIONS IN CONTROL AND AUTOMATION

APPLICATIONS DES SYSTEMES AU DOMAINE DU CONTROLE ET DE LA COMMANDE

SYSTEM-ANWENDUNGEN IN STEUERUNGEN UND BEI DER AUTOMATISIERUNG

NEW COMPUTER CONTROL SYSTEM OF 80-IN. HOT STRIP MILL

Natsuki SAIKAWA, Junzo NITTA, Takashi MINEMATSU,
Takashi MIKURIYA, Toshio TAMIYA and Toshisada TAKECHI
Kawasaki Steel Corporation, Chiba Works
Kawasaki-cho, Chiba 260, Japan

Kiyohiko YOSHIE and Hisashi EZURE
Toshiba Corporation, Fuchu Works
Toshiba-cho, Fuchu, Tokyo 183, Japan

ABSTRACT

The computer control system of a hot strip mill is discussed. The new computer control system was introduced to the 80-in. hot strip mill at Chiba Works in 1981.
The features of the hot strip mill control were considered in designing the computer control system. High-speed, large-capacity (main memory: 1M bytes) computers and high-speed (10M bits/sec) dataways are adopted to handle a large amount of real-time rolling data and to make real-time control calculations of reheating furnaces and rolling mills.
High-level data handling and high-precision process control are successfully implemented and a good control performance has been achieved in actual operations.

RÉSUMÉ

Le systemé de contrôle d'une usine de laminage à chaud est décrit. Le nonveau systemé de contrôle a été suis en service à l'usine Chiba Works en 1981. Les caractiristiques de L'application out imposé le choix de calculateurs de contrôle à grande vitesse et grande capacité (1 Mbytes) et de bus à grande vitesse (10 Mbits/s), afin de de gerér de gros flots de donneés et de pouvoir éxécuter les calculs de commande des fourneaux et des unités de laminage en temps réel. La gestion des donneés et la commande à baute précision aut été réalisées avec succés et de bonnes performances aut été obtennes à le mise en oeuvie.

ZUSAMMENFASSUNG

Diese Untersuchung beschäftigt sich mit einem computergesteuerten System für ein Warmbreitbandstraße. Das neue computergesteuerte System wurde im Jahre 1981 für das 80"-Warmbreitbandstraße im Chiba-Werk eingeführt.
Für die Auslegung des computergesteuerten Systems wurden die Kennzeichen der Steuerung des Band-Warmwalzwerks in Betracht gezogen.
Hochgeschwindigkeitscomputer mit großer Speicherkapazität (Hauptspeicher: 1M Bytes) und hochgeschwinde Datenübertragungsleitungen (10M Bits/sec) wurden verwendet, um eine große Menge von Echtzeit-Walzdaten zu verarbeiten und um Echtzeit-Kontrollrechnungen für Nachwärmöfen und Walzwerke durchzuführen.
Eine hohe Datenverarbeitungskapazität und eine hochgenaue Prozeßsteuerung wurden erfolgreich eingeführt, und im praktischen Betrieb wurde eine gute Steuerleistung erreicht.

1. INTRODUCTION

A hot strip mill is a huge installation which produces about one-half of all the rolled steels and is the most highly automated among various steel plants. It has a long history of computer control. At Chiba Works of Kawasaki Steel Corporation, research on computer control started back in 1964 and computer control, ranging from the reheating furnace exit to the coiler, was accomplished in 1969, thereby achieving effective results.[1]

Recent trends have continued toward more stringent requirements for product quality improvement and energy saving in the hot strip mill process, and the old system has become unable to cope with these new demands. Following investigation, a new computer control system was introduced to replace the old one in October, 1981 to achieve more effective improvement in process control functions.

The features of the system are given below.

(i) High-speed, large-capacity (main memory: 1 M bytes) computers and high-speed (10 M bits/sec) dataways are adopted to handle a large amount of data for real-time processing.

(ii) The heat transfer model of reheating furnaces and plastic deformation model of rolling mills are calculated in real time and the calculation results are applied to process control.

2. DESCRIPTION OF THE MILL

There are two hot strip mills at Chiba Works: No.1 56-in. mill constructed in 1958 and No.2 80-in. mill constructed in 1963. The new computer system is applied to No.2 hot strip mill which is an 80-in., fully continuous mill consisting of 4 reheating furnaces, 4 roughing mills, 7 finishing mills and 3 coilers. The general specifications of No.2 hot strip mill are shown in Table 1 and the mill layout in Fig. 1.

When rolling orders are given, slabs of various sizes and steel grades are charged into the reheating furnace and heated up to the required temperature. After reheating, they are rolled into the ordered sizes by the roughing and finishing mills, and the rolled strips are coiled by the downcoilers and sent to the weighing scale and marking machine. Various kinds of strips are rolled at intervals of about 2 minutes and the average number of coils per lot is as small as 3 pieces, so that high-speed, real-time handling of a large quantity of data and high-precision flexible process control are necessary.

3. FEATURES OF HOT STRIP MILL COMPUTER CONTROL

The features of the computer control of a hot strip mill are as follows:

(i) A variety of equipments such as reheating furnaces, rolling mills, measuring instruments, automation equipments, etc. in a long line of nearly 700 m must be concentrically controlled.

(ii) It is necessary to handle a large amount of data at high speed and with high reliability. During the rolling of the strip, process inputs/outputs of more than 2,000 points are accessed.

(iii) To roll steel strips of high quality, it is necessary to calculate complicated mathematical models on real-time basis in the control of a rolling mill and the control of a reheating furnace.

(iv) High-speed response is required in process control, operator interface and data transmission with other computers. Each of them requires response within a few seconds.

(v) Computer control is indispensable for energy saving, quality control of products and achievement of high productivity, and system failure entails the operation stop.

(vi) As the characteristic of the hot strip mill process, equipment expansion and remodelling are continually performed, and it is necessary to update the computer control system correspondingly. Thus, system maintenance should be easy, and the system should be flexible to allow future expansion.

(vii) In case of replacing the computer control system of the hot strip mill, it is necessary to replace it without disturbing the production. There is a major restriction on the construction of a new system, and the system must be constructed in a short period of time.

Table 1
General specifications of the mill

Type			Full continous mill
Productivity			
Annual production			3,600,000 tons
Slab size	Thickness		130 - 260 mm
	Width		710 - 1,910 mm
	Length		4,510 - 9,150 mm
	Weight		6.0 - 21.0 tons
Product size	Thickness		1.2 - 19.0 mm
	Width		730 - 1,900 mm
	Weight (max.)		21 tons

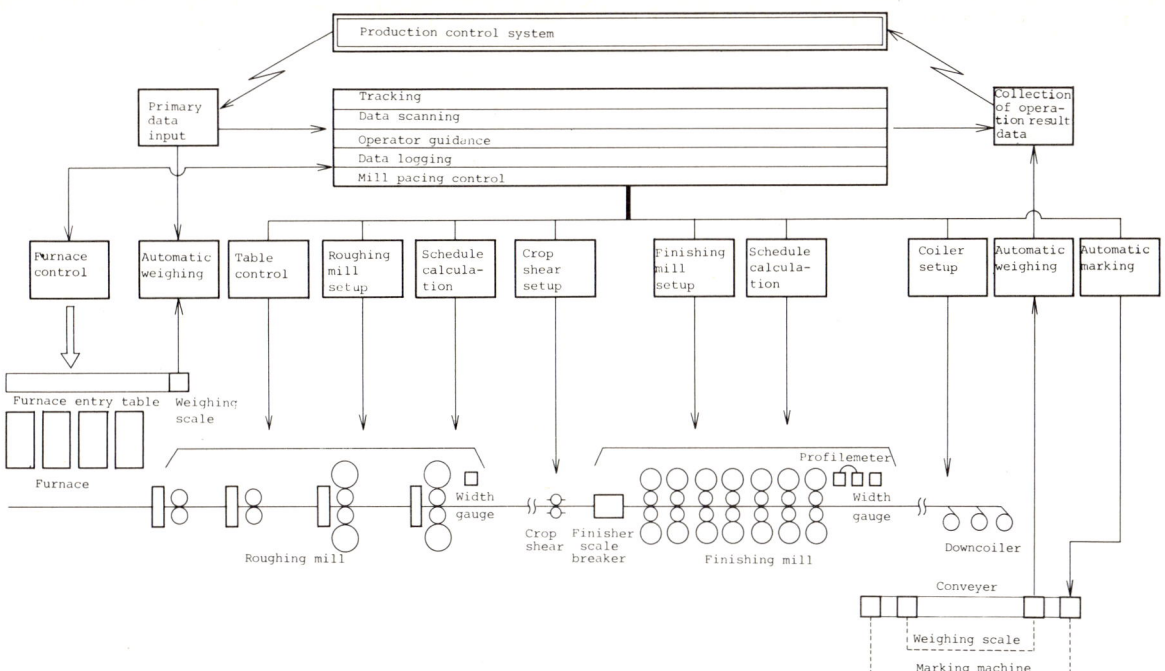

Fig. 1 Diagram of computer control as related to mill layout

4. SYSTEM DESIGN

4.1 System Constitution

For the design of the computer control of a hot strip mill it is necessary to consider the characteristics of the process control of the hot strip mill.

The constitution of the new system is shown in Fig. 2.

This system is constructed in consideration of the following:

(i) Sharing of process information among computers is promoted by employing high-speed dataways.

(ii) A distributed computer control system by a group of process computers is constructed to achieve a good control performance.

(iii) Of the two computers, one is used for mill control and the other for reheating furnace control. In case of failure of the mill control CPU, the furnace control CPU provides a backup system to enhance the reliability of the total system.

(iv) In order to start the system on-line in a short period of time, parallel run test of new and old systems has been realized by duplicating the inputs to the new and old systems.

4.2 Hardware and Software

The following are taken into consideration in the design of system hardware:

(i) High-speed dataways of 10 M bits/sec are installed to realize a distributed control system and to assure future system expansion.

(ii) Since real-time handling of a large amount of data and real-time calculations of complicated mathematical models are required, two computers of 1-M byte main memory and of 32-bit parallel architecture are employed.

(iii) As auxiliary memory units, four 4-M byte high-speed disks and one 153-M byte magnetic disk are employed. The four 4-M byte disks are duplicated and can be accessed by both CPUs. These large-scale memories support the application of higher level languages to software programs.

(iv) Since both the mill control CPU and the furnace control CPU refer mutually to the other process data, computer system linkage techniques are employed so as to allow mutual reference to data without making use of dataways.

The software of this system features the following:

(i) Software techniques, such as module structure program design, etc., are employed in order to improve the reliability and productivity of software as well as to maintain the software quality after completion.

(ii) Program compiling is possible at both the two computers. Source images of programs are stored in the 153-M byte disk, and the correction, assembling, and compiling of programs are possible on the CRT displays. The cardless system by text editor function is realized.

5. DESCRIPTION OF COMPUTER CONTROL

Fig. 1 shows the functional block diagram of the new computer control system along the line layout of the hot strip mill. After rolling orders are received from the master production

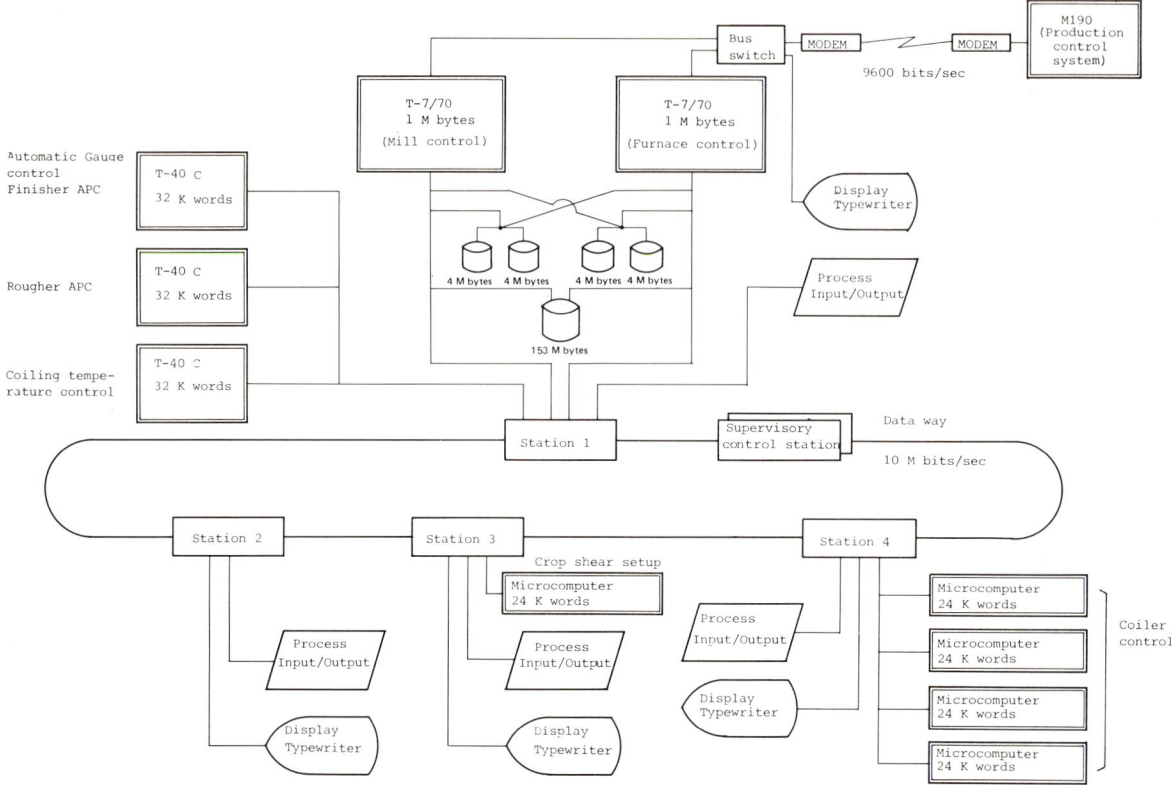

Fig. 2 System configuration of Chiba No. 2 hot strip mill

control system, automatic control and automatic operation of the hot strip mill process, ranging from reheating furnace entry to the coil marking machine on the conveyer line, are performed according to the orders.

The control functions can be divided into slab and coil tracking, data scanning, operator guidance, data logging and process control. The process control functions include mill pacing control, reheating furnace combustion control, automatic setting and schedule calculation of the roughing mill, automatic setting of the crop shear, automatic setting and schedule calculation of the finishing mill, automatic setting of coilers, automatic weighing, and automatic marking. The rolling results and quality information of the products are transmitted to the master system. Of the functions of this system, data scanning, furnace combustion control and strip profile control are described below.

6. DATA SCANNING FUNCTION

6.1 Dataways and Data Scanning

The basis of process control is the timely collection of accurate process information. In the hot strip mill, since it is necessary to handle a large amount of data at high-speed, efficient collection of process data should be contrived.

In this system, dataways are used for the computer linkage and the transmission of information of process inputs/outputs and peripheral units to promote the sharing of process information among computers.

Further, the employment of dataways greatly reduces the amount of wiring works, since information is collected from a remote station nearest to the information source. To replace a large scale computer control system such as one for the hot strip mill without disturbing the operation, the dataway is an indispensable tool.

The demerit of the dataways is that a trouble in the dataways results in failure of the total system. In this system, the reliability has been improved by duplicating the supervisory control unit (SVC) for dataways.

6.2 Data Scanning Technique

In a system in which dataways are used, both the processing time of dataways and the processing time of CPU must be taken into consideration.

The processing time for process inputs/outputs in this system is as follows:

	CPU processing time	Dataway processing time
Fixed time	1.6 msec	1.6 msec
Processing time per word	0.16 msec	0.004 msec

Therefore, in case of transmitting n words of information, the time of (3.2 + 0.164 x n) msec is required.

Since more than 2,000 points of process inputs/outputs are accessed per one strip in the computer control system of the hot strip mill, it is advantageous from the viewpoint of the load on CPU and the processing time of dataways to increase the number of points of process inputs/outputs per access.

In this system, accordingly, process inputs/outputs are divided into eight blocks to realize efficient inputs/outputs access in each block. The blocks are divided in consideration of the simultaneity of process data.

By the above-described method, the scanned data are used as the common data source of all the programs to avoid the concentration of load upon process inputs/outputs access.

Table 2 shows the specifications of data scanning function. The most characteristic feature of the specifications is that pattern recognition is employed in the coil overall-length scan at the finishing mill. Steel strips have periodical temperature variation mainly due to uneven heating in the length direction (called skid marks, arising at max. 60°C to 70°C), and selection of measuring points, when strip thickness, width, and temperature are controlled at the finishing mill, greatly affects the control accuracy. Therefore, data scan is performed at a sampling pitch of 100 msec, and pattern-recognized data are used as information for process control.

7. FURNACE COMPUTER CONTROL

Computer control of the reheating furnace is a function to set the optimum heat pattern of the reheating furnace with the aim of energy saving and mill pacing.

Fig. 3 shows the control flow. First, the furnace computer control performs tracking of slab in the reheating furnace and makes an extraction schedule from mill pacing information. Then the control calculates the atmospheric temperature in the furnace by measured temperature, and the calculation results are used to predict the temperature rise of the slab from the heat transfer model. From the heat balance of the entire furnace, the required input calorific heat is calculated to determine furnace temperature of fuel flow rate at each zone in the furnace. The control results are learnt and fed back to the heat transfer model and heat balance model.

The most important among the above-mentioned models is the slab temperature calculation model. For the slab heat transfer model, a strict solution can be obtained by solving the heat transfer difference equation and the approximation of this solution is used, so that the model can be used for calculating the temperatures of slabs piece by piece in real time.

Table 2

Data scanning of the system

Scanning type		Scanning items	No. of scanning points	Scanning period	Data processing	Use
Furnace data scan	Constant interval scan	Zone temperature Zone fuel flow rate Zone air flow rate Slab temperature O_2 percentage etc.	160	Usual 30 sec Minimum 100 msec	Noise processing Calculation of average and variance	Furnace combustion control Operation results logging
Roughing mill scan	Coil leading end scan	Rolling force Rolling current Rolling screwdown Edger current Strip temperature Strip width etc.	25	Usual 100 msec	Noise processing Calculation of average and variance	Finishing mill setup Operation results logging
	Coil full length scan					
Finishing mill scan	Coil leading end scan	Rolling force Rolling current Rolling screwdown Sideguide gap Rolling velocity Strip temperature Strip width Strip thickness Strip profile Looper current Looper height etc.	103	Usual 100 msec	Noise processing Calculation of average and variance Classification Pattern recognition	Learning control of finishing mill setup Operation results logging Quality control logging
	Coil full length scan					

The fundamental formula of slab temperature rise prediction is given below. (Stefan-Boltzmann's law)

$$q = \sigma\varepsilon[(\theta g + 273)^4 - (\theta_S + 273)^4] \quad (1)$$
$$q = \sigma\varepsilon[(\theta g + 273)^4 - (\theta_M + 273)^4](1-A) \quad (2)$$

where
- q: transfer heat by radiation (kcal/m²h)
- θg: atmospheric temperature (°C)
- θ_S: surface temperature of slab (°C)
- θ_M: average temperature of slab (°C)
- σ: Stefan-Boltzmann's constant (kcal/m²h°k⁴)
- ε: emissivity
- A: constant depending on θ_M

Fig. 4 shows the results of simulation in which a comparison is made between the results of the above-shown simplified model and the strict solution obtained by solving the difference equation.

Conditions for the simulation are given by a slab thickness of 260mm, a slab width of 1,000mm, steel grade of killed steel, a slab initial temperature of 50°C and an extraction pitch of 7-min interval.

The above-mentioned formula gives sufficient accuracy as the mathematical formula which calculates slab temperature in real time during actual operation.

The advanced computer control system for furnaces has been realized to achieve minimum fuel consumption under any circumstances.

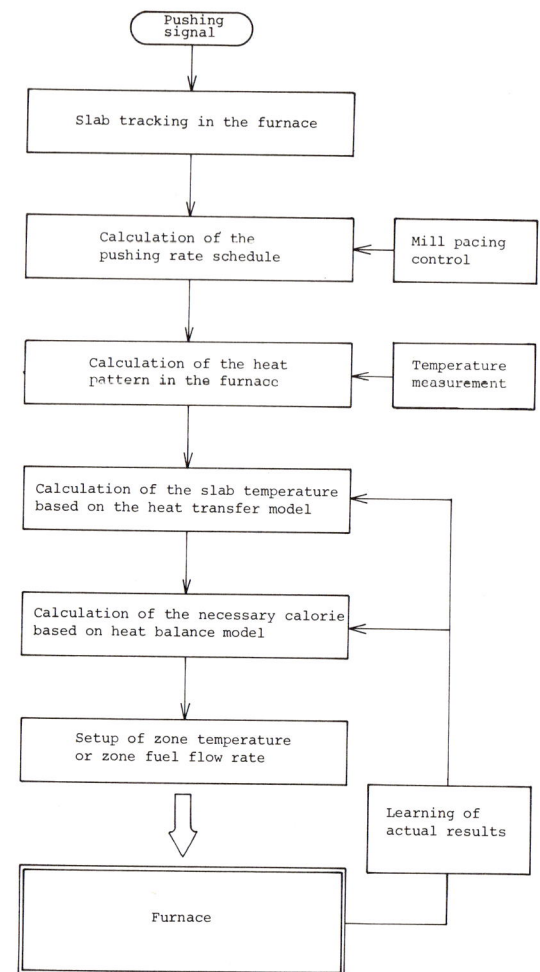

Fig. 3 General flow of furnace combustion control

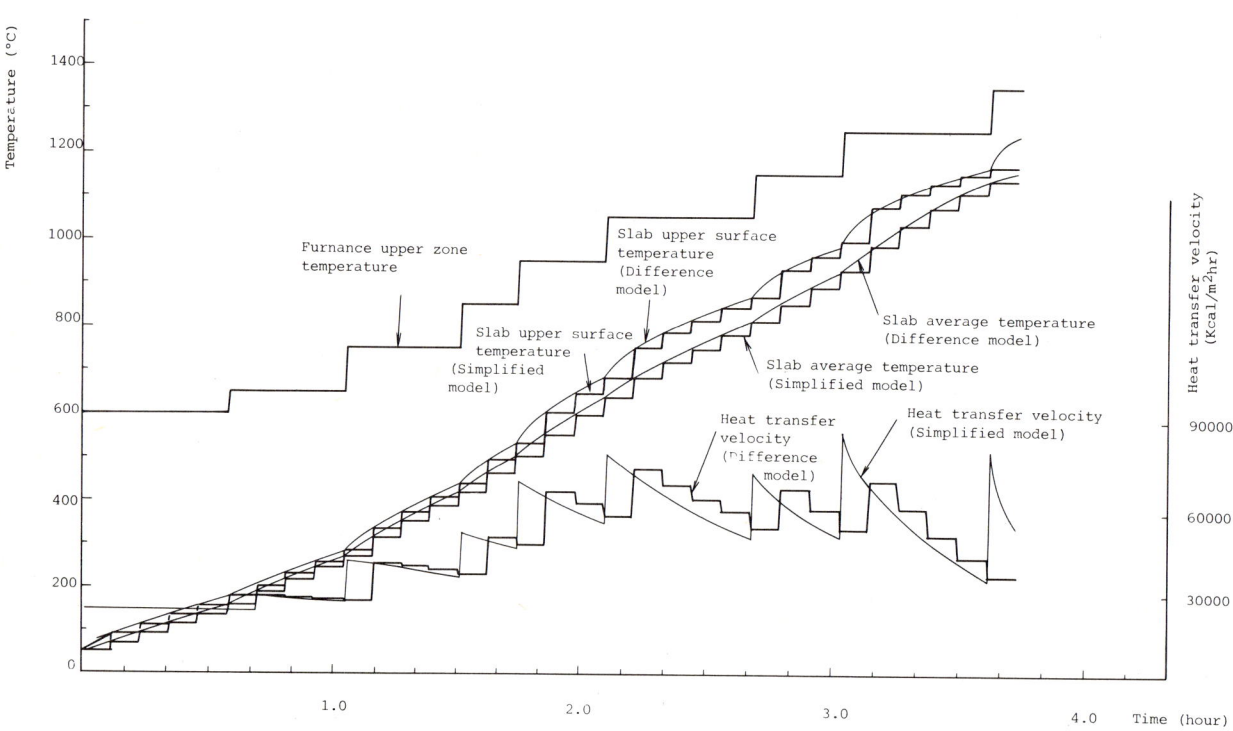

Fig. 4 Calculation of slab temperature transition

8. STRIP PROFILE CONTROL

The main function of mill computer control is the control of strip thickness. Mill computer control in the past was primarily concentrated on the strip center thickness, but recently the need has arisen for controlling the strip thickness distribution in the width direction (i.e. strip profile control).[2]

As shown in Fig. 5(a), the strip thickness changes in the width direction, and thus the center portion of the strip is usually thicker.

For each strip, the strip profile must be estimated at the finishing mill, and the load distribution of the finishing mill must be controlled, so that strip shall have the target profile and good flatness.

In order to estimate the strip profile, it is necessary to calculate the profile change of work rolls. As shown in Fig. 5(b), work rolls expand by heat from strips and also are worn by rolling. Table 3 shows the mathematical model which calculates the roll profile for each single strip.[3]

Fig. 6 shows the calculation results of the mathematical model shown in Table 3. The figure indicates how the roll profiles and strip profiles change, as rolling progresses from the first strip to the 80th strip.

These calculation results are used to modify the rolling schedule of the finishing mill for each strip to obtain the good profile.

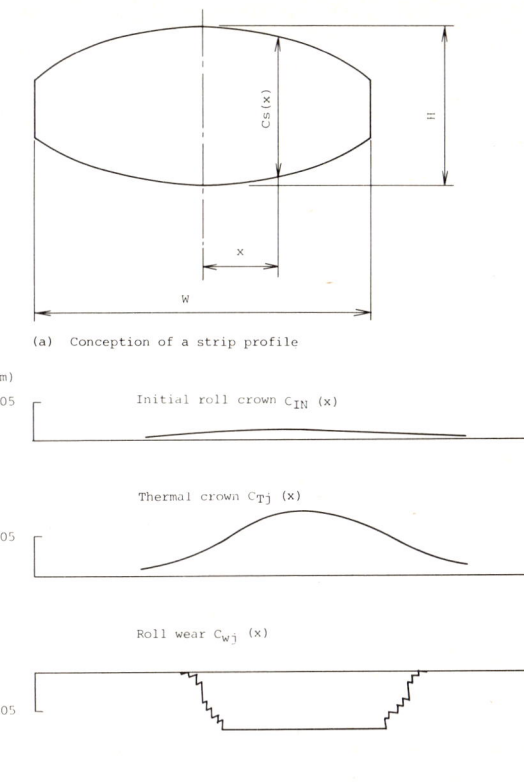
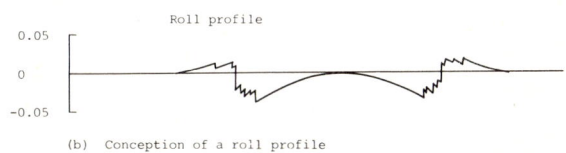

Fig. 5 Strip and roll profile of hot rolling

Table 3

Mathematical model of work roll profile[3]

Initial crown

$$C_{IN}(x) = C_{INC}[\cos(2\pi x/L) + 1]/2 \quad (3)$$

where $C_{IN}(x)$: initial crown at the position x
  C_{INC} : initial crown at the roll center
  x : any point of roll width from the center
  L : roll width

Thermal crown

$$C_{Tj}(x) = C_{Tj-1}(x) + (1 - e^{-\Delta t \cdot to})(C_{To}(x) - C_{Tj-1}(x)) \quad (4)$$

where $C_{Tj}(x)$: thermal crown at the position x after the rolling of the j-th coil
  Δt : rolling time or cooling time per one coil
  to : time constant to indicate the growth of thermal crown
  $C_{To}(x)$: steady-state value of thermal crown at the position

Roll wear

$$C_{Wj}(x) = C_{Wj-1}(x) + k_j (P_j l_j / W_j) \quad (5)$$

where $C_{Wj}(x)$: roll wear profile at the position x after rolling of the j-th coil
  k_j : wear constant, P_j : rolling load, W_j : strip width of the j-th coil
  l_j : length of the j-th coil

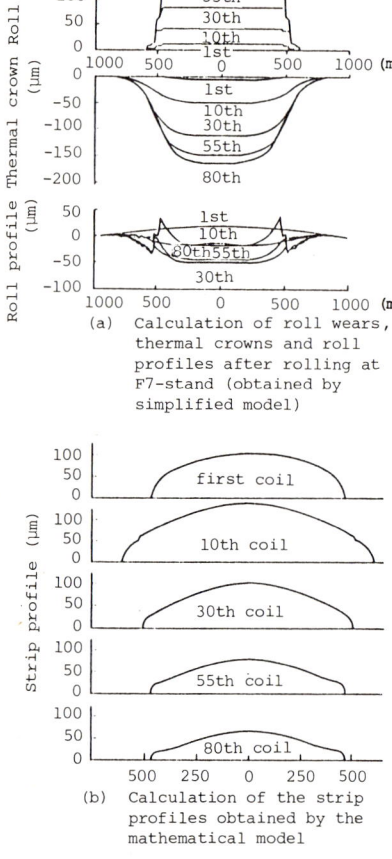

(a) Calculation of roll wears, thermal crowns and roll profiles after rolling at F7-stand (obtained by simplified model)

(b) Calculation of the strip profiles obtained by the mathematical model

Fig. 6 Calculation of strip profile

9. CPU LOAD

In a process control system in which high-speed response is required, it is important to examine the CPU load of the system.

Although many systems encounter the problem of CPU load, there is no general solution, so a symptomatic treatment in accordance with each system has to be employed occasionally.

As response delay arose in a part of programs after start of on-line operation, the CPU load was actually measured. The CPU load is defined as follows:

$$\text{CPU load} = \frac{\text{Number of monitor calls} + \text{Number of application programs calls}}{\text{Number of system timer calls}} \quad (6)$$

It has turned out that the measured value of the CPU load immediately after the start-up of on-line operation is approximately 70%, of which 25% steadily exists as the load of data scan and 15% as the CRT display.

In order to lighten this load, the following countermeasures were taken:

(i)   Reexamination of data scan programs
(ii) Modification of CRT display support programs
(iii) Redesign of integration of tasks
(iv) Reexamination of program priority
(v)   Modification of storage allocation

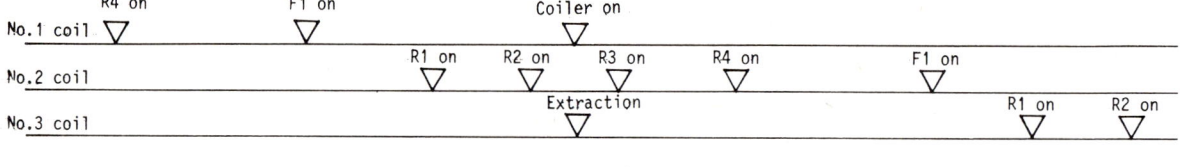

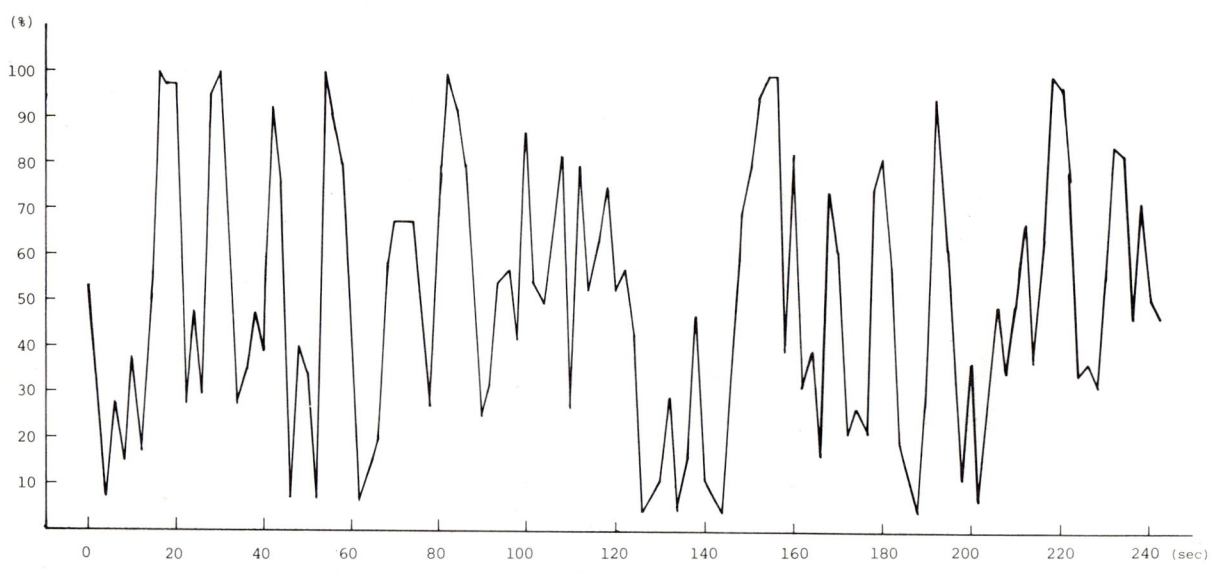

Fig. 7 CPU load of the system

As a result of the above-mentioned reexamination, the CPU load has been reduced to approximately 50%. An example of the actual measurement of the CPU load is shown in Fig. 7.

Of the above CPU load, the load of data scan is 17% in average, and the load of CRT display is several percent.

The CPU load and the control response remain to be solved in the future expansion of the system.

10. CONCLUSION

The features of hot strip mill control are considered in designing the computer control system.

High-speed, large-capacity computers (main memory: 1 M bytes) and high-speed dataways (10 M bits/sec) are adopted in the process control system of the 80-in. hot strip mill, and high-level data handling and high-precision process control are successfully implemented.

Process data are scanned at a minimum pitch of 100 msec and, after being processed, provide the process control models with timely accurate information.

On the basis of this information, the heat transfer model of the reheating furnace and plastic deformation model of the rolling mill make real-time control calculations.

The CPU load is investigated and a good control performance has been achieved in actual operations.

REFERENCES

{1} Kataoka, K. and Ushiku, Q., Computer Control of the 80-in. Hot Strip Mill, Iron and Steel Engineer. 2 (1973) 39-47

{2} Wilmotte, S., Economopoulos, M., Colin, R. and Thomas, G., New Approach to Computer Setup of the Hot Strip Mill, Iron and Steel Engineer. 9 (1977) 70-76

{3} Yarita, I., Kitamura, K., Nakagawa, K., Tamai, T., Kimishima, H. and Hamada, K., Crown Control of Hot Rolled Steel Strip by Changing of Rolling Schedules at Hot Finishing Mill, International Conference on Steel Rolling (The Iron and Steel Institute of Japan, 1980), Vol. 2, 473-484

THE CAB SOFTWARE PACKAGE

Y. Perrin, P. Scharff-Hansen, L. Tremblet - CERN
E. Barrelet, G. Fouque, K. de Kerday - Ecole Polytechnique

The CAMAC Booster, CAB, has considerable potential for widespread use owing to its unique performance (described in a joint paper) and its commercial availability. In anticipation of this, a collaboration between the Ecole Polytechnique and CERN has produced cross software, control and system software to minimize user effort for any given application. Portability, expandability and maintainability have strongly influenced the developments. This paper gives an overview of the package and of its current status, with emphasis on the techniques used to reach these goals.

Le 'CAMAC Booster', CAB, est promis à une large diffusion compte tenu de ses hautes performances (décrites dans une communication separée) et de sa disponibilité commerciale. Dans cette perspective, des outils logiciels ont eté developpés pour couvrir des besoins variés. Une collaboration entre l'Ecole Polytechnique et le CERN a produit un logiciel croisé ainsi qu'un logiciel de controle afin de minimiser l'effort de l'utilisateur pour une application donnée. Les critères de portabilité, de facilité d'extension et de maintenance ont fortement influencé les développements. Ce papier donne une vue d'ensemble du logiciel et de son état d'avancement en insistant sur les techniques utilisées pour satisfaire ces critères.

Auf Grund seiner einzigartigen Leistungsfähigkeit (siehe beiliegende Beschreibung) und der Tatsache, dass er auf dem Markt angeboten wird, kann dem CAMAC Booster, CAB, eine weitgehende Verbreitung und Benutzung vorausgesagt werden.

In Zusammenarbeit zwischen der Ecole Polytechnique und CERN entstand daraufhin eine Reihe von Programmen (Kreuz-, Kontroll- und Systemprogramme), die, für eine gegebene Anwendung, der Programmieraufwand des Benutzers auf ein Minimum zu reduzieren erlauben.

Portabilität, Erweiterungsfähigkeit und leichte Handhabung waren die wesentlichen Kriterien für diese Entwicklung.

Im Folgendem wird zunächst ein Ueberblick über das Programmpaket gegeben und der Stand der Entwicklung aufgezeigt. Der Schwerpunkt wird dabei auf die Techniken gelegt, die benutzt werden um diese Ziele zu erreichen.

1. BACKGROUND

CAB is a fast processor packaged as a CAMAC module. It is available in various configurations (crate controller, branch driver or both) under control of a CAMAC branch highway or dataway and/or IEEE 488 bus. This possibility to use CAB at various levels in a CAMAC data acquisition system combined with its ability to process data concurrently with full speed CAMAC operations makes CAB a very powerful tool for front-end data collection, event filtering/formatting.

A previous presentation at the REAL TIME DATA '79 conference in Berlin was devoted mainly to the CAB architecture and hardware. Several applications of CAB are described in a separate communication at REAL TIME DATA '82. This paper describes the software tools which have been developed for CAB.

2. DESIGN GOALS

Design of the software has been along the following lines:

- It should be portable in order to cover the wide spectrum of machines currently used in the CAB application fields.

- It must offer high-level facilities to allow the use of CAB by non-specialists (monitor calls, libraries,...).

- Provisions should be made for expandability (multi-CAB system).

- Good documentation should permit a user to benefit from these facilities and possibly to rewrite time-critical sequences, if necessary.

- Maintenance should be minimised by adopting a modular structure and source code management techniques.

3. METHODS

Adoption of standards wherever possible has been the main guideline to achieve portability. PASCAL, FORTRAN IV, the ESONE CAMAC subroutines library[1] have been used to write the program development tools.

The MUFOM FORMAT[2] has been chosen for CAB object modules to benefit from the future wide availability/easy installation of the MUFOM linker and pushers (CUFOM[3], the basis of MUFOM is used until the standard becomes official).

The technique used successfully for the maintenance of PILS[4] source code has been adopted for the CAB package. It essentially consists of working with a unique source file where all machine-specific lines are included and flagged by special comments. A simple text editor command file is used to regenerate each machine specific source from the unique source file.

4. PACKAGE DESCRIPTION

The CAB software package consists of the following tools:

- A relocatable cross assembler written in FORTRAN IV with labelled common handling to provide an HEXASCII file containing either absolute or CUFOM (MUFOM) formatted object code.

- A CUFOM pusher (cross-loader) generating, from the linker output, the CAB absolute program in an HEXASCII file. The CUFOM (MUFOM) linker handles all CUFOM outputs independently of the target machine and, as such, is not an actual part of the CAB package.

- An interactive loader/debugger written in PASCAL and running in the CAB system host permits uniform load (save) of CAB program files to (from) any CAB independently of its type and of its location in a multi-CAB system. Usual debugging tools (run/halt/single step, status, breakpoint, register and memory display/modification) are provided and enhanced by a built-in inverse assembler.

- A small monitor called Micro-Exec resides in the CAB to provide high-level facilities such as memory management (CAB data memory or any CAMAC memory module), task scheduling, I/O transfers etc. System library modules permit the Micro-Exec to be tailored to the user configuration and needs (single or multi-CAB system handling, CAB-C transparency emulation task, etc.).

- A histogram library (HCAB) based on a subset of the HMINI histogram package currently available as a standard at CERN on Norsk Data (ND-10, ND-100) and DEC (PDP, VAX) systems.

- A simulator program written in FORTRAN and presently running on CDC CYBER Series machines permits a thorough analysis of CAB program for complex debugging or whenever hardware is not available.

- A set of test programs.

5. CURRENT STATUS AND PLANS

At present the host resident software is available on the ND-100 computer and partly on other hosts as shown in Table 1. Work is being done to extend availability to other machines in cases where adaptation is simple. The different components of CAB resident software have been developed and tested separately (task scheduling, memory management, transparency emulation, multi-CAB system protocol) but integration with Micro-Exec remains to be done. Use of standards (especially PASCAL and the ESONE CAMAC library) has greatly enhanced portability. Though reducing the speed performance of the CAB, the high-level facilities offered by the Micro-Exec and the utility libraries provide efficient tools for easy writing and structuring in applications where power prevails over speed. An example of such an application is front-ending in fast data acquisition and monitoring of small events where big systems generally suffer from significant interrupt overheads. Further enhancement of the package, planned as a function of CAB users, might include:

- more user-contributed libraries,
- a menu-driven interactive loader which will allow users to generate CAB 'load modules' without detailed knowledge of the Micro-Exec and libraries structure,
- integration with the CERN standard data acquisition packages for ND-100 and VAX computers.

6. REFERENCES

[1] ESONE/NIM STANDARD CAMAC SUBROUTINES (1978)

[2] MUFOM - The MUFOM Format for Object Modules - Preliminary Document IEEE P695-81-58

[3] CUFOM - The CERN Universal Format for Object Modules, CERN DD/78/21 J. Montuelle

[4] PILS - A Portable Interactive Language System, D. O. Williams, R. D. Russell, CERN, DD-OC-82-4

TABLE 1

CAB PROGRAM DEVELOPMENT TOOLS ON VARIOUS HOSTS

Utility: Language:	ASMB Fortran	LINKER Bcpl	PUSHER Pascal	DEBUGGER Pascal	SIMULATOR Fortran
ND-10/100	√	√	√	√	-
VAX	PB	√	√	√	-
PDP-11	PO	PM	PP	PP	-
HP-1000	PO	PM	√	√	-
CDC	√	√	√	-	√
HP-9826	-	PM	√	√	-
LECROY-3500	√	PM	PP	PP	-
CAVIAR (6800) TEKTX-405X CBM-PET	-	-	-	√ in BASIC	-

√: available now
PB: adaptation in progress (byte swapping)
PO: adaptation in progress (overlaying)
PP: awaiting a PASCAL compiler
PM: awaiting MUFOM and a PASCAL compiler
- : not applicable or not planned

LE SYSTEME DE CONTROLE DU GANIL

L. DAVID, E. LECORCHE, LUONG T.T., B. PIQUET, M. PROME, M. ULRICH

GANIL, B.P. 5027, 14021 CAEN CEDEX, FRANCE

RESUME

Après une brève description du Grand Accélérateur National d'Ions Lourds, on expose la conception de son système de contrôle, qui met en jeu deux calculateurs MITRA 125, 15 microprocesseurs JCAM 10, des automates programmables APS 30-12 et un réseau CAMAC. On donne ensuite des détails sur le matériel utilisé, processeurs, interfaces, liaisons digitales et analogiques ; on explique en particulier comment sont faites les consoles, pourquoi, et comment on les utilise. On passe ensuite à l'aspect logiciel du système, en décrivant les différentes fonctionnalités qui ont dû être rajoutées au moniteur MOP2 pour gérer les consoles et contrôler les diverses parties de l'accélérateur ; quelques détails sont fournis sur la structure de la base de donnée.

ABSTRACT

The paper begins with a short description of the GANIL (Large National Heavy Ion Accelerator). Then is described the control system architecture ; it involves two MITRA 125 computers, 15 JCAM 10 microprocessors, APS 30-12 programmable controllers and a CAMAC network. Details are given about the hardware : processors, interfaces, datalinks and analog signals. Consoles design is discussed, as well as how to use them. Informations are given on the software which has been added to the MOP2 monitor to take care of the consoles and handle the various equipments controlled on the accelerator ; some details are given on the data base structure.

ZUSAMMENFASSUNG

Nach einer kurzen Beschreibung des grossen nationalen Schwerionenbeschleunigers (GANIL), wird die Anlage seines Kontrollsystems dargestellt. Für dieses System werden zwei MITRA 125 - Computer, fünfzehn JCAM 10 - Mikroprozessoren, APS 30-12 programierbare Automaten und ein CAMAC Netzwerk eingesetzt. Dann werden Einzelheiten über die benutzten Geräte gegeben : Computer, Rechnerschnittstellen, numerische und analoge Verbindungen. Ins besondere wird die Beschaffenheit des Schaltpultes deren Begründung und Verwendung erläutert. Dann geht man zum Software-Aspekt des Systems über, und beschreibt die dem MOP2 - System notwendigen funktionellen Zusätze, um die Schalpulte zu steuern und die unterschiedlichen Teile des Beschleunigers zu kontrollieren ; es werden ferner einige Details über die Struktur der Data-Basis hinzugefügt.

1. POURQUOI LE GANIL ?

L'expérience de base en physique nucléaire consiste à briser la structure du noyau en examen et provoquer des réactions nucléaires à l'aide de projectiles de masse élevée très énergiques. Les ions lourds envoyés en faisceau intense sur le noyau cible constituent des projectiles tout à fait adéquats.

On produit un tel faisceau dans un accélérateur d'ions lourds. C'est une machine complexe et puissante capable de faire croître, par processus cumulatif tirant parti de l'action de champs électriques et magnétiques, l'énergie du faisceau d'ions jusqu'à sa valeur d'utilisation dans l'expérience.

Telle est la mission du GANIL (Grand Accélérateur National d'Ions Lourds) construit à Caen (Calvados) dans le cadre d'un groupement d'intérêt économique associant le CEA (Commissariat à l'Energie Atomique) et l'I N2 P3 (Institut National de Physique Nucléaire et de Physique des Particules).

2. DESCRIPTION SOMMAIRE

Le GANIL doit accélérer les ions de tous les éléments de la table de Mendeleiev depuis le Carbone jusqu'à l'Uranium à des énergies maximales de 100 MeV/nucléon pour les ions légers et de 10 MeV/nucléon pour les ions les plus lourds. Le faisceau d'ions doit présenter une dispersion d'énergie ΔW inférieure au millième de l'énergie W.

Issu d'une source d'ions lourds, le faisceau est préaccéléré dans un cyclotron compact C0 qui lui confère l'énergie suffisante pour être injecté dans l'ensemble accélérateur principal constitué de deux cyclotrons à secteurs séparés CSS1 et CSS2 fonctionnant en cascade. (Figure 1)

L'accélération dans un CSS s'effectue grâce au champ électrique produit dans des cavités radiofréquence fournissant une tension maximale de 250 kV régulée à 10^{-4}.

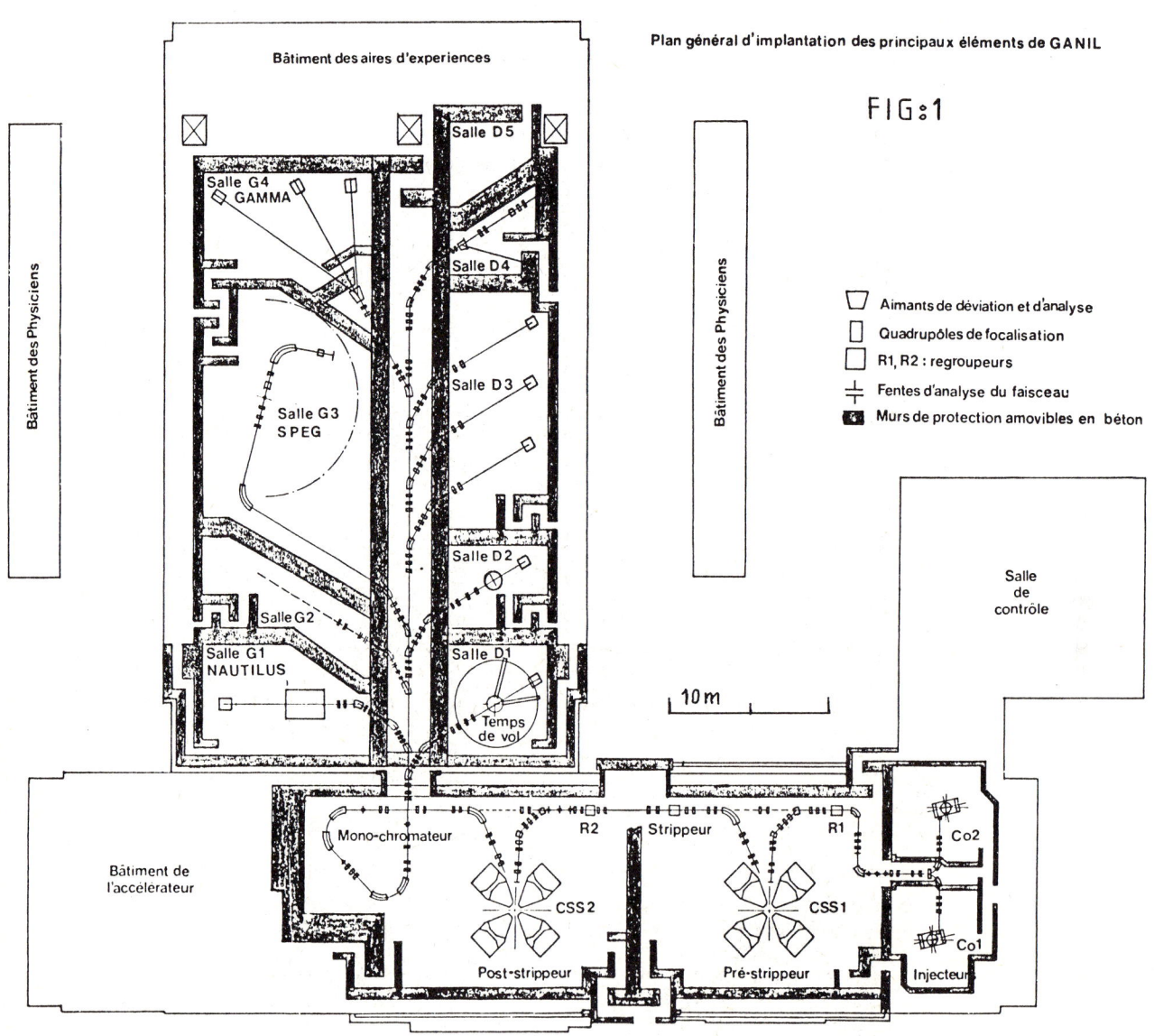

Plan général d'implantation des principaux éléments de GANIL

FIG:1

Le faisceau ainsi accéléré spirale dans le cyclotron, guidé par un champ magnétique continu dont la composante principale de 16 000 Gauss est produite par un courant de 1 850 A stabilisé à 10^{-5} alimentant un électroaimant à secteurs.

Le faisceau circule dans un espace vidé à la pression de $\simeq 6 \cdot 10^{-6}$ Pa, enclos dans une chambre à vide d'un volume de 46 m^3.

Le transit du faisceau d'ions entre l'injecteur C0 et CSS1, CSS1 et CSS2, CSS2 et les aires expérimentales est réalisé à l'aide des systèmes de transfert et de transport comportant de nombreux aimants de déviation, quadrupoles de focalisation, etc ...
Ces systèmes doivent assurer le guidage du faisceau vers l'expérience dans de bonnes conditions de conservation des caractéristiques du faisceau (énergie correcte, faisceau peu divergent, etc ...).

Des capteurs nombreux et variés servant aux diagnostics sont répartis sur l'ensemble de la machine. Ils permettent de contrôler à tout moment les composantes de la machine et de mesurer tout au long de son parcours les paramètres importants du faisceau accéléré.

3. CONSIDERATIONS GENERALES SUR LE SYSTEME DE CONTROLE (1), (2)

3.1 Généralités

La conception du système de contrôle du GANIL a été guidée par les considérations suivantes :

a) c'est une machine complexe par la variété des techniques mises en oeuvre : vide - mécanique - électrotechnique, électronique, etc ...
Environ 2 000 paramètres doivent être contrôlés et commandés.

b) le contrôle d'une telle machine s'apparente à celui d'un processus industriel
fonctionnement continu dans de bonnes conditions de fiabilité et de sécurité 24 heures par jour pendant plusieurs semaines de suite.
Avec des pauses pour changer le type d'ions accélérés, et pour un type d'ions donné changer l'énergie du faisceau.
Le système de contrôle doit pouvoir faire effectuer ces changements dans des délais les plus réduits possibles.

c) le GANIL est un instrument de recherche
En tant qu'instrument de recherche fondamentale de la physique nucléaire, l'accélérateur doit pouvoir s'adapter aux exigences de celle-ci. C'est-à-dire qu'au cours de sa vie, la machine sera amenée à évoluer plus ou moins profondément, à des échéances - a priori non prévisibles - plus ou moins éloignées.
Le système de contrôle doit être suffisamment souple pour pouvoir s'adapter rapidement à de telles évolutions sans changements fondamentaux.

d) la machine est utilisée par des chercheurs non forcément spécialistes de l'informatique.
Le système doit donc être suffisamment "transparente" c'est-à-dire simple d'utilisation pour l'opérateur.
En particulier, l'adoption d'un langage de haut niveau pour programmer les applications s'impose tout à fait.

3.2 L'architecture matérielle

Le système de contrôle du GANIL est organisé autour de 2 ordinateurs MITRA 125 : l'un servant au contrôle de l'accélérateur, l'autre destiné aux développements et à la reprise en secours du premier.

L'utilisation de microprocesseurs permet d'apporter une intelligence locale capable d'assurer des tâches ancillaires, soulageant ainsi appréciablement la charge de l'ordinateur de contrôle.

Il est à remarquer également que l'utilisation de microprocesseurs facilite grandement la mise au point de l'ensemble, chaque microprocesseur pouvant faire l'objet d'essais locaux, et n'être relié au calculateur central que beaucoup plus tard.

Des automates programmables, enfin, destinés principalement au contrôle du vide et aux sécurités de la machine, constituent le 3e type de processeur utilisé dans notre système.

Le standard CAMAC est adopté pour l'ensemble des liaisons entre les processeurs et l'accélérateur.

4. LE MATERIEL

4.1 Les MITRA

Le MITRA 125 de contrôle est principalement doté d'une mémoire de 160 K mots de 16 bits, de 2 disques de 5 Megaoctets, d'une imprimante sérielle et de 2 coupleurs CAMAC.

En cas de panne de cet ordinateur, le contrôle de l'accélérateur est repris - après commutation manuelle - par le MITRA 125 de secours qui, lui, est plus étoffé : 224 K mots mémoire, 4 disques de 5 Megaoctets, 2 bandes magnétiques de 1 600 bpi, une imprimante rapide de 600 l/min et 2 coupleurs CAMAC. Cet ordinateur de secours sert habituellement au développement du logiciel, à l'exécution de travaux par train et à des expériences en ligne (relevé des cartes de champs magnétiques dans les CSS par exemple ...).

4.2 Le CAMAC

Chacun des MITRA est équipé de 2 coupleurs parallèle CAMAC branchés sur le bus périphérique, permettant via deux modules de gestion de boucles Serial Highway Driver "Kinetic Systems" de prendre en compte 2 boucles CAMAC série fonctionnant en mode bit à bit à la fréquence de 2.5 MHz.

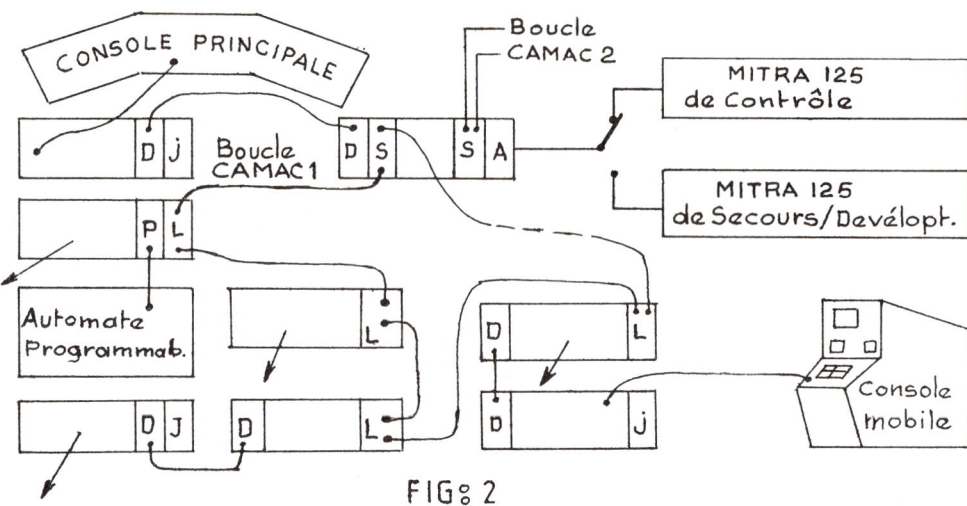

A : contrôleur de châssis type A
L : contrôleur de châssis type L
J : contrôleur autonome JCAM 10
S : module de gestion de boucle CAMAC série
(Serial Highway Driver)

D : module de liaison CAMAC-CAMAC JLS 10
P : module de liaison avec l'automate programmable
↙ : flèche indiquant les liaisons avec l'accélérateur

Fig. 2

Le système de contrôle du GANIL met en jeu environ 1 000 modules CAMAC, répartis dans 120 châssis CAMAC. On peut classer ces modules suivant divers critères.

. Suivant leurs fonctionnalités :
250 modules ont un rôle de services (contrôleurs, "dataway display", liaisons inter-systèmes)
750 modules ont un rôle de contrôle à proprement parler (consoles et processus)
. Suivant les objectifs des systèmes dans lesquels ils sont insérés :
100 sont répartis dans divers bancs d'essais de matériel, de développement de logiciel JCAM 10, de relevés de cartes de champs magnétiques.
100 sont consacrés au contrôle des consoles
100 servent au contrôle de processus, sous dépendance de JCAM 10.
700 servent au contrôle de processus, sous dépendance du calculateur central MITRA 125.
. Par familles de modules identiques :
une première famille de 200 modules est celle des modules de commande de moteurs pas à pas
une seconde famille est celle des modules ER16V, capables de contrôler digitalement les 300 alimentations stabilisées qui alimentent les aimants de guidage et de focalisation du faisceau.
D'autres familles importantes sont celles de convertisseurs analogique-digital et digital-analogique, d'interruption sur changement d'état.
Au total on compte environ 10 grandes familles pour la fonctionnalité contrôle de processus.

4.3 JCAM 10

Les microprocesseurs utilisés sont des contrôleurs de châssis CAMAC équipé d'un circuit INTEL 8080 appelés JCAM 10. Les JCAM 10 peuvent fonctionner en autonome ou être connectés à l'ordinateur central via un module CAMAC de liaison série JLS 10.

On distingue 2 types de JCAM :
× les JCAM console
Au nombre de 6 : 2 pour la console principale, 4 pour les 4 consoles mobiles, ils servent :
- à interpréter les requêtes des opérateurs et à les transmettre vers le calculateur central. Une requête complète implique en général une série de gestes de la part de l'opérateur. Le calculateur central MITRA 125 n'est alerté que lorsque la requête est complète et cohérente,
- à analyser les messages provenant du calculateur central, servant en général à des affichages sur les écrans de la console, et à les transformer en ordre CAMAC adéquats.
× les JCAM processus, s'occupant entièrement d'une mission particulière sur des équipements déterminés, ou traitant des données provenant de mesures sur le faisceau d'ions lourds. Ils sont au nombre de 8.
Dans cette catégorie rentrent 4 JCAM contrôlant chacun une ou deux cavités électromagnétiques. Le contrôle consiste en une séquence permettant le démarrage de la cavité à bas niveau, y compris la recherche de son accord ainsi que celui de son amplificateur de puissance ; puis, si certaines conditions liées en particulier à l'état du vide dans la machine sont satisfaites, en une autre séquence de montée progressive en puissance, jusqu'au niveau désiré ; une séquence spéciale de redémarrage rapide se déroule en cas de claquage.
Chaque JCAM peut fonctionner de façon parfaitement autonome, auquel cas on doit lui indiquer localement la fréquence de travail et le niveau de champ désiré, ou bien être relié au calculateur central MITRA 125 au moyen d'une liaison CAMAC-CAMAC (cf parag. 4.8), dans ce cas ses consignes lui parviennent à partir d'une des consoles via le MITRA 125.

4.4 Signaux analogiques

La transmission de l'information dans ce système de contrôle s'effectue sous forme digitale dans 80 % des cas. Toutefois certaines transmissions s'effectuent sous forme analogique, pour différentes raisons.

Une première raison est que certains signaux sont impossibles à digitaliser ; c'est le cas de signaux à très large bande (100 MHz) qui sont acheminés dans la salle de commande directement par des câbles spéciaux.

Une autre raison pouvant entrer en ligne de compte est que la digitalisation du signal introduirait un surcoût non justifié (cf nota) ou une complication inutile parce que le signal se présente à sa source sous forme analogique et qu'il doit être présenté également sous forme analogique. Dans cette dernière catégorie de signaux entrent toute une famille de mesures sur le faisceau d'ions lourds. Il s'agit soit de signaux variant très lentement et que l'on désire visualiser sous forme d'une aiguille se déplaçant devant un cadran, soit de signaux de bande passante pouvant atteindre le mégahertz, que l'on visualise sur un oscilloscope (cf parag. 4.5). L'acheminement de ces signaux jusqu'à la salle de commande se fait à l'aide d'un "système d'observation de signaux" (SOS), étudié et développé à l'origine par le CERN [3] pour ses propres besoins.

La figure 3 indique la configuration du SOS au GANIL ; 300 signaux analogiques arrivent sur des châssis SOS parcourus par 16 lignes banalisées. Chacune de ces lignes arrive en salle de commande sur un galvanomètre autocalibré ou un oscilloscope. N'importe lequel des 300 signaux peut être connecté sur n'importe quelle ligne ; la connexion est contrôlée par CAMAC. Le système a une bande passante de 1 MHz.

4.5 Les consoles

Les consoles constituent l'interface homme-machine. Contrôler et commander l'accélérateur, c'est interagir avec le système de contrôle via les consoles.

On distingue :
- la console principale installée en salle de commande et reliée directement à la branche CAMAC
- et 4 consoles mobiles pouvant être insérées dans les boucles série CAMAC pour des contrôles locaux.

De façon générale, on s'est attaché à faire en sorte que l'opérateur ait le moins de gestes possibles à faire, le moins de choses possibles à mémoriser. L'opérateur n'a en général qu'à choisir entre plusieurs options présentées par le système de contrôle, simplement en désignant du doigt ce qu'il veut.

En particulier il n'y a pas de clavier sur consoles, à part des miniclaviers purement numériques permettant de rentrer des valeurs digitales.

Les seules informations présentées spontanément par le système de contrôle, sans que l'opérateur les ait demandées, sont les messages d'alarmes, consistant en une ligne sur l'écran d'alarmes, et signant en général qu'une sécurité a fait disjoncter quelque chose.

La console principale constituée de 11 baies peut être divisée en trois zones opérationnelles (fig. 4) : la zone centrale II est réservée aux équipements non reliés à l'ordinateur, tels la source et les équipements HF et à l'écran d'alarmes. Les zones I et III sont identiques. La figure 5 donne le détail d'une zone I ou III.

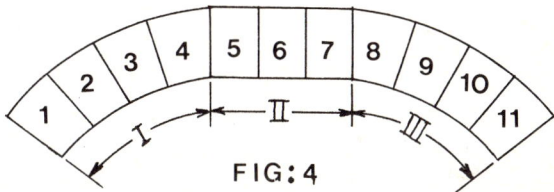

FIG: 4

Les zones α et δ de la figure 5 sont équivalentes. Une console mobile est identique à une zone α. Depuis l'une quelconque de ces zones on peut contrôler n'importe quel équipement de l'accélérateur, à l'exclusion de ceux reliés via le SOS. Ce contrôle peut s'effectuer soit à un niveau élémentaire, soit au niveau supérieur.

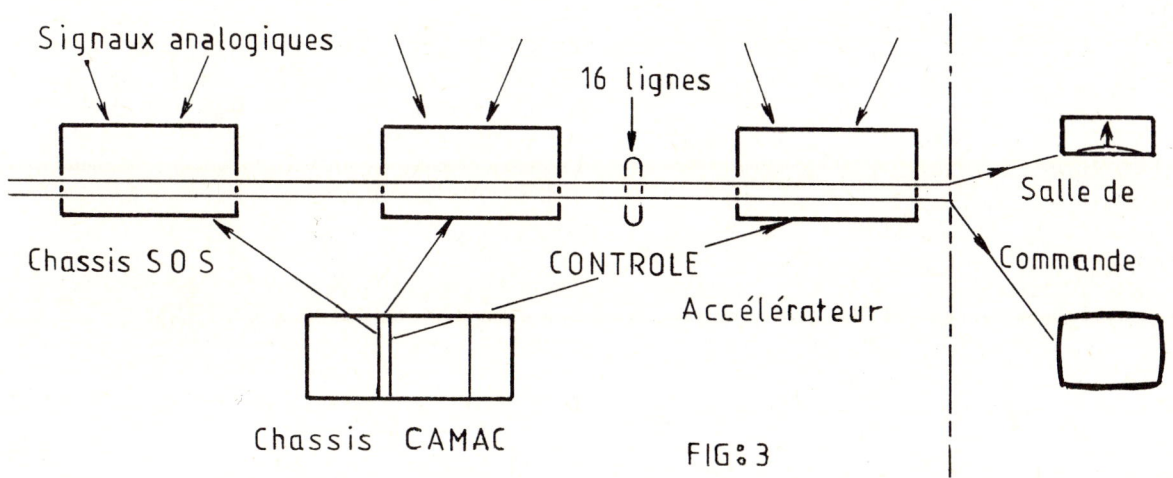

FIG: 3

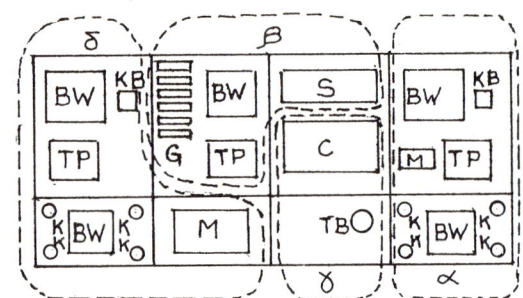

BW : moniteur de TV noir et blanc
K : pseudopotentiomètre
TP : écran à touches sensitives
S : oscilloscope
C : moniteur TV couleurs
G : galvanomètre autocalibré
TB : boule de poursuite
KB : clavier
M : mémoscope

Fig. 5

Au niveau élémentaire le système de contrôle se comporte comme un multiplexeur. L'opérateur établit une connexion logicielle entre l'un des 4 pseudopotentiomètres qu'il a devant lui et l'un quelconque des équipements de l'accélérateur, connu par son nom opérationnel.

Le nom opérationnel est une chaîne de 16 caractères, divisée en 4 champs, choisi de façon à être facile à mémoriser. Le premier champ indique à quelle partie géographique de l'accélérateur l'équipement appartient. Le deuxième champ indique la fonctionnalité de l'équipement (guidage du faisceau, système HF, etc...). Les 2 derniers champs sont spécifiques de l'équipement. Par exemple, sur la figure 6 le nom apparaissant en haut à gauche a trait au premier cyclotron injecteur (I1), à la fonctionnalité diagnostiques de faisceau (DIA) ; il s'agit de la sonde numéro 1 (SD1), et on s'intéresse à sa position (POS).

L'opérateur choisit l'équipement qu'il veut au moyen de l'écran à touches sensibles (TP sur la zone α de la figure 5) ; les noms opérationnels sont organisés suivant une structure arborescente ; la première page est le tronc de l'arbre, elle permet de choisir le premier champ du nom ; ce choix fait apparaître une autre page, branche maîtresse proposant toutes les fonctionnalités que l'on peut associer à I1, etc... Après avoir désigné une feuille, c'est-à-dire un nom opérationnel, l'opérateur doit toucher un des 4 pseudopotentiomètres (K sur la zone α de la figure 5). Alors le calculateur central est alerté, et il établit la connexion logicielle. Lorsque l'opérateur tournera le pseudopotentiomètre, le moteur pas à pas qui commande la sonde déplacera celle-ci.

A l'établissement de la connexion logicielle apparaissent sur le quart d'écran associé au pseudopotentiomètre (BW en bas de la zone α de la figure 5) des informations relatives à l'équipement connecté : sa valeur actuelle (position s'il s'agit d'un organe mécanique, courant débité s'il s'agit d'une alimentation stabilisée alimentant un aimant de guidage...)

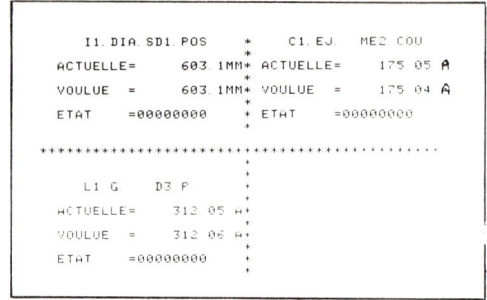

Fig. 6

sa valeur voulue, et son mot état (si l'équipement est dans un état normal, le mot d'état est à 0). Il est particulièrement important d'indiquer la valeur actuelle et la valeur voulue lorsque l'on commande des équipements qui répondent lentement ; c'est le cas pour les déplacements mécaniques, mais aussi pour les gros aimants de guidage, dont la self est importante. Il est aussi utile de disposer de ces 2 valeurs dans le cas des alimentations stabilisées où ce qu'on veut aboutit sur un convertisseur digital-analogique tandis que ce qu'on obtient provient d'un convertisseur analogique digital ; la comparaison des 2 valeurs, en principe égales après l'extinction des constantes de temps, donne une idée du bon fonctionnement des convertisseurs et des chaînes de régulation analogiques internes aux alimentations stabilisées.

C'est au niveau supérieur d'utilisation des consoles que l'on tire parti de toute la puissance du système de contrôle. C'est à ce niveau que l'on peut lancer des tâches, depuis le même écran à touches sensitives, mais sur lequel on fait apparaître un autre arbre, celui des tâches. L'écran BW du haut de la zone α de la figure 5, ainsi que le mémoscope M et le miniclavier KB sont les périphériques d'entrée-sortie des tâches, auxquelles on peut en outre affecter un ou plusieurs pseudopotentiomètres. Ces tâches, dites tâches d'application, peuvent effectuer des opérations très globales, effectuer des calculs sur les données acquises, présenter des résultats sous forme graphique, dans le but de faciliter le réglage et l'exploitation de l'accélérateur.

La zone β de la figure 5 est relative au SOS. Elle comporte un écran à touches sensitives sur lequel on peut examiner l'arbre des noms opérationnels des grandeurs reliées au SOS. Par un mécanisme analogue à celui décrit plus haut on peut visualiser la grandeur sur un galvanomètre indiqué G sur la figure 5, ou sur l'oscilloscope indiqué S.

Quant à la zone γ de la figure 5, elle est réservée à la présentation de synoptiques sur écran de TV couleur. L'intérêt de la couleur est de disposer du graphisme pour indiquer l'état d'un équipement, et de la couleur pour dire si cet état est normal ou pas. Ces synoptiques ne sont pas encore opérationnels ; ils seront surtout utiles en phase d'exploitation de l'accélérateur d'ions lourds.

4.6 Les interfaces

Dans un grand nombre de cas, l'interface matérielle entre le système de contrôle et le processus industriel qu'est l'accélérateur d'ions lourds est un module CAMAC du commerce. C'est le cas pour les convertisseurs analogiques - digitaux et digitaux - analogiques, pour les commandes tout ou rien, pour les lectures d'état.

Pour les commandes de moteurs pas à pas, on a préféré ne pas utiliser de modules du commerce, jugés peu commodes d'emploi parce que n'acceptant que des commandes incrémentales. On a donc développé un module CAMAC qui, associé à un certain codeur, se présente vu du calculateur comme un module effectuant des commandes absolues. Le codeur utilisé est celui du linac du CERN [4].

L'interfaçage des alimentations stabilisées avec une précision de 10^{-4} à 10^{-5} se fait de façon purement digitale, avec isolation galvanique. On utilise deux voies de données : l'une relie l'alimentation à un module CAMAC émetteur récepteur 16 voies (ER 16 V) ; c'est une liaison série bidirectionnelle, dans laquelle le calculateur est toujours maître : une transaction consiste en un mot de 24 bits envoyé depuis le module ER 16 V, plus une réponse de l'alimentation stabilisée. L'autre voie de donnée sert à signaler au calculateur une disjonction spontanée de l'alimentation, en générant un LAM dans un module LS de détection de changement d'état.

4.7 Les liaisons CAMAC CAMAC

Le système de contrôle du GANIL peut être considéré comme 15 ensembles CAMAC, avec un processeur dans chaque, reliés en étoile. Ces systèmes communiquent à l'aide de modules CAMAC de liaison JLS 10, repérés par la lettre D sur la figure 2. La liaison est du type série fonctionnant à 1 M baud ; pour n'imposer aucune contrainte de vitesse aux processeurs, chaque module JLS 10 est muni d'un tampon de 256 mots de 16 bits. Une transaction consiste en un message envoyé dans un sens, suivi d'un autre message en retour ; un message consiste en un nombre quelconque (mais < 256) de mots. Le processeur qui a envoyé le 1er message reste maître de la liaison pendant tout le temps de la transaction, et la libère ensuite.

5. LE LOGICIEL

5.1 Le logiciel MITRA

5.1.1 Les moyens de développement
Un logiciel appelé Ganiciel a été écrit spécialement pour le système de contrôle du GANIL. Ce logiciel a été développé sous moniteur MMT2, qui permet d'exécuter les assembleurs et compilateurs nécessaires pour créer des programmes s'exécutant par la suite en temps réel sur MITRA 125 ou sur JCAM 10.

5.1.2 Organisation du Ganiciel
Le Ganiciel s'exécute sous moniteur MOP2, et non pas sous MMT2 jugé trop compliqué et trop encombrant. L'ensemble constitue un système autochargeable : tous les programmes composant l'application sont définis et intégrés dans le système lors de sa génération. Ils sont chargés en mémoire et/ou sur disque lors de la mise en oeuvre du système.

95 % du Ganiciel est écrit en LTR, langage structuré assez comparable au Pascal, possédant en outre des instructions temps réel. Les 5 % écrits en assembleur sont relatifs soit aux opérations d'entrée-sortie CAMAC, soit à des manipulations des tables du moniteur.

On a ainsi un système à 3 couches : un noyau constitué par le moniteur à proprement parler, une couche système assurant la gestion des consoles et des équipements, et des tâches d'application, écrites par des non informaticiens, chargés de l'exploitation de l'accélérateur ; ces tâches sont toutes différées ; elles se partagent l'unité centrale quand celle-ci est disponible sous contrôle d'un dispositif distributeur qui assure qu'aucune tâche ne peut monopoliser l'unité centrale pour elle seule ; chaque tâche est assurée d'avoir un peu de temps d'unité centrale à cadence raisonnable. Ces tâches d'application utilisent les services de la couche système en appelant des sous programmes partageables lorsqu'elles désirent interagir avec les consoles ou contrôler des équipements. Les paragraphes suivants détaillent les différentes parties de la couche système.

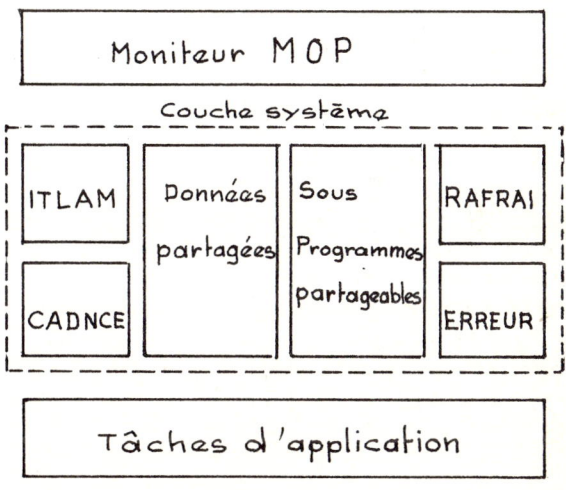

FIG : 7

5.1.3 Les tâches immédiates
CADNCE établit le cadencement nécessité par le rafraîchissement des informations affichées sur les consoles. Cette tâche est déclenchée toutes les 300 ms par un compteur matériel ; elle se borne à déclarer arrivés certains évènements sur lesquels des tâches différées sont en attente.

ITLAM traite les interruptions (LAM) arrivant du CAMAC ; de façon générale on a réduit au minimum possible l'usage des LAM ; c'est-à-dire qu'ils sont réservés aux évènements non prévisibles, émis soit par des modules LS (disjonction de quelque

chose), soit par des modules JLS (arrivée d'un message en provenance d'une console ou d'un JCAM processus). L'identification d'un LAM fait appel à des tables situées dans la zone de données partagées.
Il existe une dizaine de types de messages pouvant arriver des consoles, avec chacun sa structure propre ; ITLAM les interprète, et effectue les actions demandées, (par exemple établir une liaison logicielle avec un pseudopotentiomètre, lancer une tâche), qui impliquent en général la consultation d'une base de donnée (dans la zone de données partagées) et l'exécution d'opérations CAMAC. Les messages en provenance de JCAM processus sont en général destinés à des tâches d'application ; ils sont alors rangés dans des tampons spéciaux que la tâche destinatrice viendra lire plus tard.

5.1.4 Les tâches différées système
RAFRAI assure le rafraîchissement des consoles. Cette tâche utilise des tables situées dans la zone de données partagées ; elle peut ainsi savoir, pour chaque console, quelles informations doivent être rafraîchies, et quelles opérations CAMAC il y a lieu d'effectuer pour obtenir ces informations ; chaque information fait l'objet d'une fiche ; les fiches sont chaînées entre elles pour pouvoir être mises à jour facilement.
ERREUR assure l'édition des messages sur l'écran des alarmes.

5.1.5 La base de données et les handlers
La base de données occupe la plus grande partie de la zone de données partagées. Elle regroupe toutes les informations décrivant les équipements contrôlés. Les équipements sont regroupés par classes. Dans une même classe, les équipements qui en font partie sont contrôlés tous de la même façon, avec les mêmes types de modules CAMAC (un ou plusieurs modules ou fraction de module) ; à chaque équipement est affectée une certaine partie de la zone de la base de donnée relative à cette classe. Cette partie possède une structure, la même pour tous les équipements de la classe ; on y range les données propres à cet équipement. La figure 8 montre la structure particulièrement simple des équipements de la classe "grandeur analogique".

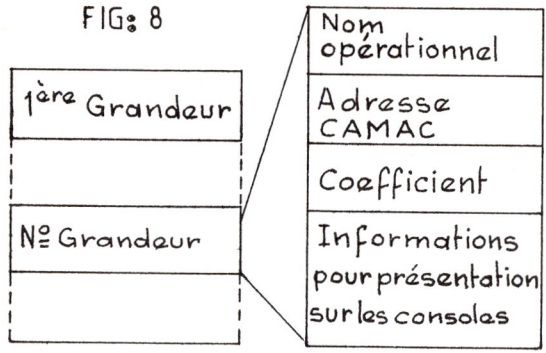

Il existe pour tout le système de contrôle une dizaine de classes d'équipements. A chaque classe est associé un handler dont le rôle est de manipuler cette classe d'équipement ; le handler reçoit des ordres émis par les tâches ; un ordre est constitué du nom d'équipement, d'un verbe indiquant ce qu'on veut faire (lire une valeur, connecter à une console, réserver l'équipement ...), et éventuellement de données numériques ; le handler retourne un compte-rendu de ce qu'il a fait, et éventuellement des données numériques. Toute action sur un équipement passe obligatoirement par le handler associé à la classe à laquelle appartient cet équipement ; le handler se trouve en partie dans la tâche ITLAM, pour les actions à effectuer pour les consoles, en partie dans les sous programmes partageables, pour les actions demandées par les tâches d'application.

5.1.6 Les noms d'équipement
Alors qu'au niveau de l'opération on contrôle l'accélérateur avec des noms opérationnels (cf parag.4.5), on manipule en fait dans le logiciel des noms d'équipement. Un nom d'équipement est constitué de 2 nombres : le numéro de la classe à laquelle appartient l'équipement, et son numéro d'ordre à l'intérieur de cette classe. Naturellement un dictionnaire situé dans la zone de données partagées permet de traduire n'importe quel nom opérationnel en nom d'équipement.

5.1.7 Les sous programmes partageables
On trouve dans cette zone, outre les handlers :
.le traducteur de noms
.des sous programmes permettant l'envoi de messages à des JCAM consoles ou des JCAM processus (messages externes) ; on a défini dans ce sens une trentaine de types de messages
.des sous programmes permettant l'envoi de messages d'une tâche à une autre (messages internes)
.des sous programmes de service : bibliothèque graphique, gestion de fichiers disque
.des sous programmes d'interface, destinés à faciliter l'écriture des tâches d'application, que ce soit pour contrôler des équipements, s'adresser à un JCAM processus ou pour dialoguer avec une console.

5.2 Le logiciel JCAM 10

Bien que l'on dispose d'un langage de haut niveau orienté CAMAC pour des essais off line de matériel, tout le logiciel on line des JCAM 10 est écrit en assembleur. Qu'il s'agisse de JCAM consoles ou de JCAM processus, le logiciel est structuré en tâches s'exécutant sous la dépendance d'un "minimoniteur temps réel"[5] rajouté au moniteur standard du JCAM 10. Ce minimoniteur assure d'une part la gestion des messages échangés avec le calculateur central, et d'autre part contrôle le déroulement des tâches, qui sont spécifiques du processus contrôlé. En connectant au JCAM une petite console on peut, sous moniteur, lancer les tâches en local, et utiliser des programmes de service destinés à aider à la mise au point.

Aux JCAM qui comportent 9K octets de mémoire sont adjoints des modules d'extension mémoire de 36 K octets. Les JCAM consoles ont en outre des modules de mémoire CAMAC, destinés à abriter 220 pages destinées aux écrans à touches sensitives.

6. CONCLUSION

6.1 L'état actuel du système

En juin 1982 une première partie du système a été mise en service ; elle permet d'accéder à environ 30 % du matériel ; cette échéance correspondait à une première phase de mise en service de l'accélérateur d'ions lourds. En novembre 1982 une deuxième partie de l'accélérateur sera en cours de tests, et le système informatique contrôlera environ 70 % des équipements. Tout le logiciel de base (rafraîchissement des consoles, lancement des tâches, échange de messages entre tâches, dialogue entre les tâches et les consoles) est opérationnel, à l'exception de certains handlers non prioritaires, et des synoptiques sur écran TV.

Un certain nombre de tâches d'application ont été écrites, dont certaines ont grandement facilité la mise en service de la première partie de l'accélérateur. De nombreuses autres tâches sont en cours de développement, et il en sera vraisemblablement encore ainsi pendant longtemps. Les premières tâches visent surtout à aider au réglage de l'accélérateur ; les tâches à venir seront principalement destinées à faciliter son exploitation.

6.2 Evaluation du système

Maintenant que ce système conçu en 1978 est opérationnel, il est difficile de ne pas se poser la question de savoir ce qui changerait si on devait en refaire la conception aujourd'hui. En ce qui concerne l'électronique digitale, la norme CAMAC apparaît encore comme un bon choix ; les modules CAMAC eux-mêmes ont évolué depuis 1978, dans le sens de la compacité, tirant parti des nouveaux circuits intégrés à grande échelle d'intégration ; certains modules CAMAC que nous utilisons seraient avantageusement remplacés par des modules ayant les mêmes fonctionnalités, mais plus simples.

En ce qui concerne l'architecture générale du système, la séparation en un processeur central et plusieurs microprocesseurs, avec une base de données centralisée, reste sans doute le meilleur choix pour notre système ; les travaux récents sur les "réseaux locaux" nous semblent s'adresser à des systèmes beaucoup plus grands que le nôtre.

Côté microprocesseurs, il est tout à fait clair que nous choisirions maintenant, au lieu de JCAM 10, des contrôleurs auxiliaires à base de 68 000. Davantage pour des raisons logicielles que matérielles. En effet, dans la réalisation de notre système de contrôle il est apparu qu'une charge importante était constituée par la programmation des JCAM 10 en assembleur. La solution idéale consiste à programmer les microprocesseurs dans un langage de haut niveau, le même que celui utilisé pour le processeur central. A cet égard le 68 000 apparaît comme très attrayant puisqu'un compilateur croisé pour LTR existe maintenant sur MITRA 125.

Ce qui a vieilli le plus dans le système de contrôle est le processeur central MITRA 125, considéré maintenant par son fabricant SEMS comme en fin de vie. Une étude économique montrerait sans doute qu'il serait bénéfique de remplacer ce calculateur qui marche parfaitement par un calculateur plus moderne, beaucoup moins cher à entretenir. Ce calculateur devrait naturellement être muni d'un coupleur CAMAC, et accepter la programmation en LTR. Ici se manifeste d'ailleurs un des avantages de la norme CAMAC, qui est d'être indépendante de quelque calculateur que ce soit ; le changement de calculateur (10 % de l'investissement global du système de contrôle) se limite au seul calculateur, et n'implique aucune autre modification.

6.3 Remerciements

Le système de contrôle du GANIL a bénéficié de nombreux apports extérieurs. L'adoption du CAMAC donne d'ailleurs d'emblée accès à toute une communauté d'idées en vigueur chez ses utilisateurs. La conception des consoles a été inspirée de celle de l'accélérateur linéaire injecteur du synchrotron à protons du CERN, qui nous a également permis de profiter de ses études relatives au système d'observation de signaux analogiques. Les modules CAMAC ER 16 V sont une conception du Laboratoire National Saturne, qui nous les a fabriqués, ainsi que les coupleurs CAMAC pour MITRA 125. Les modules CAMAC de liaison CAMAC-CAMAC, JLS10, ont été étudiés et fabriqués pour GANIL par le Département d'Electronique Industrielle et Nucléaire de Saclay.

REFERENCES

(1) M. PROME, "The GANIL Control System" IEEE Transaction on Nuclear Science, Vol NS-28 n° 3, June 1981

(2) P. BARDON and al, "The GANIL Control System as seen from the control room", Proceedings de la 9ème Conférence Internationale sur les cyclotrons et leurs applications, Caen septembre 1981

(3) S. BATTISTI and al, "A flexible wideband analog network for direct signal display on operator console", IEEE Transaction on Nuclear Science, Vol NS-28 n° 3, June 1981

(4) A. VAN DER SCHUEREN, P. TETU, J. AEBI, L. BERNARD, M. SARTORIO, "Encodeur de position", CERN/MPS/LIN/69-18

(5) M. ULRICH, H. VAN DER BEKEN, B. PIQUET, "Le minimoniteur JCAM", GANIL 81R/099/CC/16

NOTA

Il se trouve que les alimentations régulées à 10^{-3} sont des alimentations de petite puissance, peu encombrantes. Dans ces conditions il est possible d'établir une liaison relativement courte entre l'alimentation et un châssis CAMAC ; le mode commun étant limité à quelques volts, on peut tolérer une transmission analogique en différentiel ; l'interface est donc située dans le châssis CAMAC, qui fournit la tension de référence de l'alimentation.

THE CAB SYSTEM : APPLICATIONS

Etienne BARRELET, Roland MARBOT and Pierre MATRICON
Laboratoire de Physique Nucléaire des Hautes Energies, Ecole Polytechnique
Palaiseau, FRANCE

ABSTRACT

The CAB is a general purpose fast processor.

We present a variety of specific applications found for the CAB in the following areas: filtering, compacting, histogramming, signal analysis, image processing, high rate data acquisition at low cost.

These applications illustrate the main hardware aspects of the CAB. Due to its wide range of interfaces, the CAB is available in a variety of configurations : GPIB bus and/or Camac crate and/or Camac branch controller, and a memory access capability up to 80 Mbits per second.

The software tools which are now available to the CAB user are described in a joint presentation.

RÉSUMÉ

CAB est un calculateur universel rapide. Dans cet article, on présente plusicius applications possibles de CAB : filtrage, compression, histogrammes, traitement du signal, des images, acquisition rapide des données à côut réduit. Ces applications illustrent les principaux aspects matériels de CAB. Grâce à ses interfaces variées, CAB pent été utilisé daus plusieius configurations : bus GPIB et/ou creneaux CAMAC et/ou contrôleur CAMAC, et une gestion mémoire allant jusqú à 80 Mbits/s. Les outils logiciels disponibles pour les utilisateurs sont. décrits dans une autre communication.

ZUSAMMENFASSUNG

Der CAB is ein schneller Prozessor fuer allgemeine Anwendungen.

Wir stellen verschiedene Anwendungen des CAB aus den folgenden Gebieten vor ; Filterung, Datenreduktion, Histogrammbildung, Signalanalyse, Bildverarbeitung, schnelle Datenerfassung mit geringen Kosten.

Diese Anwendungen beleuchten the wesentlichen Hardware - Aspekte des CAB. Auf Grund seiner zahlreichen Schnittstellen, steht der CAB in vielen verschiedenen Konfigurationen zur Verfuegung : GPIB Bus und/oder CAMAC Crate Controller und/oder CAMAC Branch Controller bei einer Speichzugriffszeit von bis zu 80 MBit/s.

Die fuer den CAB zur Verfuegung stehenden Software Werkzeuge werden in einer getrennten Abhandlung beschrieben.

1. INTRODUCTION

The CAB is a versatile programmable processor which is much faster than Camac, [1], [2]. It is supported by software tools that are the subject of another RTD'82 paper. Now commercially available, [3], the CAB is intended to be used widely in real-time data fields, far beyond the original high energy physics domain. We schematically describe the up-to-date hardware and the applications previously found for the CAB in :
- nuclear instrumentation
- image processing
- digital video treatment
- process control
- Fourier analysis

2. ADDITIONAL BOARDS BOOST THE CAB NOW

There are three basic CAB configurations, namely the CAB G, C, B. The CAB G driven by the IEEE-488 bus is an easy way to get a powerful, low cost, autonomous crate controller. The CAB C also controls a crate Dataway but is driven by a branch highway ; moreover it can be made to act as a standard controller with different software loading. Creating a Camac branch from any network point is the job of the CAB B. Several interface boards can be added to the four basic cards. For example a CAB G or C can be implemented with a branch interface. It controls the Dataway of its own crate with its crate interface and the "sub-branch" is driven by its branch interface. A branch driver may be associated with a GPIB interface. This provides a useful way to connect a micro-computer via GPIB to debug and control a CAB branch in a Camac network. (Such a GPIB interface is called G* because it does not possess the ability to control the Dataway, as a G card does). Other cards have been made to increase the input/output facilities of the CAB.

```
IEEE-488 interface G*
Highway interface C*
Branch generation B*
Histogramming memory
Fast Access -5 Mwords/s
Ellipsometry
Gamma-camera correction
Video analysis
```

Additional boards

3. MAKE THE MOST OF YOUR DATAWAY WITH A CAB

A single crate controlled by a CAB G, and a desk computer used as a terminal, constitute a complete Camac system. This basic system takes advantage of standard modules, and in addition, is capable of handling succeeding Camac cycles at full speed while processing data. Histograms can be monitored on the screen and stored on cartridges or floppy discs. Such a setup has been used to test the RICH prototype, [4].
In another example, full compatibility is reached between the acquisition system of the experiment (including a CAB G driven by a mini-computer) and the equipment tests performed in different laboratories also with a CAB G, in the proton lifetime experiment, [5]. The one-crate system may be extended to larger, multicrate systems by putting together CAB's of all sorts to work in parallel, [6].

Digital cartoons : an example that lights up your Results

This application, [7], needs the following setup :

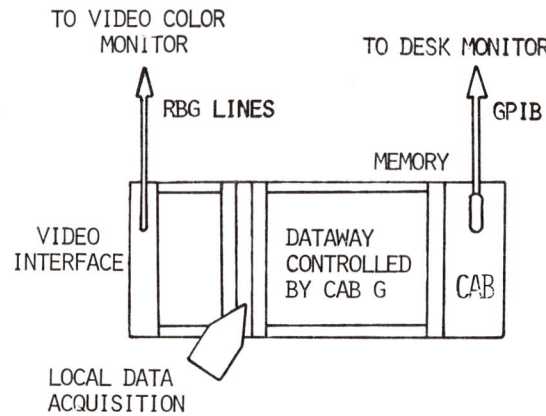

The high resolution image (512x512 picture elements) is continually being reformed by the sweep of the video signal. The data to be visualised may come from a local acquisition, an off-line processing system or a user's storage (data cartridge or floppy disc). The memory stores the processed data and the histograms.
The desk computer analyses the user's commands and transmits to the CAB the parameters that define screen areas, color ranges, zooming, fractions of histograms, axes and comments. At the present time, it takes 0.66 s for the CAB to remove the whole picture. This limit is set by the relative slowness of the one megabit video-interface memory. With a non-limiting interface, the CAB software limit would be 5 images per second.

4. USE THE CAB BUS AND SURPASS THE CAMAC SPEED

The 1μs Dataway cycle is one of the limitations of the Camac standard. Connecting specific boards directly onto the CAB bus increases the speed 5 times. Four examples illustrate this application.

4.1 High statistics in a high energy physics experiment, [8].

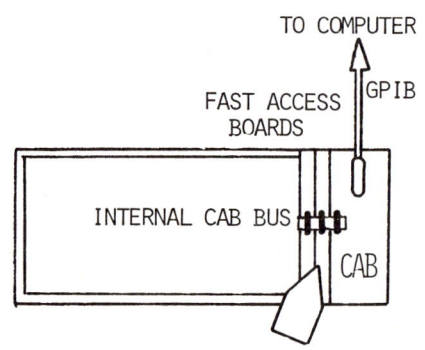

5 MWORDS/S ENCODED DATA ACQUISITION

The additional Fast Access boards contain first-in-first-out memory registers, clock and control signal generators. They allow the CAB G to get the encoded data at a rate of 5 megawords per second. The processor calculates a physical parameter through a linear combination with predetermined coefficients and takes an accept or reject decision within 3.5µs. The average computing time per event, which allows for rarer events with longer computations, reaches 5.5µs. Including the 10% duty cycle of the CERN accelerator, more than 10^9 events per day are taken, which is much above the current state of the art.

4.2 Use of the CAB for Fourier analysis.

Some solar energy physicists are studying the properties of an amorphous silicon film during its growth, [9], [10]. They have built a new type of spectroscopic automatic ellipsometer with a polarization modulation at 50 KHz ; instead of lock-in amplifiers, they use a digital method, involving a CAB G extended with the Ellipsometry board (containing an 8 bit flash-ADC, a phase locked-loop circuitry for synchronous frequency multiplication, and an accumulation register). The detected analog signal is sampled at 12.8 MHz. An on-line Fourier analysis is performed on the accumulated data by the CAB, which measures one set of Fourier coefficients within each modulation period of 20µs. The ellipsometric angles or the complex dielectric constant are then extracted from the coefficients, either by the CAB or the desk computer. In order to reduce digital errors and to improve the signal to noise ratio, a basic 5 ms sequence of 256 accumulated periods per point is chosen. At this data acquisition rate, the accuracy is 5×10^{-4}. Further accumulation over ten seconds leads to 10^{-5}.

4.3 Liquid chromatography improves with digital treatment.

A new method has been developed to separate and analyse compounds in a mixture, [11]. The thin two-dimensional sample to be analysed is lighted and passes by in front of a CCD-camera (1024 cells). The analog signal is digitised by the Video Analysis board (supporting a flash-ADC and clock generator). The CAB then analyses the sample line by line. The complete image is processed and sent to a video color monitor for visualisation and interpretation.

The same method can be used for remote measurements, parts identification, security and quality control.

4.4 On-line correction of a gamma-camera.

This example shows an intelligent "black box" inserted in nuclear medicine equipment to correct uniformity and linearity distorsions, [12], [13] :

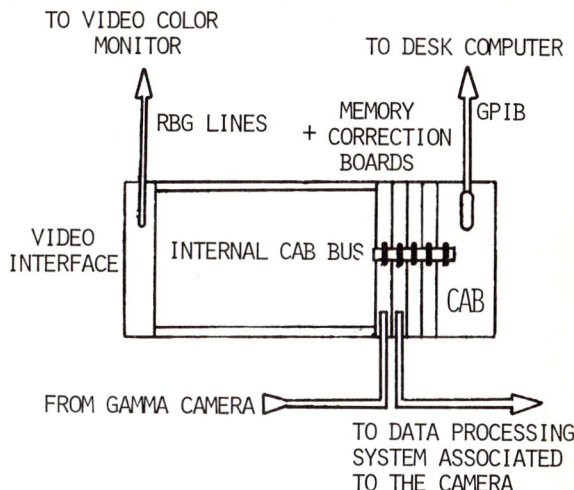

The equipment includes the Gamma-camera Correction boards and two Histogramming memories. During the calibration, the CAB accumulates histograms and determines the correcting coefficients. When operating, the analog outputs of the camera are digitised, processed with the coefficients, then returned in an analog form to the processing system associated to the camera within a few microseconds. No significant loss of time is introduced by the black box at typical illumination rates. The performance of the camera is improved and its auto-correction is of great value in an hospital environment.

5. INSERT AUTONOMOUS DATA ACQUISITION AND PROCESSING ANYWHERE IN A NETWORK

In large experiments when numerous research teams collaborate or when the apparatus is being transformed, there are certain needs which occur such as auxiliary processing units in a crate, new acquisition lines or automatic and autonomous controls. The assembly used in the experiment NA14 at CERN, [14], illustrates one of the various possibilities which can be solved by the use of a CAB.
The CAB is extended with G* and C* boards, as shown in the next drawing.

Similar setups are used in other physics experiments, [15], [16]. The specific performances of such a system rely on :
- easy insertion into a network
- autonomy
- independant and parallel processing of parts of an acquisition system
- multiple automatic controls
- acquisition line shared by several computers.

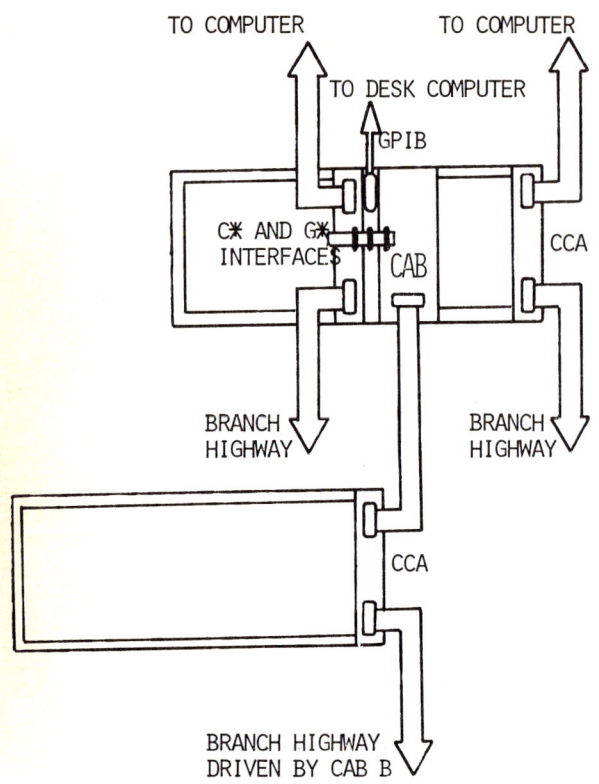

6. CONCLUSION

The CAB then can be considered as a building block set : take a CAB, connect to it additional boards, and you get a custom real-time processor for data acquisition, processing and control.
Improvements are seen in the better use of Camac Dataway, and the possibility of fast post-trigger, fast pre-processing, data reduction and formatting, local response to interrupts in a very short time.
Modifications can be added for Compex compatibility and Camac/VME interfacing.
The CAB appears as a good candidate for a powerful Camac to Fastbus interface, acting as a Fastbus master.

ACKNOWLEDGMENTS

The development of the CAB system has been partly supported by special grant from CNRS and Ecole Polytechnique. The authors wish to thank the CAB group at Ecole Polytechnique's High Energy Physics Laboratory :
R.Amailland, M.Bercher, A.Debraine, G.Fouque, K. de Kerday, J.Raguet, C.Roy, A.Simon, C.Violet, L.Zlatevski.
F.Gheno typed this text. P.Joliclercq photographed the slides for this presentation.
S.Orenstein corrected our poor english.

REFERENCES

[1] E.Barrelet, R.Marbot, P.Matricon, A versatile microcomputer for high rate Camac data acquisition, Proceedings of the Real-Time Data (1979) 77.

[2] Same authors and tittle, Proceedings of the Topical Conference on the Application of Microprocessors to High Energy Physics Experiments at CERN, Geneva (1981) 258.

[3] Lecroy Research Systems S.A., Geneva.

[4] E.Barrelet, T.Ekelöf, B.Lund-Jensen, J.Séguinot, J.Tocqueville, M.Urban, T.Ypsilantis, A two dimensional, single-photoelectron drift detector for Cherenkov ring imaging, Nucl.Instr. and Meth. 200 (1982) 219.

[5] Flash-tube read-out using a CAB microprocessor, τ_N experiment, int.report, X-Palaiseau, France (1981).

[6] C.Baglin et al., Charmonium spectroscopy at the I.S.R. using an antiproton beam and a hydrogen jet target, Proposal to I.S.R.C., CERN (1980).

[7] D.Lellouch, B.Chantalou, J.M.Chevalier, Visualisation en couleur d'histogrammes à 2 dimensions, int.report, user's guide, X-Palaiseau, France (1982).

[8] D.Lellouch, Mesure de la section efficace totale $\bar{\pi}$proton de 5 à 15Gev/c, thesis, Université Paris VI (1979).

[9] B.Drévillon, J.Perrin, R.Marbot, A.Violet, J.L.Dalby, Fast polarization modulated ellipsometer using a microprocessor system for digital Fourier analysis, Rev.Sci.Instr. 53 (1982) 104.

[10] J.Perrin, B.Drévillon, Applications of ellipsometry to in-situ study of the growth of hydrogenated amorphous films, Acta Electronica (1982) to be published.

[11] G.Guiochon, L.Beaver, H.Colin, M.F.Gonnord, A.M.Siouffi, M.Zacharia, Liquid chromatography using a bidimensional column, Proceedings of the 14th International Symposium on Chromatography, London (1982).

[12] L.Hadjeris, Correction des défauts de réponse des caméras à scintillation par un micro-ordinateur en ligne, thesis, Université Paris VI (1982).

[13] F.Soussaline, H.Nguyen Ngoc, J.Jeanjean, L.Hadjeris, R.Marbot, P.Miné, T.Hoang, Uniformity correction of a gamma-camera by an on-line microprocessor, Proceedings of the Third World Congress of Nuclear Medicine and Biology, Paris (1982).

[14] J.P.Wuthrick, J.M.Brunet, Le trigger "charges grands P_T" dans NA14, int. report, X-Collège de France, Paris (1982).

[15] Système d'acquisition de données de GANIL, Caen, France.

[16] CEA-D.Ph.N., compte-rendu d'activité (1981) 205, Saclay France.

A DATA ACQUISITION SYSTEM FOR THE VAX COMPUTERS

F. Gagliardi[1], M. Sciré[2], A. Vascotto[1], V. White[3]

A data acquisition system for VAX computers has been designed for an experiment of elementary particle physics. The system includes a VMS CAMAC driver which reads out the events, and a main data acquisition program which writes the events onto tape and distributes them to the monitoring programs. Several ancillary programs are used to control the acquisition activity.

A high degree of flexibility and a rational man-machine interface are the main features of this system that make it suitable for running today's very complex experiments.

Un système d'acquisition de données pour ordinateurs VAX a eté conçu pour une expérience de physique des particules. Le système comprend un driver CAMAC VMS qui lit les événements, ainsi qu'un programme principal qui les écrit sur bandes magnétiques et les distribue aux différents programmes d'analyse. Plusieurs programmes auxiliaires contrôlent la tâche d'acquisition proprement dite.

Une très grande flexibilité ainsi qu'une interface utilisateur bien adaptée sont les atouts essentiels de ce système et le rendent, de ce fait, parfaitement applicable aux experiences très complexes d'aujourd'hui.

Ein Programmsystem zur Datenerfassung mit VAX Rechnern wird beschrieben. Das System besteht aus einem VMS CAMAC Treiber, der die Ereignisse ausliest, und einem zentralen Datenerfassungsprogramm, das die Ereignisse auf Band schreibt und sie auf die Ueberwachungsprogramme verteilt. Mehrere Hilfsprogramme werden zur Kontrolle der Datenerfassungsaktivität benutzt.

Ein hoher Grad an Flexibilität und eine rationelle Benutzerschnittstelle sind die hauptsächlichen Merkmale dieses Systems, die es für die heutigen äusserst komplexen Experimente einsatzfähig macht.

---

[1] CERN, Data Handling Division, Geneva, Switzerland
[2] Now working at DIGITAL Milan, Italy
[3] Now working at FNAL Chicago, USA

1. INTRODUCTION

At the beginning of 1981 at CERN (the European Organization for Nuclear Research), a new particle accelerator became operational, where collisions between bunches of protons and anti-protons could be produced. One of the experiments (called UA2) set up to study this kind of reaction makes use of a large and complex detector designed and built by a collaboration of several European Universities and with the support of the CERN Laboratory. As part of this support, the Online Computing Group of the Data Handling Division took responsibility for the design and the commissioning of the computer system. The main functions to be implemented in this system were the control and monitoring of experimental apparatus and the collection of data in real time for immediate analysis at the experimental area and for more complex off-line analysis at several European computer centres.

2. OBJECTIVES OF THE PROJECT

CERN must support a large variety of high energy physics experiments. Any large investment of materials and manpower must keep account of this. One of the goals that was aimed at for a CERN data-acquisition system was flexibility, which in this case has a double aspect. On one side it should be possible to easily adapt the system to a large range of experiments' sizes and requirements; on the other, the software should be able to cope with important reconfigurations of an experiment's hardware. A high level of human engineering was also required, due to the special kind of CERN users who normally come to CERN for short periods and who need to become familiar with the data acquisition system as quickly as possible.

3. SYSTEM OVERVIEW

A data acquisition system in the field of elementary particle physics deals with units of information called 'events'. An event consists of all the data, produced by electronic equipment, which are related to a particular physical reaction occurring when a particle collision takes place. The experimental apparatus of the UA2 experiment includes several different types of particle detectors with their electronics. A typical event consists of about 10,000 to 14,000 16-bit words. The main intelligence of the system is centralised on a VAX 11/780. The computer is connected to the CERN private high-speed packet-switching network. A CAMAC interface links the VAX to the experimental apparatus' electronics.

The software running on the VAX performs two main functions: the data acquisition of events and the monitoring of the quality of the data being collected, as well as the status of the experiment in general.

The data acquisition function is performed by a program and by a CAMAC driver. The monitoring function is performed by many programs (more than 20), all written in FORTRAN 77 by the physicists of the experiment in a standard way, using a set of software utilities provided by the Data Handling Division of CERN.

4. DATA ACQUISITION

A VMS driver executes all the CAMAC functions needed to read out the events whilst a process called DAQP takes care of all data acquisition operations in general. There are also three ancillary programs: RUN, CAMLIST and PARAM.

All the CAMAC operations necessary to read out the experimental apparatus are described in a very simple language which permits the symbolical description of all the different parts of the detector. A source file written in this language is processed off-line by the program CAMLIST which performs a few basic checks and produces a list of elementary CAMAC operations in the form of a file.

The program PARAM is used to set up all the other parameters such as the maximum event size, the event buffer size, etc. This can be done independently for several users so that each one can have his private buffer of events and his own parameters. Once all these parameters have been set up, a file is created on disk to be used at the next re-start of the data acquisition program to create a shared-region which is used both by the CAMAC driver and DAQP.

The program RUN deals with the operational aspects of the acquisition system: it allows the user to control the acquisition activity by starting or stopping runs, mounting and dismounting tapes, and other similar actions. When the program RUN starts a new run, it sends a message to DAQP giving all the necessary information about this new run. DAQP, after checking the validity of the request, issues a directive to the CAMAC driver to signal that data-taking has to be started. The CAMAC driver module responsible for the handling of the events is triggered by an 'event' signal coming from the CAMAC system and producing an interrupt; it then starts the execution of the lists of CAMAC commands. These can be either very simple, just performing a DMA reading from a specific CAMAC address, or much more complicated, handling up to 15 different physics triggers, each one provoking the execution of a different sequence of CAMAC operations. The data related to these triggers are then written into all the buffers owned by those runs which have declared these triggers 'active'. Once the data are in the computer memory, they can be accessed by the program DAQP, to be analysed by a user-written routine so that some basic checks on the data quality can be performed, and to be written onto tape. Moreover, the DAQP can start to send events to all those user programs which have issued a request to sample data.

Many users can run independently of each other, each having the full control of his own stream of data, from the CAMAC read-out to the tape recording. By including an online 'filtering' routine, a user can also produce secondary output streams of selected events, which are copied to different event buffers and recorded, if needed, onto different magnetic tapes.

Data are written onto tape in a CERN standard format. Tapes are ANSI labelled and a tape-booking scheme exists which keeps track, in a disk file log, of all the tapes being written.

The data acquisition system is also able to use the tapes previously recorded as a source of data, instead of the electronics readout. In this way it is possible to fully analyse the recorded data using the same monitoring programs that are used online. This feature gives also the possibility to test the monitoring programs with realistic data.

Additionally, it is possible to read some typical events from a disk file and store them into the CAMAC hardware for a full simulation of the data acquisition process.

5. MONITORING

More than seven large programs run continuously in the VAX during data-taking, and many others are started whenever necessary, to monitor the experimental apparatus. All these programs use standard packages to interact with the users, to gather data from the data-acquisition and to display results. A major requirement in this monitoring activity is the complete independence of each program with respect to the other monitoring programs and to the data acquisition system. This is achieved by sending a complete, or partial, copy of one or more events to the monitoring program which can then process the data at its own speed without any interference from the rest of the system. The communication mechanism between DAQP and the monitoring programs is the same as that used to communicate through the CERN network. This, at the price of some increase in overhead, gives a remarkable degree of flexibility to the system. In fact, the system can be very easily reconfigured to use other computers in addition to the VAX running the acquisition, provided that they can communicate through CERNET. (This is the case for all the CERN-supported mini-computers - VAX, PDP, NORD, HP, etc.). It is also very easy, using this mechanism, to send a sample of events online to the IBM and CDC mainframes for more sophisticated analysis and to get the results back in real time.

6. CONCLUSIONS

The implementation of the system took approximately 7500 lines of macro-assembly code for the CAMAC driver (including CERNET and CAMAC lists modules), approximately 5500 lines of FORTRAN 77 code for DAQP and a total of approximately 4500 lines of FORTRAN 77 code for all the ancillary processes. Late in 1981 the system was fully operational and started to collect real data from the experimental apparatus. The conclusions we can draw from the first experimental runs are that such a system can very well cope with a large experiment like UA2, but some limitations are intrinsic to a centralised system and we observed a certain decay in performance when all monitoring functions were activated together with the full data-acquisition chain. The system is capable of acquiring, checking and storing a full event of approximately 10 Kword in 25 msec (peak rate) in the present version. The total throughput from CAMAC to tape is approximately 200 Kbyte/sec in continuous mode. Some studies are now being carried out to improve performances in the area of data-acquisition efficiency: data-taking speed, monitoring speed, etc.

SYSTEME D'ACQUISITION CAMAC
POUR TOMOGRAPHE A RAYONS X INDUSTRIEL

Jean Pierre GUERIN[*], Jacques HUET[**], Michel PAUTON[**]
[*] CEA/CEN-SACLAY/DEIN - 91191 GIF-SUR-YVETTE CEDEX
[**] CEA/Etablissement T BP N° 7 - 93270 SEVRAN

RESUME

Un prototype de tomographe industriel à rayons X de 400 kV a été développé. On décrit l'architecture du système d'acquisition CAMAC, son mode de fonctionnement et sa programmation. Cet ensemble qui comprend 3 microprocesseurs, 2 mémoires de masse et une visualisation graphique TV est très souple ; il permettra d'optimiser les paramètres de mesure en fonction de la nature et de la taille des pièces à contrôler et de chercher les limites d'application de cette méthode de contrôle non destructif. Ce système permet l'examen interne de pièces mesurant jusqu'à 400 mm de diamètre avec un volume élémentaire de $1 \times 1 \times 5$ mm^3. Suivant les dimensions des objets, les masses volumiques des matériaux constitutifs sont comprises entre 0,7 et 10g/cm^3. On espère obtenir des précisions relatives de l'ordre de 10^{-3} lors de l'examen tomodensitométrique de matériaux moulés. On indique les premiers résultats obtenus avec cet appareillage.

ABSTRACT

A 400 kV X-ray industrial scanner has been developed. The present paper describes the CAMAC acquisition system. This unit with its 3 microprocessors, 2 mass memories and TV graphic display is very flexible ; it will be used to optimise measurement parameters along with pieces to be checked and to establish application limits of this non destructive inspection method. This system allows internal inspection of objects as big as 400 mm in diameter with an elementary volume of $1 \times 1 \times 5$ mm^3 and specific weight in the range of 0.7 to 10g/cm^3. Relative precision of 10^{-3} in density for moulded objects is expected. The first results obtained are shown.

ZUSAMMENFASSUNG

Der Prototyp eines industriellen 400 kV-Roentgenstrahltomographs wurde entwickelt. Es wird die Architektur des CAMAC-Datenerfassungssystems, seine Funktionsweise und seine Programmierung beschrieben. Diese sehr flexible Einheit besteht aus 3 Mikroprozessoren, 2 Massenspeichern und einem graphischen Sichtgerät. Sie erlaubt eine Optimierung der Parameter in Abhängigkeit der Beschaffenheit und der Form der Prüflinge und weiterhin, das Anwendungsgebiet dieser zerstörungsfreien Prüfmethode zu begrenzen. Man kann Teile mit bis zu 400 mm Durchmesser und einer Volumenauflösung von 1x1x5mm^3 untersuchen. Abhängig von den Abmessungen kann sich das spezifische Gewicht der zu untersuchenden Materialien zwischen 0,7 und 10g/cm^3 bewegen. Es wird eine relative Genauigkeit von 10^{-3} für tomodensimetrische Untersuchungen eingeschmolzener Materialien angestrebt. Die ersten mit dieser Einheit erhaltenen Ergebnisse werden angeführt.

1 - DESCRIPTION GENERALE

A partir de l'expérience acquise au CEA dans la conception des tomographes médicaux, une étude de développement d'un système industriel a été lancée (1) (2) (3). Comparées à celles des appareils médicaux, les caractéristiques du prototype (fig 1) sont les suivantes :

- haute tension tube 400 kV au lieu de 150 kV,
- retour à la première génération avec un monodétecteur,
- même niveau de résolution spatiale,
- mécanique simplifiée, objets mobiles dans un flux de rayonnement fixe.

Figure 1 : prototype de tomographe à rayons X industriel

L'appareillage décrit constitue la première étape du développement de la tomographie à rayons X industrielle. Il a nécessité la mise au point au CEN-GRENOBLE/LETI d'un nouveau type de détecteur à faible traînage et d'une grande efficacité, constitué d'un scintillateur Cs I(Tl) couplé avec des photodiodes fonctionnant en régime photovoltaïque.

Ce détecteur est associé à un moniteur de flux situé en sortie du tube et constitué d'une chambre à ionisation. Les caractéristiques du détecteur et de l'électronique associée permettent d'obtenir une dynamique de mesure de 10^{+4} indispensable à l'étude de matériaux denses.

Les calculs de reconstruction d'image sont effectués pour l'instant sur un réseau indépendant du système, de même que l'exploitation des images.

2 - PRINCIPE DE LA MESURE

La tomographie permet d'obtenir de façon non destructive la cartographie interne de la densité d'un objet et sa représentation directe sous forme d'image.

Projetons sur la droite u'u l'intégrale des masses des volumes élémentaires de l'objet à tomographier parallèlement à une génératrice de direction v'v. On obtient une fonction δ (u) à θ=cste (fig 2).

En répétant l'opération pour tous les angles θ de 0 à π, puis en appliquant une méthode de déconvolution, il est possible d'établir la carte de la densité d(x,y) de chaque point de l'objet.

Le codage de la densité en niveaux de gris ou en couleurs permet la représentation de cette carte sous forme d'image.

Pratiquement, la mesure de l'atténuation d'un faisceau de rayons X qui doivent être d'autant plus pénétrants que les objets sont plus volumineux ou de plus grande densité, donne, après traitement, la valeur de l'intégrale δ (u, θ).

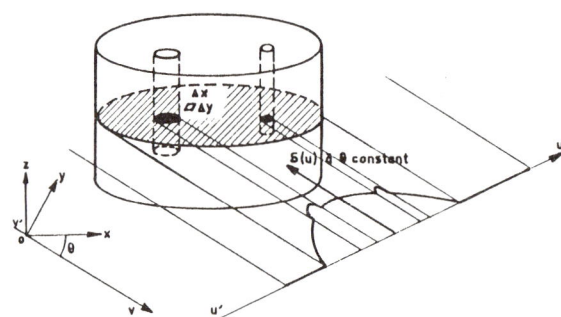

Fig 2 : la courbe δ (u) est la projection sur u'u de l'intégrale des masses des volumes élémentaires le long d'une génératrice de direction v'v.

Dans le cas de pièces industrielles, il est plus simple de faire subir à la pièce des translations à vitesse constante pour obtenir la fonction δ(u) à θ = cste que de déplacer l'ensemble générateur et détecteur. Comme le flux X au niveau du détecteur est très faible, sa mesure s'effectue par intégration des photons X en un temps correspondant au pas d'échantillonnage de la mesure. Le signal électrique délivré par l'intégrateur est alors numérisé. Après chaque acquisition le long d'une génératrice (200 à 1000 points de mesures), un moteur fait tourner la pièce d'un angle dépendant de la précision désirée et de la taille de la pièce.

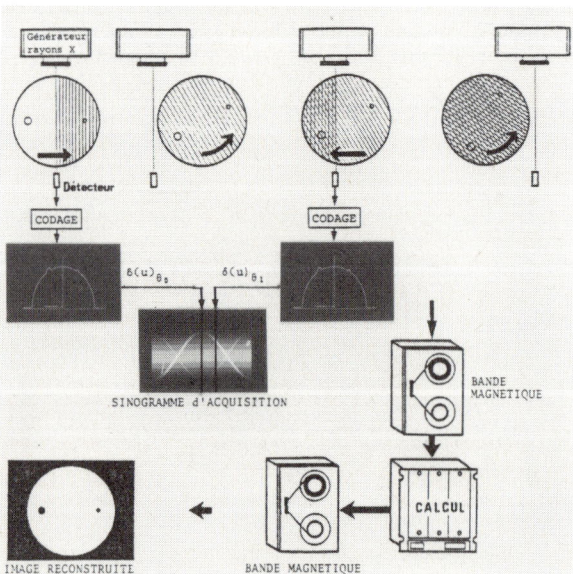

Figure 3 : principe du système

La juxtaposition des projections sur un écran graphique permet de visualiser en temps réel le sinogramme d'acquisition, le niveau de gris représentant l'atténuation (fig. 3). Lorsqu'une coupe est terminée, la pièce se déplace verticalement pour la coupe suivante. Les mesures sont stockées sur bande magnétique ou transmises par modem au centre de calcul.

3 - ARCHITECTURE DU SYSTEME D'ACQUISITION

On utilise le système CAMAC dont la conception s'adapte aisément à la conduite de processus industriels (4). L'ensemble fonctionne en mode "autonome programmable adaptatif" ; il est ordonné autour d'un contrôleur JCAM-10 (fig 4).

L'ensemble comprend trois microprocesseurs de la famille INTEL 8080 intégrés dans des unités étudiées au CEA. Le premier, situé dans le JCAM-10, assure la gestion du système. Le second assure la gestion de l'interface disquette et celle des fichiers compatibles 100 % IBM 3740 ; il sert à archiver les programmes et les résultats des étalonnages. Le troisième assure la gestion de l'interface bande magnétique 9 pistes avec un dérouleur à microformateur utilisé pour stocker les mesures dont le nombre varie de 10^{+5} à 10^{+6} par coupe suivant la taille des pièces et la résolution désirée.

Une visu graphique sur moniteur TV noir et blanc ou couleur 512 x 512 pixels, 16 teintes, sert à l'étalonnage, à l'alignement mécanique, assure la visualisation des sinogrammes et des images reconstruites.

Un seul module a été étudié spécialement pour assurer l'interfaçage, la synchronisation entre les moteurs, les règles repérant la position de la pièce et l'intégrateur du détecteur de flux X dont le signal est codé en 25 μs à l'aide d'un convertisseur 14 bits AD/ADC 1130 précédé d'un échantillonneur bloqueur AD/SHA 1144 rapide (1 μs à 3.10^{-5}).

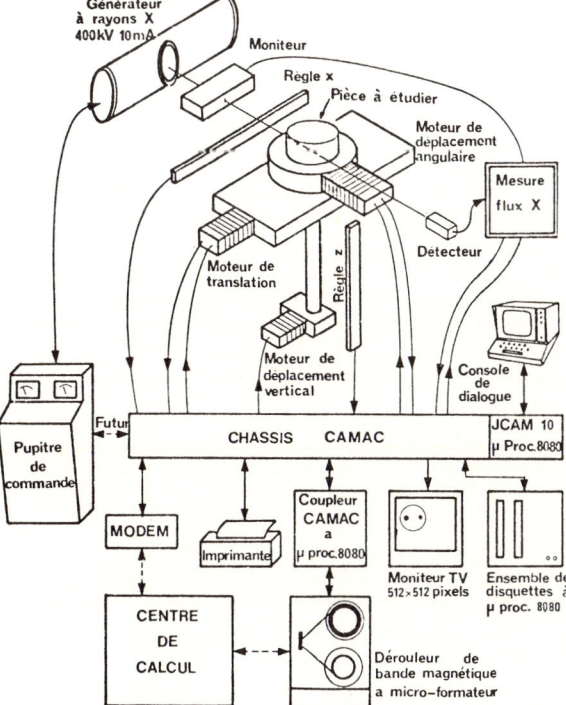

Figure 4 : Acquisition et codage de la mesure

4 - PROGRAMMATION DU SYSTEME

4.1 Le langage de programmation est le BASICAM, BASIC temps réel appliqué au système CAMAC JCAM-10 (5). Cependant, certaines parties sont écrites en assembleur pour respecter les vitesses d'acquisition de l'appareillage.

4.2 Les programmes de test et de mise au point sont utilisés pour vérifier et régler tous les périphériques et circuits du tomographe, les règles, les codeurs, les moteurs, leur vitesse et leur accélération, pour aligner le faisceau X et le banc et pour étalonner les atténuations en fonction des matériaux.

4.3 Le programme d'acquisition permet une grande souplesse d'utilisation de l'appareil. Le dialogue opérateur utilise une méthode de menus avec questions/réponses, fournit au système les paramètres de la pièce, les précisions désirées, les consignes de mesure. Ensuite, le système s'autovérifie et commence les acquisitions : commande des mouvements, codage et stockage des projections sur la bande magnétique, visualisation des sinogrammes.

4.4 Le traitement numérique de reconstitution d'images est effectué au centre de calcul. Les résultats sont mémorisés sur 16 bits par pixel sur bande magnétique.

4.5 Le programme de visualisation fournit une image reconstruite codée sur 4 bits par pixel à partir des données de la bande.

5 - PREMIERS RESULTATS

Le prototype décrit est en cours d'essais et les premiers résultats (fig 5) montrent que l'on obtient une résolution spatiale de l'ordre de 1 mm, comparable à celle des appareils médicaux.

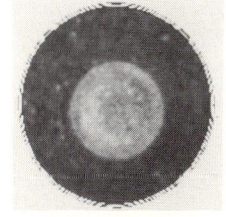

a) Tronc d'arbre.

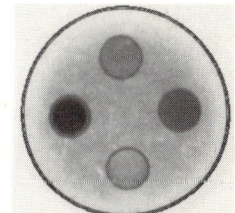

b) Inserts de tiges dans un cylindre de plastique.

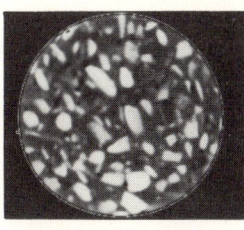

c) Cylindre de béton.

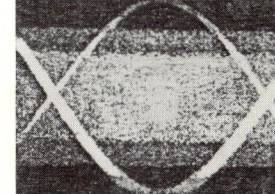

d) Sinogramme.

Figure 5 : Premiers résultats d'essais

6 - PERSPECTIVES ET DEVELOPPEMENTS

Nous poursuivons l'étude du développement de l'appareillage. Afin de réduire les temps d'acquisition, nous envisageons d'évoluer vers un modèle de la seconde génération comportant un pluridétecteur et un multiplexeur CAMAC couplé directement au minicalculateur de construction d'images.

REFERENCES

(1) GARIOD R. - ALLEMAND R. - THOMAS G. (CEA/CEN-G/LETI)
Etude et faisabilité de l'extension des méthodes de tomographie par rayons X utilisées en médecine à l'analyse non destructive d'objets.
Rapport DGRST/GR 770-799/LETI/1413/81)

(2) HUET J. (CEA/SEVRAN) - THOMAS G. (CEA/GRENOBLE /LETI)
Etude d'un prototype de tomographe à rayons X industriel. Revue pratique de contrôle industriel n° 112 - décembre 1981

(3) HUET J. - PAUTON M. (CEA/Etablissement T)
La tomographie à rayons X industrielle. Réalisation et essais d'un prototype 400 kVolts. Journées nationales du COFREND - PARIS - Janvier 1982

(4) HUET J. - PEUZIAT C. (CEA/SACLAY/SES)
Banc de contrôle automatique à rayons X géré par microcalculateur CAMAC
Real Time Data Handling and Process Control.
Berlin-Ouest - 23/25 octobre 1979.

(5) GUERIN JP (CEA/CEN SACLAY/DEIN)
BASICAM - Langage BASIC temps réel pour micro-ordinateur CAMAC. Real Time Data Handling and Process Control
Berlin-Ouest 23/25 octobre 1979.

EMPLOI DES MICRO SYSTEMES DANS LES ACTIVITES D'ESSAIS ET DE CONTROLES DU RESEAU DES LPC

S. SAVOYSKY
Laboratoire Central des Ponts et Chaussées, Paris, France

RESUME

Le développement de la micro-informatique, allié à une action de formation des utilisateurs à ces techniques, favorise la multiplication et la diversification de petits systèmes intégrés et automatiques de mesures et de contrôles dans les activités du réseau des Laboratoires des Ponts et Chaussées (Ministère du Logement et de l'Urbanisme, Ministère des Transports). Dans le même temps les architectures des systèmes de conduite de processus dans les laboratoires ou centres d'essais, organisées autour d'un seul calculateur centralisant toutes les fonctions, évoluent vers la mise en oeuvre de micro-systèmes spécialisés par fonction et organisés en réseau local.
Cette distribution de l'informatique dans un réseau d'essais et de contrôles déjà décentralisé de par la nature de ses activités, implique, en l'absence de pratiques communes, un gros risque de dispersion incontrôlée des choix techniques et du savoir faire, de défaut coûteux de coordination dans la maintenance des dispositifs et la formation du personnel. Des pratiques communes sont nécessaires ; elles sont fondées, à court terme sur des besoins de normalisation : systèmes d'exploitation, interfaces, communications, langages. A plus long terme apparaît également le besoin de méthodes et langages de spécification, pour la conception des projets et pour leur gestion après réalisation.

ZUSAMMENFASSUNG

Anwendung von Mikrosystemen bei der Versuchs- und Kontrolltätigkeit des Labornetzes der Ponts et Chaussées.
Durch die Entwicklung der Mikroinformatik und eine entsprechende Ausbildung ihrer Benutzer wird innerhalb der Tätigkeit des Labornetzes der Ponts et Chaussées (Ministerium für Wohnungs- und Städtebau, Ministerium für Transport) dis Vermehrung und Diversifizierung von integrierten Kleinsystemen zur automatischen Messung und Kontrolle gefördert. Gleichzeitig entwickelt sich der einem einzigen, alle Funktionen zentralisierenden Rechner angeschlossene Aufbau der Prozessteuerungssysteme in den Laboratorien und Versuchszentren in Richtung von funktionsspezifischen Mikrosystemen mit Ortsanschluss.
Diese Verteilung der Informatik innerhalb eines Versuchs- und Kontrollnetzes, das bereits durch die Natur seiner Aktivitäten dezentralisiert ist, beinhaltet eine grosse Gefahr der unkontrollierten Streuung bei der technischen Auswahl und dem Know-how, des kostenaufwendigen Mangels an Koordinierung bei der Instandhaltung der Einrichtungen und der Ausbildung des Personals. Gemeinsame Praktiken sind notwendig ; sie basieren kurzfristig auf dem Normungsbedarf : Auswertungssysteme, Schnittstellen, Kommunikationen, Sprachen. Langfristig zeigt sich auch der Bedarf an Spezifikationsmethoden und -sprachen zur Projektgestaltung und -verwaltung nach deren Verwirklichung.

ABSTRACT

The use of micro-computers in the testing and inspection work of the network of Road Research Laboratories.
The spread of micro-computing, coupled with measures to train users in its techniques, favours the multiplication and diversification of small integrated automatic measurement and testing systems in the work of the network of Road Research Laboratories (Ministry of Housing and Urban Development, Ministry of Transport). At the same time, the architecture of process control systems in laboratories and testing centres, organized around a single computer that performs all functions, is evolving towards the use of micro-systems specialized by function and organized into local networks.
This distribution of information processing in a testing and inspection network that is already decentralized because of the nature of its work entails, in the absence of common practices, a large risk of uncontrolled dispersion of technicial choices and of know-how and of a costly failure of coordination in the maintenance of the equipment and in the training of the personnel. Common practices are a necessity. In the short term, they will be based on the needs of standardization :operating systems, interfaces, communications, languages. In the longer term, there will also be a need for specification languages and methods for project planning and project managment after implementation.

1. Introduction

L'organisation de la recherche, des études et des contrôles au Ministère du Logement et de l'Urbanisme et au Ministère des Transports, s'appuie sur un réseau d'établissements répartis sur tout le territoire français : le Laboratoire Central des Ponts et Chaussées, les Laboratoires Régionaux, les Ateliers de Prototypes et divers centres spécialisés (LPC dans la suite du texte). C'est en 1972 que deux projets de conduite de processus furent primitivement lancés : la station d'essais de mécanique des sols du laboratoire régional de Toulouse et la station d'essais d'éléments de matériel routier de Blois. Ces stations, ainsi que d'autres projets réalisés depuis, avaient une caractéristique commune : le système d'informatique était pour chaque site, une ressource unique partagée entre plusieurs dispositifs d'essai. De telles architectures centralisées étaient imposées par le prix des matériels mis en place. Ces prix et souvent également la taille des matériels ont fortement empêché jusqu'à environ 1978 d'aborder la réalisation pratique de petits équipements autonomes. Le développement de l'emploi des microprocesseurs et circuits associés et la volonté des LPC de banaliser et décentraliser l'emploi de l'informatique sont depuis à l'origine d'une multiplication de projets et de réalisations.

2. Travaux [3,4]

Acquisition de mesures

Les essais et contrôles réalisés dans le domaine du Génie Civil sont exécutés généralement dans des sites variés, parfois sur des engins mobiles. Il est en outre souvent nécessaire de pouvoir transporter facilement, voire porter, les dispositifs et supports de mesures. Pour ces raisons, les LPC ont été amenés à développer des systèmes intégrés ou portables pour le traitement local et immédiat de mesures ainsi que des moyens d'enregistrement sur cassettes magnétiques numériques de ces mesures, pour des traitements ultérieurs et pour la constitution de bases de données techniques. Corrélativement et grâce à la micro-informatique, des lecteurs de cassettes ont pu être construits et diffusés dans tout le réseau et raccordés aux différents ordinateurs de l'Administration par les voies publiques de la télématique. Différents types d'appareils de mesures ont été ainsi équipés, éventuellement par les unités utilisatrices elles-mêmes ; par exemple dans le domaine des mesures de caractéristiques de chaussées :
- le déflectographe destiné à mesurer les déflexions ou déformations verticales de la surface d'une chaussée sous l'action d'une charge (application : étude de l'évolution du réseau routier sous le trafic, détection de zones défectueuses et contrôles des renforcements, surveillance hivernale etc...).
- le gammadensimètre mobile ("GDM 45") destiné à mesurer en continu la masse volumique d'assises de chaussées (application : étude et contrôles du compactage).

La même démarche a porté également ses fruits dans d'autres domaines : élaboration de matériels de contrôle en continu de la production des centrales de fabrication de béton ou d'enrobés bitumineux, élaboration d'appareils de relevés de données hydrologiques ou de surveillance de pollution, exploitables dans des zones éloignées de tout centre de traitement, etc...

Conduite de processus

Un exemple de réalisation de conduite de processus, parmi les plus simples, utilisant un microprocesseur est donné par l'automatisation d'une double sonde destinée à établir le profil vertical de densité humide dans un sol : deux trous étant forés parallèlement, on descend dans l'un deux une source de rayons gamma et dans l'autre, au même niveau, un compteur ; on détermine ainsi la densité d'une façon précise (puisque mesurée par diffusion directe) jusqu'à une profondeur donnée.

Le second exemple est, par contre, un projet de plus vaste envergure. L'équipement centralisé de la station automatique de Blois, précédemment citée, est progressivement remplacé par un système d'informatique distribuée mettant en oeuvre des microcalculateurs de conduite de processus spécialisés par type d'essai. Notons en outre que les liaisons de tels micro-systèmes, avec les centres de traitement de l'Administration, pour l'utilisation de SGBD, sont d'ores et déjà effectives grâce aux réseaux publics parmi lesquels, en particulier TRANSPAC.

Autres travaux

Ces techniques intéressent l'ensemble des professions du génie civil ; elles sont principalement applicables à l'automatisation des procédés de fabrication des matériaux utilisés en construction, des procédés de mise en oeuvre de ces matériaux, enfin, des procédés de construction. Les quelques exemples suivants montrent que, dans tous les cas, les objectifs essentiels sont la recherche d'une plus grande rigueur dans l'exécution des travaux et d'une peinabilité moindre pour les exécutants, améliorant ainsi les qualités économiques et techniques des résultats obtenus.
Exemples :
- des centrales de fabrication de béton prêt à l'emploi sont actuellement équipées de micro-systèmes intégrés supportant des logiciels de commande, adaptés cas par cas, aux configurations de ces centrales. Outre la conduite stricte des procédés de dosage et de malaxage du béton, garantissant la qualité du produit, les logiciels gèrent éventuellement les stocks de composants et prennent en charge, souvent, certaines fonctions administratives comme, par exemple, la facturation. L'automatisation des centrales de fabrication de produits routiers enrobés est également envisagée dans le même esprit.
- Les automatismes classiques de guidage des engins de répandage sont parfois assez primitifs ; ils sont progressivement supplantés par des systèmes numériques lesquels, prenant en compte un très grand nombre de paramètres, permettent de garantir une meilleure qualité de la mise en oeuvre par l'amélioration de l'uni de la chaussée et le respect strict des spécifications d'épaisseur.
- Les travaux de terrassement constituent enfin les sujets d'automatisation parmi les plus récents. Il s'agit de travaux pénibles et dans ce cas, toute amélioration des conditions de travail a des répercussions immédiates sur les résultats obtenus. L'une des approches les plus intéressantes dans ce domaines consiste à rechercher des processus d'apprentissage afin que l'automatisme intégré puisse gérer seul certaines séquences répétitives après seulement quelques exécutions guidées par le conducteur de l'engin.

3. Pratiques communes

Dans un réseau de laboratoires très décentralisés on peut craindre que la prolifération des produits de l'informatique et des méthodes pour leur mise en oeuvre n'entraîne à plus long terme une disparité des équipements et de leurs modes d'exploitation contrecarrant l'unicité des modes opératoires et la qualité du savoir-faire des opérateurs. Dans un environnement technique qui évolue rapidement, il est difficile par ailleurs de fixer des options définitives de choix. Nous assistons actuellement à ce que nous pourrions nommer une consommation anarchique et par conséquent parfois abusive de l'informatique. Des pratiques communes s'avèrent donc de plus en plus nécessaires. Elles concernent les points suivants pour l'essentiel.

Systèmes d'exploitation

Deux types d'actions nous semblent nécessaires :
- la définition d'un ensemble de fonctions élémentaires, spécifiques à la conduite de processus (ensemble usuellement désigné par "noyau exécutif de temps réel")[1] et dûment spécifiées non seulement dans leurs objectifs mais aussi dans leurs conditions temporelles d'exécution ;
- la définition, conjointement à la précédente, d'un langage de commande de bas niveau et le choix d'un système d'exploitation standard sanctionnant éventuellement l'existence de l'un des systèmes existants actuellement

Interfaces avec les processus physiques

Nous sommes convaincus qu'il ne peut exister de panacée dans ce domaine en raison de la grande variété des problèmes posés; nous sommes également convaincus que les quelques dispositifs, objets de standards bien connus (IEEE 488, CAMAC) sont certainement insuffisants pour couvrir techniquement mais aussi économiquement toute la gamme des besoins exprimés. Toutefois, entre la prolifération actuelle de dispositifs et une normalisation trop restrictive, il devrait être possible de caractériser les besoins les plus courants et d'en déduire des spécifications d'interface. Il serait donc hautement souhaitable que les travaux d'analyse et de documentation sur les solutions existantes soient rapidement suivis de propositions fermes pour une gamme de différentes interfaces couvrant la majeure partie de besoins. De notre point de vue, il serait en outre nécessaire d'associer à ces définitions d'interfaces, la définition de modules standardisés d'échanges. Il apparaît en effet, que dans de nombreux cas, appartenant même à des domaines totalement différents, on retrouve très souvent les mêmes problèmes d'échanges à résoudre. Nous suggérons que cette question soit examinée dans les instances de normalisation.

Communications

Il existe plusieurs niveaux de problèmes de communication ; nous nous limitons à la présentation de deux d'entre eux, les plus essentiels :
<u>Définition d'une norme de réseau local de distribution d'informations scientifiques et techniques.</u>
L'informatique distribuée en milieu industriel est une réalité ; nos stations d'essais en constituent des exemples et les principaux constructeurs de matériels d'informatique industrielle y répondent en proposant chacun leur propre conception d'architecture de systèmes industriels répartis. L'équivalent industriel d'Ethernet est donc une nécessité.

<u>Accès aux réseaux publics.</u> Les réseaux publics disponibles pour la télématique sont évidemment utilisables pour toute installation fixe ou pour toute activité de contrôle ou d'essais proches d'un de leurs points d'accès. Il n'en n'est pas toujours ainsi : c'est le cas de nombreux dispositifs de surveillance d'ouvrages ou d'environnement. Une solution envisageable est la télématique par l'intermédiaire de balises de transmission par satellites.

Méthodes de production

La programmation en langage évolué est considérée comme l'un des facteurs décisifs de banalisation de l'emploi des micro-systèmes dans nos activités, tout en ayant bien conscience des limites de cette organisation, notamment lorsque les spécifications de fonctionnement en temps réel des systèmes à produire deviennent très contraignantes. Cet objectif introduit évidemment le problème du choix d'un langage commun et des actions correspondantes de formation du personnel mais aussi la nécessité de développer des méthodes et moyens associés de production, lesquels constituent un fort investissement. L'utilisation systématique de Ada est considérée (dans les limites indiquées) comme un objectif à réaliser dans les prochaines années.

4. Conclusion

Le développement d'automatismes dans un réseau où les essais et contrôles courants sont associés à la recherche fondamentale en génie civil, conduit souvent les concepteurs de systèmes à poser des problèmes dans le domaine de l'informatique à la limite des possibilités de cette cette technique et, par conséquent, à trouver des solutions dans les résultats les plus récents de la recherche qui la concerne ; comme en outre, les activités des LPC constituent un excellent terrain d'expériences et d'observations pour les applications de la recherche en informatique, une démarche nous est fortement suggérée : participer à l'effort de recherche en informatique notamment dans les domaines qui nous semblent correspondre à nos préoccupations, et à nos moyens.

5. Bibliographie

1. BNI. Projet Sceptre. 1980.

2. BNI. Guide des interfaces physiques. 1981.

3. Emploi des microprocesseurs dans les activités des LPC .- L.C.P.C. Paris, 1979.

4. PAREY (Ch), SAVOYSKY (S) .- L'emploi des microprocesseurs dans les travaux publics .- dans : PCM, n° 4, avril 1980, pp. 33/37.

CAMAC Serial Highway Implementation
at the Joint European Torus (JET)

V. Schmidt, Joint European Torus, Abingdon UK

Abstract

The JET Control and Data Acquisition System (CODAS) uses fibre optic CAMAC Serial Highways to connect its more then 20 centralised minicomputers to the various functional subsystems of the experiment. Size and complexity of the experiment have led to the development of a sophisticated Serial Highway Driver which implements comprehensive message analysis and error recovery as well as very powerful DMA modes to achieve high data rates. This Driver is complemented by a double loop fibre optic Serial Highway providing a number of possible loop configurations for speedy detection and recovery from cable or equipment faults.

Zusammenfassung

Das Steuerungs- und Datenerfassungs-System des JET Experiments verwendet Lichtfaserkabel fuer die CAMAC Serial Highways, ueber die alle funktionellen Subsysteme des Experiments an die ueber 20 zentralen Mini-Rechner angeschlossen sind. Groesse und Komplexitaet des Experiments haben zur Entwicklung eines Serial Highway Driver gefuehrt, der sowohl eine umfassende Fehlererkennung und -korrektur durchfuehrt als auch ueber leistungsfaehige DMA-Betriebsarten zur Erzielung hoher Datenraten verfuegt. Die Serial Highways sind als doppelte Schleifen aus Lichtfaserkabeln ausgefuehrt. Die Schleifen koennen in verschiedenen Konfigurationen betrieben werden, die eine schnelle Erkennung von Kabelfehlern ermoeglichen und alternative Nachrichtenwege zu ihrer Umgehung zur Verfuegung stellen.

Résumé

Le système de contrôle et d'acquisition de donées (CODAS) du projet JET utilise des fibres optiques pour ses boucles CAMAC série connectant plus de 20 sous-ensembles. La taille et la complexite de l'expérience ont conduit au développement d'un contrôleur de boucle série très élaboré qui supporte un système de détection et correction d'erreurs et des modes efficaces de transfert direct mémoire permettant une grande vitesse de transmission. Ce contrôleur est connecté a une double boucle de fibres optiques permettant détection d'erreurs et reconfiguration rapide en cas d'incident sur les fibres.

1. General Background

JET is the major European experiment on the path to controlled nuclear fusion, nearing completion near Oxford, UK. Due to its size and complexity, control and data acquisition are fully computerised. The JET Control and Data Acquisition System (CODAS) comprises of some 30 centralised minicomputers interfacing to the experiment via some 100 CAMAC Crates distributed over the site some 500 metres in extent. CAMAC and other interface electronics are housed in 2 metre high cubicles placed near to the connected equipment which is distributed over several buildings. The structure of CODAS reflects the various functional Subsystems of the experiment. Each Subsystem Computer is linked to the experiment by means of a CAMAC Serial Highway, implemented as fibre optic cables in order to cope with the electrical problems created by the large distances between the various parts of the experiment and by the electrically noisy environment. The CODAS computers are Norsk Data, ND-100 and ND-500. The Serial Highways operate at present in Bit Mode at up to 5MHz. It is planned to increase the data transfer rate of some data acquisition loops by using Byte Mode transfer which would increase throughput by about a factor of 6.

2. CODAS Serial Highway Driver

Initial studies performed for JET by IPP Garching /1/ have shown that the standard CAMAC Serial Highway operating at its maximum speed would be able to cope with the data rates expected at JET (initially 1MWord collected from each 'pulse', ie every 10 minutes) provided the Highway was operated in 'pipeline' mode (see below). In order to achieve the fault tolerance required by JET the Serial Highway Driver (SHD) had to incorporate extensive message analysis and error recovery in the hardware. The actual SHD for JET has been developed by CAMTEC Electronics Ltd, UK making much use of a design study carried out by HMI Berlin for JET /2/. It takes the form of two PC boards that plug into the ND-100 backplane. The SHD incorporates two independant channels for Programmed I/O (PIO) operation and for DMA operation. PIO operation can be executed concurrently with all DMA operations except pipeline mode. There are essentially three different DMA modes available. In LIST Mode the SHD executes a sequence of CAMAC Commands from a list held in the computer memory. It performs its Commands in turn, and returns the results to the memory. LIST execution continues until the preset Word Count (in the hardware) reaches zero, or until a Termination Condition within each Command in the list is satisfied. CAMAC Commands in LIST Mode can be of any type and in any order. In BLOCK Mode one CAMAC Command is preloaded and remains unchanged for the duration of the transfer. The DMA channel may be set to operate in all the standard Q modes, including Q Stop Extension, ie the last valid word is accompanied by Q=0. Each transaction within the block may be optionally LAM-triggered, the SHD generating an automatic Execute Command to the appropriate LAM encoder module in order to unlock the interrupt mechanism. In BURST (Pipeline) Mode one CAMAC Command is preloaded and remains fixed for the duration of the whole transfer. The SHD transmits a burst of contiguous Command messages in a block, which will result in a corresponding number of Reply messages received back by the SHD. A burst of transactions is successfully completed when the last Command has been transmitted and the expected number of successful Reply messages have been received. If, however, any error occurred the whole burst of transactions must be repeated. Special hardware features support the Demand Handling. Correctly formatted Demand messages may be received at any time, and will be entered into a 64-word deep Demand FIFO. The SHD contains an automatic Mask/Disable facility that generates a LAM Mask or Crate Disable Command for each valid Demand received, prior to interrupting the computer. The SHD performs an Extended Message Analysis including context checking on all executed Commands, and as far as applicable, on Demands. Message analysis automatically triggers Error Recovery Procedures in all cases where such recovery is possible. Error Recovery executes Repeat Command, Read Crate Controller Status Command, or Re-Read Command depending on the type of error and type of Command that failed. Automatic error recovery may be executed for all single transaction Serial Highway cycles, ie for all transactions but BURST Mode. The 64-byte Test FIFO may be loaded with an arbitrary bit pattern which can be transmitted directly to the Serial Highway thus giving the possibility to send out deliberately corrupted Commands. A Dump Store on the receiver port continuously stores the last 256 received bytes (suppressing non-essential WAIT bytes). This last-in-first-out store is accessible from the computer for diagnostic purposes. For test purposes the SHD can be put into Echo Mode, ie connecting the transmitter output directly to the receiver input. This feature is very powerful for self-checking of the SHD.

3. CODAS Fibre Optic Serial Highway

The CODAS CAMAC Crates are connected to the central minicomputers via fibre optic Serial Highways. The components of the system have been designed and built by CAMTEC Electronics Ltd, UK based on a functional specification by JET, which in turn has used ideas from CERN-SPS /3/ for the loop configuration aspects of the design. U-port Adaptors (U-PA) 'translate' from the standard CAMAC D-ports on the Crate Controllers and on the Serial Highway Driver to the fibre optic transmission technique used on the transmission lines: Clock and Serialised Data travel on a single fibre using Manchester type coding. The system may work at four discrete frequencies: 0.5, 1.0, 2.5, and 5.0 MHz using Bit Serial transmission. The operating frequency is software selectable at the SHD. The U-PA automatically lock into any of these four frequencies. The optical Highway is duplicated, ie there is a MAIN and a BACKUP fibre optic loop connecting all U-PA. Both loops are logically equivalent. Signal flows on MAIN and BACKUP loop are opposite: the first U-PA on the MAIN loop, seen from the computer, is the last U-PA on the BACKUP loop. There are two functionally different types of U-PA in each loop. The Master U-PA interfaces the fibre optic Highway to the computer whereas Slave U-PA interface to Serial Crate Controllers. All U-PA are equipped with electronic switches which provide (1) fast switch over from MAIN to BACKUP loop, (2) program controlled loop collapsing facilities which enable the isolation of faulty cables or faulty Crates, (3) program controlled external bypassing of Crates, (4) automatic forced bypassing of Crates which lost power (battery backup maintains the fibre optic Serial Highway). The loop-switches in the Master U-PA are under direct program control. The loop-switches in the Slave U-PA automatically take into account on which loop input, MAIN or BACKUP, valid data bytes are being received. If both inputs receive valid bytes it locks to the input which first received valid bytes. This also applies to the Loop Collapse Command issued to a Crate connected to the U-PA. If the U-PA is receiving valid messages on its MAIN input it will react to the Loop Collapse Command by sending out messages on the BACKUP output. Similarily it sends out messages on the MAIN loop when receiving on the BACKUP input, if Loop Collapse is asserted. This switching mechanism can be used to cut out faulty cables or faulty Crates.

4. Software

The basic software for the Serial Highway Driver has been developed by Norsk Data, based on functional specifications by JET. It is complemented by more specific products designed and coded by JET. The low level software consists of a set of four CAMAC Monitor Calls. The normal user interface is by means of a set of FORTRAN callable Subroutines which broadly conform to the ESONE standard /4/. The subroutines which have no ESONE equivalent are mainly geared to the high performance DMA modes of the SHD hardware. The CAMAC I/O software is complemented by an extensive set of software for Test, Error Handling, and Performance Monitoring. Single CAMAC Dataway Transactions can be executed using programmed I/O (PIO). The PIO channel is protected from multiple access by an automatic booking/queuing mechanism. This booking mechanism also prevents simultaneous PIO and BURST DMA operation. The software supports all DMA modes provided by the hardware. It supports concurrent operation of PIO and DMA (except BURST). Since the ND100 is a virtual machine there exist two distinct methods for DMA execution: (1) Physical Memory Operation ie the whole required memory space is fixed contiguously, or (2) Virtual Memory Operation ie data are mapped in blocks through a reserved buffer in physical memory. This applies to all DMA modes which are available in the SHD hardware. The data throughput achieved in the various modes is as follows (16-bit words, 5MHz):

Mode	Transaction Time
Virt. Add. BLOCK Byte Mode	2.8ms + 16µs/word
Phys. Add. BLOCK Byte Mode	1.5ms + 7.3µs/word
Phys. Add. BURST Byte Mode	1.5ms + 4.2µs/word
Phys. Add. BURST Bit Mode	1.5ms + 26µs/word

The Demand Handling Software has to handle the following tasks: (1) Handling the LAM within the CAMAC Crate (ie disabling, enabling, masking), (2) Demand Allocation, (3) Error Handling. The following steps are involved: (1) A program may be linked to one or more LAM sources via a monitor call or a subroutine call. (2) When a demand interrupt is detected the LAM identifier is read and passed to the linked program. All demands for one program are queued in a software FIFO. The demand triggers the linked program to be scheduled for execution. Once the program has started it will identify the demand (by reading the queue), take care of any LAM source handling functions, and finally process the LAM. The LAM Error Handling program performs some housekeeping tasks: (1) If LAMs are detected which are not linked to any programs they are taken out of the corresponding queues, and an error message is issued. This also applies to Hung Demands as long as they are not linked to a specific program. (2) The program periodically removes from the queues any 'left-over' LAMS which are found left linked to inactive tasks. The CAMAC Error Handling programs have to deal with two basic types of Serial Highway errors: (a) the SHD detects an error condition and interrupts the ND100, (b) the SHD fails to signal completion of a CAMAC transaction and consequently the Monitor Call times out. In the case (a) the actions to be taken are: (1) recording of the error for statistical purposes, (2) if the error is fatal, the SHD is frozen. Fatal Errors are: Continuous Byte Sync Lost, Continuous No Waits, Fatal Error Flag set in the SHD, (3) if the error is not serious (typically a transient problem) the user program is informed by the returned status word and the statistics are updated, (4) if enabled, the contents of the Dump Store are logged, so recording the events immediately preceeding the error. In the case (b) we have a serious hardware fault, as the SHD even fails to generate an error message. In this case the SHD is frozen and an error message generated. The Performance Monitoring checks regularly every 60 seconds the PIO and DMA error rates. If these error rates exceed preset limits a warning message is generated. Extensive Test Software has been written which spans from functional check of the SHD boards only using the Test-FIFO combined with Echo Mode to an extensive set of Real Time programs to operate the SHD in all possible modes concurrently (as far as permissible) on a real Serial Highway. Further special programs support the initialisation of Serial Highway Loops from given parameters and the trouble shooting during installation of the Highways.

5. Status (1.10.82)

Some twenty SHDs are already in extensive use, mainly on test installations. Installation of Fibre Optic Serial Highways has started, and one Fibre Optic Serial Highway linking 3 U-PA and covering an overall loop length of over 1km is operational. It links three cubicles of the first CODAS Subsystem to the Subsystem Computer. The installation of fibres is progressing smoothly, the bulk delivery of U-PA has started and JET is confident of having all essential CODAS Serial Highways operational in Spring 1983.

6. References

/1/ Zimmermann, D, "The CAMAC Serial System and its Use in CODAS". Study performed for JET by Max Planck Institut fuer Plasmaphysik, Muenchen (September 1978)
/2/ Brehmer, W, Klessmann, H, Wawer, W, "Outline Specification of the CAMAC Serial Driver". Study performed for JET by Hahn-Meitner-Institut fuer Kernforschung, Berlin (July 1979)
/3/ Wolles, J C, Private Communication, CERN-SPS, Geneva
/4/ ESONE Committee, "Subroutines for CAMAC", Document No ESONE/SR/01 (September 1978)

Session 7

CONFORMANCE TESTING OF HARDWARE AND SOFTWARE

TESTS DE CONFORMITE DU MATERIEL ET DU LOGICIEL

KONFORMITÄTSTESTS VON HARD UND SOFTWARE

CONFORMANCE TESTING OF SOFTWARE

Roger S. SCOWEN

National Physical Laboratory, Teddington, Middlesex, UK

ABSTRACT

Software is becoming more expensive and it is natural for people to want some guarantee that any software they buy is correct. When they cannot satisfy themselves, and do not wholly trust the programmer, it is natural to seek the help of a third party who will certify the software's correctness. So far, formal certification is available only for compilers. This paper discusses some of the theoretical problems and reviews the existing compiler validation schemes. Finally there is a brief examination of other methods of quality assurance that might provide a better solution.

RÉSUMÉ

Le coût du logiciel augments et il est normal que les clients veuillent obtenir une certaine garantie de qualité du logiciel qu'ils achetent. Quand ils ne peuvent en faire eux-mêmes la vérification et qu(ils ne font pas pleinement confiance au programmeur, il est normal qu'ils fassent appel à une tierce partie qui certifiera l'exactitude du logiciel.

Jusqu'à maintenant, la certification formelle n'est disponible que pour les compilateurs. Cette contribution présente quelques uns des problèmes théoriques et les principes de validation des compilateurs existants. Ce papier se termine par un bref examen d'autres méthodes de contrôle de qualité qui peuvent apporter une meilleure solution.

ZUSAMMENFASSUNG

Software wird zunehmend teurer und dementsprechend staerker wird der Wunsch, gekaufte Software auf Korrektheit zu pruefen. Wenn der Kunde dies nicht selbst vermag, oder dem Programm-Ersteller nicht voellig vertraven kann, wird er die Hilfe einer eines dritten suchen, um die Korrektheit der gelieferten Software zu ueberpruefen. Bis heute gibt esformale Korrektheits-Pruefungen nur fuer Compiler. Die vorliegende Abhandlung betrachtet einige der damit verbundenen theoritischen Probleme und gibt eine Uebersicht veber existierende Compiler-Pruefverfahren. Zum Schluss wird eine kurze Darstellung anderer Qualitaetssicherungs-Verfahren gegeben, die moeglicherwelse bessere Loesungen ergeben.

CONFORMANCE TESTING OF SOFTWARE

Roger S. SCOWEN
National Physical Laboratory, Teddington, Middlesex, UK

> Software is becoming more expensive and it is natural for people to want some guarantee that any software they buy is correct. When they cannot satisfy themselves, and do not wholly trust the programmer, it is natural to seek the help of a third party who will certify the software's correctness. So far, formal certification is available only for compilers. This paper discusses some of the theoretical problems and reviews the existing compiler validation schemes. Finally there is a brief examination of other methods of quality assurance that might provide a better solution.

INTRODUCTION

Over the years, the scale of engineering has steadily increased. Aircraft, power stations and bridges, for example have all become more technologically advanced, complex and expensive. With any large complex system there is a problem of verifying that the finished system satisfies the design specification. How can we tell that it is working correctly? Are tests performed by an independent body or by the manufacturer himself? Does the specification include the tests that will be made? Does the manufacturer or the tester provide any form of guarantee? If the system is replicated, is each example tested to the same extent? Who pays for the verification - the customer directly, the manufacturer (and thus indirectly, the customer) or society at large (a 'free' service)? Very often the tests will have two aspects: firstly, safety (could it fail catastrophically), and secondly, performance (does it work well).

Another recent trend has been for engineering and commercial systems to depend increasingly and critically on computers, both in the initial design and later in use. It is thus important that the computers and their programs are reliable and correct. Computer hardware has improved greatly: it is smaller, cheaper, more powerful and more reliable. Computer software has lagged behind: programs are bigger, continually being altered and extended, and often understood by no single person. Without a software revolution, it seems inevitable that there will be catastrophic accidents due to program errors.

Software is labour intensive and increasingly expensive compared with other computing costs. There is therefore considerable incentive to re-use and adapt programs - perhaps on to computers with a different architecture. In practice this is simpler when programs are written in a language familiar to programmers and available on the new machine. Standard versions of many programming languages have been defined, and many users want compilers that conform to these standards, that is, process valid programs correctly. Some experienced users also like their compilers to detect non-standard programs because they regard them as erroneous and know that such programs may neither give consistent results nor be transferable.

This paper thus concentrates on testing compilers to ensure that they conform to a standard. Unfortunately there are many problems.

PROBLEM AREAS

This chapter considers in more detail some of the problems in validating compilers.

Testing a black box

Almost all current compiler validation packages treat the compiler as a black box, i.e. look only at the output from various test cases. It is well known that this method may detect some errors but is quite incapable of proving the absence of any other errors. For example there may be anomalous cases which are practically impossible to detect with random tests. Unfortunately a validation suite is the only economically viable method currently available: the source text is often a commercial secret and unavailable, and even if it were available, existing techniques could not analyse its properties. Further, validation suites can be based on past experience, e.g. test for the existence of faults that have previously occurred, and use the knowledge of how compilers are constructed. Validation suites can also be continually extended so that although there can be no certainty of detecting all errors, we can at least ensure that they do not recur.

Nevertheless any validation suite makes comparatively few tests. In particular, each part of a compiler might be satisfactory on its own, but a fault might occur because two parts interact in an unforeseen way. Similar problems may occur with compilers providing optional facilities such as listings, cross-reference indexes, runtime error checks, different character sets, or independent compilation. The options used when testing the compiler may be satisfactory when other untested cases would fail.

The need for retesting

It would be a serious fault with any compiler validation system if compilers were never retested. As fresh releases of the compiler are made and the tests altered, the original results can no longer be regarded as valid. Retesting is thus desirable and probably essential. The knowledge that an earlier version of software was valid has little value to a user: all programmers learn rapidly from experience that changing a program can introduce errors in seemingly unrelated places.

The only way to be certain that compilers are still working is to insist on complete periodic retesting, either by an independent contractor, or licensing the compiler writer to perform his own validation.

Language extensions

The pressure of the market place encourages computer manufacturers to "improve" their compilers by adding extra facilities. But testing arbitrary extensions to a language is impossible.

The US Federal Government partially solves this problem in Cobol. It has defined four levels of Federal Standard COBOL which form a sequence of nested subsets of the full ANSI Cobol Standard. As part of the validation requirements, each compiler must be able to flag on a compilation listing all uses of language features that are not included in a specified level. The implementer must provide a compile-time switch which can be set to monitor any Federal level which the compiler claims to support. The facility is validated by compiling a program (in practice several programs) four times, once for each position of the switch. On the listing, all features that are extensions of the specified levels must be flagged. No attempt is made to test the treatment of extensions beyond the standard.

This is a partial solution to a severe problem: it is very rare for a compiler to be written only for the Standard version of a language. Manufacturers almost always include extensions in order to increase the attractiveness of their computer systems. The extensions often make life easier for the programmer but are a hindrance to program portability. Usually, validation suites completely ignore language extensions. This is a serious deficiency: most users cannot distinguish extensions to the standard language and they would be very surprised to learn that even when the compiler has been certified correct, the guarantee may not apply to many parts of their programs.

It is clearly impossible for a validation package to test extensions to the implemented language. An independent validation organization cannot even provide a list of the extensions because there is no way of knowing that the programmers' manual is complete.

Besides the extensions openly described in the manual, there may be hidden extensions known to the implementer but not publicized, and accidental extensions unknown to anyone and waiting to be discovered by accident.

Thus it is important that the compiler is capable of warning a programmer whenever he contravenes the standard language. This is more easily said than done. Some extensions can easily be detected during compilation, for example the addition of complex type to Algol 60.

Other extensions can only be detected at runtime. For example, suppose the Fortran 77 language has been extended so that:
    REAL AA(1 : 0)
is not treated as an error but as an empty array. No errors will be reported unless the program tries to access an element of the array. For this extension, the compiler would have to check adjustable array declarations (i.e. those in subroutines that have variables specifying the bounds) at runtime and print a warning message when necessary. Because the array declaration might be executed many times, it would be an essential refinement that each sort of warning message is printed at most once for each run of the program. Many users are unwilling to pay the price of such checks.

Unfortunately, some sorts of extensions are practically impossible to detect. Consider the Algol 60 statement:
    xx := zz + ff(xx);
where ff is a function that has the side effect of assigning a value to zz. The statement is ambiguous and therefore illegal because Algol 60 does not define the order of evaluation for the two primaries zz and ff(xx). This sort of error cannot be detected easily at translation time or runtime; it is necessary to perform a data-flow analysis (as in DAVE [OSTE76]). In even more complicated cases (e.g. replace zz by elements of an array where the subscripts sometimes have the same value), static analysis cannot detect the error, only warn that the program is possibly invalid.

The compiler validation scheme should probably require the compiler writer to provide suitable tests for each language extension (or use previously written tests where the same extension has been implemented earlier). The extra programs should demonstrate that the extensions have been implemented correctly and that they interact correctly with standard features of the language and with other extensions. When compilers are retested, successful execution of the extra tests would give reassurance that the extensions had not been altered.

The vagueness of current standards

Contrary to popular opinion, standardizing a programming language is not just an academic exercise. The objective is usually an attempt to regularize the de facto definition of a programming language for which there are many different compilers on many different computers. Naturally each manufacturer has a large vested interest in maintaining the validity of his own software and fights strongly to ensure that his compiler meets the standard. The consequence is a compromise where the standard is defined permissively with all the faults mentioned elsewhere in this survey.

Existing standards for programming languages are written in more or less formal English. Inevitably this process has resulted in standards that are sometimes ambiguous and incomplete.

Programming language standards are not usually a model of clarity either. Some of the difficulties are deliberate because that was the only way the standards committee could agree on a definition. Precision would have caused some manufacturers' compilers to be non-standard, and enormous costs to be incurred by altering compilers and programs to comply with the standard.

More often, the ambiguity arises because programming language standards are usually written by computer experts who have had little previous training in drafting standards. This has often led to ambiguity and incompleteness. It is all too easy to make assumptions that are nowhere specified explicitly.

For example, current language standards are often vague about fundamental aspects of a language. Sometimes they say practically nothing. Thus the only information to be found in Fortran 77 [ANSI78] concerning real (usually but not necessarily floating-point) arithmetic is the following brief remarks:

"Evaluation of an arithmetic expression produces a numeric value" (page 6-1, lines 11-12);

"An integer datum is always an exact representation of an integer value" (page 4-3, lines 3-4);

"a real datum is a processor approximation to the value of a real number" (page 4-3, lines 14-15);

"This standard does not specify: ... (6) The range or precision of numeric quantities and the method of rounding of numeric results", (page 1-1, lines 33, 53-54).

There is no requirement for the result of a multiply operation ever to be even approximately correct. Thus it would probably be impossible to fail a compiler because the arithmetic operations were insufficiently accurate. At present this has not mattered because there is no agreed way of measuring the accuracy of computer arithmetic.

More effort on compiler validation will be largely wasted unless standards, as they are updated, are made more complete and precise.

The quality of a compiler

Many features of a compiler are important but untested by most validation suites because they are machine-dependent. In particular:

The compiler should be well integrated with the operating system so that, for example, source programs, data, and results can all be edited, filed and printed. The compiler should be usable in the same way as compilers for other languages with a simple, natural job-control language that is nevertheless comprehensive.

The compiler should give simple clear messages when the program contains a syntax, semantic or runtime error. The error messages should refer to the program in the language used by the programmer, and not specify machine addresses or instructions. The compiler should also warn the programmer when the program is "odd", that is it contains features which, although permitted by the standard, could indicate the presence of an error. For more details, see [SCOW77] or [SCOW82].

A compiler should be efficient with correct programs, as well as helpful with wrong ones.

Program development is also much more difficult if the turn-round time for an edit, compile and run cycle is a day rather than a few minutes. Even if the speed is adequate on one particular model, sometimes the compiler can be slower by anything up to 100 times if there is too little main store and/or backing store.

Even if a compiler is otherwise perfect, it will be most inconvenient if programs fail to compile or run because they are too big, or because the data structures are too big.

A compiler must be robust; neither it nor its compiled programs should ever lose control and end by giving a hexadecimal or octal dump.

It is important that the computational costs of a program are not excessive. But remember that costs occur not just from compiling and executing a program, but also arise from preparing, storing, loading, and amending it.

There should be good libraries of procedures available for performing common tasks. It should also be possible for programs to call them even when they are written in another language.

Tuning the compiler to the tests

Compiler writers will naturally want the tests to be public so that they know the performance necessary to pass, and know the interpretation to make in ambiguous areas of the standard definition of the language. This requirement makes it necessary to publish details of the test suite and the results required for a compiler to pass. In any case, it would probably be impractical to keep the validation suite secret because it would be run on so many different machines at so many different sites. Yet by publishing the test suite, there will obviously be a temptation to tune the compiler to pass the validation tests and make little or no attempt to discover and cure other faults. This strategem can be partially foiled by parameterizing some tests; however, completely random tests would make it impossible to confirm that an altered compiler gives unaltered results where appropriate.

Thus the validation suite must be comprehensive and programmers must realize how much (or little) is implied by any claim that a particular compiler has passed the required validation tests. Inevitably a requirement for comprehensiveness increases the cost of both writing the test suite and carrying out the tests.

Optimizing compilers

Optimizing compilers try to recognize the parts of programs that can be compiled in shorter or faster code. Compilations take longer but the result is better. Unfortunately testing such compilers poses extra problems.

Some test programs may be so simple that the compiler optimizes away the feature being tested. For example, by performing as much at compile time as possible, an optimizing Algol compiler may transform:

```
begin
    integer  ii, jj;
    ii := 3;
    jj := 2;
    if ii > jj then
        out string(' Greater_than_
            for_integers_OK`)
    else
        out string(' Greater_than_
            for_integers_fails`)
end
```

into the trivial:

```
begin
        out string(' Greater_than_
            for_integers_OK`)
end
```

This will tell us that '>' is optimized correctly at compile-time, but nothing about the runtime behaviour. The problem is serious because the printed output from the test suite cannot indicate the failure to make the required test.

On the other hand, it may happen that the programs in the test suite are so complex that the compiler doesn't attempt any optimization; if so, the optimizing facilities are virtually untested.

For some languages, e.g. Ada or Fortran, independent compilation of two subroutines may ensure a satisfactory test. Unfortunately there may be an even more intractable problem. It may happen that the optimization is perfect for simple cases, and also perfect for very complicated cases where the compiler recognizes its inability to perform optimization. However at the boundary, some cases are compiled wrongly. Unless the code of the compiler or its output is examined, only luck will discover these faults.

ALTERNATIVE STRATEGIES

It has been assumed that compilers should be judged in the future as they have been in the past, i.e. by constructing a test suite of programs which are submitted to a compiler, and examining the results to see if the compiler passes or fails. It may be that there are alternative strategies which could give the desired objective of reliable software more cheaply and easily. This section examines some of the possibilities.

A standard compiler

An old idea worthy of reconsideration is to regard the compiler as the definition of a programming language. This idea was first suggested by Garwick [GARW66]. It was then generally regarded as impractical: compilers were mostly written in low-level languages and almost incomprehensible, and there were also very many completely different computer architectures in use. However, since that original suggestion, languages such as BCPL, Pascal and RTL/2 have been designed and implemented. In each case, the first compiler was written in the language itself and almost all subsequent compilers have been based on it to a greater or lesser extent.

This approach, besides ensuring portability and a method of implementing the language cheaply and quickly, ensures that there are no gaps in the language definition. A standard compiler may even be more compact than a standard written in English; it certainly need not be more difficult to understand.

Specification languages

A programming language standard can be regarded as the specification of a compiler for the language. Recent work on formalizing the specification of computer systems could have several benefits, for example a formal specification is essential for proving the correctness of the corresponding suite of programs. Formal specifications have other advantages: processors can be written to check that the specification is:

Complete: all possibilities are specified.

Consistent: different parts of the specification do not contradict each other.

Not redundant: no requirement is stated more than once.

Unambiguous: so that there are no misunderstandings between the customer and implementer.

Not over-specified: so that the requirements do not unnecessarily constrain the implementer.

A paper [DAVI79] describes various ideas suggested for specification languages, for example:

Finite state machines define the outputs resulting from each input but become complicated when it is necessary to define synchronization.

Stimulus response sequences use a pseudo-Algol definition of the responses from user stimuli.

Petri nets (see [PETR62], [PETE77], [BRAU80]) are a graphical technique that express synchronization requirements extremely well. However their graphical nature makes them more useful as documentation output from the requirements phase rather than as an input specification.

English is unsuitable for a specification language. It is too easy to be ambiguous and for large systems it is shown in [JONE79] that English is far more prolix than the code required to implement the system.

Program proof and formalized testing

An idea that is attractive at first sight is to prove a compiler is correct; in fact some research has been done on proving programs are correct in the same way that geometrical theorems are proved in Euclid. However, the results, so far as compiler validation is concerned, are not encouraging. It is essential to have a specification and to have the text of the program available: neither requirement can be satisfied with normal commercial compilers. Although researchers (e.g. D Bjorner on Chill and Ada, Belz (TRW) on Ada in Semanol) are developing operational formal definitions (for an introduction to these ideas see [BJOR80]), a formal specification is not generally available and the text of a compiler is often a closely guarded trade secret. The task of finding techniques to analyse arbitrary programs is recognized as intractable, for example, it is not even generally solvable whether or not a program goes into an infinite loop. Strachey [STRA65] showed that there is no program which can read the text of an arbitrary program and report whether or not it terminates. All successful research in this field has been done by constructing a program to perform a specified task and then proving that it does so. But is the proof correct? An example pointing out an incorrect 'proof' is given in [GOOD75].

The theory of program testing is equally discouraging. Weyuker and Ostrand [WEYU80] point out that for every element d in the input domain of a program, there is a program which processes every element other than d correctly, but is incorrect on d. This implies that correctness can only be guaranteed by testing every possible case. Thus test cases that execute every part of the program, or even execute every program path, cannot guarantee correctness either. For an example, consider the following program which is based on an example by Goodenough and Gerhart [GOOD75]:

```
begin
    integer ii, jj, kk;
    read integer(ii, jj, kk);
    if (ii + jj + kk) = 3 * ii then
        out string(' The three
            values are equal`)
    else
        out string(' The three
            values are unequal`)
end
```

A more practical aim is to demonstrate that particular errors are absent. With this technique, assume a particular error has been made and construct test data which will disclose it. Foster [FOST80] gives a method of systematically generating test data to do this.

Grune [GRUN79] takes a more pragmatic attitude to this problem. The Algol 68 test suite has grown somewhat haphazardly rather than being systematically designed. Nevertheless he states:
"In my opinion, if a compiler processes the test set well and works well on the daily stream of average programs, it is a very good compiler. Through its unusual complexity, the test set will uncover most incorrect short-cuts, and the constant use of simple features will prevent the compiler from being too much tuned to the test set."

Other validation and certification schemes

Of course compilers are not the first complex products to be tested for conformance to a standard and for adequate performance. Before deciding on a method for judging compilers, it is sensible to see what can be learnt from the methods and experience of other industries.

This section looks very briefly at the process for granting a certificate of airworthiness for a civil aircraft and at the Ministry'of Defence's Accreditation scheme. The comparison with conventional compiler validation is marked; for example, the Civil Aviation Authority is involved during the design and manufacturing stages, and not just when the aircraft is finally complete. Certification and accreditation are also continuing processes, faults discovered subsequently must be reported and repaired.

Certificates of airworthiness

There are many other industries producing complex structures which must be proved to have acceptable performance and safety. Civil aircraft must be granted a Certificate of Airworthiness (C of A). It is instructive to compare this process with current methods of software certification and this section briefly describes the process by which a C of A is awarded. Note that this is only the barest outline of a very complex business, especially when one considers the various nuances on the subject that exist between the certification authorities and aircraft constructors around the world.

To comply with British Law as contained in the Air Navigation Order, the Civil Aviation Authority (CAA) must be satisfied, before a C of A is issued, that an aircraft is fit to fly having regard to its design, manufacture, workmanship and materials, and to the results of any flying tests. The Air Navigation Order does not specify a code of airworthiness requirements by which "fit to fly" may be judged. However, for a C of A to be issued which can claim international acceptance for flights over other countries, the Authority must satisfy itself that the aircraft complies with the "detailed and comprehensive" national code which the UK has lodged with International Civil Aviation Organization (ICAO).

The UK's detailed and comprehensive code, British Civil Airworthiness Requirements (BCAR), has been developed over the years by the Authority in conjunction with the manufacturing and operating industry in the UK (taking account also of foreign airworthiness codes) and provides a detailed basis on which the airworthiness of an aircraft may be assessed. More recently a joint European code known as JAR 25 has been developed within Europe and has been adopted by UK and some other European countries as their national code. While many requirements are simple and quantitative, a very considerable number require qualitative judgements and in many cases interpretation of the generalized intent of a requirement in relation to specific circumstances is necessary. In the UK the Authority discharges its task by investigating the aircraft constructor in sufficient depth so that it can be satisfied that proper procedures have been established by the constructor to make all the checks and balances necessary in such a complex task as the design, manufacture and testing of an aircraft. The constructor's procedures for manufacturing processes, quality control, and testing will all be examined and monitored on a continuing basis.

When the constructor embarks on the design of a new aircraft, the Authority is brought in at a very early stage. A CAA design liaison surveyor will be appointed to co-ordinate the Authority's investigation of the design and the airworthiness standards applicable in the particular case will be notified to the constructor.

The airworthiness tests, and other airworthiness requirements, are developed by the CAA in consultation with the aircraft industry. The aircraft is investigated in great depth not only part by part, system by system, but also as a total machine. The constructor is required to demonstrate adequate competence in the various aspects of design and construction.

As the design proceeds, CAA specialists visit the design organization to discuss the methods whereby compliance with the airworthiness standards will be established, to witness such tests as they consider necessary, to review the results of tests and to participate in the flying trials as necessary to determine compliance with the requirements.

An inspection surveyor is nominated to monitor the progress of manufacture, and he makes sampling checks on the quality control achieved during construction.

When the design has been finalized it is usual for an almost complete test airframe to be constructed and subjected to a range of ultimate and fatigue tests for certification purposes. This test work often continues on well beyond the date of certification in order to provide data on the behaviour of the structure ahead of the experience accumulated in service.

When the aircraft reaches the stage of having completed all necessary ground and flight tests, the constructor will certify to the Authority in writing that the aircraft complies with the appropriate airworthiness requirements.

At this stage the CAA will make a report on its investigation to the Airworthiness Requirements Board (ARB) which is an independent body representative of aviation interests. The Authority is required by the Civil Aviation Act 1971 to seek the advice of the ARB on the standards by reference to which a C of A may be granted, and whether or not a new type of aircraft complies with the standards. After receiving the advice of the ARB, the CAA will decide whether or not to issue a Type Certificate and subsequently a C of A for aircraft of the type. (Note: if the CAA decides not to take the advice of ARB it must publish its reasons publicly).

After certification of the initial aircraft of a type, all production aircraft are subjected to a less extensive flight test programme by the constructor, with CAA participation, to check that production aircraft behave in the same way as the prototype. In addition all aircraft on the UK register are flight tested periodically throughout their service lives to check for signs of deterioration in their flying qualities or performance.

If modifications to an aircraft have significant effect on its structure, aerodynamics or power plants then additional airworthiness investigation and testing is necessary. The extent of the testing will be decided by the CAA in relation to the significance of the changes.

All UK constructors are required to keep the CAA informed of defects and failures which might undermine the basis of certification as they are discovered, whether they occur on the constructor's own aircraft, a UK registered aircraft or a foreign registered aircraft. While CAA has no jurisdiction over foreign registered aircraft, and cannot compel foreign operators of UK constructed aircraft to report defects and incidents, the Air Navigation Order requires that all significant defects and incidents occurring on UK registered aircraft in service be reported to the CAA under the Mandatory Occurrence Reporting system. CAA examines all such occurrences and investigates those with significant airworthiness implications to ensure that corrective action, where necessary, is taken. Weekly listings of the incidents reported are circulated to all operators for their information and regular summary reports are published by the CAA. The data is examined at intervals to detect trends which might be developing.

CAA certification of a UK constructed aircraft is acceptable to all countries belonging to ICAO for flights over or landings in their country. Many countries have their own code of requirements (or use the American code, FAR) and in this case they may require further investigation of a UK aircraft before it can be granted a C of A on that country's register. In the UK the work by the CAA Airworthiness Division has to be paid for by the "applicant" (usually the aircraft constructor). Test work required by the CAA which the constructor has to perform is also at his cost. The incentive is simple - unless the work is done, no C of A is granted and without that the aircraft is not permitted to fly (except, of course, for test purposes).

The matter of finance varies around the world. In some cases, e.g. the USA, the Airworthiness Authority's charges are in effect borne by central government. The manufacturer still has to fund his own testing.

Accreditation

The Ministry of Defence pursue a less direct approach called quality assurance which does not test products directly, but rather confirms that a firm is competent and able to produce satisfactory products. All details of a firm will be considered: methods of production, internal quality control, the financial structure, personnel, facilities, and method of project management will all be examined.

Accreditation is known formally in MoD as the assessment of contractors to Defence Standard 05-21. Assessments are conducted by specialist members of MoD from Research and Development, Project Management, and Quality Directorates dependent upon the company's products. MoD are now cooperating with some of the nationalized industries to form joint assessment teams. This will reduce overall costs as many nationalized industries are committed to the same form of accreditation as MoD.

An assessment starts with a preliminary plan and visit. Often all branches of a firm are examined, but this depends on the firm's request and MoD's requirements. The process of conducting an assessment is defined in a manual [MOD 78a], and the matters covered by the assessment are defined in another manual [MOD 78b]. The assessment looks at many facets of the way software is produced and managed. In particular it ensures that:

(a) There is a good software management organization and a manager responsible for software quality assurance who has sufficient authority and independence to resolve problems.

(b) Adequate plans are made for all parts of software projects, i.e. design, development, testing and maintenance. Project milestones are identified so that progress may be monitored.

(c) Standards are laid down and enforced for documentation, programming, and testing. There are periodic design reviews to monitor progress, record agreed changes, and ensure the achievement of performance, reliability and maintainability. There are procedures to detect discrepancies affecting quality, and to take corrective action.

(d) Methods of testing and inspection are effective and proper records are kept.

(e) The company requires that software purchased from outside is itself inspected and tested to ensure that it satisfies the same high standards.

(f) The company periodically reviews its quality control procedures and takes steps to correct any deficiencies.

(g) There are effective procedures that ensure that copies of software are adequately identified and checked to be correct.

The result is a form reporting the findings and outlining the deficiencies or recommending acceptance. When the firm is judged satisfactory it is included in MoD's "List of Assessed Contractors" (LAC). Reassessment is normally at 2 or 3 year intervals but is actually "whenever thought necessary", for example if customers report faults, or major changes in personnel are known. Spot checks are also made.

The cost naturally varies with the size of the firm, but is typically a few man weeks or months.

There is one problem with accreditation (Contractor Assessment). Although it shows that the company is capable of controlling and developing projects to Defence Standard 05-21, it rests with the customer for each individual project to specify the Defence Standard as a contractual requirement.

Since 1976, MoD have required contractors involved with software design to meet Defence Standard 05-21. For example by October 1980, nine software houses were assessed as satisfactory. A similar scheme could be adopted for compiler validation. Instead of an independent tester running a validation suite to certify the compilers for a particular language, he would ensure that the compiler writers use the validation suite as part of their quality control process.

COMPILER VALIDATION TODAY

Several schemes already exist for validating compilers. The most important are discussed below.

Ada

Ada, sponsored by the US Department of Defense (DoD), is designed to be "a language with a considerable expressive power covering a wide application domain". Ada is intended for use in all DoD computer systems and eventually any compiler used on a DoD project must have been formally validated. The first reference manual was published in April 1979, a revised edition a year later, and a draft ANSI standard in mid-1982. The standard itself should appear in December 1982. Many related projects have been supported including the provision of an Ada Compiler Validation Capability (ACVC). This work, carried out by SofTech, is described in a long range plan [SOFT80a] and a paper [GOOD80].

The standardization by ANSI has delayed the opening of the validation service until October 1982 and no compilers are now expected to be tested until 1983.

The test suite of programs is only part of the ACVC programme, SofTech also wrote an "implementor's guide" and prepared "validation support tools" to assist in running the tests and analysing the results. These are being revised in line with the ANSI standard.

ACVO were aware of the problems of compiler validation described above and tried to overcome them. The validation tests are divided into six classes according to the expected result. The suite tests incorrect as well as correct programs, and checks both syntactic and semantic errors. Some tests indicate approximately the capacities supported by the compiler (for example the maximum number of array dimensions). The standard libraries, for example input/output, are also be tested.

The Ada Compiler Validation Office (ACVO) will organize and perform validations of Ada compilers in a similar way to those performed by the Federal Compiler Testing Center for Cobol and Fortran. One relatively unusual feature of Ada is that programs will frequently be compiled on one machine (the compiler host) and executed on another (the object host).

Sometimes a third machine (the validation host) may be required to prepare the test suite and analyse the results. Summary reports will be published every six months describing the validations that have been performed and the implementor will receive a complete report for his own compiler. Revalidation will be required, usually annually, and always within two years.

There will be a charge for validating an Ada compiler: a base charge estimated at $5000 to cover the cost of maintaining the ACVC and an additional charge to cover all incidental expenses.

Cobol

Cobol is one of the languages for which the United States Federal Government operates a compiler validation service. A paper by George Baird [BAIR79], manager of the Federal Compiler Testing Center (FCTC), describes the service and its history. A user's guide to the current test suite is available from the United States National Technical Information Service.

Out of 104,976 possible combinations of modules from the ANSI Cobol standard, the US government selected four as levels of Federal Standard Cobol. Each compiler must allow the user to specify a level of Federal Standard Cobol against which his program is to be monitored at compile time. Any syntax used in the source program that does not conform to this level must be indicated in the program listing.

The FCTC issues a Certificate of Validation, which is current for one year. It takes care to point out that this only certifies that the compiler has been submitted to its tests, and not that it provably conforms to the standard.

The programs of the test suite are organized into groups, each of which is aimed principally at one of the twelve modules in which the Cobol standard is presented. The groups are further subdivided to reflect the two levels of most modules. All the programs produce reports in the same format, using the same standard routines constructed from particularly simple and basic language elements. Apart from this, the programs in each group are as far as possible independent of each other and do not use facilities which are tested in other groups. Simple functional tests rely on a small set of basic language elements. All features included in interaction tests are also tested by themselves. Thus each part of the test suite relies only on a small set of basic elements or on those elements and others which can be tested in isolation. The tests are not, however, arranged in a single sequence. Within a test program the individual tests are generally independent and, except in a few special cases, any test may be deleted without affecting the validity of the remaining tests. The standard output from the tests is arranged to make apparent the identity of any test which has been deleted and to print the totals of tests deleted or failed for each program.

The test suite is distributed as a single file on a magnetic tape. This file contains the audit routines themselves and associated test data. There is also an executive routine, known as VP, to extract individual audit routines and tailor them to the environment in which they are to be run. The VP routine is the largest program in the validation system and any computer system which can execute it should have no size-related problems in compiling or executing the audit routines.

Most of the test programs are self-checking. They produce positive confirmation of tests passed but no details of the test. If a test fails, the actual and expected results are printed. In both cases the location of the program listing is indicated by reference to a paragraph-name.

The reporting mechanism works fairly well and the error reports stand out in the test output. However the volume of output continues to grow as new tests are devised and some way of compressing the output still further may have to be found.

The tests are intended to be comprehensive and in practice appear to exercise most syntactic and semantic areas. Nevertheless, the test suite is still being extended. The basic features of the language are fairly well covered, but there is still a lot of scope for testing interaction between them. At present the tests are almost all positive, in the sense that they check that the combination of compiler and execution system does the right thing with well formed and well behaved programs. Another line of development being considered is negative testing, exploring the effect of violating explicit prohibitions of the standard. At present the ANSI Cobol standard does not prescribe error messages, so no programs in the suite contain syntax or semantic errors. The standard does prescribe the action of a Cobol program on the detection of certain exception conditions which in some contexts would be regarded as semantic errors.

In all cases where it is possible to generate the exception condition predictably by program action, the test suite includes appropriate tests. Testing is purely functional and there is no attempt to measure any other aspect of the compilers, such as compilation speed or treatment of errors. It is Federal policy to use only the standard language, so a point is made of not exploring any possible language extensions.

The test suite has evolved over 20 years. Four people work full time maintaining and developing the Cobol and Fortran test suites. Version 3.0 of the Cobol test suite (for the 1974 standard) contained 323 programs and 218,560 lines of source code. New and revised programs were written for version 4.0 which became the official test set in January 1981.

The test suite is not formally verified, but individual programs are examined by programmers other than their authors for completeness and consistency with the relevant parts of the standard. It is always open to an implementer to object to a test which his compiler fails when he thinks it should pass.

The validation is carried out by staff members from the FCTC. In practice they are observers and the tests are run by the staff of the installation where the validation is performed. The FCTC does not have its own computers and access to a suitable system must be provided by the body requesting the validation.

The body requesting the validation pays the full cost of the validation, including all computer time used; travel and subsistence for the FCTC validation team; and the time of the validation team on work directly related to that validation. There is an additional standard charge designed to cover computer time used in preparing a copy of the population file for the validation, preparing and checking ASCII input and output files and preparing the VSR for publication. All Validation Summary Reports are published and may be obtained from the United States National Technical Information Service, who also sell the official documentation of the test suite [NTIS79]. The charge made by NTIS for copies of the VSR covers only the marginal cost of making and mailing an extra copy.

Coral 66

Coral 66 [BSI 80] was initially designed for the British Ministry of Defence in response to the need for a high-level language on small (1960s) mini-computers controlling tasks previously implemented almost without exception in machine or assembly code. Coral 66 was required to produce efficient code and some debasement of high-level language ideals was acceptable so compilers were allowed to exploit special or particular hardware features.

After Coral 66 had become the "preferred" high-level language in accordance with MoD computer policy, the Inter-Establishment Committee for Computer Applications (IECCA) took responsibility for the language. They envisaged that hardware manufacturers would provide Coral on their machines and seek approval in order to become eligible for MoD contracts. So IECCA asked the Computer Applications Division of RSRE to provide an evaluation procedure for MoD to fund the evaluation of Coral machines offered by hardware manufacturers for approval.

The evaluation considers the complete Coral 66 machine, that is the host machine, the compiler, the target machine, and the program development/support environment. However, the importance of the environment was not initially appreciated so the evaluation of a Coral 66 machine includes only a subjective report by the evaluator on the tools provided by the support system.

The test suite was originally designed by members of RSRE's Computer Applications Division and has been evolving since 1974. The tests rely on basic facilities and use these to to exercise the remaining "kernel" language features and optional features such as recursion, Coral 66 tables, Coral 66 data overlaying, Coral 66 bit manipulation. The features being tested appear in a variety of involved pieces of text (eg deeply nested conditional expressions, loops with involved "for-lists").

One test examines the compiler's ability to detect and recover from errors. A further test investigates the arbitrary limits imposed by a compiler on a Coral 66 program.

The tests establish the degree of conformity of the candidate compiler rather than a simple pass/fail - IECCA allows certain freedom in the interpretation of some Coral 66 features. A set of benchmark programs allows comment to be made on the efficiency or cost of a Coral construct.

The tests themselves have been analysed by a program derived from the research into automatic generation of compilers - the SEMSID driven test-tester confirmed that every construct appeared in every syntactically legal position.

Any hardware manufacturer must provide Coral 66 on his machine if he wishes to be eligible for MoD contracts. He therefore approaches RSRE and invites them to test his candidate compiler. RSRE then contract a third party, usually a software house, to formally test the compiler. The tests normally take about two weeks. The evaluator is also expected to satisfy himself that "extensions" are within the "spirit of Coral". The evaluator spends approximately four weeks interpreting results and preparing a report detailing the facts of the compiler's performance and his subjective opinions of the program support environment. The report is then received for discussion by IECCA - it is at this time that the manufacturer may be requested by IECCA to rectify shortcomings in his compiler or support environment. A report, once accepted, is made publicly available and the compiler placed on the IECCA standard list.

An evaluation costs about £3000, but this does not include any contribution to producing the validation suite or the indirect costs incurred by the manufacturer of the compiler.

No formal retesting of Coral 66 machines is carried out by the Ministry. However, the Ministry feels that it is at liberty to require an informal run of the RSRE test programs whenever a new version of a compiler is issued or when the underlying machine is altered. The manufacturer would be given a chance to make good any deterioration in performance, if any, noted during the informal run of the test program. Failure to repair degradations could result in the Coral 66 machine being removed from the Standards List.

Fortran

Both the US Navy and the National Bureau of Standards have produced suites of programs for validating Fortran compilers. The two suites are described in reports by Holberton and Parker [HOLB74] and Hoyt [HOYT77].

The NBS and US Navy suites differ in the way they report the results of testing a compiler. Hoyt [HOYT77] describes the differences:
> "Output results [in FCVS] produced by the execution of each routine indicate whether the code generated by the compiler passed or failed each test of the routine."

On the other hand, a major flaw in the NBS suite is that
> "all the test results were listed on a printer and required careful examination ... by the user."

The Fortran Compiler Validation System (FCVS) is based on the US Navy suite. The project was first designed in 1973 but remained in abeyance until February 1975 when the decision was made to build a self-measuring suite that would adequately test all elements of the Fortran language specified in the 1966 standard. However in March 1976 the objectives were changed to test
> "the conformance of those elements of the Fortran language which are contained in the logical intersection of the American Standard Fortran, X3.9-1966, and the elements proposed for the subset language in the draft proposed American National Standard programming language Fortran."

The FCVS (i.e Fortran Compiler Validation System)
> "consists of Fortran audit routines, their related test data, and an executive routine (EXECUTIVE) which prepares the audit routines for compilation and execution."

The EXECUTIVE reads command lines, typically from a card reader or online terminal in order to compile and run any desired set of audit routines. Facilities include the ability to insert job control instructions before, inside and after each test program, to alter the file of audit routines to correct errors in them, and to remove audit routines that failed to compile in earlier tests.

The test suite, like many others, assumes that core features of the language are correctly implemented, and uses these features to test the remaining language features one by one.

Hoyt states
> "The tests in the FCVS are 'positive' in that only statements permitted by the Standard are included. There are no 'negative' tests of incorrect statement formats which a compiler is supposed to flag as errors.

The FCVS also does not test vendor extensions to the language specifications, and does not perform an error analysis on the results of executing the Basic External Functions supplied by Fortran processors. The FCVS is not designated to measure the efficiency of the object code generated or the performance characteristics of a Fortran compiler."

Version 2.0 of the FCVS was published in September 1982 [NTIS82]. It tests the full level of Standard Fortran 77 [ANSI78] and contains 272 test cases, 143 being new for this version.

The Fortran tests are now administered by the Federal Compiler Testing Center who publish a report on each Fortran compiler tested.

Pascal

The Pascal test suite evolved from programs written by B A Wichmann (National Physical laboratory) and A H J Sale (Uinversity of Tasmania). Their initial efforts have now been extended and normalized to form the basis of a Pascal compiler validation service being run by British Standards Institution (BSI). The version of the language being tested is the British standard [BSI 82] and the test suite (version 3.1) contains 530 programs occupying 430 pages.

Included with the suite is a Pascal program, the "skeleton", which can be easily modified to insert all the necessary job control statements. However, even with a reasonably sophisticated operating system, to process and run the suite can take a week or more.

The major difference between the Pascal suite and the other suites is that it contains many programs which are correct syntactically but violate the semantic rules.

Each test belongs to one of the following six classes:

(1) Conformance Class
These tests check that programs conforming to the standard compile and run correctly. They output a pass or fail message.

(2) Deviance Class
These tests detect processors that implement an extension of Pascal by failing to check or limit some Pascal feature appropriately.

(3) Implementation Defined Class
Some sections of the Standard give implementers a choice, for example, whether or not to evaluate all operands in a boolean expression, or how big to make 'maxint'. These tests report how such features are implemented. but are not usually standard Pascal.

(4) Error Handling Class
The Standard specifies that under certain circumstances "an error shall occur". A conforming compiler need not report each sort of error but failure to detect it will produce unknown results. These tests identify such cases; they are not in Standard Pascal with respect to the feature being tested, but otherwise conform.

(5) Quality Class
These tests measure the quality of an implementation, for example, benchmarks, error recovery, or limits set by the compiler.

(6) Extension Class
These tests examine extensions approved by the Pascal Users Group.

National Computing Centre (NCC) and NPL have used the validation suite as the basis of a UK service being run by BSI to validate Pascal compilers. Two additional documents were produced as part of the project: a test suite user's guide, and a report of the procedures to be followed for formal testing.

ACKNOWLEDGEMENTS

This paper is based on a report [SCOW80]. I am very grateful to P R Brown (National Computing Centre) and T A D White (RSRE) who wrote the descriptions of the COBOL and Coral validation suites respectively, A J Heath (Civil Aviation Authority) who explained the process of granting a Certificate of Airworthiness, and Z J Ciechanowicz (NPL) who co-authored the original report. I also acknowledge the many other people who commented on the original report: A M Addyman, G N Baird, J G P Barnes, F M Blake, J W Charter, R J Cichelli, J V Cugini, M H Forrester, J B Goodenough, G Goos, D Grune, H Huenke, N Malagardis, A H J Sale, T D Wells, and B A Wichmann.

REFERENCES

[ANSI78] American National Standards Institute, American national standard programming language Fortran, ANSI X3.9-1978, American National Standards Institute, 1430 Broadway, New York, NY 10018, USA.

[BAIR79] G N Baird, Compiler validation from a functional point of view, In [NCC 80], pp 29-43.

[BRAU80] W Brauer (Editor), Net theory and applications, Lecture notes in computer science 84, Springer Verlag, 1980.

[BJOR80] D Bjorner (Editor), Abstract software specifications (1979 Copenhagen Winter School Proceedings), Lecture notes in computer science 86, Springer Verlag, 1980.

[BSI 80] British Standards Institution, Specification for computer programming language CORAL 66, British Standard BS5905:1980, 1980.

[BSI 82] British Standards Institution, Specification for the computer programming language Pascal, British Standard BS6192:1982, 1982.

[DAVI79] A M Davis, T G Rauscher, Formal techniques and automatic processing to ensure correctness in requirements specifications, In [IEEE79], pp15-35.

[FOST80] K A Foster, Error sensitive test case analysis (ESTCA9), IEEE Transactions on Software Engineering, Vol SE-6, No 3, May 1980, pp258-264.

[GARW66] J V Garwick, The definition of programming languages by their compilers, In - Formal language description languages for computer programming, (Edited by T B Steel), North Holland Publishing Company, 1966, pp139-147.

[GOOD75] J B Goodenough, S L Gerhart, Toward a theory of test data selection, IEEE Transactions on Software Engineering, Vol SE-1, No 2, June 1975, pp156-173.

[GOOD80] J B Goodenough, The Ada compiler validation capability, SIGPLAN symposium on the Ada programming language, Boston, Massachusetts, 9-11 Dec 1980.

[GRUN79] D Grune, The revised MC Algol 68 test set, IW 122/79, November 1979, Department of Computer Science, Mathematisch Centrum, Amsterdam, The Netherlands.

[HOLB74] F E Holberton, E G Parker, NBS Fortran test programs (Vol 1 Documentation for versions 1 and 3; Vol 2, Listings for version 1; Vol 3, Listings for version 3), NBS Special publication 399, October 1974, Institute for Computer Sciences and Technology, National Bureau of Standards, Washington, DC, USA, 20234.

[HOYT77] P M Hoyt, The Navy Fortran validation system, FCCTS/TR-77/18, May 1977, ADPE Selection Office, Department of the Navy, Washington, DC, USA 20736. (USGR Ref No AD A039 770).

[IEEE79] IEEE, Proceedings of Conference - Specifications of Reliable Software, April 1979, IEEE Catalog No. 79 CH1401-9C (British Library, Lending Division, Ref 79/14863).

[JONE79] C Jones, A survey of programming design and specification techniques, In [IEEE79], pp 91-103.

[MOD 78a] Ministry of Defence, Guide to contractor assessment, Book 1, The conduct of contractor assessment, 1978, Director General of Quality Assurance MOD(PE).

[MOD 78b] Ministry of Defence, Guide to contractor assessment, Book 4, Computer software QA systems, 1978, Director General of Quality Assurance MOD(PE).

[NCC 80] National Computing Centre, Language implementation validation, Proceedings of the two day workshop held at Manchester on 12-13 September 1979, edited by D J Dwyer and D I Noble, National Computing Centre, Manchester.

[NTIS79] United States National Technical Information Service, AD A036 174 - CCVS74 V3.0 User's Guide, 5285 Port Royal Road, Springfield, Virginia 22151, USA.

[NTIS82] United States National Technical Information Service, PB82-250895 - 1978 FORTRAN Compiler Validation System Version 2.0 User's Guide, 5285 Port Royal Road, Springfield, Virginia 22151, USA.

[OSTE76] L J Osterweil, L D Fosdick, DAVE - a validation error detection and documentation system for Fortran programs, Software practice and experience, Vol 6, pp473-486, 1976.

[PETE77] J L Peterson, Petri nets, Computing Surveys, Vol 9, pp 223-252, 1977.

[PETR62] C A Petri, Kommunikation mit automaten, Schriften des Rheinisch-Westfalieschen Institues fur Instrumentelle Mathematik an der Universitat Bonn, Heft 2, Bonn, W Germany, 1962.

[SCOW77] R S Scowen, The diagnostic facilities in Algol and Fortran compilers, NPL Report NAC 81, July 1977, National Physical Laboratory, Teddington, Middlesex, UK.

[SCOW79] R S Scowen, A new technique for improving the quality of computer programs, Proceedings 4th IEEE international conference on software engineering, Technical University, Munich, Sept 17-19, 1979, pp73-78.

[SCOW80] R S Scowen and Z J Ciechanowicz, Compiler validation - a survey, NPL Report CSU 8/81, December 1980, National Physical Laboratory, Teddington, Middlesex, UK.

[SCOW82] R S Scowen and Z J Ciechanowicz, Seven sorts of programs, ACM SIGPLAN Notices, Vol 17, Number 3, pp 74-79, March 1982.

[SOFT80a] Softech (J B Goodenough, J R Kelly), Ada compiler validation capability - Long range plan, Softech report 1067-1.1, Feb 1980, Softech Inc, 460 Totten Pond Road, Waltham, MA 02154, USA.

[STRA65] C Strachey, An impossible program, Computer Journal, Vol 7, pg313, 1965.

[WEYU80] E J Weyuker, T J Ostrand, Theories of program testing and the application of revealing subdomains, IEEE Transactions on Software Engineering, Vol SE-6, No 3, May 1980, pp236-246.

AN AUTOMATIC SYSTEM FOR TESTING COMPILERS

F. Federigi[1], I. Spadafora[2]

[1] ITALIMPIANTI, Genova - Italy

[2] CNUCE, Pisa - Italy

ABSTRACT

This paper presents a new, complete, language-independent system for automatically testing compilers. The approach is to stimulate the compiler with a functionally complete set of source programs, automatically generated, and to compare the trace of execution compiled programs with the trace produced by a reference interpreter based on a semantic definition of the source language. The source language syntax and semantics are defined by a unifying formalism which drives the program generator and the interpreter. This formalism extends context-free grammars in a context-dependent direction, but it still retains the structure and readability of BNF.

RÉSUMÉ

Ce papier présente un systéme complet, nouveau et indépendant du langage qui permet le test automatique des compilateurs.

Le principe consiste à soumettre au compilateur un jeu, couvrant toutes les fonctions, de programmes sources, générés automatiquement, et à comparer le parcours des programmes compilés à celui produit par un interpréteur de référence basé sur une définition sémantique du langage source.

La syntaxe et la sémantique du langage source sont définies par un formalisme commun au générateur de programmes et à l'interpréteur. Le formalisme étend les grammaires indépendantes du contexte aux formes dépendantes du contexte, mais conserve la structure et la lisibilité du BNF.

ZUSAMMENFASSUNG

Die Abhandlung beschreibt ein neues, komplettes, sprachunabhaengiges System fuer den automatischen Test von Compilern. Die Methode besteht darin, einen funktional vollstaendigen Satz von Quellprogrammen automatisch zu erzeugen und diese dann dem compiler zu unterwerfen. Der vom compiler-generierten Code erzeugte Programmablauf wird dann mit dem Programmablauf verglichen, der von einem Referenz-Interpretier-Programm erzeugt wurde, das auf der semantischen Definition der Quellsprache beruht. Syntax und Semantik der Quellsprache werden unter Verwendung eines einheitlichen Formalismus definiert, der den Programm-Generator und den Interpreter steuert. Dieser Formalismus erweitert kontextfreie Grammatiken in Richtung auf kontextbezogene Grammatiken, ohne dabei die Struktur und Lesbarkeit der BNF Notation zu verlieren.

AN AUTOMATIC SYSTEM FOR TESTING COMPILERS

F. Federigi[1], I. Spadafora[2]

[1] ITALIMPIANTI, Genova - Italy
[2] CNUCE, Pisa - Italy

This paper presents a new, complete, language-independent system for automatically testing compilers. The approach is to stimulate the compiler with a functionally complete set of source programs, automatically generated, and to compare the trace of execution compiled programs with the trace produced by a reference interpreter based on a semantic definition of the source language. The source language syntax and semantics are defined by a unifying formalism wich drives the program generator and the interpreter. This formalism extends context-free grammars in a context-dependent direction, but is still retains the structure and readability of BNF.

1. INTRODUCTION

1.1 Previous work

Control of the functional correctness of programs is a very important and actual problem in software production. Program certification is essential for complex programs widely distributed to users, such as compilers.
Although there are now many tools available for the automatic generation of compilers [1],[6], the end result must be tested as not all the components are realized automatically (e.g., symbol table construction and handling).

For this reason, different methods [4],[5],[7],[10], have been studied and developed to test compilers; however, none of these methods seem to have solved the problems completely and efficiently.
A very important work is instead present in [3] where we have described an automatic, syntax-driven generator which produces test programs that can be used to exercise a compiler.

1.2 Aims of this work

The research described in this paper is part of a project for the development of general tools for compiler testing.
The system [3], is now completed with an interpreter for these test programs. The complete testing system (see fig.1) operates in the following way:

a) test programs are automatically generated;
b) these programs are automatically interpreted and the results of this interpretation are compared with those obtained by compiling the same test programs with the compiler which is being tested and then executing them.

In [3], in order to describe the source language we adopted a formalism (the context-free parametric grammars) which is both practical to use and easy to read, and in this respect greatly resembles the widely accepted BNF grammars, and is also highly suitable to describe all syntactic aspects of programming languages, including context-sensitive ones. In addition, we designed and implemented a generative algorithm which can be applied to these grammars in order to rapidly produce a set of compilable programs which minimally cover all grammatical constructions of the source language, unless user supplied directives instruct the generator to do otherwise.

Besides, it is possible to generate incorrect programs in a controlled way, using the same algorithm; this is done for two reasons: to test the diagnostic capabilities of the compiler and to verify whether the compiler accepts all correct programs and only those which are correct.

Since our system produces compilable (or intentionally wrong) programs, the following parts of the compiler are tested:

- the lexical analyzer or scanner;
- the syntax analyzer;
- the semantic analyzer (symbol table handler);
- diagnostic and error-handling capabilities;
- limiting conditions, i.e. the maximum dimensions of compile-time and run-time tables.

This new study presents a system for the generation and interpretation of executable

---

[1] This paper has been prepared by the author under research contract signed by CNR (Consiglio Nazionale delle Ricerche - Pisa).

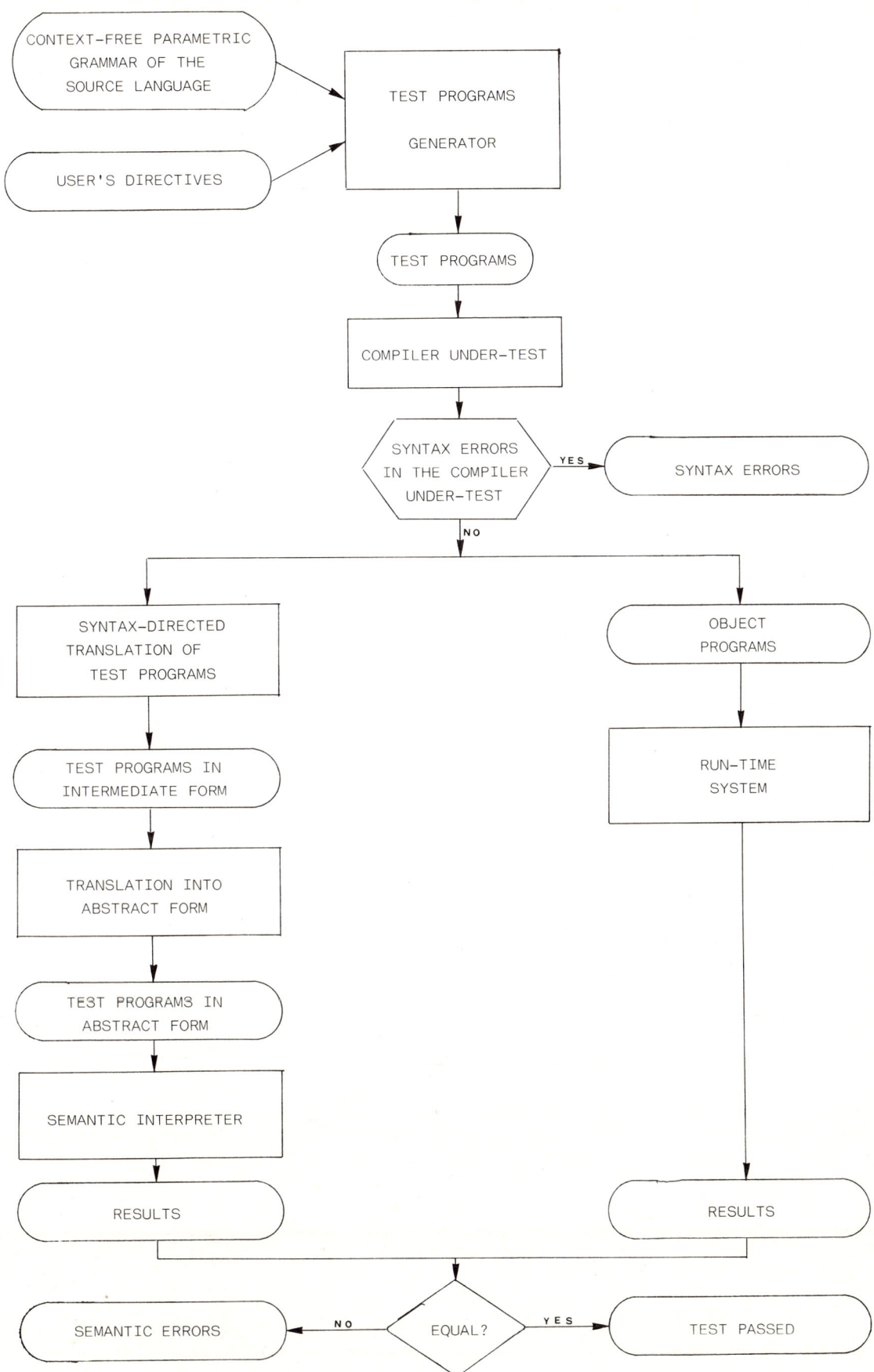

Fig.1 – Complete testing system.
Rectangular boxes stand for operations; circular boxes for input/output.

(i.e. run-time error free) programs in order to test also the code generator and run-time system. The central point of this system is to express the semantics of a programming language by a formalism which:

a) can be conveniently implemented;
b) is easily comprehensible and practical for human users.

We found it both conceptually appealing and practically convenient to use the same context-free parametric grammar formalism to describe the semantics as well as the syntax.
At the moment there is not a general, standard method to obtain from the language definition the syntax and the semantics described in the context-free parametric grammar formalism.
However the gained experiences assure us the it is always possible. Besides such task becomes more easy when all the aspects of a source language are fully described using grammatical and/or algebraic rules.

Using the test generator together with semantic interpreter, a complete testing of a compiler is thus possible.

2. FORMALISM

In this section, we briefly recall the context-free parametric grammar formalism [3]. The characterizing feature of this type of grammar is that variables (the names of languages) can be associated to nonterminal symbols; therefore, productions can be seen as context-free productions, in which some nonterminals have associated parameters, whose range is a language.
A very simple, abstract example of a context-free parametric grammar is shown below.

Example 1

Let $T = \{a,b,c\}$ be the terminal alphabet, and let

$X = \{X_a, X_b, X_c, X\}$

be the set of languages

$X_a = \{a\}$, $X_b = \{b\}$, $X_c = \{c\}$, $X = \{a,b\}$

(i.e. X is composed by non-empty sequences of a and b).
Let P be the following set of productions:

1) $S \longrightarrow Ax$
2) $Ax_a x \longrightarrow x_c Ax$
3) $Ax_b x \longrightarrow x_b Ax$
4) $Ax_c \longrightarrow x_c$
5) $Ax \longrightarrow x$

The context-free parametric grammar $G = \{X \cup \{S,A,a,b,c\}, P, T, S\}$ generates the language $L(G) = \{c,b\}^+$; this grammar acts as a sort of translator, which replaces every occurrence of a in words of x with c.
An example of generation follows:

a) in production 1) the language x appears on the right hand side but not on the left; using this production, x is replaced by any one of its elements, for example by $a^2 b$; in this way the sequence $Aa^2 b$ is generated;
b) at this point, only production 2) can be applied, replacing a with x_a and ab with x; then cAab is obtained;
c) once again, only production 2) can be applied, obtaining $c^2 Ab$;
d) only production 5) can now be applied and therefore the word $c^2 b$ is generated.

We observe that an order can be introduced between some productions; this convention does not increase the generative power of the class of context-free parametric grammars, but is only used to simplify the formalism.

3. GENERATOR EXPERIMENTAL RESULTS

In [3] the generator of test programs is completely described; now we give some generator experimental results.

We have used the system described above to test two different PLZ compilers produced by Olivetti, the PLZ/SYS [11] compiler developed by the Milan Polytechnic, the PLEO compiler developed by ITALTEL-SIT of Milan, and to examine the behaviour of four IBM 370 Pascal compilers, available for a limited time at the CNUCE Computing Center, Pisa.

PLEO is a programming language similar to PL/360 language with same Pascal aspects, e.g. type declarations.

Olivetti PLZ extends the PLZ/SYS in several directions: it introduces new types (two new real types, namely fixed-real and float-real) and it allows construction of more complex structured types (e.g., array of arrays, etc.).

The program generation of the Olivetti PLZs and PLZ/SYS was achieved as follows: first, we generated programs in a very small subset of the source language (e.g., programs with simple variable declarations only); next, in accordance with the contractors, we gradually extended this subset until test programs were generated which contained every aspect of the language. In this way, the errors are more easily detected and debugged. For each compiler, only one source grammar was written. To avoid the use of certain productions they were "masked", setting the upper bound on the number of their applications to zero.

Our principal results are summarized below.

We have no information on the detailed results for Olivetti PLZ case. We only learned that our system helped to discover a number of compiler errors and certain inaccuracy in the syntactical definition of the source language; therefore, a careful revision of the report and the language definition was necessary.

Using the PLZ/SYS test programs, both lexical, and syntactical and semantical errors were discovered: it was found that the scanner refused very long identifiers, although the report does not limit the maximum lenght of identifiers. Moreover, due to a design error, the compiler did not correctly handle the identifier scopes. Some identifier redefinition cases were not accepted into the internal scopes (local variables, record fields). This error was harder to remove because it caused a substantial change in the internal data structure of the compiler, modifications in many procedures and a more restricted definition of "scope".

As the tests covered a large number of these permitted redefinitions, they also were of great help in verifying the correctness of the modified compiler.

Semantical errors concerning the ASIZE operator (which computes the memory amount of a type or variable identifier), and initialization of variables and pointers were also discovered.

Although this compiler had already been exercised by handwritten test programs, our system detected errors which it had not been possible to spot previously.

Although the PLEO compiler had already been used by many users and its validation had been made, many imprecisions were detected by our test programs.

Besides our generator was used to test the behaviour of the following compilers:

- Stony Brook Pascal/360;
- IBM Pascal/VS; -.un 2 - Pascal 8000 IBM 360/370 Version;
- Stanford Pascal Compiler.

4. THE GENERATION OF EXECUTABLE TEST PROGRAMS

The generator produces a set of compilable, but not executable programs. For example, the generated programs can have infinite loops, not initialized variables, etc. Therefore the input grammar must also contain 'rules' to generate executable programs; these rules can be expressed by the same formalism of the context-free parametric grammars. The above semantic error can be avoided in the following way:

a) All the local variables of a block or procedure of the generated programs are initializated;
b) Loops are controlled by 'reserved' variables.

Example 2

The following program illustrates a way to avoid infinite loops.

```
    program p;
    declarations 1.1
    integer X;
    declarations 1.2
    begin
        X := 0;
        statements 1.1
    A:  statement
        statements 1.2
        if X < 5 then
        begin
            X := X+1;
            goto A;
        end;
        statements 1.3
    end.
```

The condition $X < 5$ serves to check the loop of the above program.

5. THE ABSTRACT FORM

Before the source program is interpreted, it is transformed into an abstract form; this is for two reasons:

a) to standardize the source program form, by removing the syntactical coating which differentiates identical semantic concepts in different languages. Thus the user can provide semantic definitions of concepts which are common to different languages;
b) to handle more efficiently the run-time environment (local and nonlocal) and the statements associated to a procedure or block. In source programs, the natural sequence of declarations and statements belonging to the same block can be interrupted by nested procedures and blocks. However, in the abstract form, all the declarations and statements of the same block are grouped together and can be easily found by numerical labels.

In order to translate more efficiently from concrete to abstract form, we have introduced an intermediate form. Programs translated to this intermediate form have the same static structure of the source programs but they already contain some information useful for their translation to abstract form and for their interpretation. For example, blocks are enclosed between the brackets: BLK and END, and each block contains numerical labels which uniquely identify its local and non-local environment; these numerical labels have the same function as the pointers of static chains for compilers, see the literature [9]. All the variables are initialized to undefined. To translate from concrete to intermediate form, we use the tecnique of simple syntax directed translation [2], adjusted to context-free parametric grammars.

Example 3

Consider the following Pascal-like short program:

```
program p;
type fruit=(strawberry,raspberry,bilberry);
var A: integer; B: fruit;
procedure C;
    var B: integer;
    begin
        A := 3;
        B := A+5;
    end;
    begin
        B := strawberry;
        call C;
    end.
```

The intermediate form of the above program is:

```
BLK 1 0
    A undefined; B undefined;
    proc C 2;
        BLK 2 1
            B undefined;.
            block 3;
                BLK 3 2 .
                    A := 3;
                    B := A 5 +;
                END
        END .
    block 4;
        BLK 4 1 .
            B := strawberry;
            call C;
        END
END
```

Number 2, following the procedure name C and the key word BLK, is used to uniquely identify the local environment and the statements associated to C. The second number following BLK in a procedure or block declaration identifies the environment of the procedure or block most closely nested (in the example, 1 in BLK 2 1 identifies the block BLK 1 0). Notice that BLK 1 0 always identifies the outermost block.

The declaration and statement parts are divided by a dot.

It should be clear that blocks, which are not procedure bodies, act as procedures.

The intermediate form is unsuitable for interpretation due to possible nested blocks or procedures. The program in abstract form (obtained from the intermediate form by means of a transformation using context-free parametric grammars), however, is a linear representation of the source program, and therefore easy to interpret.

Example 4

The program in example 3 has the following abstract form:

```
1 0 A undefined; B undefined; proc C 2;.
      block 4; #
2 1 B undefined; . block 3; #
3 2 . A := 3; B := A 5 +; #
4 1 . B := strawberry; call C; #
```

The brackets BLK and END have been replaced by the postfixed delimiter # which indicates the end of a block.

We observe that the static structure of the program, necessary for interpretation, has textually disappeared, but it is logically encoded by numerical labels associated with each procedure or block.

6. THE SEMANTIC INTERPRETER

This section presents the semantic interpreter in an informal way. Now we define the parameters which are used.
For singleton languages, no name is required; e.g. proc and # stand respectively for the languages {proc} and {#}.

Let N (numbers) be the integers and ID (identifiers) the language of Pascal identifiers, i.e. alphanumeric sequences beginning with a letter; we the define

V = N ∪ ID ∪ {undefined}

i.e. V (values) is the language of values which the identifiers can assume.

We indicate with ENV (environments) the language of lists made by variable identifiers with their values or by procedure identifiers with their numerical label, i.e.

ENV = ((ID V;) ∪ (proc ID N;))*

e.g. an element of ENV is:

A 3; B radish; proc C 2; D undefined;

Let STS (statements) be the language of statements which appear in the abstract form, i.e.

STS = ((ID := V;) ∪ (call ID;) ∪ (block N;))*

an example of an element of STS is:

block 2; A := tomato; B := 7; call C;

Let AF (abstract form) be the language of programs in abstract form, i.e.

AF = (N N ENV . STS #)*

the program in example 4 is an element of AF.

We indicate with ES (environment stack) and EF (environment failure) the language of the lists of environments enclosed by numerical labels: the first identifies the same environment, the second is the label of the most closely nested block; i.e.

ES = EF = (N ENV N#)*

e.g. an element of ES is:

2 B undefined; 1 # 1 A 3; B strawberry; proc C 2; 0 #

Finally, SS (statement stack) is the language of the lists of statements, i.e.

SS = (STS #)*

e.g. an element of SS is:

A := radish; block 2;#B := tomato; #

Languages ES and SS allow the representation of the flow of environments and statements of the program at run-time.

To make easy the reading, we do not give here the interpreter rules, but we describe the basic concepts by means of the interpretation of the example 3 program.

First, the interpreter pushes onto the stacks SS and ES, respectively the statements and the environment of the main program (expressed in abstract form). At any time, the tops of SS and ES contain, respectively, the statements and the local environment of the active block or procedure, and the SS top is the first statement to be executed. During the execution of a call procedure or block, the associated numerical label give all the information on their local and non local environment.

In fig.2a, the stacks SS and ES contain the statements and environment of the main program. Fig.2b shows the state of SS and ES after the execution (interpretation by means of suitable context-free parametric rules) of the statement: block 4.

Now the SS top is: B := strawberry; after the interpretation of this statement, the execution of call C pushes onto SS and ES, respectively, the procedure C statements and environment; fig.2c illustrates the interpretation of two above statements. Blocks are interpreted similarly to procedures, but they already have the numerical label to identify the statements and local environment. Fig.2d and 2e show, respectively, the execution of block 3, and A := 3, B := A 5 + (note the abstract form transforms expressions into postfix notation). At this point the SS top is: #. The symbol # indicates the end of procedure or block statements, and therefore it is written out by SS. At the same time the environment corresponding to the terminated procedure or block is popped by ES.

The program execution terminates when SS and ES are empty; therefore, the interpretation result is a sequence of environments.

7. INTERPRETER IMPLEMENTATION

The context-free parametric grammars are interpreted by an algorithm which, starting from a nonterminal and the corresponding parameter values, identifies the next production to be applied.

For the generation of the program in the concrete and intermediate forms, we use the algorithm described in [3], suitably modified to allow the parallel generation of both forms.

This algorithm could be used also to transform the program to the abstract form and to interpret it; but, for this case, a simpler algorithm can be used. In fact, the algorithm of [3] produces the shortest possible derivation using all the productions of the grammar at least once (unless user supplied directives instruct the generator to do otherwise), and, at some steps of program generation, different productions may be non-deterministically applied.

Instead, in the case of program transformation into the abstract form and interpretation, each step is completely determined by the initial value of parameters (respectively, the program in intermediate and abstract form) associated with the first production of their grammars.

The interpreter implementation makes use of a stack containing terminal strings and pairs (M,L), where M is a nonterminal and L is list of actual parameter values.

Initially the pair (S,ε) (S is the starting symbol of the grammar and ε is the empty word) is pushed onto the stack; successively the following steps are repeated:

a) According to the pair (M,L) on top of the stack identify k the next production to be applied, and pop (M,L).
b) Apply production k in two stages:
   b1) the right part of production k, starting with the right-most symbol, is pushed onto the stack in the following way: if the symbol is a terminal, it is placed in the stack. If the symbol is a nonterminal M, let L be the actual value of the parameters associated to M; the pair (M,L) is pushed onto the stack;
   b2) if the stack is empty, the algorithm

finishes. If the symbol at the top of the stack is a terminal, it is written out and popped; otherwise, step a) is applied.

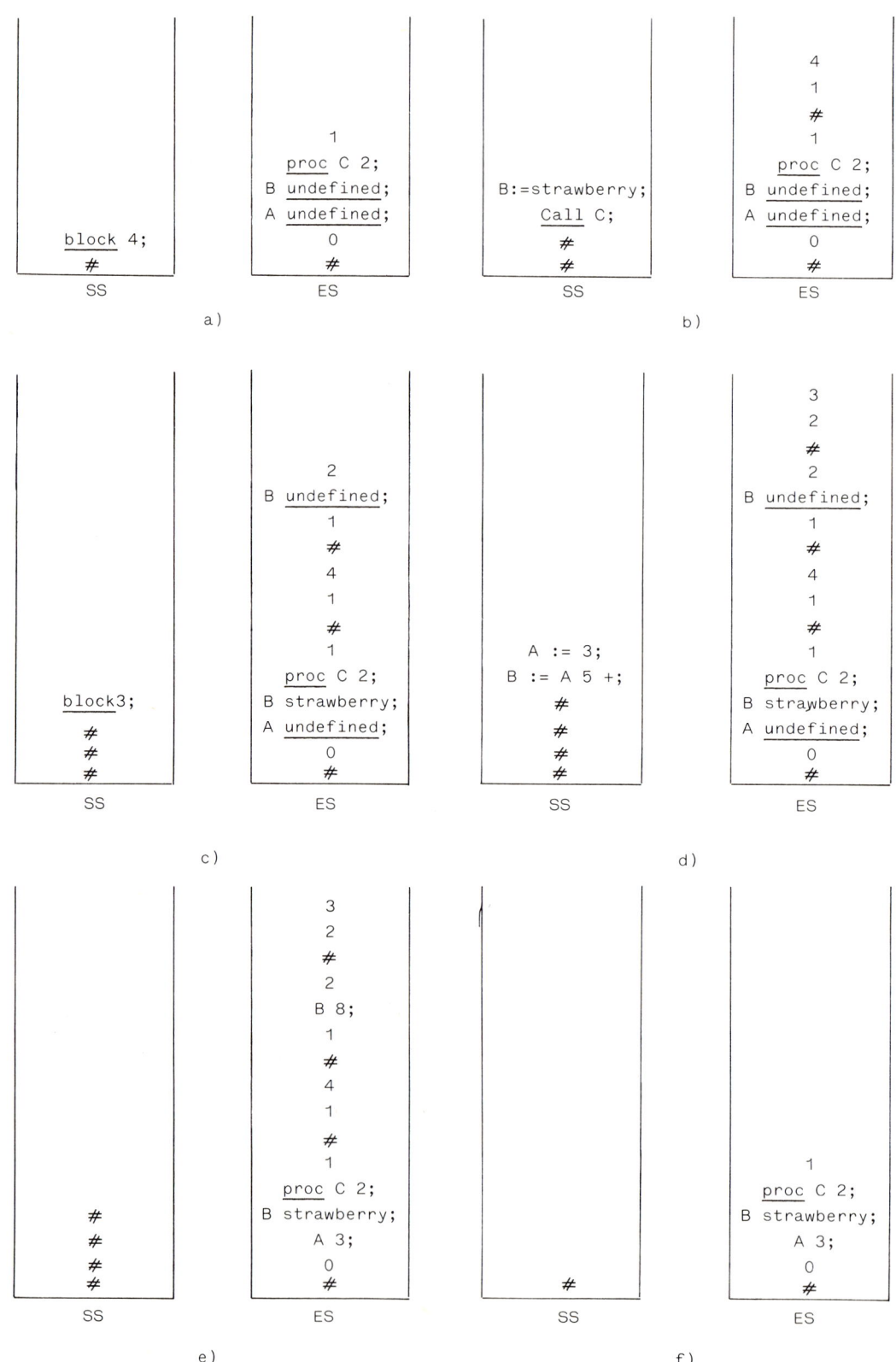

Fig. 2 –

8. CONCLUSION

The interpreter described is implemented in PL/1 on an IBM 3033 at CNUCE Computer Center of Pisa. A limitation of the present system is that the results produced by the semantic interpreter and by the tested compiler must be compared by hand. The next step (which is now being realised) will be to replace this human intervention by a program which compares the traces of computations.

ACKNOWLEDGEMENTS

This work is partially supported by CNUCE, an Institute of the Italian National Research Council, in Pisa. In addition, we are very grateful to Prof. S. Crespi-Reghizzi for his constructive comments and suggestions during the research.

REFERENCES

[1] A.Aho, et al., Principles of compiler design (Reading Mass., Addison Wesley, 1978).

[2] Aho, and Ullman, The theory of parsing, translation, and compiling, Prentice-Hall, New York 1973, vol.2, 730-736.

[3] Bazzichi, and Spadafora, An automatic generator for compiler testing, to appear in IEEE Transaction on Software Engineering, (July 1982).

[4] A.Celentano, et al., Compiler testing using a sentence generator, Software Practice and Experience, vol.10 (Nov. 1980) 897-918.

[5] S.Crespi, et al., Un generatore di test per la verifica dei compilatori e delle specifiche sintattiche dei linguaggi, Tecnologie Sistemi Informatica Olivetti, n.10 (Dec.1978) 81-107.

[6] D.Gries, Compiler construction for digital computers (J.Wiley E.S., New York, 1971).

[7] K.V.Hanford, Automatic generator of test cases, IBM System Journal, vol.9 (Dec. 1970) 242-257.

[8] J.C.Huang, An approach to program testing, ACM Computing Surveys, vol.7 (Sept. 1975) 113-128.

[9] Pratt, Programming languages: design and implementation (Prentice-Hall, New York, 1975).

[10] P.Purdom, A sentence generator for testing parsers, BIT, vol.12, n.3 (July 1972) 366-375.

[11] T.Snook, et al., Report on the programming language PLZ/SYS, New York, Springer-Verlag, 1978.

[12] N.Wirth, The programming Language Pascal, Acta Informatica, vol.1 (1971) 35-63.

LES PROCEDURES DE VALIDATION DES COMPILATEURS

Jacqueline SIDI

BNI : Domaine de Voluceau - Rocquencourt
BP 105 - 78153 Le Chesnay Cedex
SEMA : 16,18 Bd Barbès 92126 Montrouge Cedex

ZUSAMMENFASSUNG
Die Gültigkeitserklärung den Compilers hat als Zweck, die schnelle Zunahme den Compilers zu entmutigen, die unangemessen den Normen sind : PASCAL/SOL Standard mit der ISO Norme und den Zuschläge der SOL Spezifikationen ; das ADA Gerätehandbuch. In der Tat wird die Übereinstimmungskennung garantiert, daß die Programmen tragbar sind, die mit Hilfe den für gültig erklärten Compilers geschrieben wurden.

Die Gültigkeitserklärung den Compilers wird heutig durch Testprogrammen gemacht. In diesem Artikel vergleichen wir die ADA und PASCAL Methoden, um Testkartensätze auszufürhen und Ergebnisse zu analysieren. Das Thema "die manuellen und automatischen Testausführung" wird auch erörtert.

ABSTRACT
The aim of compiler validation is to discourage proliferation of compilers that do not conform to the standard : PASCAL/SOL standard that includes ISO and SOL specifications, ADA reference manual. In fact, the conformity label will be a garantee for portability of programs written with validated compilers.

Compiler validation is at present done with test programs. In this article we compare ADA and PASCAL approach for realizing test suites and analyzing results. Manual and automatic generation of test programs are also approached.

RESUME
Le but de la validation des compilateurs est de décourager la prolifération des compilateurs non conformes aux normes : Standard PASCAL/SOL comprenant la norme ISO et les compléments de spécifications SOL, Manuel de Référence ADA. En effet, le label de conformité sera un garant de la portabilité des programmes écrits à l'aide des compilateurs validés.

La validation de compilateurs se fait actuellement à l'aide de programmes de tests. Nous comparons ici l'approche utilisé pour la validation des compilateurs ADA et PASCAL au niveau de la réalisation des jeux d'essais ainsi que de l'analyse des résultats. Le problème de la réalisation manuelle ou automatique des tests sera aussi abordé.

1. INTRODUCTION

Dans le cadre de la mission qui lui a été impartie lors de sa création par la DIELI, le BNI participe à la création du Centre de Validation SOL des compilateurs PASCAL et assurera cette validation dès la mise en place effective de ce service /1/. En collaboration avec le GMD (Gesellschaft fur Mathematik und Datenverarbeitung) et le NPL (National Physical Laboratory), le BNI a été chargé par la CCE (Commission des Communautés Européennes) de promouvoir et mettre en place un Centre de Validation Européen pour les compilateurs ADA.

2. LES JEUX D'ESSAIS

Que ce soit dans le cadre de la validation PASCAL/SOL ou celle d'ADA, la validation se fait à l'aide d'un jeu de programmes de tests : environ 600 pour PASCAL /2/ et 1700 pour ADA /3/. On retrouve dans les deux jeux à peu près les mêmes catégories de tests. En particulier :
- la classe des tests qui doivent se compiler correctement, puis s'exécuter ;
- la classe des tests correspondant à des programmes non conformes à la norme ;
- la classe des tests permettant de vérifier ou de mesurer la valeur de certains paramètres.

2.1 LES CLASSES QUI TESTENT LA CONFORMANCE A LA NORME

Cette classe contient des programmmes qui testent les spécifications de la norme.

PASCAL : classe "conformance"
Cette classe comprend les programmes qui doivent être compilés puis exécutés. Lors de leur exécution ils impriment un message indiquant si le résultat se trouve être conforme à ce qui était prévu.

ADA : classes A et C
La classe A contient les programmes qui doivent se compiler correctement.
La classe C contient des programmes auto-controlables qui doivent se compiler et s'exécuter correctement.

2.2 LES CLASSES DES PROGRAMMES NON EXECUTABLES

Ces classes regroupent les programmes qui en principe ne doivent pas s'exécuter car ils contiennent des éléments interdits par la norme.

PASCAL : classes "déviance" et "détection d'erreurs"
La classe "déviance" a pour but de détecter si le processeur traite une extension du langage, ou s'il permet une manipulation erronée des traits du langage, ou encore s'il présente quelques unes des erreurs communes à des processeurs déjà existants.
La classe "détection d'erreurs" vérifie les points explicitement définis comme erreurs dans la norme. Il s'agit dans ce cas d'utilisations incorrectes de constructions légales.

ADA : classes B et L
La classe B contient les programmes devant produire une erreur à la compilation.
La classe L contient les programmes qui même s'ils n'ont pas produit d'erreurs à la compilation doivent être rejetés au niveau de l'éditeur de liens.

2.3 CLASSES PRODUISANT UNE INFORMATION

Ces classes permettent de calculer la valeur des paramètres qui peuvent varier d'une implémentation à l'autre.

PASCAL : classes "qualité" et "défini par l'implémentation"
La classe "qualité" ne statue pas sur la conformité du processeur vis-à-vis de la norme mais tente de cerner ses caractéristiques.
La classe "défini par l'implémentation" teste la conformité du processeur au manuel de référence produit par le constructeur. Si ce manuel ne précise pas la valeur pour un de ces paramètres, le test essaiera d'en obtenir la valeur.

ADA : classes D et E
La classe D teste la valeur de certains paramètres qui produisent des limitations de capacité.
La classe E vérifie le choix des options prises par l'implémenteur et la façon dont les ambiguités ont été traitées.

3. GENERATION DES TESTS

La réalisation de jeux d'essais peut se faire de plusieurs façons : on peut les écrire manuellement ou bien utiliser des outils automatiques tels que les générateurs de jeux d'essais. Le choix de la méthode dépend du type de problème à résoudre.

Les procédures automatiques donnent la possibilité de produire assez simplement et rapidement, un grand nombre de lignes de programme. En utilisant un générateur on peut, ainsi, tester le bon fonctionnement du compilateur et même sa qualité dans le sens de son degré de résistance à l'erreur.

Au contraire, les tests manuels écrits par des personnes qui ont une expérience en conception de compilateurs permettent de mieux cerner les déviances, ajouts ou erreurs vis-à-vis des spécifications.

L'utilisation de générateurs de tests tels que celui de B. HOUSSAIS /4/ ou de F. BAZZICHI /5/ permet d'obtenir un ensemble de programmes assez longs : 200 à 2000 lignes. Chaque programme teste, dans des contextes variés et avec des complexités pouvant être importantes chaque type d'instruction. On peut ainsi produire assez rapidement un nombre important de programmes pour tester le compilateur.

En écrivant manuellement des tests, on peut ainsi tester la conformité stricte à la norme ou au standard du langage sous forme de programmes compilables ou exécutables. L'inconvénient de cette méthode est l'effort humain considérable nécessaire au développement de ces tests. En contrepartie, l'utilisation en est facilitée grâce aux outils de dépouillement automatique qui peuvent être utilisés pour la plupart des tests.

4. OUTIL DE DEPOUILLEMENT

Les tests ont été écrits de façon à faciliter leur dépouillement. Mais en fait à l'heure actuelle il existe différentes conceptions de l'interprétation des résultats.

PASCAL : les tests considérés comme étant refusés après dépouillement automatique sont analysés manuellement. Dans le cas de la classe "déviance" ce sont les tests considérés comme acceptés qui subiront une inspection manuelle afin de vérifier si l'erreur prévue est celle obtenue.

ADA : tout test conduisant à un constat d'échec est considéré comme ayant de toute façon produit un effet contraire aux spécifications, il s'agit donc d'un échec. L'automatisation des résultats est maximun :
- les tests exécutables produisent eux-mêmes un message FAIL en cas d'échec. Donc à moins d'un plantage ou d'un message FAIL, ils sont considérés comme réussis quel que soit le message produit.
- les tests non exécutables contiennent un commentaire indiquant l'endroit approximatif où doit se produire une erreur. Un outil d'analyse vérifie si une erreur est bien produite dans cette zone quelle qu'en soit la cause.

5. CONCLUSION

La validation de compilateurs est maintenant reconnue comme étant un besoin. Les outils développés bénéficient de l'expérience acquise dans des contextes de recherche. Aujourd'hui les axes de recherches pouvant profiter à ce domaine sont ceux qui permettront de pouvoir certifier (c'est-à-dire prouver formellement) la conformité aux spécifications.

6. REFERENCES

/1/ M. GIEN, J. SIDI, SOL project and validation, Conférence on Pascal validation, NPL, fév. 1982

/2/ J. SIDI, Procédures de validation de logiciel et application à la validation SOL, Rapport BNI n° 39, sept. 1982

/3/ J.B. GOODENOUGH, The Ada compiler Validation Capability, IEEE computer janv. 1981

/4/ B. HOUSSAIS, Un générateur de tests commandé par les grammaires, publication IRISA n°119, juil. 1979

/5/ F. BAZZICHI, I. SPADAFORA, An automatic generator for compiler testing, IEEE vol SE-8 n°4 juil. 1982

IDA - TESTS DES LOGICIELS TEMPS REEL ASSISTES PAR ORDINATEUR

G. LAMARCHE et P. TAILLIBERT

Electronique Serge Dassault (ESD)
55, Quai Carnot 92214 SAINT-CLOUD
Tél.: 602.70.17 / 602.50.00

RESUME : Ce document décrit les résultats d'une étude ayant contribué à définir un ensemble de moyens permettant d'informatiser les opérations de test des logiciels temps réel. Ces moyens comprennent un langage de test dont les principales caractéristiques sont tout d'abord présentées. Après avoir justifié l'existence d'une bibliothèque d'outils, l'article décrit ensuite l'un de ces outils plus particulièrement adapté au test des logiciels comportant des processus parallèles.

MOTS CLES : Test de logiciel, langage de test, logiciel temps réel,
test en temps réel, processus parallèles, réseaux de Pétri.

ABSTRACT : This paper describes the main results of a study whose aim was to define a set of means to computerize real-time software testing. These means mainly encompass a test language whose characteristics are first presented. After exposing the reasons leading to a test tools library, this paper describes one of these tools especially designed to concurrent process checking.

KEYWORDS : Software Testing, Test language, Real-time Software, Real-time testing concurrent
processes, Petri nets.

ZUSAMMENFASSUNG : Dieser Article beschreibt die Ergebnisse einer Entwicklung, welche mitgewirkt at eine Sammlung von Mittels zu definieren, die die Prüfungsoperations der Echzeit Software zu informatisieren ermöglichen.

Die Mitteln beinhalten eine Test Sprache, deren Haupteingenschasten zu nächt vorgeschaft sind.

Nachdem die Existence einer Werhzeugs bibliothek begrünten worden ist, beschreibt der Article eines davon, das besonders zum testen der Software geeingnet ist, die parallel durchgeführte Process beinhalten.

Schlüsselworten : Software Prüfung, Prüfung Sprache, Echtzeit-Software, Parallel Process,
Petri nets.

I - INTRODUCTION

Parmi l'ensemble des techniques de validation de logiciel, l'analyse dynamique et plus particulièrement le test des programmes jouent encore un rôle privilégié [HOW 78]. Il est certain que l'utilisation de techniques d'analyse statique, appliquées tant au programme lui-même qu'à ses divers niveaux de spécification, tend à se généraliser.

Cependant ces techniques ne sont pas en elles-mêmes suffisantes pour atteindre les exigences de qualité requises de la part d'un nombre croissant de logiciels et plus particulièrement des logiciels temps réel intégrés à un équipement. Il en résulte donc une phase de test des programmes longue et coûteuse où l'on tente de placer le logiciel dans le plus grand nombre possible de configurations. Cette étape est d'autant plus longue que l'impossible exhaustivité du test ne permet pas d'en déterminer de bornes.

Il ne nous semble pas que cet état de fait doive évoluer radicalement dans les années à venir et ce malgré les efforts considérables entrepris tant pour formaliser les spécifications et donc y appliquer des contrôles automatiques [TEI 77] [ALF 80] [BAR 82] que pour améliorer les techniques d'évitement ou de tolérance aux fautes [LIS 77] [CII 80] [AND 76].

Il nous semble donc justifié de proposer un ensemble de moyens destinés à améliorer les conditions dans lesquelles se déroulent les opérations de test des programmes.

Trois critères principaux ont été retenus pour la conduite de l'étude :

- la généralité, supposant l'indépendance des moyens de test par rapport aux langages de programmation des programmes testés et à la machine sur laquelle ils s'exécutent.

- la portabilité, nécessitant que les outils soient définis hors de toute hypothèse d'implantation autre que celle de pouvoir s'exécuter sur les mini-ordinateurs classiques.

- la simplicité, évitant l'introduction de contraintes d'utilisation par rapport aux conditions actuelles de déroulement des tests.

II - PRINCIPAUX RESULTATS

2.1. Machine de test distincte de la machine sous test

L'étude se place dans le cas où les tests sont commandés à partir d'un autre calculateur que le le calculateur cible ; cette approche permet la conduite des tests en temps réel, c'est-à-dire sans perturbation du programme à tester ;
elle présente également l'avantage d'offrir la "puissance informatique" nécessaire à la conduite des tests (mémoires de masse, dispositifs d'impression ...) ;
ces possibilités sont en effet souvent absentes des ensembles informatiques intégrés dans un équipement.

2.2. Langage de test

Il permet de décrire de manière formelle et standardisée les opérations qui sont habituellement exécutées lors du test d'un programme. Ces opérations peuvent être regroupées en trois grandes fonctions :

- commande du programme testé, permettant l'exécution de tout ou partie de celui-ci sur un jeu de stimuli d'entrée.

- mesure du programme testé par observation directe ou enregistrement des valeurs produites par ce programme.

- vérification par comparaison des valeurs mesurées aux valeurs de référence ou plus généralement, du comportement observé à un comportement de référence.

2.3. Bibliothèque d'outils de test :

Elle sert de structure d'accueil à un ensemble d'outils de tests réalisés à l'aide des primitives du langage. Elle comporte en particulier de puissants outils d'enregistrement ou de modélisation du comportement du programme sous test.

Le chapitre 3 présente les caractéristiques principales du langage de test ; après avoir présenté la bibliothèque d'outils le chapitre 4 décrit l'un d'entre eux, prévu pour permettre le contrôle d'un programme par rapport à un modèle décrit à l'aide d'un réseau de Pétri.

III - LE LANGAGE DE TEST

Un langage de test doit offrir des possibilités algorithmiques adaptées aux traitements les plus fréquemment rencontrés dans les opérations de test ; mais, et c'est là un aspect important, il doit permettre la description et la manipulation d'objets extérieurs au programme de test lui-même (données et procédures du programme à tester).

3.1. Description des objets du programme à tester

Les objets du programme à tester se répartissent en deux catégories : les variables et les points de contrôle.

a) Les variables sont décrites par la structure de la donnée mais également par les indications nécessaires pour y accéder ("adresse" et procédé de lecture ou d'écriture). Pour cela, un type "variable-testée" permet d'indiquer :

- le type de la donnée,

- une procédure d'accès en lecture décrivant les opérations nécessaires pour acquérir la valeur de cette variable,

- une procédure d'accès en écriture.

Un paramètre supplémentaire (attribut adresse) peut être précisé à la déclaration d'une telle variable et référencé dans les procédures d'accès. Ainsi par exemple, toutes les variables entières basées par rapport à l'adresse "PROC1" peuvent être décrites par le type ci-dessous :

```
type  entier-proc-1 is    tested-var
      integer (32) ;
      reading is
              -- Procédure permettant l'
              -- accès en lecture aux
              -- objets du type
      end   reading ;

      writing is
              -- Procédure d'accès en
              -- écriture
      end   writing
end tested-var ;
```

Un objet de ce type est déclaré par :

VAR at 200 : entier-proc-1 ;

-- 200 représente "l'attribut adresse" de l'objet déclaré.

b) Les points de contrôle : un point de contrôle permet de décrire certains points particuliers de la structure de contrôle d'un programme (étiquette, début de procédure, de bloc, numéro d'instruction ...) . Il peut être utilisé comme point de lancement, point de surveillance ou point d'arrêt d'un programme ; dans ces deux derniers cas, le passage du programme sous test devant un tel point peut engendrer l'activation d'un événement dans le programme de test ou l'arrêt du calculateur sous test.

La description de ces objets comporte un prologue décrivant la séquence d'opérations à réaliser lorsque le point de contrôle est utilisé comme point de départ (y compris le passage éventuel de paramètres au programme sous test) et un détecteur précisant les opérations nécessaires pour utiliser ce point de contrôle comme point de surveillance ou d'arrêt (action sur le matériel ou modification du code du programme sous test).

3.2. Génération des déclarations

Un système de test "réaliste" ne doit pas obliger son utilisateur à redéclarer tous les objets du programme à tester ; cette opération ayant été faite lors de l'étape de codage, il est souhaitable que ces objets soient connus "implicitement" dans le programme de test.

Pour aboutir à un tel résultat tout en restant indépendant du langage de programmation utilisé et de ses processeurs, un préprocesseur a été défini afin d'aider l'utilisateur à déclarer les objets du programme sous test. Ce préprocesseur est muni d'opérations d'entrée-sortie lui permettant d'accéder aux différentes tables produites par les processeurs du langage de programmation et si nécessaire au texte source du programme à tester. Il autorise la génération d'instructions représentant les déclarations souhaitées établies à partir des informations recueillies dans les tables.

Il est ainsi possible pour un utilisateur particulier (soit homme système, soit simplement premier utilisateur) de définir pour chaque couple langage - processeur un ensemble de procédures que l'utilisateur final pourra utiliser pour déclarer les objets et les manipuler avec pratiquement la même facilité que si l'existence de ces objets était implicite.

3.3. Instructions du langage adaptées au test des programmes

Nous ne décrirons que les éléments les plus caractéristiques.

a) Expressions et conversions

Le langage de test permet d'écrire des expressions où figurent indifféremment objets du programme sous test et objets du programme à tester.

Pour en faciliter l'écriture, le langage prévoit la possibilité de redéfinir la sémantique des opérateurs (+, -, *, /, =) pour les nouveaux types éventuellement introduits.

Si cette opération s'avère trop coûteuse, il est possible de définir une fonction de conversion qui, afin d'en faciliter l'utilisation, porte le nom du type cible (l'objet à convertir étant passé en paramètre).

b) Itérateur

L'observation des techniques utilisées pour le test des programmes fait apparaître l'importance de la structure de contrôle répétitive.

Cette constatation a conduit à prévoir un constructeur spécial appelé "itérateur" [LIS 77], permettant de disposer d'une succession de valeurs sur lesquelles porte l'itération sans être contraint de stocker ces valeurs au préalable dans une structure appropriée.

c) Interface standardisée

Afin d'assurer la portabilité des systèmes de test, l'interface entre machine de test et machine sous test a été standardisée. Elle se compose de procédures, définies par leur spécification externe, que chaque implémentation doit réaliser en tenant compte du matériel de l'installation.

d) Gestion du temps

Le langage de test étant principalement destiné à la vérification de logiciels temps réel, il est nécessaire qu'il dispose d'outils pratiques de gestion du temps ; ce besoin se fait plus particulièrement sentir lorsque l'utilisateur désire commander ou observer l'environnement de la machine sous test. La précision des outils habituellement implémentés peut s'avérer insuffisante compte tenu du fait que ces opérations sont soumises à l'aléa du "scheduling" logiciel. C'est pourquoi les constructions agissant sur le temps peuvent être utilisées en deux modes distincts :

- le mode "normal" où la synchronisation est réalisée par les mécanismes habituels de scheduling.

- le mode "précis" où au contraire l'opération est réalisée après une attente active garantissant ainsi une meilleure précision.

Une instruction du langage permet de préciser :

- la date de début de l'action associée
- sa périodicité
- une clause de fin de répétition (durée ou condition)
- une clause de précision

e) Gestion des actions parallèles

En plus de la structuration des programmes en tâches asynchrones et d'un mécanisme de communication par évènement bien adapté à l'implantal'implantation d'outils de test exploitant la détection de points de contrôle, le langage de test permet de spécifier l'exécution parallèle de quelques actions élémentaires sans avoir recours à un découpage en tâches.

Il est muni pour cela d'un moyen de description de "blocs parallèles" formés d'un ensemble d'instructions qui, au lieu de s'exécuter séquentiellement engendrent autant de tâches s'exécutant en parallèle ; seule l'initialisation des instructions s'effectue dans leur ordre d'apparition dans le bloc. Une telle construction est équivalente à la création d'autant de tâches que d'"instructions parallèles", mais permet une formulation beaucoup plus souple et immédiate des opérations de test.
Dans l'exemple suivant l'instruction INST1, le bloc B1 et l'instruction INST3 s'exécutent parallèlement.

<u>Parallel</u>
   INST1 ;
   B1 : <u>begin</u> INST2, .. <u>end</u> B1 ;
   INST3 ;
<u>end parallel</u> ;

IV - BIBLIOTHEQUE D'OUTILS GENERAUX

4.1. Présentation

Elle a pour rôle de fournir un certain nombre d'outils réalisés à l'aide du langage de test et couvrant une bonne partie des besoins les plus courants. Elle permet ainsi de réduire notablement le temps à consacrer au développement des programmes de test. Elle sert également de de structure d'accueil aux outils plus spécifiques d'une méthodologie ou d'une application.

La bibliothèque standard comprend en particulier :

* <u>une macro-instruction d'exécution</u> autorisant le lancement du programme sous test en un point de contrôle particulier ou au début d'une procédure (il est possible dans ce cas de fournir des paramètres effectifs à la procédure). Cette macro-instruction permet de spécifier un invariant sous la forme d'une expression qui est évaluée avant puis après l'exécution et dont la variation entraine un diagnostic d'erreur. Enfin, il est également possible de demander le calcul d'une variation et de préciser une durée limite.

* un outil de <u>simulation d'instruction</u> utilisable lorsque certaines instructions machine ne peuvent pas être exécutées (entrées-sorties non câblées) ou que certaines procédures ne sont pas encore au point. Il est possible, grâce à cet outil de substituer aux instructions ou procédures absentes l'exécution d'une procédure du programme de test.

* <u>un enregistreur</u> ayant pour rôle de mémoriser pendant l'exécution du programme un certain nombre d'informations. Le test consiste alors à dépouiller l'enregistrement ainsi obtenu. L'utilisateur doit préciser <u>quand</u> enregistrer (point de contrôle, accès à un domaine ..), <u>quoi enregistrer</u> et <u>quand</u> cesser l'enregistrement.

* <u>un outil de modélisation</u> permettant de contrôler le comportement d'un programme par rapport à un modèle décrit à l'aide d'un réseau de Pétri. Cet outil est décrit ci-dessous.

4.2. Utilisation de la technique de l'observateur pour le test des programmes temps réel

4.2.1. Objectif

Afin d'atteindre le double but de formalisation et d'automatisation des opérations de test fixé à l'origine de l'étude IDA, il est rapidement apparu nécessaire de disposer de procédés permettant de contrôler <u>sans perturbation du programme sous test</u> :

- le comportement des processus parallèles (synchronisation, partage des ressources...)

- les contraintes de date ou de durée d'exécution.

Dans l'état actuel de disponibilité des outils de test, le second point peut être partiellement pris en compte grâce à l'outillage utilisé habituellement pour les tests du matériel (analyseur logique, oscilloscope,...).

Par contre, en ce qui concerne le test du comportement des processus parallèles, le programmeur se trouve réellement démuni. Ainsi la simple vérification du fait que deux séquences ne s'exécutent jamais simultanément ne pourra être effectuée qu'après une mise en oeuvre laborieuse d'un équipement mal adapté (analyseur logique "performant").

4.2.2. Principe

Il s'appuie sur le concept d'observateur [AYA 79] et est schématisé par la figure 1 ; il met en jeu :

- le programme sous test, inchangé par rapport au programme définitif ;

- un modèle décrivant le comportement de référence faisant l'objet du test ;

- un ensemble de connexions entre programme et modèle, réalisé par détection de points de contrôle par l'intermédiaire de l'interface matérielle ;

- un "contrôleur" chargé de faire évoluer le modèle parallèlement au programme tout en s'assurant de la validité de son évolution.

Figure 1

Les possibilités offertes par cette méthode nous ont conduit à développer deux types de modèles :

- le "modèle parallèle" orienté vers le test de processus asynchrones ;

- le "modèle séquentiel" destiné au test de la structure interne des programmes.

Seul le modèle parallèle est décrit dans ce document.

4.2.3. Le modèle parallèle

Ce modèle est construit à partir d'un réseau de Pétri et est donc particulièrement adapté au contrôle des synchronisations et des exclusions.

Afin de simplifier les modèles d'une part et de permettre l'introduction du temps d'autre part, nous avons admis la possibilité que, dans le modèle parallèle, le tir d'une transition ne soit plus de durée nulle.

Cet aménagement ne modifie en rien les possibilités de vérification statiques faites sur les réseaux de Pétri dans la mesure où une transition à tir non instantané est équivalente à une séquence transition-place-transition où les tirs des transitions seraient instantanés.

L'intérêt de cette modification est d'une part de diminuer le nombre de places et de transitions d'un modèle, d'autre part d'associer à chaque transition un traitement et à chaque traitement une durée minimum et une durée maximum, le contrôleur devant s'assurer que la durée effective de la transition respecte bien la fourchette indiquée.

Chaque transition doit alors être connectée au programme sous test par deux points de contrôle, l'un correspondant au début d'exécution du traitement associé à la transition et l'autre à la fin.

La figure 2 représente un modèle permettant de contrôler que les séquences S1 et S2 ne s'exécutent jamais simultanément.

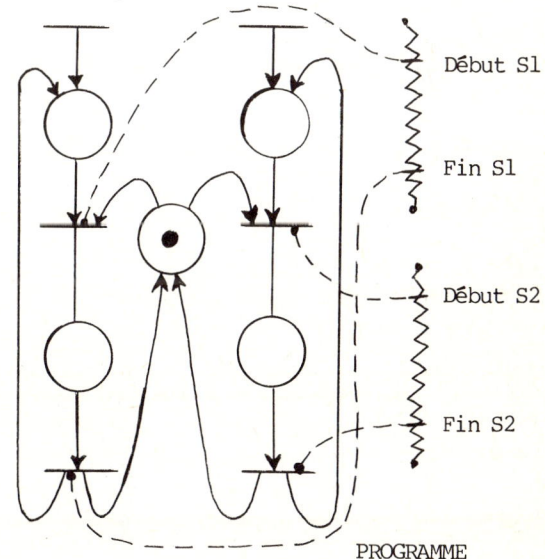

Figure 2

Remarques :

* Le modèle parallèle conserve la possibilité de transition à tir instantané (durée maximum nulle et points de contrôle identiques).

* Il est également possible d'acquérir à la fin d'une transition la durée effective de celle-ci et de pouvoir la traiter à l'aide de prédicat et d'actions définies sur chaque transition.

4.2.4. Implémentation

Afin de ne pas perturber le programme sous test, l'interface entre machine de test et machine sous test se limite à une observation du Bus interne de la machine testée et des quelques signaux permettant d'identifier les informations y circulant.

Une telle interface permet d'effectuer malgré ces restrictions :

- la détection des points de contrôle et leur datation ;

- la détection des modifications intervenant dans certains emplacements mémoire désignés au préalable.

Cependant, la prise en compte de tels événements par la machine de test n'est pas possible en temps réel. (Ceux-ci pouvant apparaître ponctuellement de manière très rapprochée) ; une telle prise en compte n'est d'ailleurs pas utile, les contrôles pouvant être effectués en différé pourvu que la chronologie d'apparition des événements soit respectée.

En conséquence machine de test et machine sous test sont désynchronisées ; une file d'attente contenant les événements observés (points de contrôle ou modifications mémoire) et leur date d'apparition permet d'"adapter" la charge de travail de la machine de test (figure 3).

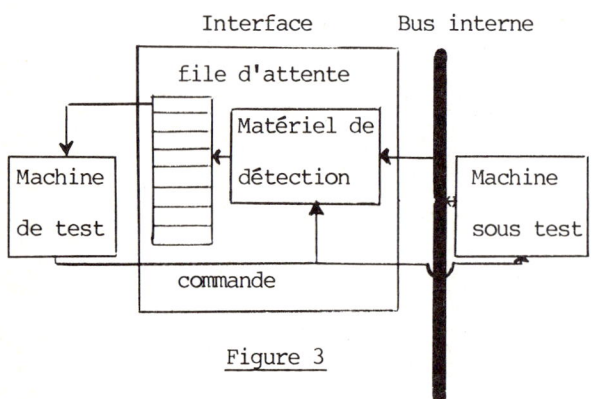

Figure 3

Le "matériel de détection" est réalisé grâce à un comparateur dynamique formé d'une mémoire RAM d'une capacité d'adressage égale à celle de la machine testée et d'une capacité de mémorisation de 1 bit.

Une première phase - avant le test lui-même - consiste à programmer cette mémoire aux adresses pour lesquelles une détection est souhaitée. Les techniques actuelles d'intégration permettent d'envisager un tel comparateur avec un minimum de matériel.

Remarque

Dans le cas où la machine sous test est multiprocesseur chaque processeur est muni d'une interface de détection, la chronologie des événements enregistrés étant restituée grâce aux datations mémorisées dans chaque file d'attente.

4.2.5. Utilisation

L'outil qui vient d'être décrit permet d'effectuer des tests de niveau élevé et cela sans perturber le programme testé.

La puissance des tests est liée directement à la puissance des modèles proposés. L'absence de perturbation est, pour sa part, due au fait que seules des observations fugitives de la machine testée permettent de mettre en oeuvre la méthode.

L'utilisation de cet outil, et en particulier de la modélisation par réseau de Pétri, pourra être rendue aisée par une interface homme machine bien adaptée (interface graphique dans la mesure du possible). D'autre part les réseaux à décrire restent simples compte tenu du fait que seul le comportement à tester doit être modélisé et non pas l'ensemble de l'application comme cela est le cas pour effectuer des analyses statiques sur ces modèles.

Enfin on peut remarquer que cet outil peut rendre de grands services pour la mise au point des programmes lorsque, une erreur ayant été détectée, de longues manipulations sont nécessaires pour en trouver la cause.

V - CONCLUSIONS

Les "moyens de test" qui viennent d'être présentés, résultats d'une étude menée avec l'aide de l'agence de l'informatique, permettent d'envisager la conduite des tests des logiciels à haute sécurité sous un jour nouveau :

- pratique systématique des tests de non-régression,

- application de jeux d'essais longs et complexes grâce à l'automatisation introduite par les outils,

- contrôle de comportements difficilement observables jusqu'ici grâce à la puissance de modélisation disponible.

L'indépendance de ces moyens par rapport aux langages de programmation et aux calculateurs cibles permet de minimiser les coûts d'adaptation à chaque utilisation particulière.

Un interpréteur d'un sous-ensemble du langage de test, limité aux tests non temps réel, a été réalisé à l'Electronique Serge Dassault (système LOTUS [VIE 81]) et est utilisé pour le test des logiciels MIRAGE 2000.

Enfin, une baie de mise en oeuvre, munie d'une interface matérielle remplissant les fonctions décrites précédemment, est en cours de développement pour un calculateur de l'ESD.

REFERENCES

[AND 76] T. ANDERSON - R.KERR,
"Recovery Blocks in Action",
Proceedings Second International Conference on Software Engineering,
IEEE, 1976

[LIS 77] LISKOV-SYNDER-ATKINSON-CHAFFERT
Abstraction mechanisms in CLU
Communications of the ACM
Aout 1977

[TEI 77] D. TEICHROEW - E.A. HERSHEY
"Computer - aided technique for structured documentation and analysis of information processing systems
IEEE Transactions on software Engineering - Vol. SE 3 n° 1
Janvier 1977

[HOW 78] W.E HOWDEN
A survey of static Analysis Methods
A survey of dynamic Analysis Methods
Tutorial : Software testing and Validation techniques
IEEE 1978

[AYA 79] AYACHE-AZEMA-DIAZ
Observer : a concept for on-line detection of control
errors in concurrent systems
9 th International symposium on fault tolerant computing
MADISON JUIN 79

[ALF 80] M.W. ALFORD
"Software requirements engineering methodology (SREM), At the age of four";
Proceedings COMPSAC 80.
1980

[CII 80] CII-HONEYWELL BULL
Reference Manuel for the ADA programming language
Juillet 1980

[FAI 80] FAIRLEY
Ada Debugging and Testing Support Environments
ACM, Mars 1980

[VIE 81] G. VIENNET - J.C. SEGUIN - M.ESTEVENY
Manuel de référence du système LOTUS
Document EMD
JUIN 81.

[BAR 82] P. BARDIER - S. CHENUT-MARTIN - F. DOLADILLE
DIAO : Définition de logiciel assistée par ordinateur
Premier colloque de génie logiciel
AFCET
PARIS - Juin 1982

MONITEUR DE TESTS

UN OUTIL POUR L'ECRITURE RAPIDE DES TESTS DE CONFORMITE

Christian TRIOLAIRE
SEMS
1 Rue de Provence 38130 ECHIROLLES

ZUSAMMENFASSUNG :

Die bei der Zuverlässigkeit des Softwäres Kontrolle getroffene Schwierigkeiten werden hier besprochen : das grösste Problem liegt an der schnellen Prüfungs programme Schreibung.

Mehrere Programme werden leicht bei Prüfungsprogramme Generatoren entwickelt, die bringen aber keine hervorragende Lösung für die Untersuchung der Erzeugnisse.

Das Prüfungsmonitor ist ein Mittel irgendwelche Prüfungskombinationen zu generieren. Der Testprogrammer gibt einige Einrichtungen und dann die gemeinsame Prozesse werden automatisch ausgeführt, das geprüfte sofwäre gerannt, die Erzeugnisse geprüft und eine Diagnose verlegt.

Irgendwelches Softwäre kann damit geprüft werden und besonders wenn die Einrichtung der Prozessen auf seinem Betragen übt.

ABSTRACT :

Difficulties involved in software reliability control are discussed : a major problem is to write correct test programs quickly.
Test program generators provide many programs easily but do not give a satisfactory solution for result checking.
The test monitor is a tool which allows any software test combination. The test programmer provides a few instructions from which common operations are automatically run, tested software launched, results checked and a diagnosis printed.
Any software can thus be tested especially when the operations scheduling has effects on its behaviour.

RESUME :

Les difficultés rencontrées au cours du contrôle de la fiabilité du logiciel sont présentées : un des problèmes essentiels est l'écriture rapide de tests sûrs.
Les générateurs de programmes de tests fournissent facilement de nombreux programmes mais n'apportent pas de solution satisfaisante pour la vérification des résultats.
Le moniteur de test est un outil permettant n'importe quelle combinaison de test de logiciel. Le programmeur de test donne quelques instructions et les opérations communes sont déroulées automatiquement , le logiciel testé est lancé, les résultats vérifiés et un diagnostic édité.
Tout logiciel peut être ainsi testé, surtout lorsque l'ordonnancement des opérations influence son comportement.

MONITEUR DE TESTS
UN OUTIL POUR L'ECRITURE RAPIDE DES TESTS DE CONFORMITE

Christian TRIOLAIRE
SEMS

1 Rue de Provence 38130 ECHIROLLES

1 - INTRODUCTION

Les meilleures techniques de production de logiciel ne permettent pas d'espérer obtenir un logiciel sans erreur. Les preuves de programme sont limitées à des programmes de petite taille. Les coûts de maintenance évalués entre 40 et 80% du coût total du logiciel [1], imposent le contrôle de sa qualité avant toute diffusion. Constructeur d'ordinateurs, la SEMS doit contrôler des logiciels importants présentant des interfaces complexes avec les programmes appelant. C'est dans ce contexte qu'est écrit cet article.

2 - LES PROGRAMMES DE TEST

L'analyse du contrôle qualité d'un logiciel débute dès que ce dernier a été défini. Les programmes de test sont donc écrits pendant le codage du logiciel à tester.
Les concevoir avant que ce logiciel n'existe amène le responsable du contrôle qualité à faire des hypothèses sur son comportement.
Ne pouvant assurer un contrôle exhaustif, il fait des paris sur les contrôles à effectuer. Certains de ces paris échouent et l'on est obligé d'écrire de nouveaux programmes de test pendant la phase de contrôle qualité. Leur mise au point est difficile car il y a toujours doute sur l'origine de l'anomalie : logiciel testé ou programme de test ? ceci prolonge la durée du contrôle qualité et retarde la date de disponibilité du produit. Les programmes de test seront plus facilement écrits et exploitables si la méthode d'écriture, la présentation, l'affichage des résultats sont les mêmes. On a donc intérêt à définir un standard. Ils seront plus efficaces s'ils décrivent le comportement du logiciel testé au fur et à mesure, et émettent un diagnostic en fin d'exécution. C'est ce que nous appelerons le diagnostic automatique.
Les tests non retenus pour les contrôles ultérieurs de "non-régression" n'ont qu'une vie éphémère : ils sont joués une seule fois car pour la plupart ils ne mettent pas en évidence d'erreurs. Cela incite leur concepteur à négliger l'aspect documentation de ses programmes. Cette négligence rend définitif leur abandon car on préfère bien souvent en écrire un autre plutôt que d'utiliser l'existant qui nécessiterait un examen jugé plus coûteux que la réécriture. L'idéal est l'autodocumentation : le programme décrit au moment de l'exécution ce qu'il fait, ce qu'il teste, comment l'arrêter, le relancer, etc...

3 - LES GENERATEURS DE TESTS

A partir d'une grammaire affixe ils produisent automatiquement des ensembles de tests [3] et assurent une bonne couverture des cas possibles d'utilisation du logiciel à tester.
Quelques règles de grammaire suffisent pour obtenir plusieurs dizaines de programmes. Mais la production d'un grand nombre de tests conduit à un dépouillement long et fastidieux. On peut le réduire en fournissant des programmes dont les résultats sont proches du diagnostic automatique et de l'autodocumentation.
Pour s'assurer de la "non-régression" d'une nouvelle version du logiciel testé on doit réutiliser les jeux complets des tests générés, mais le coût du contrôle de "non-régression" est alors élevé.
Lorsque l'ordonnancement des évènements extérieurs détermine le comportement du logiciel à tester (moniteur d'exploitation, logiciel temps réel, etc...) il faut générer les tests des différents cas de simultanéité et de synchronisation. L'explosion combinatoire du nombre de tests obtenus rend difficile leur utilisation.

4 - UNE AUTRE SOLUTION : LE MONITEUR DE TEST

On se propose d'automatiser le plus possible les opérations communes à tous les tests :

- initialisation de variables ou de structures,
- activation du logiciel testé,
- test des variables et du compte rendu,
- analyse et élaboration de diagnostic,
- édition des variables de l'interface mise en cause lors d'une erreur.

Chacune de ces opérations peut-être paramétrée par l'utilisateur, voire inhibée à l'aide d'un langage de commande du moniteur que nous appelerons langage de test.
Contrairement aux générateurs de tests, l'utilisateur à la responsabilité de l'enchaînement des opérations. Il indique pour chaque test la succession de requêtes du logiciel à tester qu'il souhaite activer, les synchronisations entre séquences, et les contrôles à effectuer. Le moniteur est interactif, toutes les commandes données par l'utilisateur sont analysées et exécutées immédiatement, ce qui accélère la mise au point [2]. Un mécanisme de génération par défaut permet d'omettre l'écriture de commandes, de paramètres..., ainsi, avant d'activer le logiciel à tester, l'utilisateur doit initialiser les variables de l'interface. S'il ne le fait pas, le moniteur leur donnera des valeurs aussi pertinentes que possible.
Le moniteur autorise une analyse immédiate du test en affichant toutes les informations sur le comportement du logiciel testé, et une analyse à postériori en fournissant les mêmes indications horodatées sur un listage. Si l'utilisateur souhaite conserver son test, il demande son archivage. Il pourra être rejoué lors des contrôles de "non-régression".

4.1 Le langage de test

A - Les directives :

Elles décrivent le comportement automatique du moniteur. L'utilisateur peut donner la liste des informations à éditer en cas d'anomalies, les conditions d'interruption d'un test, celles d'abandon des boucles, les méthodes d'initialisation des variables ou des buffers à chaque utilisation,...

Exemple : directive de vérification de résultats. REPORT <liste-variables>, <liste-buffers> ON <liste-condition>.

Les conditions portent soit sur la comparaison du compte rendu à une valeur ou plage de valeurs, soit sur la comparaison de buffers ou de variables entre eux.

Le moniteur vérifiera après chacune des commandes ultérieures si les conditions spécifiées dans la directive sont satisfaites. Dans l'affirmative le compte rendu et les buffers mentionnés, ainsi que toutes les variables utilisées par la commande seront éditées.

B - Les commandes :

Un ensemble de commandes, dites générales, activent les fonctions communes à tous les tests :

- modifier les valeurs par défaut des variables et buffers,
- incrémenter, décrémenter, comparer des variables ou buffers,
- itérer une séquence de commandes, sortir d'une itération,
- mesurer le temps (évaluation de performances)
- exécuter en parallèle les séquences de commande, les synchroniser.

Pour chaque logiciel à tester un ensemble de commandes, dites spécifiques, est créé. Elles ont le même aspect que l'interface que l'on souhaite activer.

Par exemple, si l'on souhaite tester un logiciel de gestion de fichiers :

- avec l'interface assembleur : le moniteur connait autant de tables d'interface que l'utilisateur le souhaite, il connait les variables de chacune d'elles et les initialise avec des valeurs par défaut. L'utilisateur dispose de commandes pour chaque action du logiciel de gestion de fichiers permise par l'interface assembleur. Ces commandes ont pour paramètres le numéro de la table d'interface à traiter et éventuellement le nom de chacune des variables de cette table dont on désire modifier le contenu.

- avec l'interface COBOL : le moniteur connait autant de fichiers que l'utilisateur en a décrit à la génération, il autorise la modification des paramètres de la "file définition" (FD) de chacun et peut activer tous les verbes COBOL sur une commande de syntaxe semblable à celle de COBOL.

4.2 Architecture

Elle est schématisée par la figure 1

Le module "dialogue - analyse syntaxique" reçoit les commandes de l'utilisateur et les codifie dans un langage intermédiaire. Il édite les résultats.

Le module "répartiteur" lit un ordre du langage intermédiaire et en demande l'exécution au module réalisant l'action correspondante. Il assure également la cohérence entre les modules en cas d'abandon du test, d'erreurs,...

Le module "fonctions générales" réalise toutes les actions liées aux directives et aux commandes générales : initialisation des variables, test de compte rendu, préparation et exécution de boucles, préparation des éditions,...

Les modules "test interface" assurent l'activation du logiciel à tester. Ils sont écrits dans le langage dont on souhaite valider l'interface (ASSEMBLEUR, COBOL, PL, FORTRAN...).

La multiplication de ces modules fait courir le risque de rendre le moniteur inutilisable pour des raisons de complexité et d'encombrement. L'utilisateur peut l'adapter à son environnement en indiquant à un "générateur" les interfaces qu'il souhaite tester. Le moniteur résultat est alors constitué des modules : fonctions générales, dialogue, répartiteur, et de tous les modules d'activation des interfaces déclarées. L'analyseur syntaxique, paramètré par une grammaire, permet d'assurer que le dialogue reconnaitra uniquement les commandes exécutables avec cette version du moniteur. Lorsqu'un nouveau logiciel (ou une nouvelle interface) doit être testé, il suffit d'écrire le module de gestion du nouveau service et d'ajouter les règles de grammaire nécessaire. Une documentation d'utilisation est éditée pour chaque moniteur.

5 - CONCLUSION

Une première version du moniteur a été réalisée pour le logiciel d'entrée-sortie du SOLAR. Elle est écrite en PASCAL. L'introduction d'une nouvelle gamme de disques a pu être qualifiée exclusivement à l'aide de cet outil. La rapidité d'écriture des programmes de test (de même que celle de la reproduction des cas d'anomalies signalés par les utilisateurs) nous fait envisager de l'étendre à d'autres logiciels, et à d'autres gammes de machine. Ces extensions permettront de confirmer l'hypothèse sur les actions communes à tous les tests. L'adoption progressive du moniteur permettra alors de définir un standard de tests de conformité.

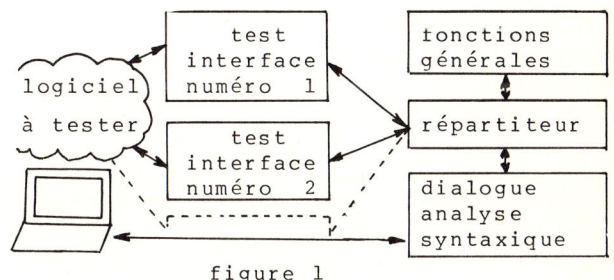

figure 1

BIBLIOGRAPHIE :

[1] BOEHM, B. W., Les facteurs du coût du logiciel. Présenté par E. GIRARD, Technique et Science Informatique vol 1 n° 1 (1982) 5-24.

[2] BROWN, P. J. Dynamic programm building, Software practice and experience vol 11 (1981) 831-843.

[3] HOUSSAIS, B., Production systématique de tests commandés par une grammaire, thèse de docteur en troisième cycle, Rennes, (décembre 1976).

ABOUT THE SECOND GENERATION OF BUS STRUCTURES

E.V. Chernykh
Joint Institute for Nuclear Research, Moscow, USSR

It was concluded that bus structures being developed at present may be classified as the second generation of programmable buses. Parameteres of these buses are enumerated and divided into five groups in accordance with its functional components. A timing structure of single and interrupt operations affecting the throughput is briefly considered.

RÉSUMÉ

On a conclu que les structures de bus actuellement développées peuvent être considérées comme constituant la deuxiéme génération de bus programmables. On énumère dans ce papier les paramètres associés à ces bus et on les répartit en cinq groupes d'après leurs composants fonctionells.

On y présente brièvement une structure d'opérations - simples ou avec interruptions - modifiant la capacité de traitement.

ZUSAMMENFASSUNG

Busstrukturen, die zur Zeit in Entwicklung sind, koennen als programmierbare Busstrukturen der zweiten Generation betrachtet weiden. Parameter dieser Busstrukturen werden dargestellt und in fuenf Gruppen gemaess ihren funktionellen Bestandteilen eingeteilt. Das Zeitverhalten fuer einfache und Interrupt-gesteuerte Betriebsweise und sein Einfluss auf den Daten durchfluss werden betrachtet.

1. BUS SYSTEMS GENERATIONS

From the early eighties high energy physics installations: spectrometers, accelerators and test benches /1/ demand much more sophisticated data acquisition systems. Computer systems having distributed structure are being developed to meet these requirements. A distributed computer system consists of crates connected by a bus but a performance of a single local control centre is much more than earlier due to present-day powerful microprocessors using.

With a sufficient number of processors the performance of these systems strongly depends on interconnection bus. For this reason an investigation and development of bus structures have become quite important task. Parameters of bus systems being developed at present - E3S (Eurobus), VME, P896, Fastbus and many others - make possible to qualify them as the second generation of programmable buses (Table 1). This paper is based on the consideration of standard bus systems and candidates to standard ones for high energy physics installations. As basic principles are common to all second generation bus systems the models under discussion can be considered as representatives of generation as a whole.

2. BUS SYSTEM PARAMETERS

A bus system as a system component is characterized by a vector of characteristics /2/. These characteristics are functions of parameters of bus system and other system components and system features as well. It is possible to use the formal procedure of bus system selection if only these functions can be obtained /3/. At present they can be hardly obtained so a numerical evaluation of the characteristics is usually used /2/. A number of parameters has been discussed in the paper /4/.

Parameters of the second generation bus system can be divided into five groups in accordance with its functional components:
1. Parameters of address and data, arbitration, interrupt buses.
2. Parameters of auxiliary series bus.
3. Mechanical and electrical parameters.
4. Elements of standardized software and their parameters.
5. Specific parameters.

Parameters in the first group fall into parameters of backplane and intercrate buses. The first subgroup comprise following parameters: ADP-matrix (Address, Data, Cycle Phase), bus communication technique (asynchronous or synchronous), set of operations, addressing modes and methods of address space extension, means for bus operations checking. ADP-matrix for every bus presents in a short form an interrelation of two degree of a designer freedom: space and time (Table 2). Parameters in the second subgroup are the following: limitation imposed on a system volume and architecture, type of interconnection unit, maximum distance between these units.

Parameters of arbitration bus are: a number of arbitration levels, technique of requests processing (centralized, distributed), technique of response generating (serial, parallel, radial), functional independence of an arbitration bus from an address and data bus.

Parameters of interrupt bus are similar to those of arbitration bus: number of interrupt levels, technique of requests transmitting (radial, wired OR, along the data bus), technique of requests processing (centralized, distributed - with active or passive interruptor).

Parameters of auxiliary serial bus are: bus communication technique (asynchronous or synchronous), clock frequency, number of priority levels, number of registers addressed, data exchange possibility.

Mechanical and electrical parameters are: number of stations in crate, type of mechanics, types of main and auxiliary connectors and number of pins, available power voltages, bus coupler implemented in LSI.

Elements of standardized software and their parameters: control and data registers; algorithms of operations, initialization and so on; control language, data base, operational system.

Specific parameters of bus system: number of companies supporting, level of the standard (international, national and so on), independence of specific microprocessor type and others.

3. OPERATIONS TIMING STRUCTURE

The system performance depends on the basic bus structure characteristics - its throughput. The throughput of first generation bus structures has been evaluated by single or block operations /2/ because of a limited number of operations. It seems necessary to evaluate the performance of the second generation bus structures by an operation mix similar to computer evaluation. To compose the mix, it is necessary to determine the duration of different operations and the frequency of their usage in high energy physics tasks. Further in this paper we discuss categories of operations and a qualitative dependence of the duration of interface operations on the parameters of system components.

An operation is defined as a logically completed series of actions for information exchange between a program source and one or several sources (sinks) of information. From the functional point of view operations can be divided into data exchange and interrupt operations. Data exchange operations are the following: single, zero, block, mixed, multiple. During the operations use is made of geographical and logical modes of addressing. There are two methods of address space extension: group and extended addressing. Interrupt operations may be executed by passive or active interrupt source.

Each operation consists of a subset of a set of microoperations. The electronic components and structure mainly affect the minimum duration of microoperations[5]. All bus structures can be devided into two groups depending on electronic components used: ECL (Fastbus) and TTL (others).

4. CONCLUSION

With a sufficient number of processors the performance of distributed computer system strongly depends on interconnection bus. Bus system investigation on a basis of analytical models is essential. An attempt of qualitative analysis of a bus system is undertook in this paper. A formal model of relationship between characteristics and parameters must be found in future.

REFERENCES

1. B.Zacharov, Distributed processor systems, In: CERN School of Computing Proceed., Geneva, 1976, CERN-76-24, Dec. 1976, p.3.
2. I.F.Kolpakov, Selection criteria of standard interfaces, In: Proc. of the 1st European Symp. on Real Time Data Handling and Process Control, Amsterdam, 1980, p. 261.
3. M.J.Gonzalez, Jr., and B.W.Jordan, Jr., A framework for the quantitative evaluation of distributed computer systems, IEEE Trans. on Comput., 1980, vol. C-29, p.1087.
4. R.Patzelt, Interconnection problems, unified hard-software solutions, In: Proc. of the 1st European Symp. on Real Time Data Handling and Process Control, Amsterdam, 1980, p. 501.
5. E.V.Chernykh, Standartnye interfe'sy dlya programno-modulnykh mnogoprotsessornykh sistem (Obzor), Pribory i tekhnika experimenta, No.4, 1982, p. 5-35.

Table 1
Parameters of second generation bus systems

Bus characteristics	Advanced parameters defining corresponding characteristics
1. Performance	Extended set of operations, including those for synchronization of multiple processors work. Extention of a memory area accessible during a single cycle. Overlapping of data exchange and arbitration operations. Auxiliary serial bus.
2. Modularity: Mechanical aspect	Set of card sizes, possibility of several modules interconnection beyond a crate.
Functional aspect	Possible extension of the data and address words, operations to speed up the scanning of large number of similar modules, software standartization.
Structural aspect	Bus coupler implemented in LSI.
3. Reliability	Asynchronous information exchange protocol, means of bus operations checking, means of deadly embrace exclusion, physical integration of buses.

Table 2. ADP matrix of multiprocessor bus systems

Info.exchange	Type of bus system	ADP matrix	Cycle dur., μs
Parallel	1. GEC-Elliott system crate	$\|A_{24}\|$	1.4
	2. COMPEX	$\|D_{24}\|$	0.7
	3. VME	$\|A_{23(15)} D_{16(8)} A_8 D_{16(8)}\|$	0.15
Serial/ /parallel	1. Kinetic Systems syst.crate	$\|A_{24}A_{24}D_{24}D_{24}\|$	6.4
	2. Eurobus	$\|A_{26(18,10)}D_{24/8(16/8,8)} A_6 D_8\|$	0.5
	3. P896	$\|A_{32}D_{32/16/8}\|$	0.4
	4. FASTBUS	$\|A_{32}D_{32}\|$	0.1

Session 8

NETWORKING AND DATA HIGHWAYS

RESEAUX ET TRANSMISSION DE DONNEES

RECHNERNETZE UND DATENÜBERTRAGUNGSLEITUNGEN

ON STANDARDS, PRE-STANDARDS AND DE-FACTO STANDARDS IN DATA COMMUNICATION

R. Popescu-Zeletin

Hahn-Meitner Institut fur Kernforschung Berlin FRG

ABSTRACT

The fast technological developments in the field of data communication have been materialized in a variety of products which claim to conform to a certain standard.

On the other hand the standardization activities have been supported by different organizations and the standards proposed are often contradictory or mutually exclusive. The result of this situation is that the user community has a difficult task in developing strategies to get unscathed through the product jungle of the market.

The paper outlines the characteristics of the main standardization bodies in the field of data communication, their output and impact on the data communication market.

RÉSUMÉ

Les développements technologiques rapides en matieré de communications digitales ont débouché sur une variété de produits annoncés conformes à certains standards. Par allieurs, les activités de standardisation soutennes par plusieurs organismes et les standards proposés sout souvent contradictoires ou s'excluent les uns les autres. Le résultat est que l'ensemble des utilisateurs ont d'énormes difficultés pour échaffauder des stratégies leur permettant de cheminer indemnes à travers la jungle des produits disponibles. Cet article décrit les aspects marjeins des organismes de standardisation concernés, leurs résultats et leur impact sur le marché des produits.

ZUSAMMENFASSUNG

In der Dokumentation entsteht im Zuge der technologischen Entwicklung eine Vielfalt von Produkten mit dem Anspruch, daß diese bestimmten Standards folgen. Auf der anderen Seite werden Standardisierungsgremien durch unterschiedliche Organisationen unterstützt, so daß die resultierenden Vorschläge oft widersprüchlich sind oder sich gegenseitig ausschließen. Für den Benutzer ergibt sich daraus die oftmals schwierige Aufgabe, sich mit seinen Interessen dem fast unüberschaubaren Markt anpassen zu müssen. Der folgende Beitrag zeigt Charakteristika der wichtigsten Standardisierungsgremien auf, beschreibt die resultierenden Ergebnisse und untersucht die Konsequenzen auf dem Gebiet der Datenkommunikation.

1. INTRODUCTION

Not very long ago, the interface between the user's devices and the modem was generally regarded as the line dividing data communication and data processing. But over the past few years the network has crept through that connector and has infiltrated in terminals, frontends and main-frames. On the other hand the users have precipitated this invasion by calling for increased connectivity, higher reliability, lower costs in networking and support for interconnecting heterogeneous equipment /1/.

In order to support the requirements of standard communication features in heterogeneous environment and for different applications tremendous research and standardization efforts can be witnessed in the past years.

The Oxford dictionary defines a standard as: "a thing serving as basis for <u>comparison</u>".

Very often people involved in data communication are faced with statements like: "- this product is a 99%, X25 standard interface - or - that protocol is a standard teletex protocol with some new options - or - the network architecture is similar to the standard ISO-OSI reference model".

A closer analysis of these statements based on <u>comparison</u> of a certain product with a certain standard shows that the benefit of the standardization for the user is null. The scope of standards in data communication is a rigid scope, and can not be a comparison but a conformity issue, which has to be provided in order to provide interconnection of heterogeneous products.

Since standards are vital in the field of data communication a closer look at the different standard producers and their output will ease the understanding of their impact in the next future in this field.

We may distinguish three different types of standards:

- standards which are issued by organizations which are created on a national and international level to represent the interest of both users and suppliers (ISO, AFNOR, DIN, ANSI, BSI, etc.).

- pre-standards which are created in technical committees representing a certain group of or ganizations with a common interest (ECMA, CCITT, NBS etc.).

- de-facto standards which are the result of a wide dessimination of a certain product and usually are the result of a certain good solution for a market gap or a good vending strategy of a certain manufacturer (DIX, IBM etc.).

Certainly, there are interrelations and influences between these aspects of the standardization work and a closer analysis will help the understanding of the technical, political and commercial implications of the standardization efforts.

The first part of this paper gives an overview of the different organizations and some comments on their style of work and implications as seen by the author.

Part two gives a short overview of the main characteristics of the actual networks and highways under standardization, and on the actual stage in their development.

Part three, which is based on a cross-reference between the different standardization documents outlines some gaps and necessary items to be addressed in the next future.

2. STANDARDIZATION ORGANIZATIONS

The need for standards in data communication which are world-wide recognized is everywhere stated in terms which range from firm belief to passionate commitment /2/.

Two aspects have to be taken into consideration:

- the conflict between the user needs and the manufacturer marketing policies,

- the time factor, which gives to a standard its practical value.

We will try to give an overview of the standardization bodies considering the above two aspects.

2.1 International Standard Organization (ISO)

The representation at ISO is from both users and manufacturers through the national standardization bodies. There is a formal liaison between ISO and other associations interested in certain common items of work.

Since the representation of interests is very wide, usually, an indigestible flood of contributions on technical and political decisions slows down the progress towards a standard at the right time.

On the other hand, to achieve consensus of all interested parties in the world at every stage in the development of a standard is a Sisyphus enterprize.

The result of this policy may be outlined by reading, for example, the 1980 report of ISO/TC97 which lists 129 projects in the field of information technology standards. 117 projects out of 129 had no target date shown - the remaining 12 had all missed their target /2/.

In each member body of the ISO, representing the national standardization organizations, the internal ISO structure in subcommittees, working groups etc. is usually replicated; i.e. to produce an international standard first a national consensus has to be established (task which usually is difficult enough) in order to be able to start the process of getting an international agreement with other member bodies and liaisons.

The result of this <u>iterational</u> process and other constraints imposed by the ISO rules (mailing, translations, fluctuation of experts etc) is that ISO produces standards which have a usual timing range between late to too late /2/.

Nevertheless, ISO is probably the only body which is intitled to name its output __standards__ with international relevance.

In the field of data communications which is covered by the ISO/TC97/SC16 and SC6 even if some of the above considerations are also true, the situation is probably better.

This is motivated by the heavy engagement of the specialists involved in this subcommittee. The output of this subcommittee is the Open Systems Interconnection reference model and the associated work on the definition of standards for services and protocols for different layers within the model. The model has two main objectives:

- to provide a logical framework to bring order into a very complex domain,

- to separate functions characterizing a certain layer in order to provide independence between standards at different layers.

The reference character of the ISO-OSI model has and will have a dominant role in the future of standards and products in data communication since it imposes a rigorous discipline for the development of products and systems to achieve compatibility.

2.2 C.C.I.T.T.

It represents the association of common communication carriers and national PTTs administrations. They are mainly involved in producing standards for long-haul networks.

CCITT issues recommandations which have a de-facto standard character in the common communication carriers community and a pre-standard character for ISO.

Historically, CCITT was involved in the lower layers of the ISO-OSI reference model but they are increasingly concerned with the provisions for information processing services in the upper layers and their integration.

Technically, there is a good communication between CCITT and ISO since CCITT had recognized the implications of the OSI reference model.

The policy of CCITT in order to produce recommandations in a short period of time is the use of top specialists in a certain restricted area of work. Since the consensus has to be found between a well defined group of "travelling standardizers" concentrating on an answer to a well defined question, the CCITT standardization effort may be characterized as fast and effective. Another aspect which helps towards their effectiveness is, probably, the common philosophical understanding that the main quality of a standard does not reside in being the ideal for all applications but in allowing communication at the right time for the majority of users.

The closed interest representation and the restricted area of aspects addressed by CCITT are materialized in a lot of valuable recommendations with de-facto standard character in public networks (X25, class 0,1 transport layer etc.)

2.3 European Computer Manufacterer Association (ECMA)

It represents most European computer manufacturers and some American representatives residing also in Europe.

ECMA is not a trade organization but tries to produce suppliers' standards for the European environment.

ECMA standards are in fact pre-standards which are mainly intended to the ISO work for further processing.

They are rapidly produced and usually influencial, even if their final standards are very often subject to change.

ECMA is a closed organization which gives the user the oportunity to have an observer status (i.e. user groups can not influence ECMA decisions since they have no vote).

The ECMA working groups (TC24,TC25) are deeply involved in the ISO-OSI working groups supporting the work on the technical level.

Two aspects have to be taken into consideration:

- the impact of an ECMA standard on the ISO work. Does ECMA have enough influence to promote its standards to a de-facto standard by European manufacturer implementations?

- the reaction of each specific European manufacturer to an ECMA standard versus an ISO standard since the financial effort for implementation is not negligeable /3/.

2.4 International Electrotechnical Commission (IEC)

This organization, even if affiliated to ISO, acts and feels independently. Its work in the field of information technology has been fairly limited. IEC and ISO have a common responsability in defining standard interfaces but the models on which these interfaces apply are contradictory. The IEC group involved in data communication is IEC65 which continues to work on its PROWAY instrumentation bus, dealing with the lower layers of their own model.

2.5 Institute of Electrical and Electronic Engineers (IEEE)

Its activity in data communication is mainly represented by the project 802 focussing on the definition of local area network (LAN) standards.

The project is supposed to define a LAN which will reflect the current available technologies. This work is in progress for about two years with a very ambitious and rigorous time schedule.

It tries to accomodate requirements coming up from different environments like real-time, digitized voice and office automation towards an integrated standard.

Since also political aspects have influenced this work the actual result is out of the time schedule and consists in a variety of standards and options which are often mutually exclusive. A practical question to the IEEE approach is: "Which benefit has a standard from the integration aspect of different mutually exclusive options and solutions?"

2.6 Activities supported by CEC in data communication

Recognizing the trade-off in data communication technology the CEC have started to support different activities in the framework of the multi-annual plan in informatics.

This support may be classified in:

- support at a managerial level (WGS) where coordination at European level between national standardization bodies is promoted,

- support at a technical level which is performed mainly in the COST-11bis projects and EWICS.

The COST-11bis program is intended to support multilateral technical cooperation and collaboration between European research institutions actively involved in their own data communication. The projects granted by COST-11bis focus on items of the OSI reference model which are studied and implemented in order to gain practical and theoretical draw-backs for the standardization work. The main areas outlined by the projects are: message systems, local area networks management and internetworking, distributed database systems, file access and transfer and satellite communication).

Some technical committees (TC5 and TC10) in the European Workshop for Industrial Computer Systems (EWICS) are also deeply involved in data communication standards for local area networks and real time applications.

Their main tasks are technical contributions to the international standardization bodies (ISO, IEC, IEEE) and developments of technical solutions for those items of works which are qualified as for further study in the program of work of the standardization bodies.

2.7 DEC-INTEL-XEROX (DIX)

The agreement reached between the three companies is of enormous importance for the market penetration and for the establishment of a de-facto standard.

Even if the solution proposed is not covering the entire range of requirements of all applications, at least for the next 5 years this de-facto standard will be predominant.

This is motivated by the fact that it stresses a cheap and easy product covering a large variety of applications for office automation and light industrial applications.

The DIX specification have been succesful in persuading IEEE, ECMA, NBS to adopt the ideas and mechanisms of this specification.

The solution has also influenced a lot of companies to buy licences (NIXDORF, OLIVETTI, SIEMENS, ZILOG, TEKTRONIX etc.) or other to provide network interfaces for most widely accepted host computers.

Also, probably, a major technical factor for promulgating this specification towards a de-facto standard is the anouncement of VLSI DIX-controller chips (AMD, FUJITSU, INTEL, MOSTEK, etc.).

2.8 National Bureau of Standards (NBS)

This organization is in charge of the production of a set of Federal Information Processing Standards (FIPS) for the American governmental organizations and other interested groups (NUA-network user association). An exception is the Department of Defence, probably for political and historical reasons.

In the field of data communication a lot of valuable contributions may be witnessed in the last four years.

The reports produced are detailed specifications of services and protocols of the different OSI layers.

They have a pre-standard status and are thought as an input for the ISO/OSI effort but also as a reference for the user implementations. The reports cover the whole area of data communication starting from LANs to satellite communication and from public networks to interconnection of networks of different technologies.

3. SOME TECHNICAL CHARACTERISTICS OF THE PROPOSED STANDARDS

Since the organizations described above have defined their main aim in standards for a certain environment or application field, a rough classification may be introduced depending on these aims.

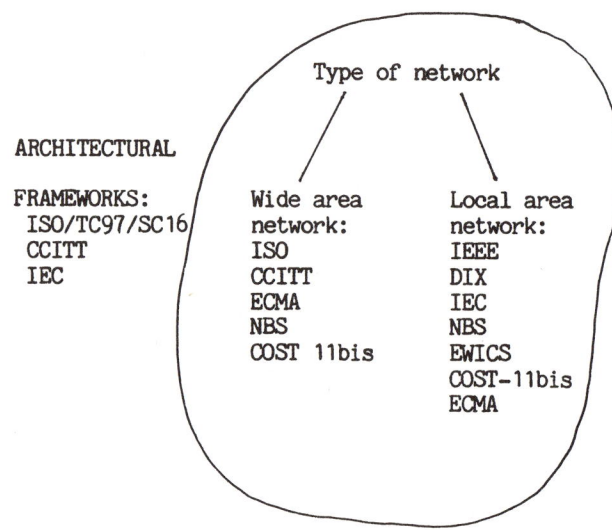

The differences between the specifications can be analyzed with respect to the lower layers (1-3) in the ISO-OSI architecture since proposals for standards in the layers above have been produced only by ISO, CCITT, ECMA and NBS and usually the result is a compromise between the different proposals towards a common standard.

Another argument for concentrating on the lower layers is the fact that by definition and probably in most cases in practice the upper layers will not be influenced by the adopted network technology.

3.1 Wide area networks

Primarily the work in ISO was influenced by the de-facto standard public networks introduced by the common communication carriers. The standardization work in the lower layers was influenced by the fact that public X25 networks are available and operational in different countries (DATAPAC, TRANSPAC, DATEX-P, DN-1 etc.).

These are meshed networks spanning over large distances and may be characterized by a point-to-point full duplex access mechanism between host and network. Layer 1 is coered by the CCITT recommendation X21bis for a bit serial full duplex transmission. Layer 2 is a standard HDLC-LAP B procedure. The usual throughput is max 48 kB/sec and flow-control, error detection and correction are provided in layer 2 based on a 16 bit CRC.

The services provided at layer 3 are connection oriented.

3.2 Local area networks

3.2.1 PROWAY (IEC)

The PROWAY standard is oriented to the process control and real time environment. The whole system architecture has not been finalized and the actual results concentrate in the specification of the coupler and the path unit which encompass the first two OSI layers from the point of view of functionality.

The PROWAY is designed for distances up to 2km as a passive multidrop bus system interconnecting up to 100 stations. The transamission media has a throughput of 1 to 2 MB/sec over a baseband coaxial cable.

It uses at the link level the standard HDLC frame with a 16 bit CRC. It provides a deterministic, guaranteed access time to the media (2ms). An important characteristic is the fact that it provides a priority mechanism for expeditious communication and a jabber control (watch-dog timer) is used for iterruption of a station which transmits longer than it is allowed.

The prefered access mechanism is token passing and three types of services are provided to the higher levels:

- datagram (send and pray),
- send and acknowledgement of delivery,
- send and acknowledgement of receiving.

3.2.2 ETHERNET (DIX)

The main area of applications addressed by the DIX specification is office automation. The network is based on a passive bus (coax-cable) with a data rate up to 10 MB/sec.

The access method is non-deterministic and based on the carrier sense multiple access with collision detection (CSMA/CD).

A trunk spans over 500m distance which can be enhanced by repeaters between trunks up to 2500m.

The DIX collision technique is a compromise, as there is a speed-distance limitation, i.e. an interdepency between minimum packet length maximum transmission distance and data bit rate. Any change of one of the parameters will influence the overall system function /4/.

It covers the layers 1 and 2 of the ISO-OSI architecture providing error detection using a 32 bit CRC field.

Error recovery and flow control is left for the upper layers called client layers.

The type of services provided are unconfirmed datagrams.

The maximum configuration of a network (including repeaters between trunks) comprises 1024 stations.

The addressing capabilities provided are station to station and multicast.

3.2.3 IEEE 802

The intended network standard has probably the widest set of functional requirements attempting to provide:

- conformance with ISO/OSI model
- no limit in distances and number of connected stations,
- no intermediate nodes,
- support of different network technologies (baseband and broadband),
- allow datagram and peer acknowledged communication at link level.
- support for individual, multicast and broadcast addressing.

The network will provide different data rates from 1 to 40 MB/sec in baseband and broadband technology.

Many options are provided and most of them are mutually exclusive but the services provided by the Logical Link Control (LLC) will isolate these differences from the higher layers.

The access mechanism is CSMA/CD for the bus topology but also token passing is supported for bus and ring topologies. The types of services provided are connection and connectionless oriented.

The frame format of IEEE 802 is similar but not identical to the DIX frame.

The area of application is indended to cover commercial (office automation) using the CSMA/CD option and industrial (real time) applications using the token access technique (to a bus or ring topology).

4. REMARKS AND CONCLUSIONS

Table 1 gives an overview of the main characteristics of the presented networks:

	WIDE AREA NETWORKS ISO-CCITT	PROWAY IEC	DIX	IEEE 802
STATUS	100%	40%	100%	50%
DISTANCE	unlimited	2km	500m,...2500m with repeters	depends on options
TOPOLOGY	meshed	bus	bus	bus and ring
REF. TO OSI MODEL	1 to 3	1-2	1-2	1-2
TRANSMISSION MEDIA	full duplex bit serial	passive coax	passive coax	different base+broadband
RATE	max. 48 Kb/sec	1-2 MB/sec	10 MB/sec	1-40 MB/sec
ACCESS MECHANISM (L2)	multi-master	multi-master token passing	multi-master CSMA/CD	multi-master CSMA/CD and token passing
ERROR DET. RECOVERY	yes 16 bit CRC	both 16 bit CRC	only error det. 32 bit CRC	depends on opt. 32 bit CRC
FRAME FORMAT	standard HDLC	standard HDLC	DIX own	IEEE own
TYPE OF SERVICE	connection oriented	connectionless three options	connectionless	connection and connectionless
PRODUCTS & PRICES	available medium	not available	available low	not available

Table 1

By an analysis of the status in the development of the different proposals one may conclude that there is room and need for harmonization of the different approaches.

A major factor in this process is probably not a formal liaison but an intensive coordination and collaboration work between the different organizations (a good example is the ECMA-NBS work or those specialists attending actively technical meetings in different standardization bodies).

The required harmonization addresses the different models and terminologies (ISO,IEC,IEEE).

But also there is a need of a common approach in the definition of the types of services provided by the different approaches at the boundary of the lower layers. This is a crucial issue in order to provide isolation of functions and layers. If a layer does not provide a certain function this has to be done in the layers above which makes one of the main objectives of the ISO-OSI reference model - independence of the layers - obsolete.

Even if the standardization work has already started on issues like connectionless services, a rapid integration in the model is required.

This is not only motivated by the fast developments in the local area networks but also by the developments in the satellite communications.

Inside a layer, for the same type of applications and the same technology and access mechanism, we think that there is no technical reason for having different solutions, frame formats etc. (e.g. CSMA/CD in DIX and IEEE), if cheap products for large dissemination is one of the scope of the standardization work.

A major technical issue in the next future is the interconnection of networks of different technologies. A special attention has to be paid to the definition and to the architectural implications in the network layer in all standardization bodies (proposals in this area have been already made by ECMA, NBS and COST-11bis project on LAN).

For a wide acceptance of networks other important issues for network operation are provisions for network management.

These aspects have architectural and technical implications on the present proposed standards. Since the conformity aspect is vital in the interconnection of heterogeneous devices, the compliance of a certain product to a certain standard (services and protocols) will lead to requirements of certification procedures associated to each standard.

The standardization process in the data communication field is a complex and iterative process. Since the technological developments in this area are very fast an imperative factor in this process is the time.

The aim of having a right standard at the right time can be only achieved by an intensive, close cooperation between all involved organizations using probably new procedures in the standardization process.

5. REFERENCES

/1/ Mier, Edwin E., High level protocols, and the OSI reference model, Data Communications, July 82.

/2/ O'Connor, R.M., Information technology standards: Priorities for Europe, CEC Report WGSN70 1981

/3/ Lenzini, L., Popescu-Zeletin, R., and Vissers, C., State of the art study on standardization of level 4 of the ISO-OSI reference model, CEC Report 1981

/4/ Klessmann, H., Local area networks for distributed intelligent systems, IAEA Advisory Group Meeting 1982 IAEA Austria.

IMPLEMENTATION OF A POWERFUL LOCAL AREA NETWORK
ON AN FIBER OPTIC LOOP

IMPLEMENTATION EINES LEISTUNGSFÄHIGEN DATENNETZES
MIT RINGSTRUKTUR AUF EINEM OPTISCHEN MEDIUM

E. Querasser, M. Lindner, H. Preineder, F. Buschbeck
Austrian Research Centre Seibersdorf Ltd
Lenaugasse 10, A 1082 WIEN, AUSTRIA

This contribution describes the implementation of a LOCAL AREA NETWORK in the AUSTRIAN RESEARCH CENTRE SEIBERSDORF. This future oriented Implementaton differs from other available Local Area Networks especially in the following aspects:

- timeslotted system
- defined transmissiondelay
- no Store and Forward
- no Local Area Network specific protocol
- optical medium in a loop configuration

This Local Area Network has a capacity of about 500 full duplex connections (channels), each working up to 9600 bit/s. If higher datarates are desired, the number of channels decreases proportional. The Network is CONNECTIONORIENTED and transparent to any transmissionprotocol.

Der Beitrag beschreibt die Realisierung eines lokalen Datennetzes zur Kopplung der örtlich verteilten Prozessdatenverarbeitungsanlagen untereinander und mit einer zentralen Datenverarbeitungsanlage im Österreichischen Forschungszentrum Seibersdorf. Diese zukunftsorientierte Implementierung unterscheidet sich von anderen Datennetzen besonders durch

- Timeslotted System
- definierte Übertragungsverzögerung
- kein Store and Forward
- Protokolltransparenz
- Verwendung eines Lichtleiters als Übertragungsmedium in Ringstruktur

Das Netz hat eine Kapazität von etwa 500 gleichzeitig aktiven, bidirektionalen Datenkanälen mit einer Datenrate von je 9600 bit/s. Wenn höhere Datenraten gefordert werden verringert sich die Anzahl der Kanäle proportional. Das Netz ist VERBINDUNGS-ORIENTIERT und protokolltransparent.

Cet article décrit la realisation d'un Réseau Local au Centre de Recherche Seibersdorf en Autriche. Les caractéristiques particulierés à ce réxau seut les suvantes:

- multiplexage temporel
- délai de transmission spécifié
- aucun stockage intermédiaire
- aucun protocole spécifique aux reseaux locaux
- boucle optique

Ce réseau local a une capacité de 500 canaux full-duplex de 9600 bits/s chacun. On pent obtenin des dibits plus élevés en réduisant le nombre de canaux en proportion. Ce réseau est orienté Connexion et transparent aux protocoles de transmission.

1. PROBLEMSTELLUNG

Die fortschreitende Entwicklung der Prozessdatenverarbeitung zu vielen kleinen, lokal verteilten Systemen mit gemeinsamen zentralen Datenbanken und die Notwendigkeit, auch mit weit verstreuter Peripherie zu kommunizieren, führte schon vor einigen Jahren zu nahezu unlösbaren Verkabelungsproblemen. Mit der bis dahin üblichen, individuellen Verkabelungstechnik konnte die gerade in einem Forschungszentrum wichtige Flexibilität nicht befriedigend gewährleistet werden. Aufgrund dieser Erkenntniss wurde die Entwicklung eine Datennetzes in Angriff genommen.

Ausgehend von der Struktur und Verteilung der Datenverarbeitungsanlagen im ÖFZS wurden folgende Anforderungen an ein Datennetz formuliert:

a) Standardschnittstellen an den Geräteinterfaces (z.B.: V24 asynchron)

b) Hohe Flexibilität und leichte Konfigurierbarkeit

c) - Implementation von Standleitungen, die keiner Setup-Procedur bedürfen.
   - Möglichkeit eines raschen, freien Verbindungsaufbaues ohne Unterbrechung des Netzbetriebes für temporäre Verbindungen

d) - Viele verfügbare Datenkanäle mit mittlerer Datenrate
   - Implementierbarkeit von High Speed Kanälen
   - Kein Store and Forward Betrieb

e) Frei wählbare Datenrate an den Geräteschnittstellen

f) Protokolltransparenz
   jedes end to end Protokoll soll im Datenring implementierbar sein

g) - Optisches Medium (galvanische Entkopplung)
   - aber auch Möglichkeit ein anderes Medium einsetzen zu können (auch gemischter Betrieb)

h) Fehlererkennung, Fehlerortung und Diagnose bis zum Geräteinterface, sowie zentrale Dokumentation aller Fehler und Betriebsstörungen

i) Hohe Zuverlässigkeit und kleine Restfehlerrate

j) Steuerung der Network Setup Prozedur
   - durch zentralen Vermittlungsrechner
   - durch intelligente Geräteinterfaces
   - manuell von jeder Station aus

2. TECHNISCHE REALISATION

2.1 Struktur der ÖFZS Datennetzes

Aufgrund der guten Erfahrungen mit seriellen Systemen und den am Markt erhältlichen Fiber Optic Systemen wurde eine ringförmige Struktur gewählt(Abb.1). Hierbei ergeben sich, verglichen mit anderen Konfigurationen (z.B. : Sternkonfigurationen), die kürzesten Verkabelungslängen. Ebenso ergibt sich dadurch eine leichte Erweiterbarkeit der Konfiguration (Einfügen von zusätzlichen Stationen).

Die Information wird am Datenring im Übertragungsmedium in einem Zeitmultiplexverfahren mit festem Rahmenformat (Abb.2) übertragen. Die Datenrate am Medium beträgt 6.144 Mbit/s. Ein Rahmen besteht aus 512 Timeslots mit einer Breite von je 10 bit. Dieser Rahmen wird pausenlos sequentiell übertragen, d.h. nach Slot 511 folgt unmittelbar Slot 0.

Jeder dieser Slots ist geeignet, einen auf Grund der Ringstruktur bidirektionellen, fullduplexfähigen Kanal mit maximal 9600 bit/s zu bedienen.

Einem logischen Kanal wird im Falle einer Standverbindung fix, ansonsten temporär ein (oder mehrere) Slot zugewiesen. Von den 512 im Rahmen enthaltenen Slots werden 18 für die Steuerung und Überwachung des Ringes verwendet (3 zur Synchronisation und 15 für Ringmanagement, Verbindungsauf-und -abbau, Errordocumentation,...), die restlichen 494 Slots stehen zur Datenübertragung - volltransparent - zur Verfügung.

Die Zuordnung der Interfaceports zu einem freien Slot erfolgt:

- im Falle der Standverbindung durch starre Zuordnung über einen PROM-Baustein lokal in jeder Station. (z.B.: um eine Standleitung zwischen Station 1 Interface 16 und Station 4 Interface 3 zu definieren, muß in Station 1 ein (oder mehrere) Slot (z.B.: Slot 37) dem Interface 16 und in Station 4 derselbe Slot (37) dem Interface 3 zugeordnet werden.

- im Falle der temporären Verbindung wird diese Zuordnung durch Zusammenspiel der lokalen und zentralen Ver-

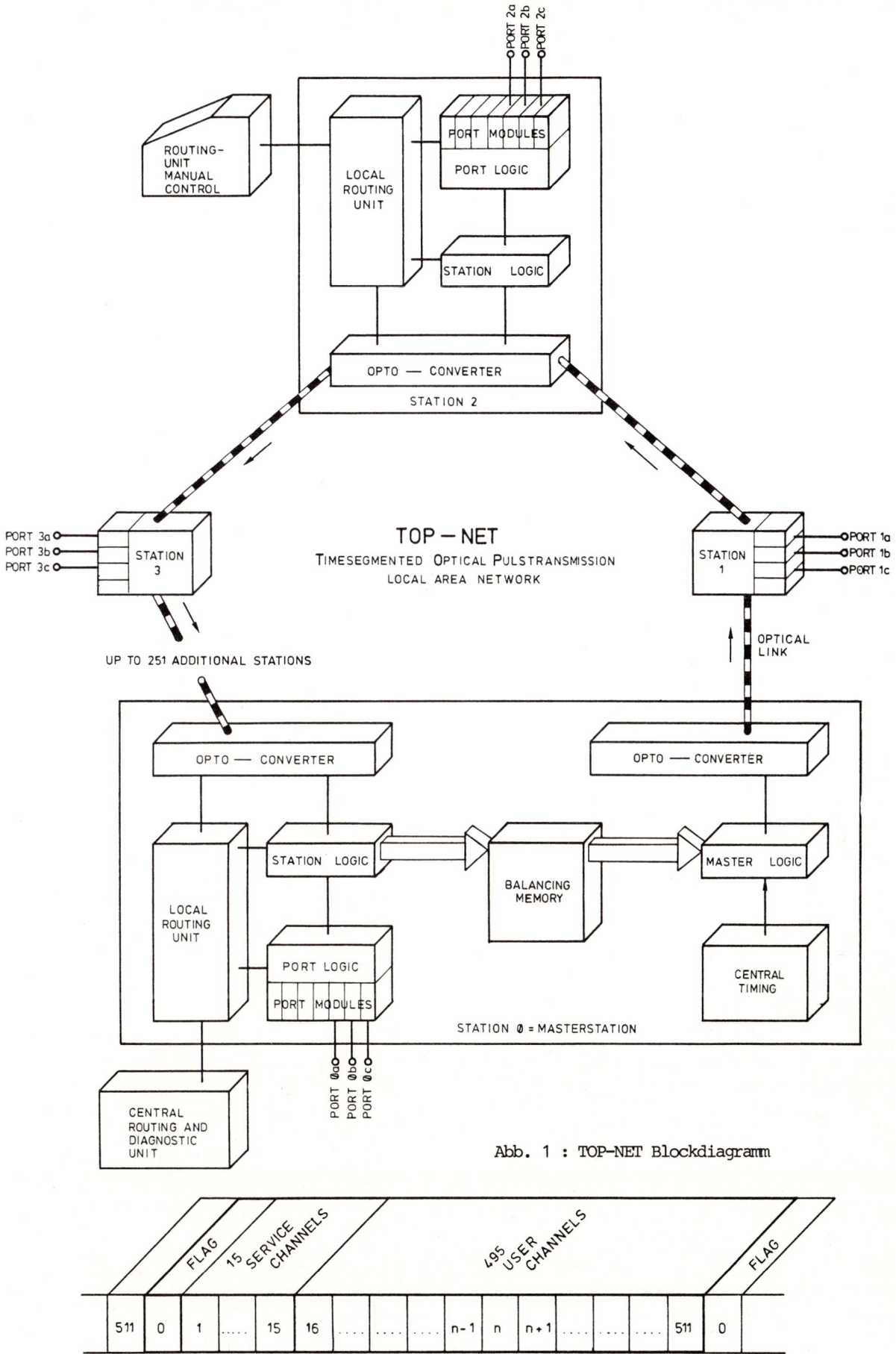

Abb. 1 : TOP-NET Blockdiagramm

Abb. 2: Rahmenformat

mittlungseinheiten durch einen Vermittlungsrechner hergestellt. Hier kann aber auch symbolisch adressiert werden. (z.B.: verbinde DRUCKER1 mit RECHNER5) Die Eingabe der Verbindungsanforderung kann entweder manuell über eine Eingabeeinheit oder über ein intelligentes Interfaceport (protokollgesteuert) oder vom zentralen Vermittlungsrechner aus gestellt werden.

Die maximale Übertragungszeit für ein 10 bit- Wort beträgt etwa 1ms, und entspricht der Umlaufzeit eines Rahmens.

Die Darstellung des Rahmens am Medium ist stark von diesem abhängig. Im Falle des Lichtleiters wird zur Kodierung der Rahmendaten der Biphase-Code verwendet. Dadurch wird die Rückgewinnung des Bittaktes stark vereinfacht und die Möglichkeiten der Fehlerüberwachung erweitert. Die für das Connection-Setup benötigte Zeit ist stark von der Art der Verbindung und von der gewünschten Dokumentation und Validitätsprüfung abhängig. Im einfachsten Fall, einer Point to Point Verbindung ohne Dokumentations- und Prüfungsanforderungen, kann mit Aufbauzeiten zwischen 4 und 10 ms gerechnet werden.

2.2 Synchronisation

Zur Synchronisation des Rahmens wird Slot 0 herangezogen. In diesem Slot, wird ständig ein spezielles Bitmuster übertragen. Jede Station vergleicht dieses Muster auf 8 bit Länge mit einer Referenz. Tritt keine Übereinstimmung zum erwarteten Zeitpunkt ein, so liegt mit Sicherheit ein Fehler in der Übertragung oder in der Stationslogik vor.

In diesem Fall wird sofort jede weitere Datenübertragung zu den Kanalinterfaces unterbunden und dort der Fehler "unsynchron" angezeigt. Zur Neusynchronisierung wird ebenfalls dieser spezielle Slot herangezogen. Diese Resynchronisierung nach einem Fehler wird von der zentralen Masterstation gesteuert. Diese gibt danach den Ring wieder zur Datenübertragung frei.

2.3 Folgerungen aus den Anforderungen an das Datennetz:

a) Standardschnittstellen an den Geräteinterfaces :

Alle Schnittstellen zu den Geräten sollen gebräuchliche Standardschnittstellen sein. Es sollen möglichst viele Standards implementierbar sein.

Daraus ergibt sich die Forderung nach einem modularen Aufbau der Stationen. Für die Schnittstelle zu den einzelnen Interfacemodulen ist ein einfacher interner Standard Bus vorhanden.

Durch die Definition dieses 'Zwischenstandards' ist es möglich während der Übertragung eine Konvertierung des Datenformats vorzunehmen (z. B.: Verbindung von einem Gerät mit V24- Interface zu einem Gerät mit 8 bit Paralellinterface) ohne spezielle Konvertierungshard- und software. Dies ist jedoch nur soweit möglich, als nicht auch eine Protokollkonvertierung erforderlich ist.

In der ersten Ausbaustufe sind folgende Standardinterfacemodule verfügbar:

V24/RS232C asynchron bis 19.2 kbit/s
8 bit paralell I/O
X.25   bis 32 logische Kanäle

In weiterer Folge ist an die Implementierung von IEEE488/IEC625 und Sprachübertragungsinterfaces gedacht.

Ebenso ist die Verwendung von applikationsabhängigen Interfaces möglich (Analogsignalübertragung,...).

b) Hohe Flexibilität und leichte Konfigurierbarkeit

Durch die Wahl eines extrem modularen Aufbaues kann die Flexibilität nicht nur zu den Geräteinterfaces, sondern auch gegenüber dem Medium gewährleistet werden. Es ist somit durchaus möglich innerhalb des Ringes andere Medien bzw. andere Codierungsmethoden zu verwenden.

256 Standverbindungen sind derzeit nicht in das Vermittlungssystem eingebunden, wodurch sich deren Setup-Proceduren nach Power-On erübrigen.

Durch das intelligente Vermittlungssystem besteht auch bezüglich der Arten von Verbindungen hohe Flexibilität (Point to Point, Broadcast, User defined, ...).

Weitere Stationen können im Falle einer Konfigurationsänderung ohne lange Betriebsunterbrechung eingefügt werden. Bei der zusätzlichen

Instalation eines Interfaces in einer bereits bestehenden Station ist mit keiner Betriebsunterbrechung für die anderen Kanäle zu rechnen.

c) Implementierung von Standleitungen und temporären Verbindungen

Bei der Abschätzung des Bedarfs an Konfigurationsänderungen bzw. der Häufigkeit der Modifikation der Verbindung stellte sich heraus:

- etwa 50% des Verbindungsbedarfs sind reine Standverbindungen und bedürfen nur sehr seltener Konfigurationsänderungen, erfordern dafür aber eine hohe Verfügbarkeit (Rechnerkopplungen, entfernte, aber fix zugeordnete Peripherie, .. .)

- etwa ebensoviele Verbindungen benötigen häufig Modifikationen. In den meisten Fällen benötigt man diese Verbindungen trotzdem über längere Zeiträume (Terminals, Meßanordnungen,...).

- Nur in sehr wenigen Fällen wird ein betrieblich rasch wechselnder Verbindungsweg benötigt (X.25, manche Prozessteuerungselemente).

Es liegt also auf der Hand, bei der Auslegung des Netzes auf Standleitungen und länger bestehende Verbindungen (hier spielt die Zeit des Verbindungsaufbaues keine Rolle, aber es ist meist eine Protokollierung und Validitätsüberprüfung erforderlich) besonderes Augenmerk zu legen.

d) Viele verfügbare Datenkanäle mit mittlerer Datenrate

Bei der Wahl einer Standarddatenrate (sie stellt die maximale Datenrate für die Verwendung eines einzelnen Slots dar) wurde von der im ÖFZS häufig verwendeten Peripherie ausgegeangen. Dabei ergab sich, daß nahezu 90% der Verwendeten Geräte mit Datenraten bis maximal 9600 bit/s arbeiten. Für die Installation des Ringes wurde daher eine Konfiguration 512 Slots zu 9600 bit/s gewählt. Für die schnelleren Geräte können entsprechend mehr Slots zusammengefaßt werden.
Grundsätzlich sind auch Broadcast Übertragungen möglich, aber derzeit nicht implementiert.

e) Frei wählbare Datenrate an den Geräteschnittstellen:

Die Datenrate mit der günstigsten Kanalauslastung ist durch die Standarddatenrate von 9600 bit/s beziehungsweise einem ganzzahligen Vielfachem davon gegeben. Es lassen sich jedoch auch alle anderen (auch nichtgenormte) Datenraten bis zu 960 kbit/s realisiern. Für langsamere Datenraten besteht keine Begrenzung (es können auch statische Signale übertragen werden) Die Verwendung bzw. Zuweisung von mehreren Slots zu einem Datenkanal erfordert keine Änderung der Interface Hard- oder Software, da sie ausschließlich vom Vermittlungssystem bewerkstelligt wird. Eine Interfacebaugruppe kann daher für ein weites Spektrum an Datenraten verwendet werden.

f) Protokolltransparenz:

Die meisten am Markt befindlichen lokalen Netzwerke verwenden an den User-Schnittstellen ein vorgegebenes, netzwerkspezifisches Protokoll. Es ist daher oft mit großen Schwierigkeiten verbunden auf so einem Netz ein "fremdes Protokoll" zu verwenden.

Ein weiterer Punkt ist dabei das Problem der Uebertragungsverzögerung. Die Mehrzahl der bekannten Netze können keine maximale Zeit hierfür angeben.

Im Gegensatz zum beschriebenen System werden in den herkömmlichen Systemen die Daten blockweise übertragen. Daraus folgt, daß mit der Datenübertragung in vielen Fällen erst begonnen werden kann, wenn die ganze Informationseinheit übergeben wurde das heißt nach einer von der Blocklänge und Übertragungsgeschwindigkeit abhängigen Zeit (z.B.: bei 9600 bit/s kann mit der Übertragung eines 128 byte Blockes im Netz erst nach 133 ms begonnen werden. Unter Vernachlässigung der Übertragungsverzögerung und des Protokolloverheads steht der Datenblock beim empfangenden System erst nach weiteren 133 ms vollständig zur Verfügung.

Diese Tatsachen führten zu dem Verlangen nach:

- Kein Store and Forward:
  Es soll keine zusätzliche Verzögerung bei der Übertragung entstehen

- Protokolltransparenz:
  Es soll in dem Datennetz jedes übliche Netzprotokoll implementiert werden können.

In der beschriebenen Implementation verhält sich das Netz nach Verbindungsaufbau wie eine getaktete, vollduplexfähige Leitung, daher kann darauf jedes Protokoll abgewickelt werden.

Unter Einbeziehung des intelligenten Vermittlungssystems und spezieller Interfacemodule sind auch Netzprotokolle wie z.B.: PROWAY, P802, ETHERNET-type, .. Implementierbar bzw. können simuliert werden. Es können daher auch mehrere gänzlich unterschiedliche Protokolle gleichzeitig im Datenring betrieben werden, ohne sich gegenseitig zu beeinflussen. Ebenso können daneben z.B.: V24-Terminalverbindungen - ohne jegliches Protokoll - betrieben werden.

g) Optisches Medium

Für die Verwendung eines optischen Mediums sprechen vor allem:

- Unabhängigkeit von der elektrischen Umgebung des Mediums

    Ein Lichtleiter ist unempfindlich gegen Einstreuungen elektrischer oder magnetischer Natur. Es muß daher darauf keine besondere Rücksicht bei der Wahl des Verkabelungsweges genommen werden.

    Da die Fehlerrate in einem optischen Medium mit der Fehlerrate eines abgeschlossenen Logiksystems vergleichbar ist, verringert sich der Aufwand und somit auch der Overhead für die Überwawachung und Sicherung der übertragenen Daten auf Leitungsebene.

- Elektrische Entkoppelung der Stationen und galvanische Trennung der Geräteinterfaces. Auch bei Verbindungen von einem Gebäude zum nächsten müssen keine Blitzschutzmaßnahmen getroffen werden.

- hohe Datenrate über große Entfernungen ohne Verwendung von Trägerfrequenz.

- moderne, zukunftsorientierte Technologie

- Trotz der Festlegung auf ein optisches Übertragungsmedium soll die Möglichkeit der Verwendung jedes anderen Mediums mit genügender Bandbreite nicht ausgeschlossen sein. Darüberhinaus wird auch noch die Verwendung unterschiedlicher Medien in einer Konfiguration unterstützt.

h) Fehlererkennung, Fehlerortung und Dokumentation:

Aufgrund der räumlichen Ausdehnung des Netzes wurde eine zuverlässige Fehlererkennung, -ortung und -dokumentation sowie Diagnosemöglichkeiten für unumgänglich erachtet. Die Dokumentation und Registrierung der Fehler und Betriebsstörungen erfolgt zentral.

Es konnten 4 verschiedene, grundsätzliche Fehlerursachen identifiziert werden:

- Ausfall des Mediums bzw. dem Medium zugeordneter optischer und elektrischer Komponenten

- Ausfall, bzw Störung der Stationslogik

- Fehler die das Vermittlungssystem betreffen (keine Beeinträchtigung der Standverbindungen)

- Fehler in den Interfacemodulen

Daraus resultieren die folgenden Maßnahmen:

- Bei Fehlern am Medium:
  Die in Übertragungsrichtung nächste Station am Ring erkennt den Ausfall und generiert eine Fehlermeldung an die Masterstation. Dazu muß jede Station fähig sein, selbstständig einen Rahmen zu generieren.

- Ausfall einer Station (z.B.: Ausfall der Synchronisation):
  Die entsprechende Station wird auf leitungsnaher Ebene überbrückt. Das Fehlen dieser Station wird durch regelmäßiges Polling des Vemittlungsystems erkannt und gemeldet.

- Fehler im lokalen Teil des Vermittlungssystems (Erkannt z.B.: durch "watchdog"):
  Die jeweilige Einheit versucht einen geordneten Restart und danach eine Fehlermeldung an die Masterstation abzusenden. Läßt sich der Fehler so nicht beheben, so versetzt sich die lokale Vermittlungssystemeinheit in einen passiven Zustand. Das logische Fehlen der Einheit wird durch zyklisches Polling des Masters erkannt und gemeldet.

- Fehler in den Interfacemodulen können bei Verwendung intelligenter Module über das Vermitt-

lungssystem gemeldet und protokolliert werden.
Ebenso können hier eventuell Maßnahmen zur Eliminierung des fehlerhaften Interfaces getroffen werden, sowie Diagnoseinformationen erfaßt werden.

Zu Servicezwecken können auch über das Vermittlungssystem einige Fehler simuliert und Diagnoseabläufe ausgelöst werden.

i) Hohe Zuverlässigkeit und kleine Restfehlerrate:

Die geforderte Zuverlässigkeit wird durch die optische Datenübertragung die zentrale Fehlererfassung und durch Verwendung moderner Technologien (z.B.: PAL- Bausteine) und die damit verbundene Reduktion der Bauteilanzahl angestrebt.
Die Fehlereinstreuungen in ein optisches Übertragungsmedium ist im Vergleich zu elektrischen Übertragungsleitungen vernachlässigbar klein, da keine elektrischen oder magnetischen Einflüsse zu Fehlern führen können.
Durch die Verwendung des Biphasecodes in Verbindung mit der Synchronisationsmethode erreicht man schon an der Schnittstelle zu den Geräteinterfacemodulen eine Hammingdistanz von 2. Diese läßt sich durch Einführen eines Paritätsbits auf dieser Ebene noch auf 4 erhöhen. (Auf Grund des Biphasecodes und der Synchronisation wird schon bei der Dekodierung des Rahmens jede ungerade Anzahl an Fehlern pro Rahmen erkannt. 2 Fehler auf dieser Ebene können aber nur höchstens einen Fehler in einem Datenwort eines Kanals erzeugen.)

Bei der seit etwa 2 Jahren im ÖFZS installierten Version des Ringes konnte während des nahezu kontinuierlichen Betriebes bis jetzt kein auf die Ringfunktion oder das Medium zurückzuführender Fehler registriert werden. (Ein Datenkanal wurde dauernd betrieben und überwacht)

3. REFERENZEN

(1) Hewlett Packard, Application Note 1000: Digital Data Transmission With the HP Fiber Optic Systems (11/78)

(2) Digital Equipment Corp, Intel Corp, Xerox Corp, The Ethernet, Version1.0 (September 30, 1980)

(3) IEEE 802 Local Network Standard, Draft B (Oct.19, 1981)

A Q-BUS INTERFACE AND AN RSX-11/M DRIVER FOR THE
LOCAL NETWORK DANUBE

V. Tschammer, K. Emmelmann, W. Wawer
Hahn-Meitner-Institut für Kernforschung Berlin GmbH, Berlin (West)

ABSTRACT

The paper describes a Q-Bus interface an an I/O driver for the DBC-100 DANUBE controller, being the basic equipment for the connection of our standard station, an LSI-11/23 with an RSX-11/M operating system, to the local area network. The DANUBE network at the HMI Berlin is used to support various types of investigations and developments, most of them being concerned with real time applications of local area networks and with the lower four layers of the OSI architecture.

RÉSUMÉ

Cet article décrit une interface Q-Bus et un gérant d'entrée-sortie pour le controlêur DBC-100 du réseau Danube. Ces éléments constituent l'équipement essentiel pour la connexion de notre station standard, un LSI-11/23 sous RSX-11M, au réseau local. Le réseau Danube est utilisé au HMI de Berlin conitue support d'experiences, la plupart d'entre elles étant de type temps réel et portant sur les quatre niveaux inférieurs du modéle de référence ISO-OSI.

ZUSAMMENFASSUNG

In dieser Arbeit wird als Grundausrüstung zum Verbinden unserer Standardstation - einem LSI-11/23 mit Betriebyssystem RSX-11/M - mit dem Ortsbereichsnetz eine Q-Bus-Schnittstelle und ein E/A-Treiber für die DANUBE-Steuereinheit DBC-100 beschrieben. Das DANUBE-Netz in HMI Berlin wird zur Unterstützung verschiedenartiger Untersuchungen und Entwicklungen benutzt, die meistens Echtzeitanwendungen von Ortsbereichsnetzen und die unteren vier Schichten der OSI-Architektur betreffen.

1. Introduction

The DANUBE local area network is a development of the project KAYAK at INRIA, France. It has been installed at the HMI Berlin to support various types of investigations and developments, most of them being concerned with real time applications of local area networks and with the lower four layers of the OSI architecture. The DANUBE hardware is fairly well documented, commercially available and has been easily adapted to our different problems.

2. DANUBE Network and Controller

DANUBE is an ETHERNET-type local area network /1/, which uses a coaxial cable as the transmission medium to connect up to 255 stations in a bus configuration. The network may be extended to a maximum length of 1 km and offers a transmission rate of up to 2 Mbits/sec. The access of the stations to the common bus is accomplished by CSMA/CD algorithms.

The logical architecture of DANUBE conforms to the ISO Reference Model for Open Systems Interconnection (cf. figure 1). The functions of the Physical and the Data Link Layer are covered by a DANUBE controller (DBC-100), which is available from Bertin & Cie., Aix-en-Provence.

In general, such a controller is required to provide a transmission medium independent communication facility to the higher layers, often collectively referred to as the "Client Layer".

The main tasks of the DBC-100 are the encapsulation/decapsulation of transmit and receive data, the link management for channel allocation and contention resolution, the filtering of received data packets according to the physical address of the station and the logical name of the client, and the detection of errors.

3. Q-BUS Interface

The DBC-100 has a dual port 4 Kbyte RAM (Random Access Memory), which forms the interface to the client layer, containing status variables and several buffers for the exchange of DANUBE data packets /2/. This RAM is normally interfaced via the MULTI-BUS. To make it accessible to the LSI-11/23, a special interface to the Q-Bus has been developed at the HMI.

This interface provides two address windows, which map Q-Bus addresses into the RAM. The windows of 256 and 3x256 bytes respectively may be relocated independently within the RAM via a Page Control Register. In this way, a client entity residing within the LSI-11/23 may easily access the status and the buffer region of the RAM in parallel. The status region contains flags for the buffer management and for error and status indications, and the buffer region is used for the transfer of data packets, i.e. DANUBE messages as well as control information for the DBC-100 itself.

The access to the RAM is controlled by a status variable. The update of this variable is signalled to the LSI-11/23 by an interrupt.

4. RSX-11/M Driver

The I/O driver for DANUBE /3/ is built as a loadable driver with a loadable data base for mapped RSX-11/M operating systems. Other versions, e.g. for RSX-11/S systems, may easily be derived from this implementation.

The driver handles two device units, one unit being reserved for input and the other unit for output. This method provides a full duplex data transfer via the driver.

A client task running on the LSI-11/23 may use the standard IO.WLB and IO.RLB functions for data transfer, and IO.KIL to cancel outstanding operations. The data presented to the driver are copied directly from the LSI-11/23 memory to the controller RAM or vice versa by programmed data transfer.

Status returns include the total number of bytes transferred successfully, or several error reports, indicating the different error conditions, which are signalled by the DBC-100 or the Q-Bus interface.

5. Transport Station

In general, a client task should not access the driver directly. A better and more common practice is that the client task uses the "services of a higher quality" offered by the Transport Layer.

Our DANUBE network is supported by a transport station program running on the LSI-11/23 /4, 5/. It offers both a datagram and a virtual connection service, the latter providing error and flow control as well as in sequence delivery.

6. Further Activities

The DANUBE network at the HMI Berlin is expected to support various types of further investigations and developments:

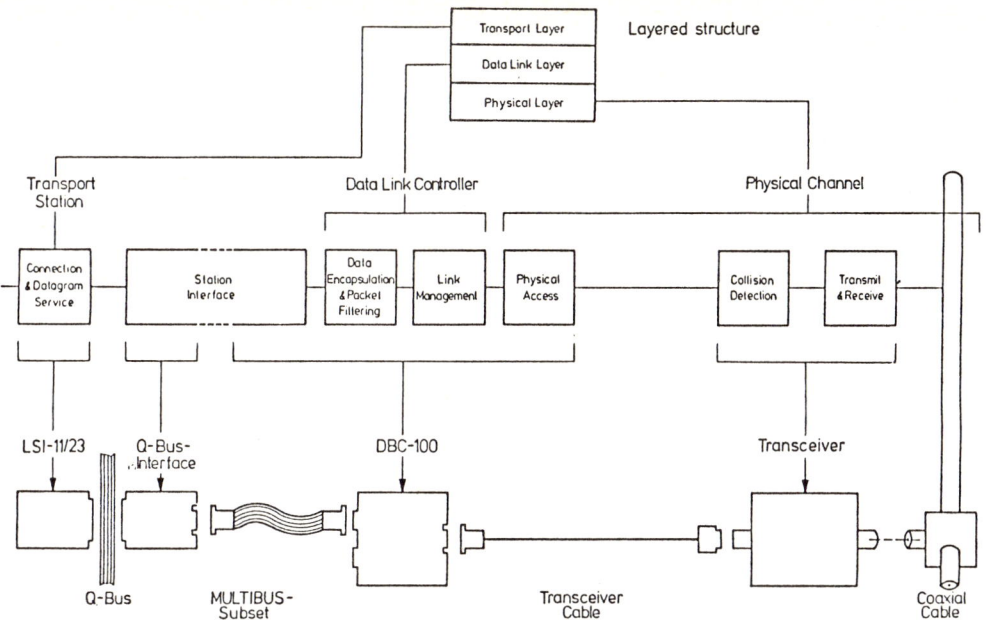

Fig.1: DANUBE Architecture and Implementation at HMI Berlin

a) Realization of TOPAS, a Low Level Token Passing Scheme using a source addressed token with time slices. This deterministic and contention free bus access method will be implemented as an add-on to the ETHERNET-type hardware for use in real time applications /6/.

b) Implementation of a network monitor for the observation and presentation of network activities and statistics at the Data Link and Transport Level.

c) Performance measurements under operational and test conditions, with special emphasis on different load patterns, including burst and periodic load.

d) Development and implementation of a gateway to operate DANUBE as a subnet of the existing local X.25-Net (HMINET II).

7. Acknowledgements

The authors thank M. Mercier-Laurent (INRIA), Mme. Vatton and M. Quint (IMAG Grenoble) for their support to the development of the Q-Bus interface, the driver and the transport station.

8. References

/1/ Description foncionelle du réseau expérimental DANUBE, INRIA Projet Pilote KAYAK, REL 2.514.1 (Juill. 1980).

/2/ Martin, M., Mercier-Laurent, C., Santourian, R., Spécification d'interface avec la carte coupleur Bertin DANUBE (BD), INRIA Projet Pilote KAYAK, REL 2.542 (Juin 1981).

/3/ Tschammer, V., An RSX-11/M Driver for DANUBE, an Ethernet-type Local Area Network, to be published as an HMI-Bericht (October 1982).

/4/ Martin, M., Naffah, N., Quint, V. Scheurer, B., Proposition d'un protocole de transport pour réseaux locaux, INRIA Projet Pilote KAYAK, REL 2.504.2 (Septembre 1980).

/5/ Rousset, X., Vatton, I., Interface de la station de transport sur MICRO-1, IMAG Laboratoire d'informatique et de mathématiques appliquées de Grenoble, version 6 (Septembre 1981).

ANALYSIS OF THE COMMUNICATION AND FAULT ASPECTS OF
PATH AND HIGHWAY PROTOCOLS FOR PROWAY NETWORKS

W. Ansaldi, M. Olobardi - ANSALDO, Genova (Italy)
A. Faro, O. Mirabella - Università di Catania (Italy)

ABSTRACT

The aim of this paper is to analyse the path and highway layers of the PROWAY architecture proposed by IEC, pointing out how it is possible to extend the HDLC protocol to perform such layers. Both communication and fault aspects are analysed and a formalization of such functions is given in order to clearly show the data highway behaviour.

RÉSUMÉ

Le but de cet article est d'analyser les niveaux "path" et "highway" de l'architecture Proway proposeé par l'IEC, indignant comment il est possible d'étendre le protocole HDLC pour geren de tels niveaux. Les aspects communication et fantes de fonctionnement sont étudiés et formalisés, afin d'étudia le fonctionnement des niveaux de transmission.

ZUSAMMENFASSUNG

Das Ziel der Abhandlung ist es die Path-und Highway-Ebenen der vom IEC vorgeschlagenen PROWAY-Architektur zu untersuchen, wobei darauf hingewiesen wird, in welcher Form das HDLC Protokoll erweitert werden Koennte, um die erwaehnten Ebenen zu bedienen. Datenuebertragung und Fehlerverhalten werden untersucht, wobei eine formalisierte Beschreibung angegeben wird, die eine klare Beschreibung der Verhaltens des Datenweges ermoeglicht.

ANALYSIS OF THE COMMUNICATION AND FAULT ASPECTS OF
PATH AND HIGHWAY PROTOCOLS FOR PROWAY NETWORKS

W. Ansaldi, M. Olobardi - ANSALDO, Genova (Italy)
A. Faro, O. Mirabella - Università di Catania (Italy)

The aim of this paper is to analyze the path and highway layers of the PROWAY architecture proposed by IEC, pointing out how it is possible to extend the HDLC protocol to perform such layers. Both communication and fault aspects are analyzed and a formalization of such functions is given in order to clearly show the data highway behaviour.

1. INTRODUCTION

In the last years, distributed systems for process control have been more and more studied. This has determined a great development of local networks for the interconnection of measurement and control devices. Thus the need arises of standardizing both protocols and services of such networks in order to facilitate both the connection of different devices (networking) and the connection of different networks (internetworking).

At present different architectures have been proposed with this aim. For example Ethernet (developed by Xerox, Digital and Intel) and IEEE 802. The first seems to become de facto a standard for both data acquisition and office automation because of its diffusion on the market. The second, on the contrary, is developed essentially as a standard but it seems still to be in an evolutive phase.

Both these architectures present general purpose characteristics. On the contrary the need arised of having specialized standard architectures which take into account both the particular work environment conditions and the special requests of the applications.

For this reason IEC (International Electrotechnical Committee) is developing a standard data highway architecture called proway which is a communication HIGHWAY between devices which form a distributed process control system. Because of the particular environment where it has to operate and the applications to be covered, proway must provide very reliable communications also in the presence of electromagnetic interferences. Data transmission will be serialized over a single electrical transmission line and characterized by a short access time, thus allowing real time (event driven) communication processes.

Proway may provide "connectionless" and "reliable datagram" services. As known connectionless service performs data transfer of single messages without opening a connection between the interacting communication processes and without acknowledgment from the destination of the received message. Also reliable datagram service performs data transfer without opening a connection, but it uses message pairs so that each message is followed by the acknowledgment sent from the destination to the source. In the proway terminology the source process is called active primary (initiator) whereas the receiver is called secondary unit and it can behave as a responder and/or a listener. Of course if the secondary unit is a responder the proway provides a reliable datagram service otherwise it provides a connectionless service.

The proway service is suited for process control applications where short question and answer messages are often exchanged by the units, but not for exchanging very long files. It is instead designed for communications between input, output and control devices for man/machine interfacing and for service or support equipment connections.

The network can support up to 100 stations. This depends on the required performances in terms of delay, because a high number of stations increases the access time; in addition when one increases the number of the stations, it is necessary also to increase the cable length to avoid that the station distance becomes critical, and consequently one is obliged to reduce the transmission bit rate for maintaining the signal quality.

The proway foresees at least 2 supervisor units which arbitrate, in real time, contentions among candidates (e.g. demanders or initiator candidates) to acquire the active primary token and at least 2 manager units which monitor the line performances and choose a supervisor unit among the canditates. The presence of at least 2 supervisor units and 2 manager units enhances system reliability with a management level redundance.

The proway architecture has a layered structure almost universally accepted for the organization of the communication among cooperating systems (fig.1). In such a structure, each layer provides a set of communication services to the upper layer masking how these services are performed, so dividing the total problem in smaller pieces. According to IEC, the architecture of a communication system for distributed process control may be splitted into 5 levels (some of which constituted of 2 or more sublevels). The first three levels (line, path, highway) implement a low level communication system that IEC is developing as a standard, while the other two (network and application) perform functions necessary for correct connection and interface with the network users (transducers, measure instrumentation, operators, etc.). Fig.1 puts also in relation the proway architecture with the 7-layer architecture proposed by ISO for Open Systems Interconnection (OSI).

The aim of this paper is to analyze the path and highway layers of the proway architecture by pointing out how it is possible to extend the HDLC protocol to perform such layers. Both communication and fault recovery aspects are analyzed and a formalization of such functions is given in order to clearly show the data highway behaviour.

Sect.2 discusses the path layer, while sect.3 presents the highway layer. The line coupler foreseen in the proway standard is not treated in this paper because its purpose is only to convert data from their logical representation to signals compatible with the adopted transmission line. Sect.4 analyzes communication and fault aspects of the proway, whereas sect.5 presents the formal description of the highway layer.

2. PATH PROTOCOL

The path level protocol of the proway draft standard is responsible of parallel to serial data conversion and of lower level error handling. The characteristics of this protocol level are almost entirely implemented in hardware using integrated chips executing basic HDLC functions.

So the following functions must be surely implemented inside the path layer:
 - serializing and deserializing frames - adding and removing frame synchronization patterns
 - recognizing frames addressed to a designated station - generating and monitoring error detecting code - handling highway frames of widely different lengths - detection of frame size errors

Other remaining functions can be provided by the path level protocol and can be obtained by ad hoc circuitry or by software. The first solution is, when appliable, to be preferred leading to a more efficient implementation. These functions could be the following:

- preventing the station from transmitting without pause for an excessive time: may be realized using a one-shot flip-flop connected to the electrical interface between the interface board and the transmission line.

- switchover to a redundant transmission line: may be implemented inserting circuitry into the electrical interface to monitor the line electrical continuity and the switch to the redundant line. If the network is organized around two different transmission lines (information line and control line) the messages travelling on the broken line could be routed to a third back-up line or to the remaining active line, at the cost of a degradation of the network performance.

In case of detection of errors or failures the path level protocol should inform the higher level protocol to allow the start of adequate recovery procedures.

3. HIGHWAY PROTOCOL

The highway unit controls and manages the data highway operations including error recovery and control line access. For this aim, the highway functions are organized into the ranks which are below briefly explained from the lowest to the highest.

- Listener

It accepts the correct frames addressed to its station

- Responder

It accepts and responds with either an ack or a data frame at the reception of a correct frame addressed to its station.

- Initiator

It responds to the supervisor poll with a request, if any, for access to the data highway and sends data (to either listener or responder stations) when it is allowed from the supervisor. The initiator starts a suitable recovery action when it detects an error condition.

- Demander

It can perform an unsolicited request to access the data highway and become Active Primary (AP). This request may be transmitted over a dedicated line (in this case the reservation bus will be separated by the data bus) or by means of special states on the data bus.

- Supervisor

It controls the access to the line by arbitrating contentions among demanders and candidate initiators. The supervisor initiates suitable recovery actions when it detects fault conditions. For example when the active primary keeps the line for an excessive time.

- Manager

It controls the data highway performance and assigns the line control to a supervisor by arbitrating contentions among candidate supervisors. The manager ensures data highway continuity when the active supervisor fails.

4. COMMUNICATION AND FAULT ASPECTS

Two principal problems arise for the communication and management in a data highway, that is the data transfer protocol, the reservation protocol and the fault management. Two different bus structures are possible for this aim: single bus based solution offers hardware simplifications but decreases the network performance, while the double bus based solution shows high reliabity and performance but needs more expensive hardware.

However both the data transfer and reservation protocols are not influenced by the bus structure. In fact in case of a double bus structure, the data transfer and reservation messages pass through different busses so allowing parallel activities; while in case of a single bus structure, the data transfer and the reservation messages share the same bus sequentially. In the following we propose to use a modified form of HDLC as highway protocol and moreover we analyze the error and recovery problems.

4.1 Use of HDLC for Highway protocol

The proway standard can be implemented without inventing new protocols, by extending the functions of the HDLC protocol. This choice depends on both the HDLC diffusion and potentiality in governing the data exchange between two interacting stations. However since the proway does not need opening and clearing of virtual channels between the stations, we propose of utilizing the HDLC messages relative to these phases for the bus reservation, leaving unchanged the meaning of the messages relative to the data transfer phase. In particular in order to make the HDLC adequate to the needs of the proway, it is necessary to add a source field in the header of the HDLC frame; in fact the HDLC was conceived for governing the interaction between two fixed stations while in our case it has to be used for the communication among several stations. A way for obtaining a protocol for controlling the bus access is that of using the opening and clearing messages of the HDLC in the following manner:

- the message SABM with P/F =0 as a message exchanged among the candidate initiators and Supervisor with Token Passing tecnique for Data Highway reservation;

- the message SABM with P/F=1 as a message sent from the Supervisor to a demander or a candidate initiator to transform it in Active Primary;

- the message DISC as a message sent from an active unit to another of higher rank in order to communicate that it has finished its funtion,or as a message sent from an active unit to another in order to impose the end of the lower rank unit function;

- the message UA as a message sent from an active unit to another of higher rank in response to SABM or DISC received from it.

The use of the poll/final bit also during the opening phase is to distinguish the polling messages from the reservation or nomination messages.

The HDLC allows us to obtain also a data transfer flow controlled with window mechanism so increasing the performance because this avoids that the responder must send an ack after each received message. At present the proway proposal foresees an unitary window, but it is envisaged that the proway can support also non unitary window to enhance the throughput in case of long message transfer. In addition the HDLC can be used for error recovery due for example to loss or misordering of frames because of line noise. In fact this protocol presents a reject function by which the destination informs the sender on lost or misordered data.

4.2 Error detection and recovery

The high data integrity and fault tolerance requested by the PROWAY draft can be obtained only considering all error sources and carefully analyzing every possible alteration of normal functioning conditions. This can be obtained examining the different kinds of information flowing to and from every network node and also analyzing the protocol state diagram,in order to detect possible deadlock situations.As a result of this analysis ,a comprehensive set of recovery procedures (acknowledgements,retransmissions,diagnostic messages,time outs)can be outlined, thus allowing the requested level of fault tolerance characteristics. Three kinds of error sources can be identified:

1)Transmission error, caused by electromagnetic noise during the transmission of a message.The noise can modify one bit (spike error) or more consecutive bits (bust error);

2)Transmission line break;

3)Node malfunctioning, caused by H/W error on the line interface or in the equipment connected to it(power down ,etc).

For every identified error source the following procedure will lead to the identification of suitable error recovery tools:

a) identify possible effects of the considered error source on the normal functioning conditions of the protocol;

b)for every effect identified in a),select adequate detection procedures;

c)for every malfuntion detected by the procedures identified in b), select adequate recovery procedures;

d)repeat steps a,b,c in order to deal with the new error effects connected to the newly detection and recovery procedures added in phase b and c. The outlined procedure should be repeated for every rank a highway protocol level can be organized with. As a result of the overall process a set of six tables (one for each rank) should be produced as shown in fig.2. Note that every error source can cause different effects in different working conditions and that different detection and recovery procedures can cover the same effect. On the other hand, the same detection and/or recovery procedures can cover more than one effect (dotted lines in fig.2).

In order to be shure that the detection and recovery procedures previously identified are able to exhaustively deal with every possible error, verification tecniques can be used. A survey of existing verification tecniques and pratical examples can be found in bibliography.

A complete description of the previously outlined procedure is out of the scope of this paper. In the following, a brief survey will be given of the main error detection and recovery tecniques used in the protocol design, for the three considered error sources.

Transmission error

Depending on the kind of electromagnetic noise over the line, a transmission error can cause the following effects on the message transmitted:

1) the messages arrives to the right destination. The transmission causes the modification of one or more bits and the error is detected by the Frame Check Sequence (FCS) appended to the message by the path level protocol;

2) the error causes the modification of the address field of the message and the message itself is lost;

3) same as 2) but the modified address field causes the reception of the message by a node other than the intended destination. The FCS indicates that the message has been damaged.

The detection and recovery tecniques to be introduced in order to deal with the error stated above depend on the rank of the transmitting and receiving nodes.

A hierarchical approach based on the rank machanism is used as a general rule for the detection and recovery strategy: if a message is exanged between two nodes of different ranks, the detection and recovery activity is carried out by the higher rank node. The lower rank node is only responsible of the integrity of the equipment connected to it and thus performs only a filtering action on the incoming messages.

For every rank, the recovery procedures needed depend on the kind of message transmitted. The most common tecniques include: acknowledgement messages and retrasnsmission, time outs, monitoring of the line status.

The selection of the right tecnique depend on many factors, such as: characteristics of the current protocol implementation, characteristics of the transmission medium, typical error rate, characteristics of the application, performances needed.

Acknowledgement + retransmission guarantees prompt detection and recovery, at the cost of high time overhead and low throughput.

Through the use of different mechanisms, lower overhead can be obtained but the recovery time will of course increase. Such mechanisms are:

1) sending acknowledgement messages every N (N>1) packets received (window mechanism)

2) sending acknowledgement messages only upon reception of a correct packet. If the received packet has a wrong FCS, no acknowledgment is sent; the error is then detected by an ad-hoc time out on the sending node

3) no acknowledgement is sent. In this case a node other than the sender or the receiver will detect the error, for instance monitoring the line activity and verifying that no station is transmitting.

No solution is optimal for every case but must be selected by the designer, depending on the actual needs. Possibly, the designer could include in the protocol more than one mechanism and allow the user to select, during the system configutation phase, a particular detection procedure.

Recovery procedure consists, normally, in retransmitting the message and, after N retransmissions, starting a reconfiguration procedure.

It's worth noting that different detection and recovery policies must be used if the control frames travel on the same line with the information frames or over a dedicated line. In the former case, care must be taken in order to avoid collisions between control and information frames caused by the loss of some previous packet. An example of such situation can be found when a supervisor node sends an activation message to a candidate Initiator but the corresponding acknowledgement message is lost. At this point the initiator node begins transmitting information frames while the Supervisor node retransmits the activation message.

Transmission line break

The detection of a transmission line break is responsibility of the path level protocol. The data and/or control flow is then switched to a redundant line, when provided (see section 2)

Node failure

Two cases can be distingued:

1) the failure (break or power down) is located on the interface board.

2) the failure (break or power down) is located on the equipment connected to the interface board.

In case 1) the interface failure can be detected in one of the following ways: a) when the node fails to respond to N consecutive messages; b) when no message is travelling on the line for a certain time; c) when the node fails to respond to a polling message during a configuration procedure.

When the node failure is detected, the information is forwarded to the Supervisor node that sends an update message to all the active nodes.

In case 2), the equipment failure can be detected by the corresponding interface. Upon detection of the failure, the interface sends a special management message to the network Supervisor and then logically disconnects itself from the network, discarding any message addressed to it.

5. FORMALIZATION OF THE HIGHWAY PROTOCOL

In order to clearly represent the data highway functions, the various units can be formalized by using a generalized finite state graph. Such a graph consists of:

- nodes representing the principal states of the unit;

- arcs which determine the transitions from a state to another.

The transition is activated:

- by the reception of a message from a peer unit or a command from a upper layer unit;

- by internal conditions, e.g. internal time outs.

During the transition, the unit executes the procedure associated to the arc. This procedure produces different outcomes depending on the other secondary internal states and on input parameters.

Let us note that the arcs which contain procedures, are labelled by inputs before the procedure and by outputs after the procedure, whereas the arcs which do not contain procedures, are labelled by an input/output pairs.

In the following figures the commands received from the upper layer are denoted by n, whereas the commands given to the upper layer are denoted by c. In addition the internal mechanisms are denoted by I.

Appendix shows the list of the commands, the internal mechanisms, the procedures and the special I-Frames used in the formalization.

5.1 Listener formalization

Fig.3 shows the listener formalization. The listener is configured by the Supervisor by means of a SABM message with P/F bit=1 whose text contains the list of the stations from which the Listener can accept data. To disconnect a listener the Supervisor sends a DISC message. The Listener responds UA to either SABM or DISC received from the Supervisor. If the Listener has not the possibility of responding UA, it must be configured manually from the users because this cannot be performed by the supervisor in a reliable way. Note that the supervisor considers out of service a listener which does not respond UA within a fixed timeout. During the data transfer phase a listener simply passes data to the upper layer if CRC and source test is positive, otherwise the arc labelled by (1) is executed in fig.3.

5.2 Responder formalization.

The responder behaviour is represented in fig.4; it is actived and configured through a SABM,P=1 message from the supervisor node. It accepts by responding UA. The supervisor considers out of service a responder which does not respond UA within a fixed timeout. When responder receives an I-frame, it checks data correctness, transfers the data to level 3, then sends an ack to the active primary node. If the data is not correct, the arc labelled by (1) is executed in fig.4.

5.3 Initiator and Demander formalization.

The initiator and demander behaviour is represented in fig.5. When a candidate initiator receives a SABM,P=0 message from Supervisor or from another candidate initiator, it updates the reservation list in the message and transfers it to following candidate primary, then it passes in wait state. When it receives a SABM,P=1 message in this wait state, it is called Active Primary (initiator), then it firstly performs a recovery action whose consequences may be:

1) to remove from transmission buffer data previously acked.

2) to send a message to supervisor if it has not received an ack from a responder after n-retransmissions, then it becomes AP, sends data and waits for ack.

Let us note that if the candidate initiator does not receive SABM,P=1 message in the above wait state within the timeout I3, then it passes in active primary candidate state. Whereas if it receives a SABM,P=1 in this last state, then it conseders this nomination as an error and informs the supervisor on this situation.

When Active Primary receives an ack during the data transfer phase, it sends again a data, if any, while if it receives a DISC from supervisor or the internal mechanism I1 acts because this station does not have data to transmit, AP goes respectively in AP candidate state or in wait for data.

In fig.6 we have formalized a station which perform at different times the active primary and listener functions. Note that the listener functions not only concern the CRC and source test but also the updating of the window mechanism or the predisposition of the reject procedure (see procedure Reception window control in the appendix). Thus I-Frame, ack and reject messages (concerning data previously transmitted by the station) are received only when the station is a listener, whereas I-Frame, ack and possible reject messages (for requesting the retransmission of non correctly received data) are sent only when the station becomes active primary. Such a behaviour is not actually foreseen for the proway but it could be useful in the case in which many messages have to be transmitted by the same station. The graph is similar to the previous one with a difference regarding the possible transmission of acks and/or rejects to the previous active primary (which is now a listener of the above type) according to the "reception flow control" procedure. Since we do not foresee the possibility of opening virtual channels between the stations. The case represented in fig.6 needs static virtual channels between all the potentially interacting stations, but this could require a great memory capacity of the stations. So it seems to be preferable that a station can have a limited amount of channels whose access rights are controlled by the supervisor during the nomination phase as indicated in sect.4.1.

5.4 Supervisor formalization

The supervisor behaviour is represented in fig.7. In particular when a candidate supervisor receives in the idle state SABM,P=1 from the manager, it responds UA and becomes active supervisor. Then it starts the procedure to nominate the active primary. The supervisor returns in idle state when it does not receive the reservation train after an N' attempts or when it does not receive UA from any nominated AP after N" attempts. In these cases it informs also the manager. Of course the supervisor nominates another AP after a DISC message from the preceding AP. When the communication system presents high reliability it can be useful that the supervisor does not wait for UA from the nominated AP, in this case in fig.7 after the "choice of the active primary procedure" the arc labelled by (2) is executed otherwise the arc labelled by (1).

5.5 Manager formalization

The manager behaviour is represented in fig.8. In particular when a candidate manager receives SABM,P=1 from the network director, it responds UA and becomes active manager. Then, if necessary (i.e. I10 = 1), it performs the procedure for nominating a supervisor. Note that in this figure we introduce the input "ERROR" to represent any error signalled to the manager either through a message coming from lower rank stations or by internal mechanisms. During the "ERROR" procedure the manager tests the network as described in sect.4.2. On the basis of this test, the manager disconnect the supervisor or re-nominates the same supervisor by communicating to it the new network configuration.

6. CONCLUSIONS

In this paper a data highway proposed by IEC called Proway was analyzed. Some aspects have been deeped and an extension has been proposed to allow a more efficient use of the proway when a station has to send a sequence of many messages. Very powerful management and supervisor functions characterize the link layer. This determinates a high implementation effort but permits a controlled access to the line depending on the user needs. This is not possible with CSMA-CD or token passing mechanisms which produce respectively a random or sequential access to the line. Therefore the proway seems to be particularly adapted for real time distributed systems. However this aspect and the others analyzed in the paper need further studies to obtain a standard proway.

APPENDIX

a) The following internal mechanisms are used in the highway layer formalization:

- I1: no data to transmit
- I2: presence of an urgent request of becoming active primary.
- I3: timeout associated to the "wait for active primary nomination" state
- I4: timeout associated to the state "wait for UA".
- I5: maximum number of SABM,P=1 repetitions without UA from the nominated AP Manager formalization.
- I6: maximum number of unsuccessfully repetitions of SABM,P=0.
- I7: timeout associated to the state "wait for reservation".
- I8: impossibility of nominating a preselected candidate primary or demander as active primary.
- I9: absence of candidate initiator or demander requests.
- I10: need of nominating a new supervisor.
- I11: impossibility of nominating any candidate supervisor as active supervisor.
- I12: timeout associated to the state "wait for UA".
- I13: impossibility of nominating a preselected candidate supervisor as active supervisor.
- I14: timeout associated to the ack reception.
- I15: the active primary must release the bus for either the presence of an urgent access request or the firing of the timeout associated to the bus keeping.

b) The following commands are exchanged between the highway and its upper layer:

- η1: request of sending data
- η2: send data
- c1: notification that this unit was con-

figured as listener
- c2: data indication
- c3: notification that this unit is not configured
- c4: notification that this unit was configured as responder
- c5: data sent
- c6: request of new data to be transmitted.

c) The following procedures are used in the formalization:

- CRC AND SOURCE TEST, which controls the CRC and if the source is one of those stated in the configuration phase.
- UPDATE THE RESERVATION LIST, which updates the reservation list in the reservation train.
- RECOVERY, which removes from the transmission buffer data previously acked, retransmits data non-acked within a prefixed timeout and sends a message to the supervisor after n-time repetitions of the transmission of the same non-acked data.
- RECEPTION FLOW CONTROL, which sends, if necessary, ack and reject messages relative to the received I-frames.
- TRANSMISSION WINDOW CONTROL AND COPY, which sends data if their sequence numbers are inside a transmission window.
- RECEPTION WINDOW CONTROL AND COPY, which accepts data if their sequence numbers are inside the reception window, otherwise it ignores these I-frames and predisposes the reject procedure for the appropriate actions. Note that when this procedure receives ack or reject messages, it respectively updates the window mechanism or predisposes the reject procedure.
- FAULT RECOVERY, which sends a DISC to the nominated AP when a fault arises.
- ACTIVATION POLICY, which determines if there are requests of becoming AP. Note that when the supervisor receives $DISC_1$ from the AP, it passes this message to the manager and goes in the idle state.
- CHOICE OF THE ACTIVE PRIMARY, which selects the AP on the basis of the reservation table.
- CHOICE OF THE SUPERVISOR, which selects the supervisor between the two candidates.
- ERROR which tests the network stations to find if the signalled error situation depends on the supervisor or on other units.

d) Special messages

- IF_1, which represent the I-Frames retransmitted by the AP because they are non-acked within a certain timeout.

- $DISC_1$, which represents a DISC message containing the code of the error which determined the emission of this disconnection message.

7. BIBLIOGRAPHY

/1/ IEC : Process data highway for distributed process control Systems. Part I°: general description and functional requirements, February 1981.
/2/ ISO: Reference model of OSI - TC97/SC16 N.7498 August 1981
/3/ W.Ansaldi et alii : Definition and development of a protocol for an industrial PLANT CONTROL NETWORK - Proc. Symp. on LANs, Firenze 1982.
/4/ G. Le Moli : The theory of colloquies - Alta Frequenza N.10 1973.
/5/ A.Faro : Formalization of protocols and interfaces in view of the theory of colloquies. - To appear on Computer Networks and Simulation, S.Schoemaker ed., North Holland, 1982.
/6/ G.V. Bochmann, C.Sunshine : Formal methods in communication protocol design - IEEE Trans. on Communications, n.4 April 80, 624-631

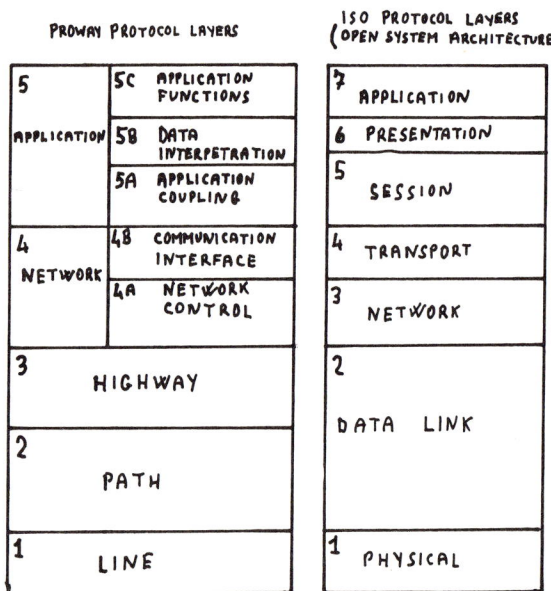

Fig.1 - Proway and OSI architectures

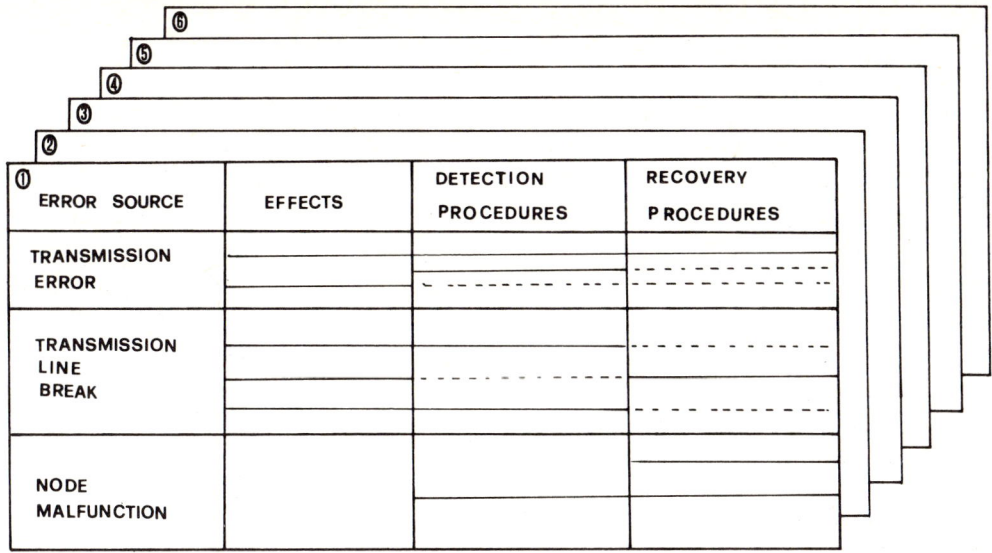

Fig.2 - Rank modules for fault management

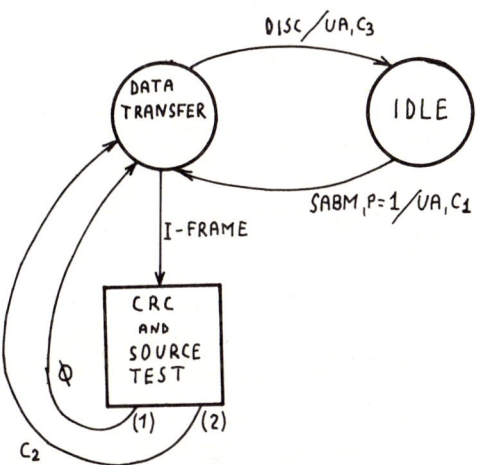

Fig.3 - Listener formalization

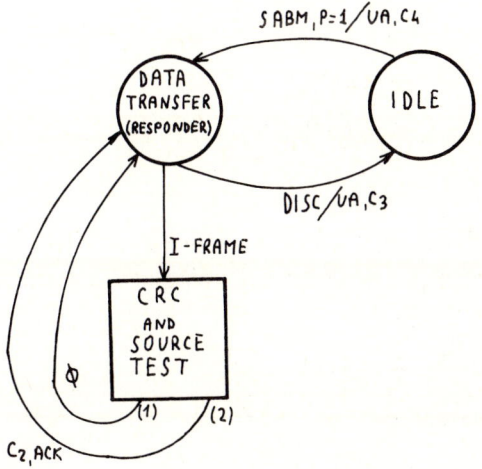

Fig.4 - Responder formalization

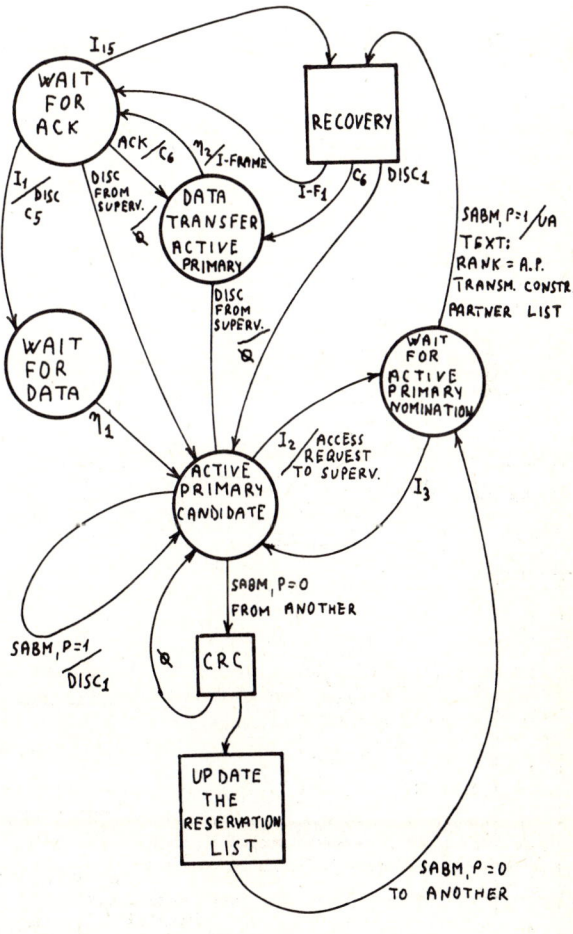

Fig.5 - Demander and initiator formalization

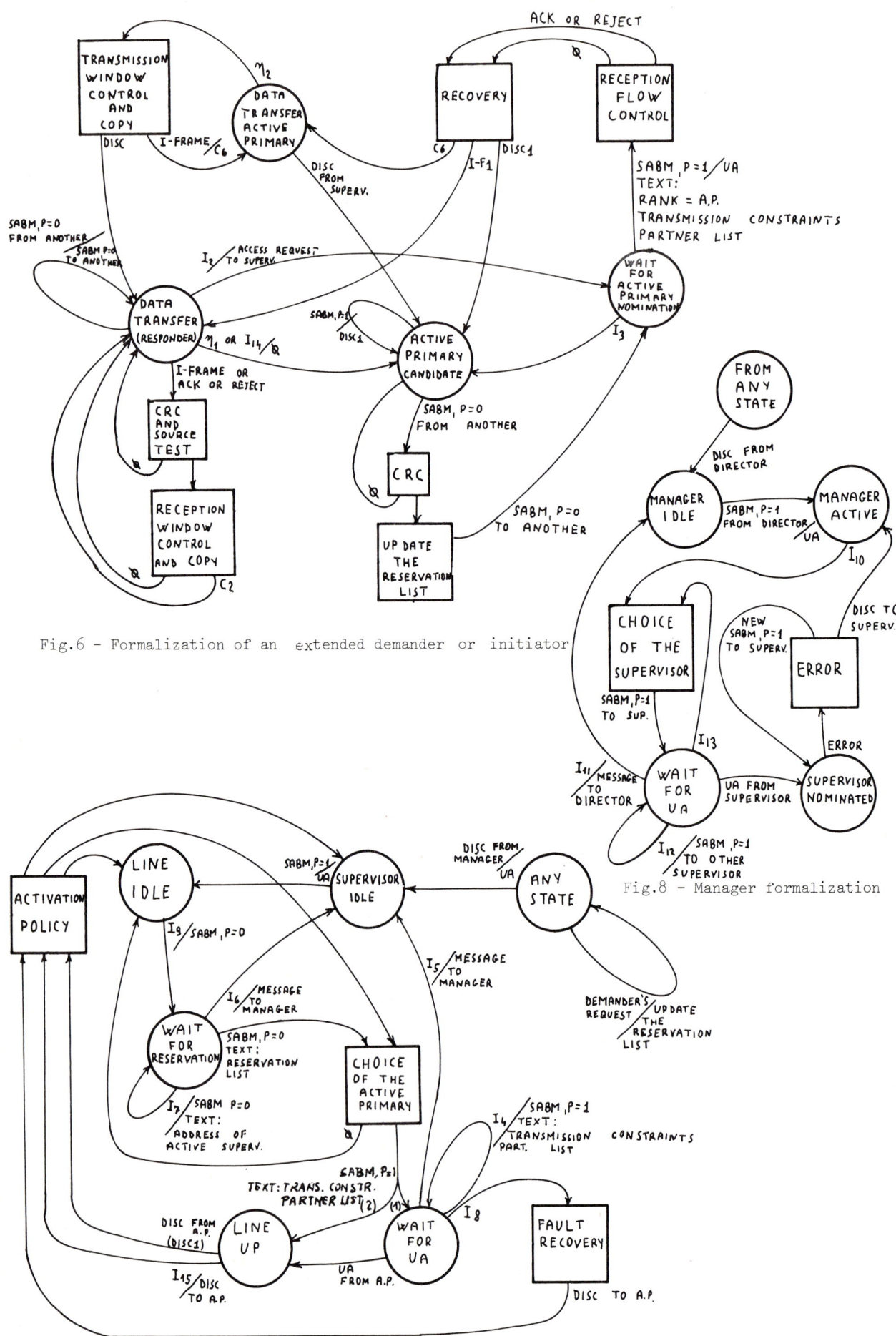

Fig.6 - Formalization of an extended demander or initiator

Fig.7 - Supervisor formalization

Fig.8 - Manager formalization

FASTBUS, A HIGH-SPEED MULTI-SEGMENT BUS

H. MUELLER EP-Division
CERN
1211 Geneva 23
SWITZERLAND

ABSTRACT

FASTBUS is a general bus system being developed, primarily, for high speed data acquisition and processing in the next generation of large physics experiments. It is built up from backplane segments housed in crates, linked together by cable segments for inter-crate communication. Each segment supports multiprocessors, and independent segment operation permits a high degree of parallelism. Handshake bus protocols, uniform system-wide, ensure reliability and both high- and low-speed devices can be accomodated. A synchronous mode provides for data block transfers at maximum speed.

RÉSUMÉ

Fastbus est un bus général, developpé principalement pour les applications d'acquisition et de traitement des données à haute vitesse destinées aux futures expériences de Physique. Ce bus est construit à partir de segments fond-de-panier en creneaux, reliés par des câbles pour la communication inter-créneaux. Chaque segment supporte des multiproccesseurs et le fonctionnement autonome des segments pesmet un degré élevé de parallélisme. Des protocoles "handshake" uniformes assure une bonne fiabilité et des équipements de vitesses variees sont connectables. Un mode de transfert synchrone permet les échanges de donnees à vitesse maximum.

ZUSAMMENFASSUNG

FASTBUS ist ein Universalbussystem, das hauptsachlich für die Hochgeschwindigkeits-Datenerfassung und -verarbeitung bei den Großversuchen der Physik der nächsten Generation entwickelt wird. Es ist aus in Kästen untergebrachten Rückwandsegmenten aufgebaut, die für Kasten-Kasten-Kommunikationszwecke durch Kabelsegmente miteinander verbunden sind. Jedes Segment unterstützt Multiprozessor-betrieb und die unabhängige Segmentbetriebsweise gestattet ein hohes Maß an Paralleilität. Mit systemweit einheitlichen Busprotokollen für den Quittungsaustausch wird Verläßlichkeit gewährleistet und ermöglicht, daß Geräte hoher sowie niedriger Geschwindigkeit aufgenommen werden können. Mit einem Synchronmodus sind Datenblockübertragungen mit höchster Geschwindigkeit durchführbar.

1. INTRODUCTION

FASTBUS [1,2,3,4] is a parallel BUS STANDARD of general applicability in both COMPUTER-and INSTRUMENTATION-type environments. The ability to communicate system-wide through high speed CRATE SEGMENTS and long BUS CABLE SEGMENTS meet the requirements of High Energy Physics experiments but are in no respect restricted to use in this domain.

System wide PROTOCOLS on wide area TOPOLOGIES and very high bandwidth CRATE SEGMENTS are outstanding features of FASTBUS. Nevertheless each segment is primarily an independent communication medium for its own attached DEVICES, which can be close to each other in a CRATE or distributed over tens of metres on a CABLE SEGMENT [5].

The FASTBUS ARCHITECTURE is independent of technologies or manufacturers. Though high performance technologies are generally made use of in the SYSTEM MODULES any kind of slow or simple electronics can be interfaced to FASTBUS [6,7]. The POWER and COOLING requirements are in direct proportion to the required performance and technology of the devices.

2. BUS ELEMENTS

FASTBUS consists of terminated 32 bit BUS ELEMENTS (SEGMENTS) multiplexed for ADDRESS and DATA information. Various timing, control and information lines add up to 60 essential BUS lines. CRATE SEGMENTS use a 130 line BACKPLANE with power and ground lines, position code lines, daisy chain and diagnostic lines as well as ample reserved lines. Additionally an optional 195 pin PRIVATE BACKPLANE connector in the CRATES gives maximum flexibility of system design.

Any combination of MASTERS or SLAVES or SEGMENT-INTERCONNECTS and INTERFACES can reside on a SEGMENT. Each SEGMENT needs TERMINATORS and Arbitration Timing and Control (ATC) logic as well as a detector circuitry for geographical addresses (EG-generator).

3. SLAVES

FASTBUS SLAVES [8] connect to BUS MASTERS when they recognize their address within PRIMARY ADDRESS CYCLES. Their actions then are determined by the decisions imposed from the MASTER. However, the FASTBUS PROTOCOL allows SLAVES to indicate difficulties or special situations by use of the WAIT (WT) line and by a 3-bit response-code (SS) accompanying every cycle. In this way simple SLAVES can operate together with sophisticated or high-speed MASTERS.

The functional components FASTBUS SLAVES are

(a) Basic set of primary address recognition circuitry; bus receivers/transmitters; CSR register(s) and NTA-pointer.

(b) Responder logic to ADDRESS LOCKED OPERATIONS.

(c) USER or SYSTEM implementations.

(d) Implementation dependent decoding.

Components (a) and (b) which perform a coupling between the bus and the implementation-logic are reproduced in more or less similar form in any SLAVE. The production of standard chip-sets for these functions is being discussed.

4. MASTERS

MASTERS use their internal ARBITRATION LOGIC in order to compete with other MASTERS to use a common SEGMENT. The common ATC logic controls the ARBITRATION PROCESS.

The responsibility of MASTERS is: not to make any logical errors concerning BUS MASTERSHIP and the responses received by the SLAVES. Further MASTERS have to react in unique ways in critical situations, for example if SLAVES do not respond or if the communication is lost.

The completion of cycles initiated by a MASTER normally done asynchronously is at the arrival of the SLAVES handshake signals which allows high-speed MASTERS to operate with slower SLAVES. MASTERS can also enter a synchronous mode pipelining a block of data without handshakes. This allows MASTERS to cycle at their intrinsic speed and to increase the information transferred per cycle.

MASTERS also behave as SLAVES when addressed. This allows the System Controller to identify and control their parameters. The functional components built around a DECISION LOGIC in a MASTER are:

(a) ARBRITRATION LOGIC (any MASTER)

(b) BASIC SET of Bus drivers/receivers; CSR register(s) primarily address generation and recognition logic; NTA-pointer.

(c) USER/SYSTEM implementations

(d) Implementation dependent decoders.

The DECISION LOGIC in a MASTER can be classified as:
- hard wired (sequencer) [9, 10]
    Fixed-Task MASTERS, simple, high speed.

- microprocessor-driven [11]
    Programmable-Task MASTERS, very flexible, medium speed, sophisticated devices.

- interfaced from outside [12, 13, 14]
    Host-system advantages, information links, efficient synchronisation and mapping required.

5. MASTER-SLAVE CONNECTIONS

Having gained the BUS by an arbitration cycle the BUS MASTER demands a connection with the required SLAVE(S). This locking condition is established by a PRIMARY ADDRESS CYCLE and consists of the MASTERS Address Assert (AS) signal and the SLAVE'S Address Acknowledge (AK) signal. The operations within this two-signal envelope exclusively affect only the connected DEVICES and they are called ADDRESS-LOCKED-OPERATIONS.

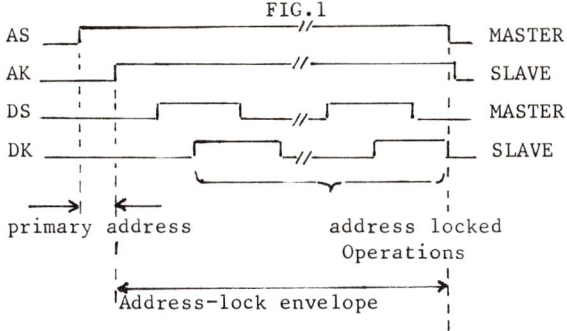

FIG.1

6. OPERATIONS

The common way to exchange information within an ADDRESS LOCK consists of a two-step process repeated n times: (a) Definition of the INTERNAL ADDRESS for the following DATA TRANSFER (b) SINGLE or BLOCK DATA TRANSFER. Both kinds of cycles are provided by the standard FASTBUS PROTOCOL. However, ADDRESS - LOCKED OPERATIONS in general include any sequence of (meaningful) data cycles. SPECIAL BROADCAST address cycles as for example the conditional assertion of a position encoded bit on the bus is one example of a different procedure.

Frequently BUS MASTERS want to operate sequentially with several SLAVES without interruption which is possible by maintaining the Grant Acknowledge (GK) signal. This kind of OPERATIONS are called ARBITRATION LOCKED.

7. SLAVE INTERNAL ADDRESSES

The definition of the INTERNAL ADDRESSES make use of a Next Transfer Address (NTA) pointer in the SLAVE which can both be loaded or read within a SECONDARY ADDRESS CYCLE. The validity of the value is verified and indicated by an SS RESPONSE. SINGLE DATA TRANSFERS provide 32 bit information exchange according to the direction defined by the READ (RD) line between the locations pointed at by the hardware - NTA in the SLAVE and a pointer in the MASTER. BLOCK TRANSFERS automatically increment the NTA until a boundary value is reached which is indicated by the END OF BLOCK SS-code. Fig. 2 shows the general flow of decisions to be made by a FASTBUS SLAVE according to the protocols.

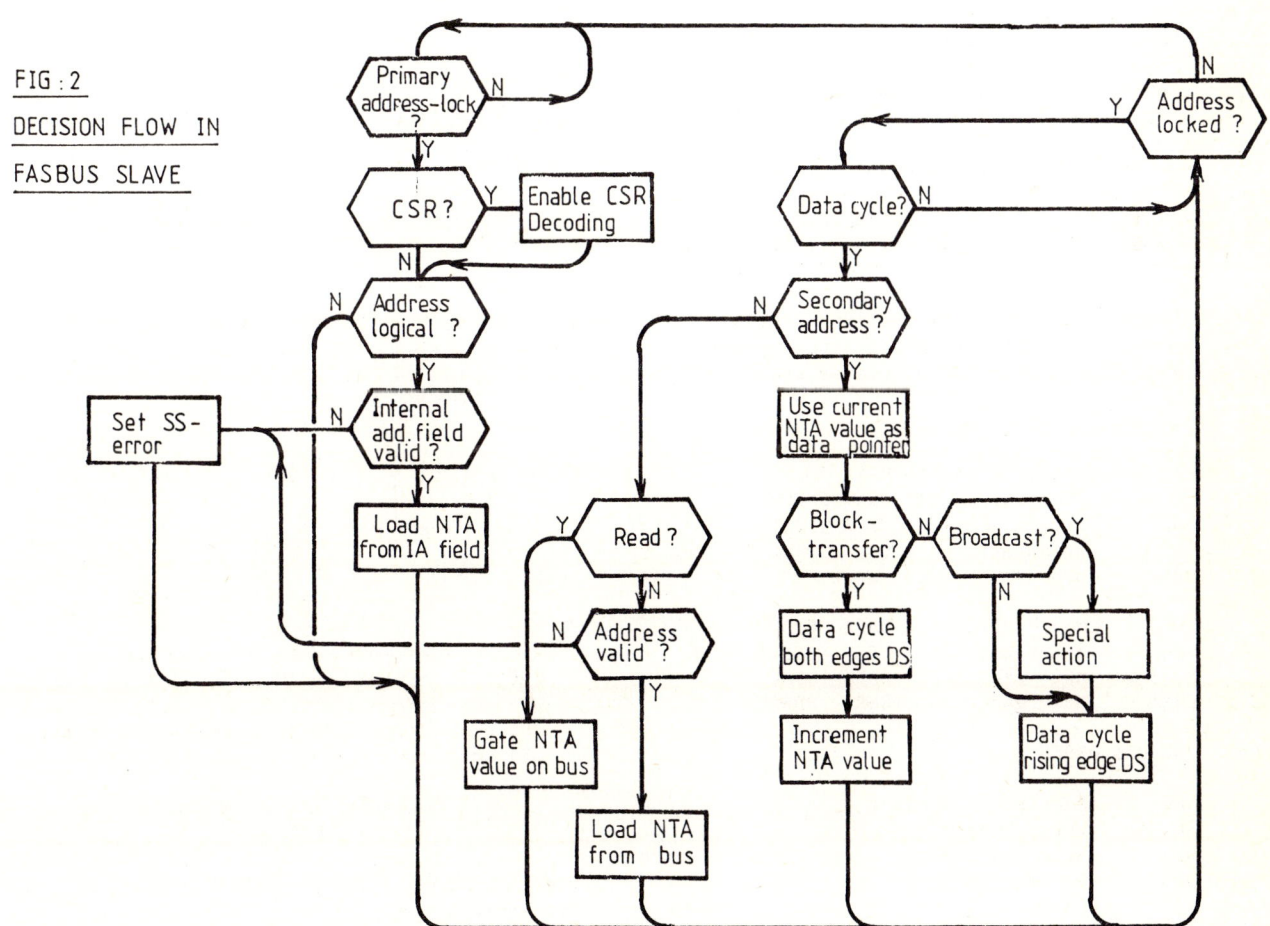

FIG:2 DECISION FLOW IN FASBUS SLAVE

8. PROTOCOLS

The procedures to obtain MASTERSHIP, to connect MASTERS with SLAVES and to operate within connections are defined in the FASTBUS SPECIFICATION [1] by detailed protocols. The FASTBUS PROTOCOL provides most commonly used ADDRESS-LOCKED OPERATIONS which are specified by a 3 bit MODE-SELECT (MS) code from the current BUS MASTER along with a DATA STROBE (DS). Slaves must acknowledge within a maximum time (can be stretched by use of the WAIT line). Along with the DATA ACKNOWLEDGE (DK) signal the SLAVE provides a 3 bit response code (SS). PIPELINED TRANSFER PROTOCOLS work without DK-handshake and partially out-of-phase SS-responses.

8.1 FASTBUS PROTOCOLS OF ADDRESS LOCKED OPERATIONS

MS0	random data
MS1	handshaked Block Transfer
MS2	secondary address
MS3	pipelined transfer
MS4..7	reserved

8.2 SLAVE SS RESPONSES

SS0	valid action
SS1	busy
SS2	end of block
SS3	user defined
SS4	reserved
SS5	reserved
SS6	Data Error (reject)
SS7	Data Error (accept)

The PROTOCOLS of PRIMARY ADDRESS CYCLES use the same MODE SELECT LINES together with a geographical indicator (EG) to specify the protocol which is valid at primary-address-assertion time. At the same time the decision is made to select wither the USER-DATA or the CONTROL and STATUS (CSR) SPACE. There are three categories of PRIMARY ADDRESS PROTOCOLS:

- geographical: physical position is address
- logical   : 32 bit system wide address
- broadcasts   : classes of slaves addressed

8.3 PRIMARY ADDRESS PROTOCOLS

MS0	specific device	DATA SPACE
MS1	specific device	CSR SPACE
MS2	broadcast	DATA SPACE
MS3	broadcast	CSR SPACE
MS4..7	specific device	reserved

EG=0	MS0	logical	DATA SPACE
EG=0	MS1	logical	CSR SPACE
EG=1	MS0	geograph.	DATA SPACE
EG=1	MS1	geograph.	CSR SPACE

8.4 SS RESPONSES

SS0	Address recognized
SS1	Network busy
SS2	Network failure
SS3	Network abort
SS4	Reserved
SS5	Reserved
SS6	Invalid IA Address rejected
SS7	Invalid IA Address accepted

Every SLAVE must support the geographical access to the CSR SPACE. This allows unique access in order to initialize and identify the SLAVE.

9. ARBITRATION

A parallel, high speed ARBITRATION SCHEME [15] is used (Fig. 3) Masters issue an Arbitration Request (AR) in order to request the use of their SEGMENT. On receipt of the Arbitration Grant (AG) from the ATC-logic all participating MASTERS start to compare their 6 bit ARBITRATION VECTORS with the value on the bus. Detecting higher values than its own, a MASTER removes its value from the BUS. After a short settling time the highest vector remains on the BUS and the corresponding MASTER indicates his MASTERSHIP by asserting Grant Acknowledge (GK). AN ASSURED ACCES PROTOCOL allows low - priority MASTERS to get the bus subsequent to higher - priority MASTERS. The ARBITRATION PROCESS can take place in parallel with other FASTBUS OPERATIONS since the new MASTER is excluded from using the BUS for the duration of the current ADDRESS LOCK.

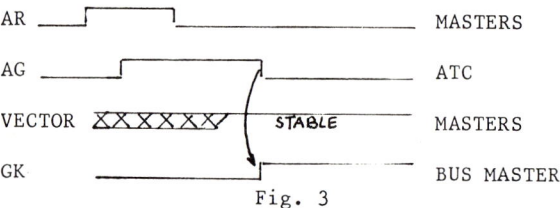

Fig. 3

10. ADDRESS SPACE

An ADDRESS FIELD of up to 32 bits can be used to connect SLAVES within a PRIMARY ADDRESS CYCLE. For internal use in the SLAVES, FASTBUS provides a dual 32 bits range for every SLAVE: the Control and Status Register (CSR) space for protected access and the normal Data - or USER space.

The PRIMARY ADDRESS FORMAT for LOGICAL ADDRESSING contains the GROUP NUMBER of the SEGMENT, the DEVICE ADDRESS and a part of the 32 bit INTERNAL ADDRESS RANGE.

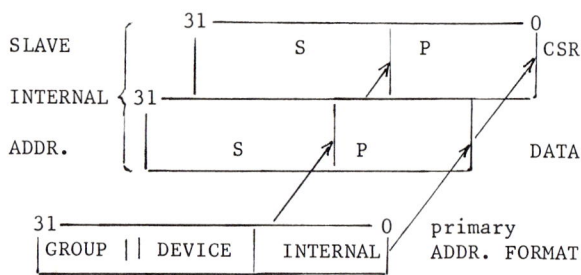

S: only accessible through Secondary addr.
P: accessible from Primary address cycle.

PRIMARY LOGICAL ADDRESS
Fig. 4

The INTERNAL ADDRESS RANGE which is not available during PRIMARY ADDRESS LOADING can be accessed through SECONDARY ADDRESS CYCLES which are standard ADDRESS LOCKED OPERATIONS and allow full 32 bit loading.

BROADCASTS and GEOGRAPHICAL ADDRESS FORMATS have no DEVICE - and INTERNAL ADDRESS FIELDS

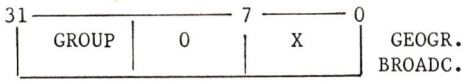

Fig. 5

The information X in the low byte specifies either the position of the DEVICE on the SEGMENT (geographical) or the condition or type of the BROADCAST. In both cases INTERNAL ADDRESSES have to be specified by SECONDARY ADDRESS CYCLES.

11. CONTROL AND STATUS (CSR) SPACE

The access to CSR SPACE is enabled at PRIMARY ADDRESS TIME whenever the MS1 or MS3 code is specified. The 32 bit CSR SPACE, of internal addresses, is divided into 4 sections of equal size:
* NORMAL
* PROGRAM
* PARAMETER
* USER

A variety of registers in the normal section and the meaning of their bit-positions are defined by FASTBUS. The register CSR (0) is mandatory for any SLAVE since it contains a unique MODULE IDENTIFIER and important functions such as the enable/disable of logical address recognition. Freqently used FUNCTIONS can be set or reset SELECTIVELY by writing a ONE to the corresponding bit-position and ZEROES to all other positions. (Fig. 6) CSR registers are mainly used for IDENTIFIERS, SYSTEM FUNCTIONS, USER FUNCTIONS, INTERRUPTS, ARBITRATION VECTORS, LOGICAL ADDRESS STORAGE and for the SEGMENT INTERCONNECTS.

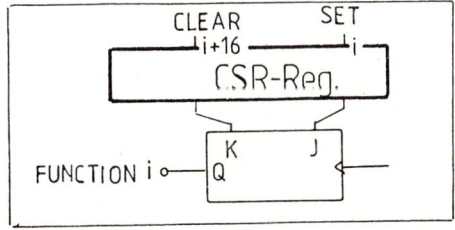

11. PARALLEL PROCESSES

FASTBUS provides a fast communication path between logically addressable MODULES of a processing system. This can consist of several PROCESSOR MODULES, shared MEMORY MODULES and IO MODULES on one CRATE SEGMENT. The efficiency of such a system is enhanced by the parallel action of the ARBITRATION, the very-high BANDWITH (200 MHz for ECL-backplanes), the ability of FASTBUS MECHANICS to power and house self-contained PROCESSOR-DEVICES (up to 350 IC in one DEVICE), and the 32 bit architecture. The concept of segmented busses additionally allows distributed parallelism in the FASTBUS System.

12. TOPOLOGY

The FASTBUS SYSTEM consists of 1 to n terminated BUS ELEMENTS called SEGMENTS. CRATE SEGMENTS provided a HIGH-SPEED COMMUNICATION PATH for its attached DEVICES of up to 500 Mb/s, while BUS CABLE SEGMENTS stretch over distances of tens of metres to interconnect DEVICES. CRATE and CABLE SEGMENTS are connected by SEGMENTS INTERCONNECTS [16]. These allow complete separation of two adjacent BUS SEGMENTS, or linkage to form one single BUS.

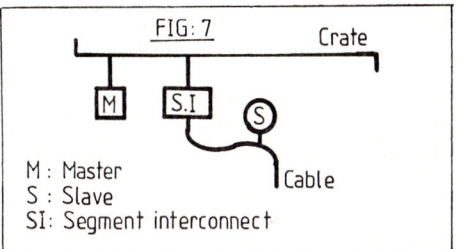

Since SEGMENTS can have several SEGMENT INTERCONNECTS, complicated TOPOLOGIES can be built up.

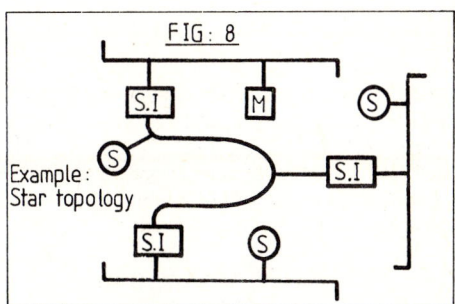

13. INTER SEGMENT COMMUNICATION

SEGMENTS are disconnected or linked together by the SIs within PRIMARY ADDRESS CYCLES according to the high order part of the address field. This GROUP FIELD (recommended as 8 bits) is used as a key to pass the SEGMENT boundary.

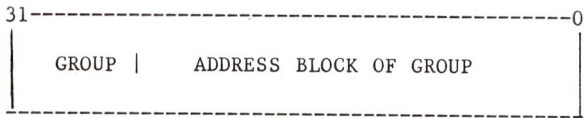

Primary address format
Fig. 9

The SIS contain ROUTE TABLES to match with the GROUP FIELD information. They can be initialized to the address groups to be passed to the FAR SIDE SEGMENT. The entries in the ROUTE TABLES thus define the TOPOLOGY of the BUS SEGMENTS to be linked together. An SI seeing a "PASS ADDRESS" will act as a SLAVE who has recognized the address but who delays the ACKNOWLEDGE signal by asserting WAIT. On its FAR SIDE the SI acts as a MASTER who arbitrates for the BUS SEGMENT and, being BUS MASTER initiates a new PRIMARY ADDRESS CYCLE. If this is not yet the DESTINATION SEGMENT the next SEGMENT is linked in the same way. The ACKNOWLEDGE of the SLAVE on the DESTINATION SEGMENT is propagated back to the originating MASTER together with a removal of WAIT. It there is no ACKNOWLEDGE only WAIT is removed and the MASTER times out. Because of the step by step ARBITRATION for every new SEGMENT the setup time for the connection through segments may be considerable. The ADDRESS LOCKED OPERATIONS however are possible at much higher speed, especially if PIPELINED TRANSFERS are used.

14. FASTBUS IN PRACTICE

Prototyping efforts during the last years in several laboratories in both Europe [17] and the USA [18] have produced a series of standard SYSTEM DEVICES and INTERFACES being in use in several FASTBUS systems. A FASTBUS PILOT PROJECT has been successfully built at CERN [19, 20] using a mixture of 100K ECL SYSTEM MODULES [9, 21, 22].

CONCLUSION

The FASTBUS STANDARD defines a very flexible high performance bus which is technology and manufacturer independent. Its regular ARCHITECTURE on the basis of BUS SEGMENTS with system wide PROTOCOLS supports variable and modular SYSTEM DESIGNS in both COMPUTER and INSTRUMENTATION fields. Dynamically variable BUS TOPOLOGIES allow distributed processors to communicate efficiently on the basis of a 32 bit system wide address. A PILOT PROJECT has proven the reliability and logical consistency of the standard.

REFERENCES

[1] FASTBUS SPECIFICATION/US NIM Committee, September 1982.

[2] THE FUNDAMENTALS AND STATUS OF FASTBUS, E.M. Rimmer, CERN-DD, August 1982.

[3] INTRODUCTION TO THE FASTBUS STANDARD DATA BUS. R. Larsen, SLAC-PUB 2859.

[4] FASTBUS FUNDAMENTALS, P.J. Ponting, CERN in EP-Electronics Note 82-03.

[5] CABLE SEGMENT, engineering Note, L. Pregernig, CERN-SPS.

[6] FASTBUS COUPLER PROPOSAL, G. Fremont, CERN EP-Electronics.

[7] FASTBUS MASTER AND SLAVE INTERFACING, H. Mueller CERN-EP September 1982.

[8] FASTBUS SLAVES, R. Downing, IEEE, Trans. Nucl. Sci. Vol NS-28 No 5.

[9] SDS-CONTROLLER SPECIFICATION, L. Pregerning, CERN-SPS.

[10] FASTBUS SEQUENCER, G. Fremont, CERN EP-Electronics.

[11] 68000 general purpose master (GPM) H. Mueller, CERN-EP prelim. SPECIFICATION.

[12] REGISTER TO FASTBUS INTERFACE (IORFI) C. Logg and L. Paffrath, SLAC.

[13] UNIBUS PROCESSOR INTERFACE TO FASTBUS (UPI), E. Barsotti, FNB008.1/FNAL September 1981.

[14] AN INTELLIGENT INTERFACE BETWEEN FASTBUS AND A MICROCOMPUTER BUS, A.W. Booth, CERN-DD.

[15] CONTENTION RESOLVING CIRCUITS FOR COMPUTER INTERRUPT SYSTEMS/ Proc. IEEE Vol. 123 No 9, September 76, D.M. Taub et al.

[16] THE SEGMENT INTERCONNECT, R. Downing IEEE Trans. Nucl. Sci. Vol.NS-29 No 1.

[17] STATUS OF FASTBUS IN EUROPE, H. Foeth, IEEE, Trans. Nucl. Sci Vol. NS29 No 1, February 1982.

[18] Several papers in IEEE, Trans. Nucl. Sci. Vol. NS-28 No 5 and Vol. NS-29 No 1.

[19] TRIGGER SYSTEM IN PILOT PROJECT, L. Gustafson in CERN-EP-Electronics Note 82-03.

[20] THE FASTBUS PILOT PROJECT, R.W. Dobinson in CERN EP-Electronics Note 82-03.

[21] FASTBUS DISPLAY AND MEMORY (FDM), H. Mueller, CERN EP Specification July 1982.

[22] FASTBUS ROMULUS INTERFACE, L. McCulloch CERN EP-Electronics Specification.

AN IMPROVED ETHERNET FOR REAL-TIME APPLICATIONS

R. Hainich

Hahn-Meitner-Institut fuer Kernforschung Berlin GmbH

Germany

ABSTRACT

In the area of office communications, Ethernet is likely to become a standard because of its flexibility, reliability and ease of use, properties also desirable with real-time applications. Unfortunately, Ethernet does not guarantee an ordered scheduling of tranmissions and has no means for insuring functional security at overload.
We will propose some simple modificatons to the Ethernet protocol, providing for nearly deterministic round-robin message scheduling while maintaining interconnection compatibility with normal Ethernet interfaces.
The improvements are achieved by varying the random delays following collisions, in dependency of the individual message waiting times and the number of active nodes, which is estimated from the number of collisions taking place. Results from computer simulations involving up to 1000 nodes are shown.

ZUSAMMENFASSUNG

Ethernet wird wahrscheinlich ein Standard in der Bürokommunikation werden, wegen seiner Flexibilität, Zuverlässigkeit und einfachen Anwendung, Eigenschaften, die auch in der Prozeßdatenverarbeitung von Vorteil wären. Leider erlaubt Ethernet keinen hinreichend geordneten Nachrichtentransport; ebenso existieren keine Vorkehrungen, um eine Minimalfunktion bei Überlast sicherzustellen.
Wir wollen hier einige Protokollmodifikationen vorstellen, die eine nahezu deterministische Zugriffszuweisung erreichen und trotzdem die volle Kompatibilität zu normalen Ethernet-Interfaces bewahren.
Die Verbesserungen werden erreicht durch Variation der zufälligen Verzögerungszeiten nach Kollisionen, in Abhängigkeit von den individuellen Nachrichten-Wartezeiten und der Anzahl der aktiven Stationen, welche nach der Anzahl der auftretenden Kollisionen geschätzt wird. Wir zeigen Resultate von Computersimulationen mit bis zu 1000 Stationen.

RÈSUMÈ

Pour les applications de bureautique, Ethernet est en passe de devenir un standard grâce à sa flexibilité, sa fiabilité et sa facilité d'emploi, propeétes "egalement recherchées dans les applications temps réel. Malheureusement, Ethernet ne garantit pas d'ordonnancement particulier des transmissions et ne peut se camporta de façon prévisible en cas de surcharge. On propose des modifications simples au protocole d'Ethernet, assurant un ordonnancement quasi-détermininiste des messages tout en maintenant la compatibilité avec les interfaces standard Ethernet. Les améliorations sout obtennes en faisant varier les délais de déférence en fonction des délais d'attente et du nombre de noeuds actifs, estimé d'après le nombre de collisions. Des resultats de simulation pour un reseau de 1000 noeuds sent fournis.

INTRODUCTION

Ethernet has a simple, reliable protocol with no need for initialization (which is a weak point with many other networks). The absence of special protocol messages and complicated procedural dependencies guarantees for high throughput and deadlock-free operation.

In terms of office communications, Ethernet is fast. In most cases, its limited message length and high data rate result in short response times.

With real-time applications, however, a secure upper limit of response time (message delay) is needed, especially under overload conditions.

Because it is normal with such applications that several nodes are functionally linked by a common process, there is no way to exclude overloading especially in delicate situations. In this case, waiting times have extremely wide deviation because Ethernet gives all nodes arbitrary chances of getting their messages transmitted, to the point of discriminating messages with longer waiting times. This may cause delays more than ten times longer than necessary.

TRANSPARENCY

We concentrated most of our investigations on overload simulations (see also 'Simulation Model'). This 'worst case' is especially worth of consideration, because there is no way to prevent it, except for extremely oversizing the network capacity.

Overload will always cause long message queues within the nodes; however, if every node can always send its currently most important message within a short, predictable time, this essentially improves safety and simplicity; we call such a net 'transparent' to the upper system layers, meaning that it guarants for a minimum transport ability in every situation.

It should be possible to insure that of n nodes, none would have to wait for significantly more than n-1 messages of others, until it may transmit at least one message of its own. Our suggested improvements are especially aimed at this goal.

We will not consider the internal message queueing within the nodes; we assume that they promote their most important ones. We also do not use information that is not available at the Link Layer, e.g. global priorities /5/.
We will not try to make any improvements of contention time or number of collisions; however, the maximum throughput of the network does not suffer from our enhancements, while the worst-case response times at high load are reduced at least by a factor of 10.

The improved strategy was developed from a CSMA/CD strategy we call CSMA-B; after a few modifications, we were also able to obtain a result that is plug-in compatible with the Ethernet standard /2/.

ETHERNET

Although Ethernet is described thoroughly in /9/, /10/, and especially in /2/, we will highlight some points important to us. We will ignore all aspects of data representation but are interested in what we call the Ethernet 'protocol', a CSMA-scheme (Carrier Sense Multiple Access) with collision detection.

Ethernet nodes, when having a message to send, wait until they detect the channel to be idle and then start transmitting. Because of the signal propagation delay, some nodes might start transmitting simultaneously, an event called 'collision'. It might take twice the delay time from one end of the cable to the other until all of them sense this by watching for distortions of their own transmission. In /2/, the worst case round trip delay (including some safety margins) is called a 'Slot'.

In case of a collision, the nodes cease transmission for a random number of Slots and then try again as soon as the channel becomes idle. The node with the accidently smallest random delay then gets access while all others keep waiting. Using random delays minimizes the probability of repeated collisions.

Because the shortest message length with Ethernet is not much longer than one Slot, several messages of other nodes may pass during the random delay time. With our first protocol variations, we assumed that the nodes finish delaying as soon as they sense a signal (but still wait until the net becomes available) because this faciliates queueing control by the nodes themselves.

With low traffic, collisions will be rare; following the end of a transmission, however, they are to be expected more frequently, because other nodes might have got ready for another access attempt or have provided new messages meanwhile. If lots of messages are provided, queues will develop and several nodes will collide each time the net becomes available. If the nodes suspend delay counting as long as they sense a signal on the net, this 'First Collision' may be avoided /8/. As to /2/, Ethernet does not use this method.

OPTIMUM DELAYS

For our considerations, some rules on choosing optimum random delay times will be of interest:

1) It is reasonable to choose delay times that are integral multiples of one Slot (/2/, pp.13).
2) The time values should be equally distributed between 0 and a maximum value D_m equal to the number of nodes involved in the contention /10/.
If we further try to avoid the initial collision when the channel gets available, by delaying before the first transmission attempt, the total contention time (i.e. all subsequent delays and collisions until one first single node may transmit successfully) converges to e-0.5 (2.22..) Slots as node numbers go to infinity (assuming worst-case conditions about

cable length and node positioning on it) /4/. With First Collision, up to one more Slot is wasted at high load. The contention time can keep that short because among many nodes, there is probably always one which computes a short delay time.

There are deterministic Protocols that might need as few as 1 Slot if all nodes are involved /7/, but error recovery, switching nodes on and off and especially flexible addressing is still a problem with this.

B E B

In order to avoid excessive delay times as well as too many collisions, Ethernet starts up with an instant transmission attempt and then uses delay times redoubled at every collision: $D_m = 2 \uparrow C - 1$, where C is the number of collisions that a station experienced since first trying to transmit its actual message. This strategy is called 'Binary Exponential Backoff' (BEB). While collisions are very likely while D_m is short, the chance of resolving the conflict rapidly increases when it becomes longer. Therefore, this works with any number of nodes, making a deadlock impossible. In /2/, this Strategy is modified to 'Truncated BEB', where D_m is limited to a maximum value of 1024. This may help to avoid uselessly long delay times in some cases, but may also increase contention times when some hundred nodes are competing (see fig.1).

ETHERNET WAITING TIMES

Because the Collision Counter of an Ethernet node is reset after its message has been transmitted, but nowhere else, and because delay times are always completed regardless of any events on the net, some bad effects on access assignment take place.
At high load, nodes often finish delaying while some message is being transmitted. They wait until the net becomes idle and then try to transmit. If there are more nodes or the one just having transmitted still has another message, a collision occurs; the node that just had access would win this contention almost for sure, because its C was set to 0 (resulting in zero delay), while all others get their C's increased, lowering their future chances.
By this mechanism, a node once having access may send as many subsequent messages as it wants, leaving almost no chance for the others ('Blocking' the net). This might be useful when long documents are to be transmitted, splitted in many packets, but is dangerous with real-time applications.
It also has the advantage of dramatically lowering the contention times (but only if every nodes really wants to send several messages). In practice, nodes are not likely to send more than a few messages in immediate series, due to the slowness of their software and their limitation in data volumes; however, nodes newly becoming active, also have C-values of 0 and might therefore alternately block up others that have waited longer. This implies that message waiting times may at best have a distribution like with random queueing (fig.2);

however, especially with load patterns that are good for queueing (i.e. that inhibit Blocking), Ethernet's contention times tend to be too long because its BEB increases the delay values too fast.

SIMULATION MODEL

The empirical studies we have undertaken could only be done by means of simulation. Simulation of very large systems requires careful modeling to keep computing times as short as possible.

We decided to use the Slot as the smallest time raster; Ethernet's message lengths are therefore approximated to be from 1 to 24 Slots. Because this is nearly accurate for the shortest message length, no finer resolution was required.

The simulation routine was of course event-driven. Message generation was allowed only during transmission time, i.e. no new demands arose during contention. This is suitable in the infinite-load situation upon which we concentrated in this research, as well as with long messages.

At high load, messages will queue up within the nodes causing a number of them to be constantly competing.

Because the number of nodes actually competing is statistically dependent on the number of connected nodes and the present system load (lower load means that fewer of them are involved at the average), considerations about strategy behaviour in this situation, with different node numbers, will also apply to limited periods of time under medium load.

In order to avoid Blocking with the Ethernet strategy, we also ran simulations where a total of n nodes out of 1000 nodes were revolvingly kept active by random message generation.

Our simulation program had a runtime of O(nodes*messages). We ran several 1000 simulations, with about 400 Million nodes*messages.

ETHERNET AT OVERLOAD

With Ethernet, overload may have diastrous consequences: If it continues for some time, the Collision Counters in the nodes approach 16 after about 4000 Slots, causing an error message and possibly suspending the node's activities, a catastrophic event with many real-time applications (if there is overload, a node that accidently experienced some collisions, will likely not get access because of its high C-value but increase it until 16 in 'First Collisions').

In order to get some reasonable results for comparison at high load, we assumed that the Ethernet nodes' Collision Counters are simply reset to 0 when approaching 16; this will then occur frequently (see also fig.3).

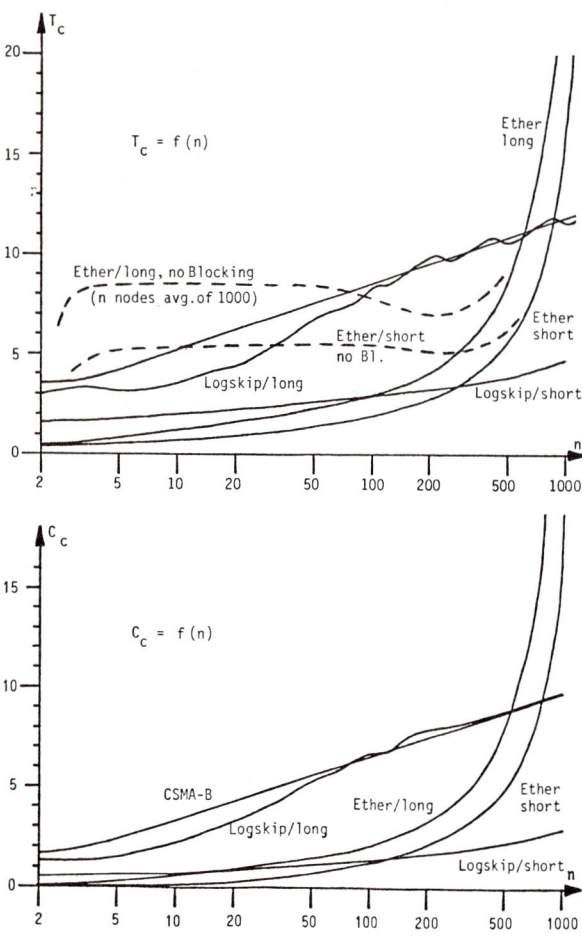

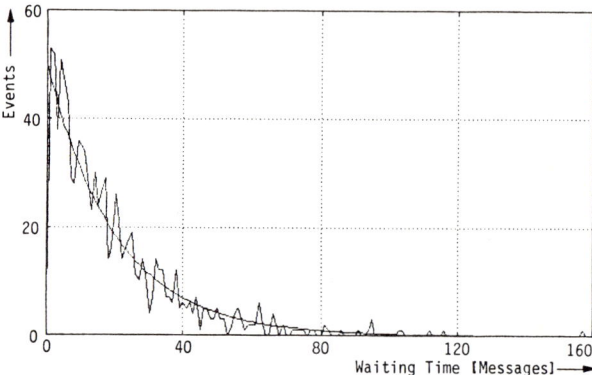

$$r = \frac{m0}{N} e^{-\frac{m}{N}}$$

Where N : number of active nodes,
       m : waiting time in avg. message-lengths,
       m0: total messages processed,
       r : no. of messages that were waiting for m others.

Fig.2: The theoretical r-curve for 20 nodes after 1000 messages, together with a simulation result.

EXPECTED MAXIMUM

r gives the distribution of m0 discrete events; we could split the area under the curve into areas of the content 1, each representing a single event. We further might think of those events like being placed exactly in the middle of the areas, between two adjacent areas of the size 1/2. With this, the area right of the longest waiting time to be expected, mx, should also be 1/2:

$$\int_{mx}^{\infty} r \, dm = \frac{1}{2}$$

Leading to
$$mx = N \ln 2 \, m0$$

This is very close to our simulation results. For a system with accurate round-robin queueing, the corresponding formula would write like:

$$mx' = N - 1$$

Although the m e a n transmission waiting time of a n y protocol always has to be N-1, simple CSMA will easily produce maxima 10 times longer. Ethernet with Blocking shows maxima more than 100 times above the average; however, we have to admit that few messages will wait for longer than the duration of the longest congestion phase occuring (during an idle interval, they have a good chance of being tranmsmitted).

Fig.1 shows the mean contention time (Slots), Tc, and the mean number of collisions per contention, Cc, for various strategies at infinite load (n nodes constantly competing). Ethernet and the compatible Logskip strategy are shown at short and long messages (1 resp. 24 Slots). Because Ethernet's BEB is truncated at 1024, contention times with Blocking explode above 200 nodes. If Blocking is prevented (see also 'Simulation Model'), contention times are always higher than necessary.

WAITING TIMES WITH RANDOM ACCESS

The term 'Delay' often used in communication protocol analysis does not sufficiently characterize the performance with real-time applications. First, it is a mean value, while here we are interested in maxima; secondly, it normally includes queueing within nodes, which we cannot consider to be part of a protocol defining the two lowest ISO-layers only. Of primary interest to us are the waiting times of messages already offered to the net until their transmission, especially their deviation and expected maxima.

A simple random access scheme without any influences on queueing may illustrate this. Because each message has the same chance at every contention phase, the waiting time distribution (no. of messages transmitted above waiting time, expressed in message lengths), is given by the Poisson distribution:

STRATEGY DESCRIPTION

The most important points of the strategy variants to be presented will be shown in tables like the following:

```
S T R A T E G Y  : ETHER
Collis./Involved : C:=C+1, Dm:= min(2^C-1,1024)
        Actives  : -
Success/Winner   : C:=0,   Dm:=0
        Losers   : -
```

This describes the Ethernet strategy by the action taken on strategy variables (C, the collision counter, and Dm, the maximum delay time) at certain events: We distinguish between Collision and Success (one node gets access).
At Collision, there are Involved (colliding) and Active nodes (having messages pending, but delaying).
At Success, the Winner node transmits, while the Losers keep waiting.
Nodes having no message pending (Passives) are not subject to any of the strategy variants.
The strategy formulas are ordinary assignments, and the leftmost expression is to be computed first.
A random delay D between 0 and Dm is computed with all strategies; this is therefore not mentioned.
We can now easily identify the BEB at the Collis./Involved line, where C is incremented and exponentiated.

With Ethernet, C and Dm are reset at Success and not changed at Losing. This is exactly the place for improvements.

INITIAL MODIFICATIONS

The most natural way to avoid exessive waiting times for network access is, to establish some mechanisms of round-robin access assignment among the competing nodes. Because Ethernet nodes will only react on their own collisions, some changes are required before we can think of introducing any form of distributed queueing control; the following strategy we used as a base for improvements:

```
S T R A T E G Y  : CSMA-B
Collis./Involved : C:= C+1, Dm:= 2^C-1
        Actives  : C:= C+1, Dm:= 2^C-1
Success/Winner   : C:= 0,   Dm:= 0
        Losers   : C:= 0,   Dm:= 0
```

Here, all active nodes sense every collision and all successful transmissions. They may sense collisions they are not involved in, because the resulting Collision Fragments are always shorter than valid messages (/2/,pp.14). All delays are ended and all collision counters reset at Success. In addition all C's are incremented at every collision on the net (but only in nodes competing for access); we call this behaviour 'Global Consensus' (GC). The BEB starts over again at each contention phase, which of course increases the number of overall collisions; however, there are no overflowing C's, and every node has at least an equal chance at each contention phase. The overhead resulting from contention times can be kept sufficiently low.

QUEUEING CONTROL

Most interesting with CSMA-B is, that the average number of collisions during one contention phase, Cc, is logarithmically dependent on the number of competing nodes, with a standard deviation of about 1 (fig.1), /4/:

$$Cc \simeq log2(Na) \quad \text{(binary logarithm)}$$

Therefore, we can estimate Na from Cc and use this knowledge in order to vary the maximum delay times Dm of the nodes, giving them different chances for access according to their different waiting times.

At ideal round-robin access assignment, each node should wait for exactly one transmission of all other competing ones before it claims access itself; its waiting time then equals the number of competitors minus one (itself):

$$Qopt = Na - 1$$

In order to know the right time for an access attempt, the node neeeds to know its own waiting time Q as well as Na.

A simple counter for successful transmissions by others, started when an own message gets pending, will deliver the right value for Q.

Knowing Q and, approximately, Na, is sufficient for variating Dm appropriately. We want some algorithm that tends to decrease Dm until zero in the longest waiting node and lets it remain higher in the others. If more nodes happen to calculate short delay times because of the variability of the Na-estimation, this is no matter of concern because the conflict may easily be solved by maintaining the BEB.

We will maintain the strategy formula $Dm=2^C-1$. Then, we must vary C in order to get the right Dm-values for queueing control. We will still count collisions with C, so that it increases by Cc at every contention phase; however, after the contention, the loser nodes shall decrease their C by a value derived from the waiting time Q. Because now C does more than simply counting collisions, we will re-christen it 'Collision Weight'.

Idealizing, we can define a minimal condition that C should have the value 0 in the longest waiting node and the value 1 in the second longest waiting one following to a contention.

This would cause the C of the former second longest waiting node to increase to Cc+1 (averagely) at the next contention. The longest waiting node we assume to transmit; its C rises to Cc. Our new longest-waiting node should get its C down to zero by subtracting some function of Q:

$$Cc + 1 - f(Q) = 0$$

With

$$Cc \simeq log2(Na)$$

we get

$$f(Q) \simeq log2(Na) + 1$$

We assume round-robin queueing

$$Q \simeq Na - 1$$

and get

$$f(Q) \simeq \log_2(Q+1) + 1$$

INTEGER LOGARITHM

With regard to simple implementation, we did not consider the use of floating-point logarithms. Instead, we use an integer logarithm that can simply be derived from a priority encoder:

$$f(Q) = ld'(Q)$$

where
$ld'(0)=0$, $ld'(1)=1$, $ld'(2,3)=2$, $ld'(4,5,6,7)=3$ etc..

Simulations showed that more 'accurate' implementations of $f(Q)$ would produce no better results.

We may summarize: The Collision Weight is incremented at every collision (as was), but then decreased by $ld'(Q)$ after the contention. We will see that this simple approach really works.

THE WINNER NODE

We still need an appropriate strategy for the Winner node. We know, that the shortest contention times may be expected when $Dm=Na$. Therefore, with

$$Dm := 2\uparrow C - 1$$

we get

$$C \simeq \log_2(Dm+1) \simeq \log_2(Na)$$

Because we said that in the winner node, C should have approached

$$C \simeq Cc \simeq \log_2(Na),$$

it should already have the most reasonable value; however, this worked out poorly with low node numbers. We therefore prefered to set C explicitly:

Assuming

$$Q \simeq Na$$

we may define:

$$C := ld'(Q)$$

This leads to lower contention times with few nodes; however, one may doubt if the additional effort is worthwhile, because few nodes simply can't produce high load for long, due to the slowness of their software.

LOGLOG

The strategy is now complete:

```
S T R A T E G Y  : Loglog
Collis./Involved : C:= C+1, Dm:= 2↑C-1
        Actives  : C:= C+1, Dm:= 2↑C-1
Success/Winner   : C:= ld'(Q), Dm:= 0, Q:= 0
        Losers   : Q:= Q+1, Dm:=0, C:= C-ld'(Q)
```

The difficulties in the theoretical analysis of this strategy led us to the use of simulation in order to investigate its performance and to test the effects of changes in the formulas. The conclusion of this investigation was that performance was good and no changes were recommendable except for some minor additions.

C-LIMITATION

A theoretical analysis of the development of the C-values within single nodes under equilibrium conditions, as well as simulations, showed it to be reasonable to limit C to a maximum of about 16. We also found that the C-values would react very fast if we changed the number of active nodes; the strategies' dynamic behaviour at load changes is excellent.

C-RANGE

An implementor should know that with Loglog, C-values down to -8 are to be supported, although they are to be interpreted as 0 when it comes to calculating Dm (see fig.5)

DELAY TIMEOUT

Given the strategy formula $Dm=2\uparrow C-1$ with a maximum C of 16, it becomes obvious that Dm could theoretically become very large. Although this could only result from an instantaneous load decrease by violently discarding nodes or messages, it could sometimes become a problem.

A simple modification will solve this: D (not Dm !) should be limited to about 16 Slots.
If the net has been idle for this time, C should be lowered or set to 0 (we preferred the latter), and D should be recomputed (we call this 'Delay Timeout'). If more nodes are still competing, all are doing this simultaneously; the instant access attempt (D=0 !) results in a contention phase that serves for regaining reasonable C-values. An arbitrary node gets access, but the Losers then decrement their C's according to their waiting times, preparing the next messages to be sent in the right sequence.

Because Delay Timeouts are extremely rare, other strategies than setting C to 0 in this case will yield no practical advantage.

With the Ethernet strategy, Delay Timeout could replace the present Dm-limit of 1024.

QUEUEING PROPERTIES

Message waiting times, in our terms, begin when the message is at the head of the node's output queue and last until transmission. We are not concerned with the waiting times within the node's internal queues. 'Queueing' therefore takes place between the foremost messages or, so to say, between the nodes themselves.

We will show queueing behaviour by the waiting-time distribution: The number of messages having waited for some other messages until transmission, are plotted above those times, that are denoted in (average) message-lengths. The sample plots shown are representative.

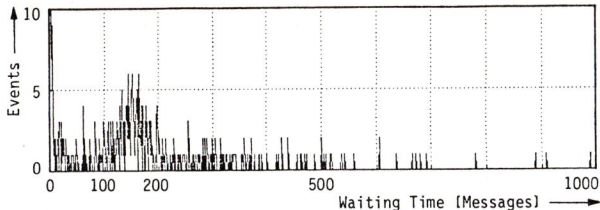

Fig.3 shows the distribution of the waiting times (in message lengths) with an Ethernet of 100 nodes after 1000 messages of maximum length have been transmitted (with Blocking). Note that one message had waited for almost all others. The maximum around 150 messages ((24+3)*150≈4000 Slots including contention time) is caused by the Collision-counters approaching 16 and being reset after about 4000 Slots at the average; this temporarily increases the node's chances at contention. If Blocking was inhibited, the distribution would more resemble to fig.2.

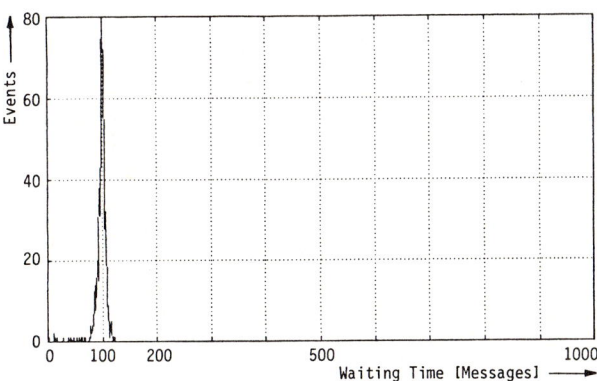

Fig.4 is drawn from a Loglog-simulation with the same assumptions like in fig.3 in order to show the difference in queueing behaviour.

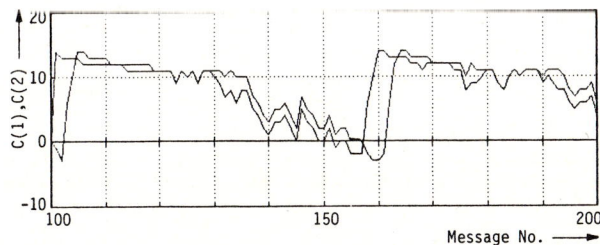

Fig.5 shows the Collision Weights of 2 Loglog-nodes out of 60, at long messages. The Collision-Weights leap upwards after transmission and then stick to their limit until they go down again when the time for the next transmission has come. Negative values are interpreted as zero; This causes some extra collisions, but it is also good for proper queueing. Throughput is not seriously affected, because with the Ethernet-compatible strategy Logskip (see 'Compatibility'), we can save contention time with short messages (fig.1).

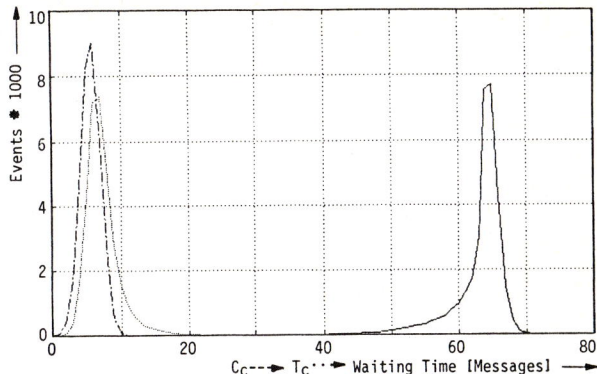

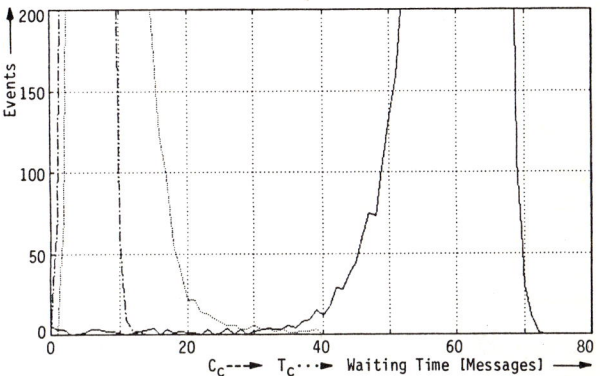

Fig.6 was drawn from a long-term simulation with 64 Loglog nodes over 30000 messages. In the second drawing, we have enlarged the vertical axis, in order to make single events visible. The waiting time distribution (solid line) is well limited; no single message had to wait for more than 72 others. The conclusion out of several very long simulations was, that maximum waiting times do not increase by more than one single message (at any number of nodes) if the number of processed messages is multiplied by a factor of ten. Also shown in the above diagram is the number of collisions per contention (dashed line) and the contention time in Slots (dotted line).

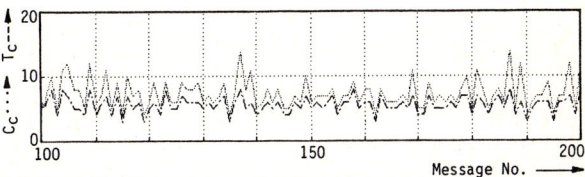

Fig.7 shows the contention times and collisions associated with each of 100 subsequent messages in a net of 60 Loglog nodes (like fig.5). The Collision numbers (dash-dotted line) have only a narrow deviation; an extra Collision Counter could be used for error detection, with a threshold of 16 like with Ethernet. The Contention Times (dotted line) show some minor spikes; this is similar to Ethernet.

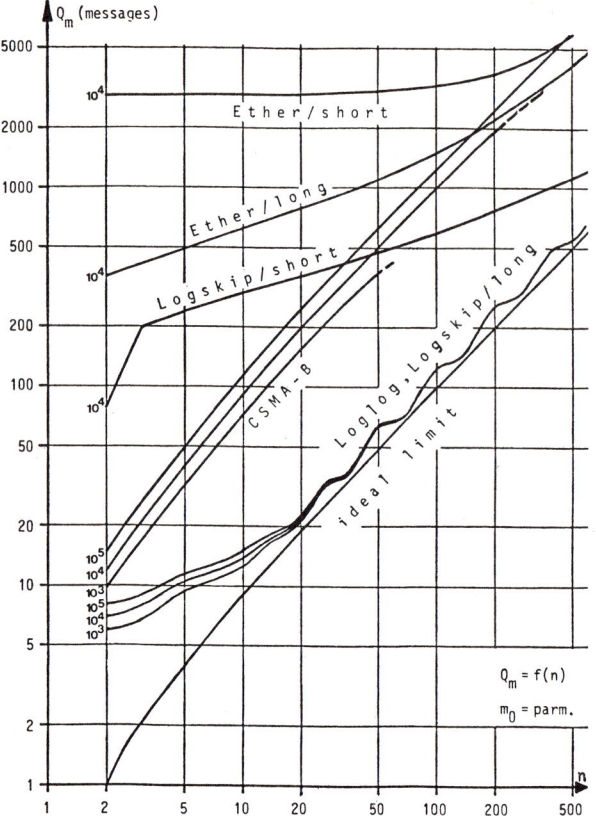

time happens to 'skip' over the duration of a short message by others. This even allows for some 'Blocking'.

Everything else remains the same. During Contention, Global Consensus is maintained. With messages of 16 or more Slots in length, delay ends during the transmission, but this is not considered to be a Timeout, because the net has not been idle;
therefore, at message lengths of 16 up, Logskip is absolutely identical to Loglog.

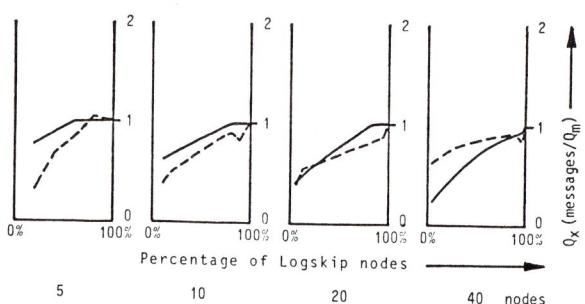

Maximum expected message delays for the Logskip nodes
(normalized, message length ---1, —— 24 Minislots)

In fig.8, we have collected the information about waiting times. The maximum message waiting times Qm (in message lengths) to be expected with the different strategies are shown as a function of the number of active nodes N (load=1), for message lengths of 1 (short) and 24 Slots (long) and for different observation times (total messages processed, m0). Numerous simulation runs were necessary to derive these curves. The vacillation of the Loglog curves is due to the use of ld' (integer logarithm) in the strategy formulas, which makes performance somewhat dependent on the number of nodes with respect to the next power of 2. If Blocking is inhibited, Ethernet's maximum waiting times may be lower.

COMPATIBILITY

During our investigations, we found that interfaces using different CSMA/CD schemes could join the same net, provided that their retransmission attempt probabilities during contention are approximately alike. Loglog and Ethernet nodes are not compatible; however, little modifications of the Loglog strategy are sufficient to achieve this. Because we also desired to improve Loglog's throughput at short messages (i.e. to achieve less collisions), we tried the following change:

A node finishes delaying only if its delay time ends naturally (still limited to 16) or if it senses a collision. The node only consideres itself to have been 'Losing', if its delay time ends during the transmission by another one. It therefore does not increment its Q and does not take part in a First Collision if its delay

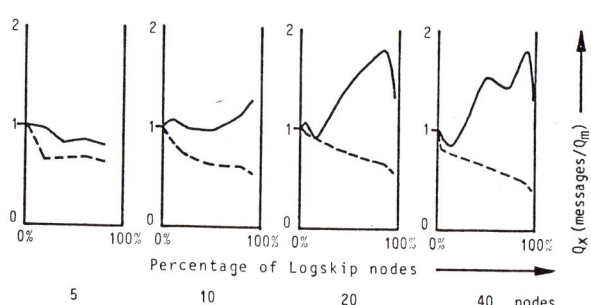

Maximum expected message delays for the Ethernet nodes
(normalized, message length ---1, —— 24 Minislots)

Fig.9 contains detailed information about the performance of mixed systems: it shows the relative message delays to be expected with different percentages of Logskip nodes in Ethernet systems of 5,10,20 and 40 nodes.
The general conclusion is, that there are no substantial disadvantages, if especially the more demanding nodes are equipped with Logskip. We have found that the Ethernet nodes may get a smaller relative share of the channel capacity if they represent less than 1/4 of all nodes. This is, however, no serious problem, because the Ethernet nodes, as said, would already show undesirable behaviour under extreme load situations. We have indications that throughput and compatibility might be further improved with some compromises in queueing behaviour; however, queueing showed to be the most important key to short waiting times. We therefore decided first to concentrate on thoroughly investigating the Logskip strategy, which, in its present form, is adequate for upgrading exitsting netorks without problems.

Because with short messages Q is not always incremented, the nodes cause less collisions on the cost of queueing.

The worst-case waiting time for access (most important with real-time applications) remains unchanged because it depends mostly on the behaviour with long messages. Even with short messages, queueing is still better than that of Ethernet with long messages (see fig.8).

An important point is, that Logskip is indeed fairly compatible with Ethernet (fig.9). Channel utilization (given by the contention time / message length relation) with Logskip is, all in all, no worse than with Ethernet.

PRIORITY CONTROL

Sometimes there is a demand for processing certain messages faster than others. This might be achieved by presetting the Q-values within certain node interface to values other than zero at message generation.
Care must be taken to keep all waiting time demands within the net consistent. Some of the Q-values therefore should begin below zero, interpreted as zero in the strategy formulas.

This type of priority control is very interesting for its security, because simulations showed that Loglog and Logskip nodes would recover almost instantly if we were setting any of their strategy variables to inconsistent values.

RESPONSE TIMES

We get the worst-case message waiting time by taking the maximum message length of 24 Slots, adding the contention time (fig.2) and multiplying this with the maximum waiting time (in messages) from fig.8.
Given 50 microseconds Slot length and 1024 nodes, we get 40 seconds for Ethernet with Blocking (unrealistic, because the nodes can't generate messages fast enough), about 30 seconds for Ethernet without Blocking and 1.8 seconds for Logskip.

Because the absolute minimum (no contention time, perfect queueing) would be about 1.3 sec., Logskip is already near optimum. If some application requires time limits below this, one would have to reduce the maximum message length, the maximum node number, or even the Slot time.

CONCLUSION

We have shown a CSMA/CD strategy with controlled queueing behaviour, maintaining the security advantages of a decentralized, probabilistic protocol.
Priority control may be achieved in a very fault-tolerant way. Interfaces using the proposed strategy may work together with Ethernet interfaces on the same cable, allowing for easy and fully compatible upgrading of existing networks.

ACKNOWLEDGEMENTS

This work originated from a Diploma Thesis /3/ done at the Institut fuer Prozessdatenverarbeitung of the Technical University of Berlin. The author also wishes to thank the Process Control Center (Prozessrechnerverbundzentrale-FSP/PV) of the Technical University of Berlin for providing computing time for the simulations.

R E F E R E N C E S

/1/ Abramson, N., 1970. The ALOHA System-Another Alternative for Computer Communications. AFIPS Conf.Proc. vol.37, (1970)

/2/ BLUE BOOK, 1980. XEROX Corporation, OPD Systems Development, 3450 Hillview Avenue, Palo Alto, CA 94304

/3/ Hainich, R., 1980. Problemanalyse und Entwurf eines dezentralen, lokalen Kommunikations- Systems fuer die Realzeit-Datenverarbeitung. Diplomarbeit am Inst. f. Techn. Informatik der TU Berlin

/4/ Hainich, R., 1983. Acess Mechanisms for CSMA/CD with Real-Time Applications (to be published)

/5/ Iida, Ishizuka, Yasuda, Onoe, 1980. Random Access Packet Switched Local Computer Network w. Priority Function. NTC'80, IEEE 1980 Nat. Telecomm. Conf., Houston,TX, USA, 30Nov-4Dec

/6/ Johnson, D.H., O'Leary, G.C., 1979. A Local Access Network for Packetized Digital Voice Communication. NTC'79, IEEE 1979 Nat. Telecomm. Conf., Washington, DC, USA, 27-29 Nov 1979

/7/ Kleinrock, L., Scholl, M.O., 1980. Packet Switching Radio: New Conflict-Free Multiple Access Schemes. IEEE Trans. Comm. 7/80

/8/ Labetoulle, J., 1980. Etude Analytique du Reseau DANUBE. (Communication no.204). INRIA Rocquencourt B.P.105, 78150 Le Chesnay, France

/9/ Metcalfe et al., 1977. Multipoint Data Communication System w. Collision Dtection. U.S. Patent No. 4.063.220, 13.12.77

/10/ Metcalfe, R.M., Boggs, D.R., 1976. Ethernet: Distributed Packet Switching for Local Computer Networks. Comm. ACM, 7/76

/11/ Shoch, J.F., Hupp, J.A., 1979. Performance of an Ethernet Local Network. Proc. Local Area Netw. Symp., Boston, May 79

/12/ Tobagi, F.A., Hunt, V.B. 1979. Performance Analysis of Carrier Sense Multiple Access with Collision Detection. Proc. of the LACN Symposium, May 1979

/13/ Yemini, Y., Kleinrock, L., 1979. On a General Rule for Access Control or Silence is Golden. Flow Control in Comp. Netw., ed. J.L.Granse & M.Gien, North Holland 1979

Session 9

SYSTEM AVAILABILITY, RELIABILITY AND MAINTAINABILITY

DISPONIBILITE, FIABILITE,   **VERFÜGBARKEIT, ZURERLÄSSIGKEIT**
MAINTENABILITE DU SYSTEME   **UND INSTANDHALTUNG VON SYSTEMEN**

SAFETY PROBLEMS IN ADVANCED CONTROL

J.R. Taylor
Risø National Laboratory
DK-4000 Roskilde, Denmark

ABSTRACT

Hardware developments in the 1970s make possible "truly advanced control" in which not just normal operation, but abnormal and emergency operation is effectively controlled. There is a need for this, arising from plant complexity and the difficulties for operators of dealing rapidly with integrated plant. The problems of achieving advanced control lie in developing design procedures which can take account of abnormal operation and in the problems of program and specification error. To make advanced control work, such problems need to be solved, and the techniques of safety and risk analysis need to be incorporated into the process of system design.

RÉSUMÉ

Les développements matériels des années 70 rendent possible un contrôle vraiment avancé et effectif non seulement d'un fonctionnement normal, mais aussi d'un fonctionnement anormal. Il y a un besoin dans ce domaine créé par la complexité des machines et la difficulté pour les opérateurs de se débrouiller rapidement avec les machines intégrées. Les problémes de contrôle avancé résident dans les procédures de développement qui permettent de prendre en compte le fonctionnement défectueux et dans les erreurs de programmation et de spécification. Pour parvenir à ce contrôle avancé, ces problèmes doivent être résolus et les techniques d'analyse de sécurité et de risques incorporées au processus de conception du système.

ZUSAMMENFASSUNG

Des Fortsdiritt der Hardware-Entwicklung in den siebziger Jahren haben den Bau "echt fortgeschrittener Steuerungssysteme" ermoeglicht. Diese umfassen nicht nur die Steuerung der normalen Ablaeufe sondern auch die Steuerung abnormaler Ablaeufe und von Notfaellen. Die Notwendigkeit hierfuer hat sich aus der Komplexitaet der Anlagen sowie aus den Schwierigkeiten fuer die Operateure ergeben, sehr schnell auf komplexe Vorgaenge zu reagieren. Die Probleme einer solchen fortgeschrittenen Steuerung Liegen darin, dass Entwurfsverfahren entwickelt werden muessen, die auch abnormale Betriebssituationen beruecksichtigen, sowie in den durch Programm - oder Spezifikations fehler hervorgerufenen Problemen. Um fortgeschrittene Steuerungen tatsaechlich zu installieren, muessen die genannten Probleme geloest werden. Techniken fuer Sicherheits - und Risikoanalysen muessen in den Systementwurf einbezogen werden.

SAFETY PROBLEMS IN ADVANCED CONTROL

J.R. Taylor
Risø National Laboratory
DK-4000 Roskilde, Denmark

Hardware developments in the 1970's make possible "truly advanced control", in which not just normal operation, but abnormal and emergency operation is effectively controlled. There is a need for this, arising from plant complexity and the difficulties for operators of dealing rapidly with integrated plant. The problems of achieving advanced control lie in developing design procedures which can take account of abnormal operation and in the problems of program and specification error. To make advanced control work, such problems need to be solved, and the techniques of safety and risk analysis need to be incorporated into the process of system design.

INTRODUCTION

Since the middle of the 1950's, there has been a trend in many industries towards larger unit sizes, more integrated production, and tighter control, allowing operation closer to safety limits.

This can be seen in the process industries, where large scale integrated plants have replaced smaller batch plants. It can be seen in air transport, where ever larger aircraft land with ever shorter separation intervals. Automatic train control systems have allowed train speeds to be raised and intervals between trains to be lowered. More integrated power systems are operated, with less reliance on reserves.

Computers have been used to take over control in an ever wider range of automation and safety functions. It is seldom realised just how much responsibility has been encoded in these devices. As an example, some of the areas where computers are already responsible for personal safety are listed below:

- control of traffic lights
- control of elevators
- control of cranes
- raising oil rigs and holding them in position
- fire fighting system control on oil rigs
- air traffic control (with manual back up)
- automatic landing systems (full responsibility on the computer, if used)
- aircraft flight surface control
- railway signalling and switching
- train speed control and braking
- nuclear power plant shutdown (experimental)
- chemical plant shutdown
- control of medical life support systems
- robot control

The problems of hardware reliability with these systems have largely been solved. The problems of software reliability have certainly not been solved, for a range of reasons which will be discussed later.

IRONIES OF AUTOMATION

That computers are used at all for safety functions is not so much a question of equipment economics. Rather, it is because they work better than the <u>man</u> they are replacing, or because they offer completely new safety facilities.

In train speed control for example, the systems are designed to work with the driver, and to prevent him driving past red signals. In nuclear systems, the devices were at first applied to completely new safety functions such as departure from nucleate boiling monitors (measures the fraction of bubbles in boiling water). In chemical plant, equipment functioning can be continuously monitored. In aircraft landing, landings in fog are made possible.

As time goes on, the advantages of using computers here become more and more that they can perform <u>more complex functions</u>, using more complex criteria for shutting down plants, reducing load, or smoothing disturbances. The borderline between safety and control actions gradually disappears.

The increased capability is not just an optional benefit obtained by the use computers. For some applications, increased capability (and hence complexity) is a <u>necessity</u>.

Automation makes larger, more integrated plants possible. It gives the operator the possibility to control or supervise more units as an integrated whole and to improve plant performance as a result. But if things go wrong, his increased scope of activity has several effects. He has a more complex plant to diagnose. Even if he can find the fault, he has a more difficult job in deciding what to do. He has less experience and "feel" for the individual plant units, partly because he has more of them, partly because a good deal of their control has been taken over.

The automation devices too introduce problems themselves. As more instruments and information processing devices are fitted to the plant, it becomes more likely that any failure will be due to an instrument rather than the plant itself. The operator needs to understand the automation system as well as the plant.

These problems, among others, Bainbridge has called "ironies of automation". They have led to an increasing transfer of safety responsibility from the operator to the designer. It is now common to see plant instructions with "five minute" or "half hour" rules. These mean that operator assistance must not be required in any incident for at least five minutes, or half an hour. In some instances rules forbid operator interaction during major incident.

Such rules are a direct result of the increasing difficulty for the operator to act correctly and rapidly in an emergency. This does not necessarily leave the operator unemployed. He has more than enough problems, in keeping the complex plants running effectively, in optimising their operation and in performing diagnoses after accidents.

ADVANCED SAFETY AND CONTROL SYSTEMS

Some of the modern, though by now fairly conventional facilities, which have helped to keep safety problems under control in integrated plant are

- much more extensive safety shutdown systems, with more shutdown criteria, and sequential control of shutdown.

- load reduction systems which serve as a first line of defence, before shutdown becomes necessary.

- much more extensive interlocks on controls, so that simple operator errors are prevented.

- sequential controllers which perform complex startups and shutdowns without error.

- automatic data logging.

- extensive alarm systems.

- mimic displays of plant status.

The *performance* of normal control has been improved by so-called "advanced control theory", with multivariable control and adaptive control.

Not all of these developments have been an unmitigated success. Automatic data logging works, producing miles of data print out, but does not always communicate much information to plant operators and managers. Extensive alarm systems work well during normal operation. But in real emergencies, when they are needed most, they flood the operator with an excess of information. To quote from one operator, after the Three Mile Island incident "we wished we could switch the alarm system off".

"Advanced" control systems too suffer from problems. They modify plant performance, so that an operator has little chance of understanding the way the plant works. In the case of failure, the operator is faced with the problem not only of solving the plant problem, but with either fighting the control system, or taking over from it.

Some progress *has* been made towards solving this problem. "Robust" controller design theories have been suggested and used. However, the range of disturbances to which they can respond are at present minimal - only a few kinds of sensor failure (see fig. 1).

There are some developments in progress which help to solve some of these problems.

- Selective data logging systems such as "Anticipation" (MUN) which record and present only unusual plant information.

- Alarm analysis systems (AND) (FEL) which can group alarms, set priorities on them, present a range of possible causes, present probable consequences, and even suggest possible disturbance reducing actions.

- Integrated disturbance analysis and operator support systems (EPR) which provide the operator with just the information he needs during a disturbance, in a graphic form (trends, mimic diagrams) which help to support understanding.

These developments have not been completely successful as yet. Overenthusiastic application has sometimes led to disappointment. The reason is that the interaction between complexity of the operation task and the complexity of the engineering analysis task required has not always been appreciated.

As an example of this, the failure analysis of a plant required for an

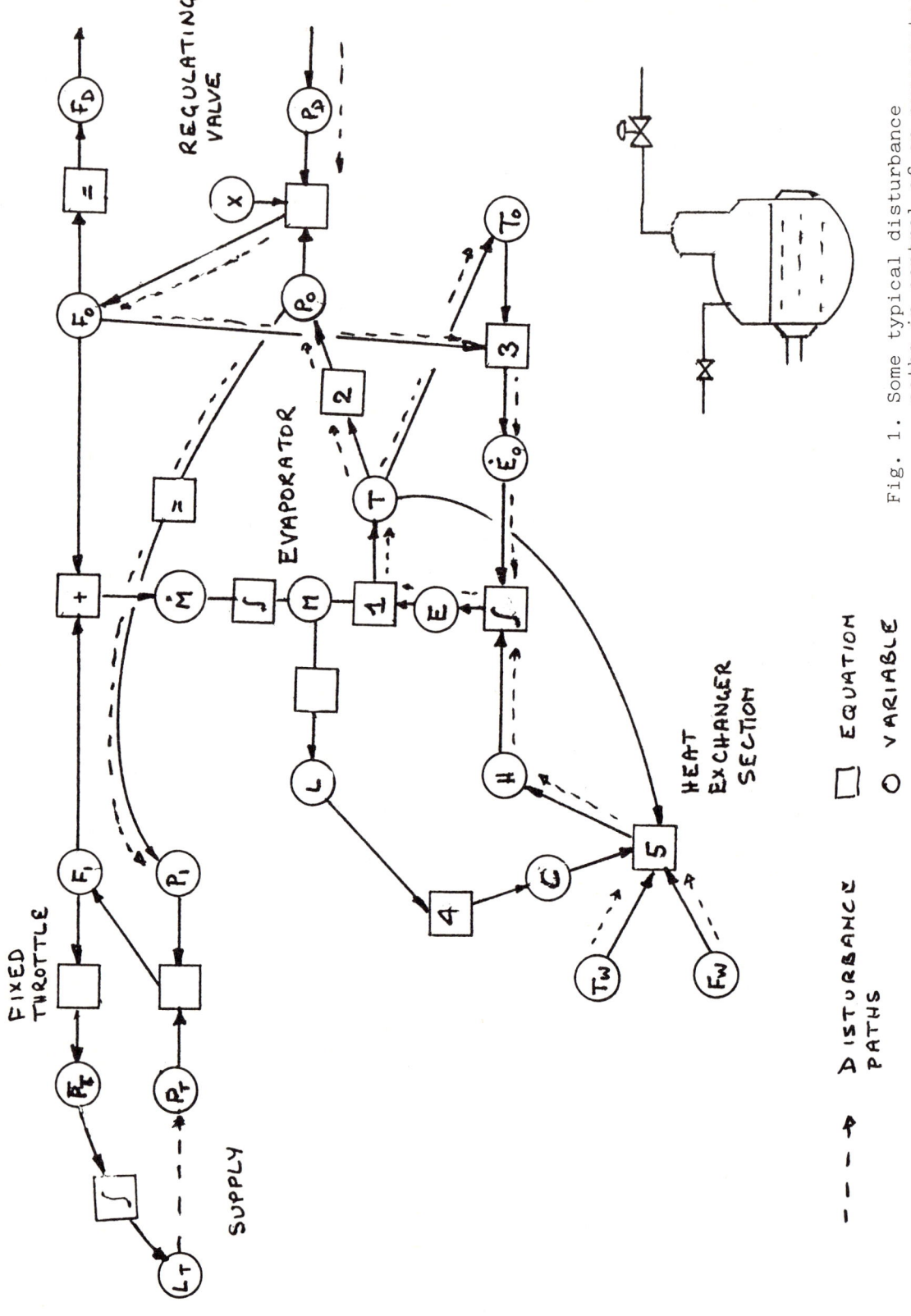

Fig. 1. Some typical disturbance paths in control of an evaporator

alarm analysis, system can take tens of man years. Just keeping such a system up to date can be a problem. Also one must recognise the possibility of errors in analysis. The theory necessary to determine what is a correct failure analysis of a plant has only recently been developed (TAY 82).

If these problems are to be solved, then we need

- a much more direct route from the engineering design process to the actual implementation. Theories which take a plant description in physical terms are much better than ones which require detailed coding.

- a better theoretical understanding of what is required. As an axample, many earlier systems used a simple functional model of plants, ignoring the fact that energy systems must be modelled in terms of two way effects, and not just functionally (DE 82).

- a multilevel approach to understanding of plant behaviour, which is not just hierarchical, but takes account of the way some systems fulfill various purposes.

As an example of developments taking account of the first need could be mentioned alarm analysis systems based directly on plant models (HOL) (AND).

As an example of an approach fulfilling all the needs can be mentioned the multilevel flow modelling of Lind (RAS) (LIN).

Looking forward, one can see that existing computer hardware already can provide the possibilities for what might be called "truly advanced control". This would provide

- highly optimised multivariable adaptive control during steady state operation.

- "reasonable" or "robust" control responses to failures in the plant or the control system, if these are at all possible, by continuous variation of control variables.

- automatic sequential control of equipment to reduce disturbances, when this cannot be done by normal regulation.

- fully automated safety systems, in those cases where the operator simply cannot be expected to respond quickly enough or correctly.

- dialogue systems (text, voice (?) and graphic) to support scheduling of plant production, planning, and condition monitoring of the plant. (Help the operator understand plant performance).

- dialogue systems to help diagnose failure or poor performance in those cases where neither the operator nor the computer alone have sufficient information.

The hardware to allow this is in fact available now, and moreover quite cheaply. What is needed is the necessary theory, engineering design and analysis techniques, and software techniques.

SOFTWARE PROBLEMS

While hardware reliability problems can be reduced to a minimum, by use of distributed systems, a high degree of circuit integration, and redundancy, the situation is not so good in the case of software errors.

Distributed software systems to some extent reduce the problems of software reliability, by keeping modules small, but there is a limit to this. For high capability modules, programs tend to have several thousand lines of code.

Consider a system with some 10,000 lines of program. It will, if coded conventionally, contain about one error per 10 lines. If a systematic design approach is taken, this can be reduced to one per 30 lines, after syntax analysis, or about 300 errors. Ordinary testing will remove about 50% of these errors. Inspection may remove a further 60% or so, so we end with about 50 errors or so.

A really thorough on line test might involve some 10,000 on line tests. The program itself might typically have some 10^{10} paths, so it might seem that there is little chance of detecting an error (10^{-6}). However, the errors generally lie on many paths for example typically 10^4 paths each. We say that the "path size" of the errors is about 10^4 (RAM). So we now have some 50 errors distributed over 10^6 path groups, to be detected by some 10,000 tests. There is a reasonable probability of detecting some of the errors, but certainly not all. (There is no guarantee that an error will be detected, even if the test activates a path on which an error occurs).

Looking at the probability of path execution, this typically follows a pareto distribution (fig. 2). We find that the probability of there being an error in any one path is about $50/10^6 = 5 \times 10^{-5}$. We also find that they are distributed on the less probable paths, with a path execution probability of less than 10^{-4}. (The more probable paths will already have been debugged).

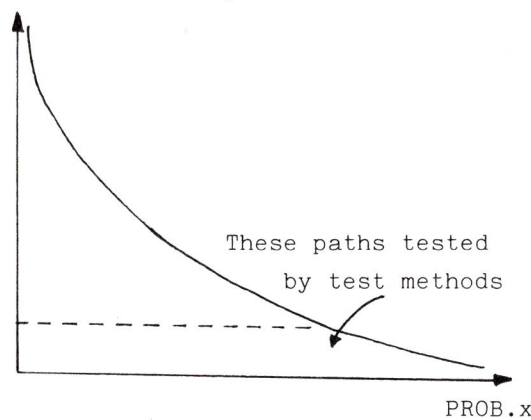

Fig. 2. Distribution of path execution frequencies in program testing.

There is now a probability of meeting an error in an execution of 5×10^{-9}, - giving a very respectable reliability per execution.

If we now perform one execution every millisecond and executions are repeated continuously, we have $1000 \times 3600 \times 10,000 = 3.6 \times 10^{10}$ executions per year. This gives about 180 failures per year.

This is unacceptable in a safety monitoring system, and is probably unacceptable in most control systems.

The parameters of this example can be changed enormously, and still leave the conclusion unchanged. The problem is that testing only detects errors on the most probable program paths. Indeed, the problem is worse that this, because the distribution of paths executed in testing is generally different from that in on line use.

Looking at what we could do to improve this, we could

- improve testing and debugging efficiency by a few orders of magnitude (incredible).

- use diverse redundancy, that is, two programs made by two groups (a good practical technique, but difficult to apply to advanced systems).

- increase the number of tests by a few orders of magnitude (very expensive and gives diminishing returns).

- cut the number of paths in programs by using few branches and no loops (very effective, but drastic).

- make path coverage of tests more uniform (how?).

- use a theoretically based approach.

If we look at the theoretically based techniques actually used today in this area, we have:

- Proving correctness. This has suffered from a setback in that many published "proofs" have contained errors. The technique is having a resurgence currently in the form of multilevel formal specifications.

- Automated random testing - this requires that the controlled plant is simulated so that test results can be evaluated automatically. Or that two diverse programs are tested in parallel. Ehrenberger (EHR) and his colleagues have shown that this approach is prohibitive in terms of numbers of tests required, unless combined with an analytical approach, such as program analysis, or with redundant programming.

- Path and path domain testing approaches. This suffers from the so-called "missing path problem", and by problems of combinatorial explosion, but can be applied to programs of a few thousand lines (BOL. All approaches suffer from the problems of specification error.

SPECIFICATION ERROR

Specification errors arise when the task which is explicitly stated as a requirement does not match the implicit requirement. The specification may be incomplete, inconsistent within itself, or inconsistent with the world it is desired to control. A frequent cause is that the designer's model of the real world is incomplete or inconsistent.

Specification errors are very common. A large part of the software production and testing process involves exploring performance, investigating aspects of the specification, and revising specifications. Nevertheless, about half of the problems arising with computer systems after they have been put into operation are observed to be due to specification errors.

One way of avoiding such errors would be to document specifications and purpose of computer systems more carefully, and integrate specifications more directly into the design process. There is a problem here, in that the language used by computer engineers is very different from that used by plant engineers. Plant engineers are notoriously poor at documenting the details of their requirements, and remembering all the exceptions and constraints on

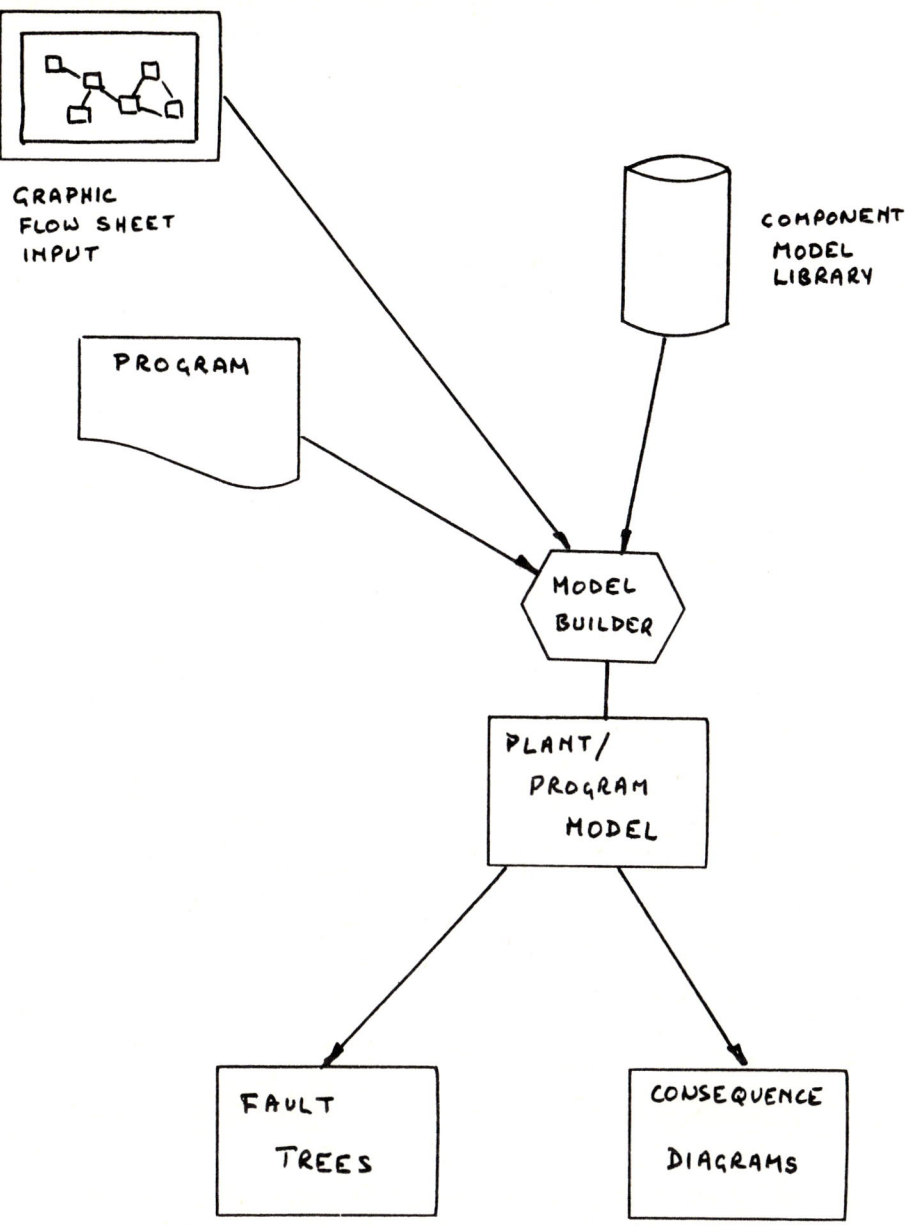

Fig. 3. Automatic fault tree and consequence analysis for plant control software

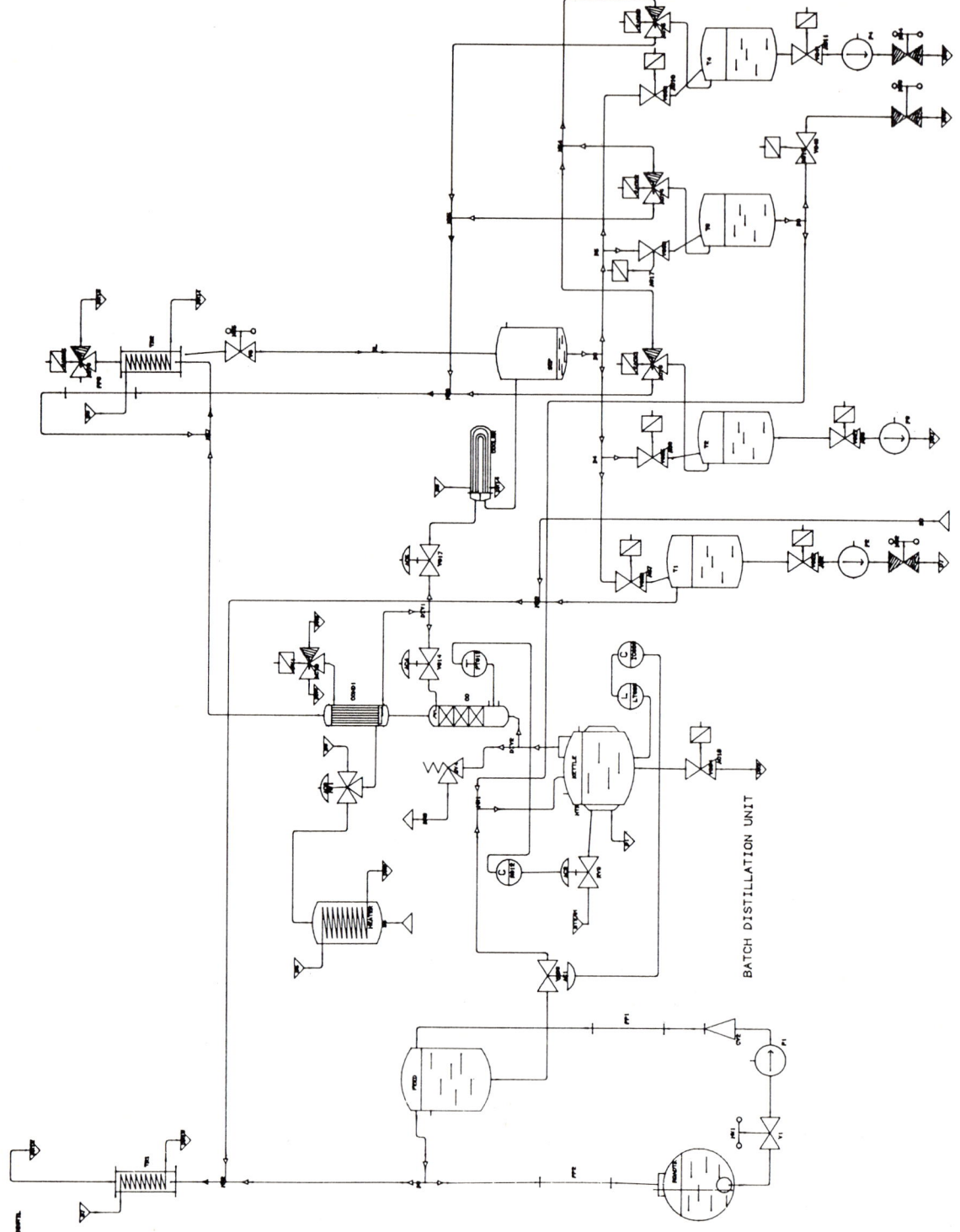

Fig. 4. Input to control system analysis programs

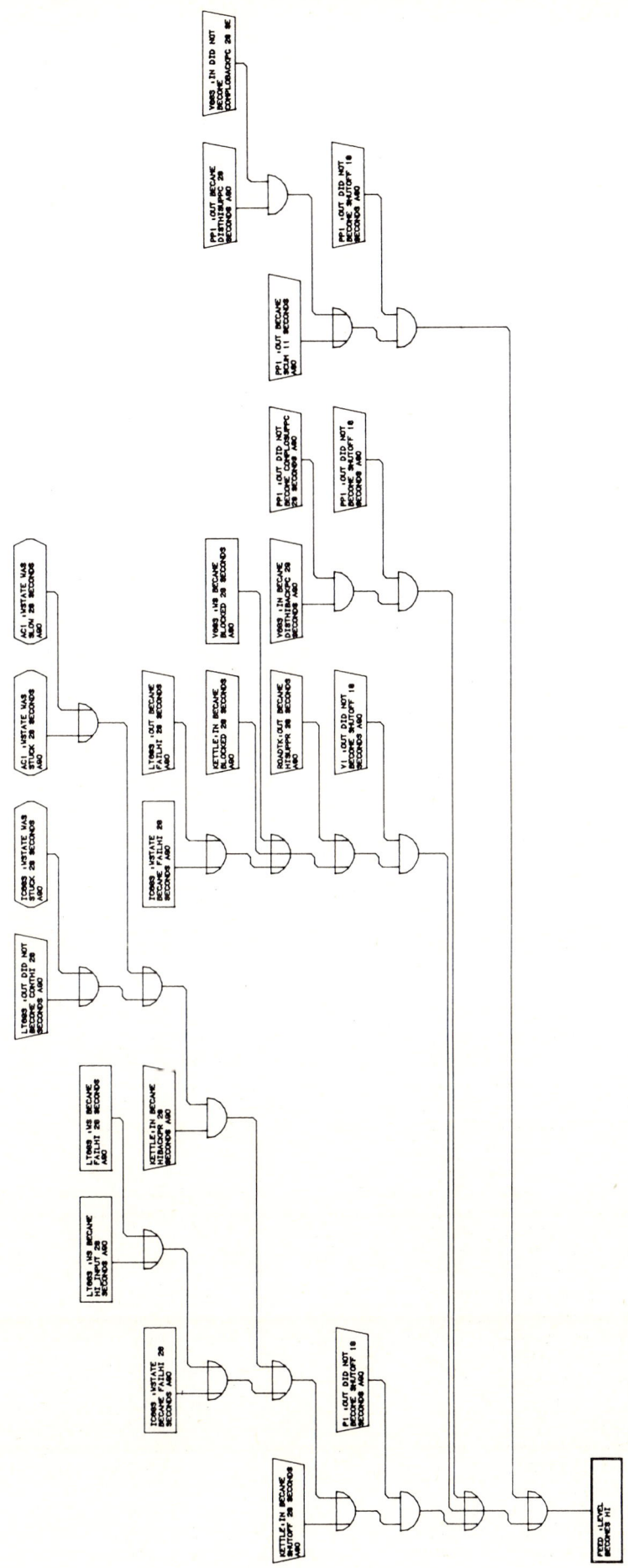

Fig. 5. Partial fault tree resulting from hardware/software analysis

their needs. Perhaps as more experience is gained, and as the number of engineers with experience in both fields increases, the problem will be reduced. Certainly development of documentation forms which are understood by all will help. But the primary problem is that of complexity.

These problems are not peculiar to computer systems. Earlier studies of design errors in process plant (TAY 75) (FAW) showed that up to 30% of safety related incidents arose as a result of design error, and about 2% from system design error. But the complexity allowed by computer systems certainly makes matters worse.

An approach which bypasses some of these problems is to analyse the safety of the plant and the computer system together, using one technique. Software fault tree analysis is one such technique (TAY 82b) (fig. 3). Here a model of the plant is provided in the form of a piping and instrumentation diagram, and a set of standard plant component models. Starting with a "top event", some accident in the plant, both failure events in the plant and errors in the computer system which can lead to accidents are traced.

This approach bypasses the design process in discovering what kind of plant or computer behaviour is dangerous. Its standard of safe behaviour is the safety or risk analysis model, which can be made independently of the design process.

If safety validation is to depend on such models, however, the models had better be correct. To test this, several comparisons of model results with "hand made" risk analyses have been carried out (TAY 82c). The results identify the areas where the automatic techniques are superior, and where there are loop holes and locunae. The results also show that manual safety analysis techniques are much less effective than is generally realised.

CONCLUSIONS

Significant improvements in control are now possible in hardware terms. "Truly advanced control", in which systems take account not only of normal operation, but also plant component failure and control system failure, seem within reach. The obstacles are the design theories and techniques at present in use.

If the promise is to be achieved, then computer system design needs to be incorporated properly into plant design. Languages are needed (probably graphic?) which bridge the gap between plant engineer and computer specialist. To integrate both in one person is one approach, but for large projects is practically impossible. Plant engineers, though need to face up to the complexity of the requirements that the make, and the problems of specification error. They need to understand their own design process better. And if systems are to respond to plant failure, then the techniques of failure and risk analysis need to be built into design procedures.

REFERENCES

(MUN) G. Munday, On line hazard identification during chemical processes, Loss Prevention in the Process Industries, Heidelberg 1978, (Dechema).

(AND) P. Andow, Process plant alarm systems: general considerations, In: Loss Prevention and Safety Promotion in the Process Industries, 1975, (Elsevier).

(FEL) C. Felkel and H. Roggenbauer, The "star" concept, International meeting on thermal nuclear reactor safety, Chicago 1982.

(EPR) On line power plant alarm and disturbance analysis system, EPRI report NP-1379, April 1980.

(TAY 82) J.R. Taylor, An algorithm for automatic fault tree analysis, IEEE Trans. Reliability, June 1982.

(DE) M.K. De et al., A functional design approach to PWR safety, International meeting on reactor safety, Chicago, 1982.

(HOL) E. Hollo and J.R. Taylor, Use of dynamic fault tree construction as a basis for alarm analysis, Enlarged Halden Group Meeting, Frederikstad, 1977.

(RAS) J. Rasmussen and M. Lind, Coping with complexity, Report Risø-M-2293, 1981.

(LIN) Generic control tasks in process plant operation, European Annual Manual, Bonn, June 1982.

(RAM) C.V. Ranmamoorthy and F.B. Bastani, Software reliability - status and perspectives, IEEE Trans. Software Eng., vol. SE1-8, no. 4, July 1982.

(EHR) W. Ehrenberger and K. Plogert, Statistical verification of reactor protection software, Proc. Int. Symp. Nuclear Power Plant Control, Cannes, 1978.

(BOL) S. Bologna and J.R. Taylor, Validation of safety related software for a nuclear reactor, IAEA specialist meeting on computerised control and safety systems in nuclear

	power plants, Pittsburg, 1977.
(TAY 75)	A study of failure causes based on US nuclear reactor abnormal occurrence reports. In: Reliability of Nuclear Power Plants, IAEA, Vienna, 1975.
(FAW)	S. Fawcett, The design of nuclear power plant for high reliability.
(TAY 82b)	An integrated approach to the treatment of design and specification errors. In: Reliability in Electrical and Electronic Systems, Eurocon conference, North Holland, 1982.
(TAY 82c)	J.R. Taylor, Evaluation of cost completeness and benefits for risk analysis procedures, Int. Symp. Risk and Safety Analysis, Bonn, July 1982, (Springer).
(TAY 83)	J.R. Taylor, A background to risk analysis, vol. 1 to 4, to be published. Draft available from author.

AUTOMATION EINES WASSERWERKES FÜR UNBEMANNTEN,
VOLLAUTOMATISCHEN WERKSBETRIEB MIT EINEM
PROZESSRECHNER UNTER BESONDERER BEACHTUNG ELEKTRISCHER
UND VERFAHRENSTECHNISCHER FEHLERTOLERANZ

Ulrich Rüdiger
SCS Scientific Control Systems GmbH, Essen
B. R. Deutschland

ZUSAMMENFASSUNG

Der Systemanalyse lag die Aufgabenstellung zugrunde, bei Auftreten von Störungen jeder Art die Wasserversorgung der Stadt möglichst lange aufrechtzuerhalten. Dazu wird in Störfallbehandlungsvorschriften die verfahrenstechnische Redundanz des Wasserwerkes ausgenutzt. Diese vom Normalbetrieb des Wasserwerkes abweichenden Betriebsweisen nehmen ein wesentlich größeres Volumen in der Programmierung ein als die Prozeßsteuerung im Normalbetrieb. Aus der Kenntnis der verfahrenstechnischen Zusammenhänge sind Plausibilitätsprüfungen möglich. Die als fehlerhaft erkannte Erfassung von Gerätezuständen führt in jedem Fall zu einer Nachricht an den Betreiber, da aber das Wasserwerk unbemannt gefahren wird, ist auch eine direkte Beeinflussung der Steuerungen notwendig. Maschinen, Anlagen und Meßgeräte, die nicht einwandfrei arbeiten, werden im Rahmen der verfahrenstechnischen Möglichkeiten umgangen. Dadurch ist die optimale Betriebsweise des Wasserwerkes eingeschränkt, jedoch kann die Betriebssicherheit der Gesamtanlage und damit die Sicherheit der Wasserversorgung erhöht werden.

ABSTRACT

The most important assumption for the system analysis was the necessity of maintaining the city water supply as long as possible even in the event of a major disturbance. In order to do this the operating instructions for handling such disturbances take into account the technical redundance in the water works.
A much greater portion of the programming effort has to be invested to handle such nonstandard operating conditions as does the programming for the normal process control situation. Once the technical interrelationships are known, it is possible to make plausibility checks. Alarm messages are always transmitted to the operator. But since the water works are normally unmanned, it is necessary to provide for automatic interaction with the process control. If at all possible, machines, installations and measuring equipment which are not working properly, are taken out of operation.
Thus even though an optimal mode of operation is not possible, the security of the entire installation and hence the security of the water supply is increased.

RESUME

Automation d'une usine à eau en vue de son exploitation automatique, sans personnel, à l'aide d'un calculateur de processus industriels, compte tenu d'une tolérance d'erreur électrique et technologique.
L'analyse des systèmes avait pour but de maintenir aussi longtemps que possible l'alimentation en eau de la ville, quelles que soient les perturbations intervenant. A cet effet, on utilise dans le règlement concernant la méthode à suivre en cas de perturbation, la redondance du procédé technologique de l'usine à eau. Ces modes d'exploitation qui s'écartent de la marche normale de l'usine à eau, occupent dans la programmation un volume beaucoup plus grand que la commande automatique du processus en marche normale. La connaissance des rapports du processus technologique permet de procéder à des contrôles de plausibilité. Lors du recensement des états du matériel, toute détection de défectuosités se traduit, dans tous les cas, par une communication à l'exploitant, mais comme l'usine à eau est exploitée sans personnel de service, il est aussi nécessaire d'intervenir directement dans les commandes automatiques. Les machines, installations et appareils de mesure qui ne fonctionnent pas parfaitement sont "court-circuités" dans le cadre des possibilités offertes par le procédé technologique. Cela limite, certes, le mode d'exploitation optimal de l'usine mais permet d'accroitre la fiabilité de l'ensemble de l'installation et, par conséquent, la sureté de l'approvisionnement en eau.

1. Einführung in das Verfahren

Bild 1 zeigt eine Übersicht über die verfahrenstechnische Auslegung des Werkes. Bis zu 16 Grundwasserbrunnen mit einer Gesamtförderleistung von rd. 8 Mio cbm pro Jahr fördern Rohwasser aus entfernt liegenden Wassergewinnungsgebieten mit Tauchpumpen. Am Werkseingang sind 4 Entsäurer vorhanden, gefolgt von 3 zweistufigen Schnellfiltern mit den zugehörigen Rückspüleinrichtungen und Schlammwasserbehandlungsanlagen. Ein Reinwassertiefbehälter mit einem Nutzvolumen von 5600 cbm ermöglicht Vorratswirtschaft und eine gleichmäßige Betreibung der Rohwasserressourcen. Das Puffervolumen des Reinwasserbehälters dient auch dazu, die Rohwasseraufbereitung energieoptimal und mit wenigen Schaltspielen zu betreiben. Die dazu erforderlichen Werkssteuerungsstrategien und die Aufstellung von Verbrauchsprognosen mit den dazu erforderlichen langfristigen Statistiken werden vom Prozeßrechner ausgeführt.

Es schließt sich das Reinwasserpumpwerk an, das mit einer Förderleistung von max. 4000 cbm/h bei 100 % Reserve netzdruckgeregelt arbeitet. Der konstante Netzdruck von ca. 4 bar wird durch je eine geregelte Drehstrompumpe gehalten, ein statischer Wasserdruck besteht nicht. Die Anordnung und die Verrohrung der einzelnen Werksteile gestattet es, die wesentlichen Aggregate in unterschiedlichen Kombinationen zu betreiben. Damit ist ein redundanter Betrieb des Werkes in Störsituationen möglich. Die Energieversorgung des Wasserwerkes erfolgt aus einer eigenen 10 kV-Umspannanlage mit zwei getrennten Einspeisungen und Notstromdiesel. Auch die Steuerung dieser Anlage obliegt dem Prozeßrechner, der mit der zugehörigen Steuerungstechnik aus einer unterbrechungsfreien, batteriegepufferten Stromversorgung gespeist wird.

Bild 1: Verfahrenstechnik des unbemannten vollautomatischen Wasserwerkes

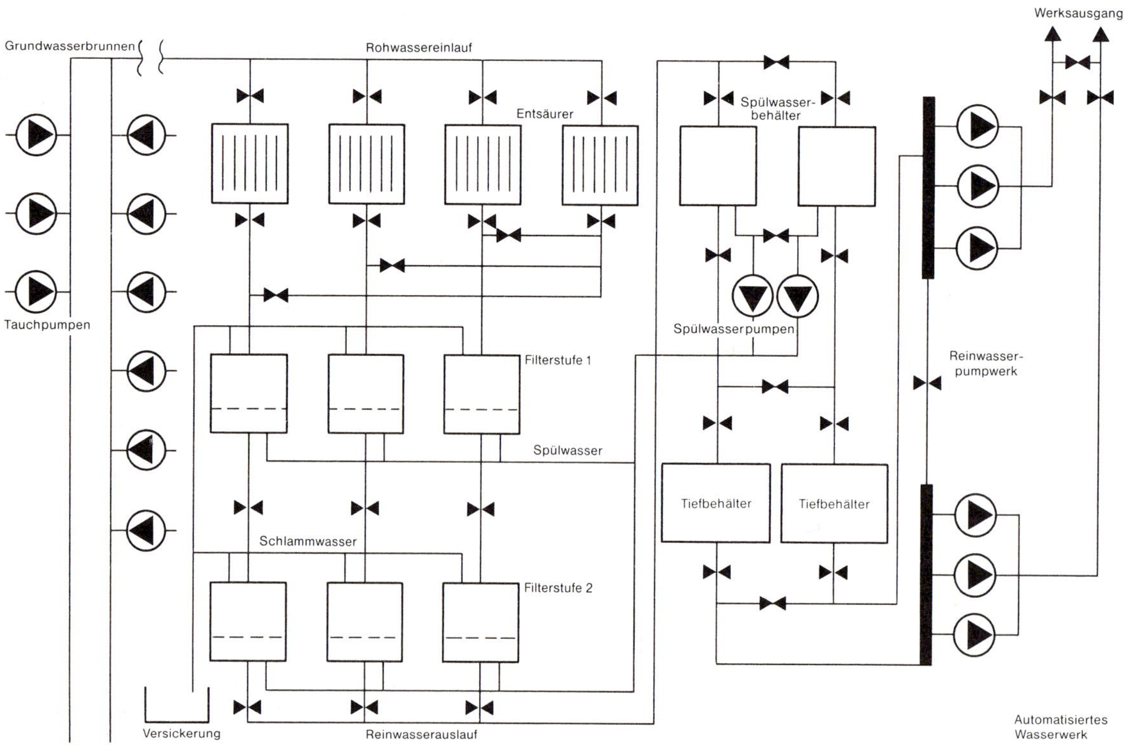

2. Automatisierung des Prozesses

<u>Bild 2</u> zeigt die gerätetechnische Ausstattung für die Automation.

Die Niederspannungsschaltanlage ist sowohl mit dem Prozeßrechner, als auch mit einem konventionellen Anzeige- und Schaltpult verbunden. Eine Betriebsartenumschaltung kennt folgende Zustände:

Handbetrieb	Rechner außer Betrieb
Pultbetrieb	Rechner protokolliert und optimiert, steuert aber nicht.
Automatikbetrieb	Rechner steuert

Neben der Betriebsartenumschaltung ist ein vollautomatischer Systemrestart implementiert, der zur Beseitigung unklarer Werkszustände dient. Er enthält umfangreiche Werkszustandsanalysen und Schaltprogramme, um einen definierten Werkszustand über alle Bereiche einzustellen. So können z. B. die Schieber und Klappen der Filter aus einer beliebigen Position automatisch richtig gestellt werden.

Die Prozeßperipherie umfaßt:
- 96 Analogeingänge
- 96 Zählereingänge
- 640 Digitaleingänge
- 656 spontane Digitaleingänge
- 64 Unterbrechungseingaben
- 624 Digitalausgaben

Mit dieser Ausrüstung kann die gesamte Verfahrenstechnik gesteuert und überwacht werden. Die Zentralwarte erhält zur Überwachung des Betriebes das Ereignisprotokoll, sowie Statistikdaten zur weiteren Auswertung auf einem getrennten Rechner.

Bild 2: Elektrische Ausrüstung zur Automation des Wasserwerkes (Hardware)

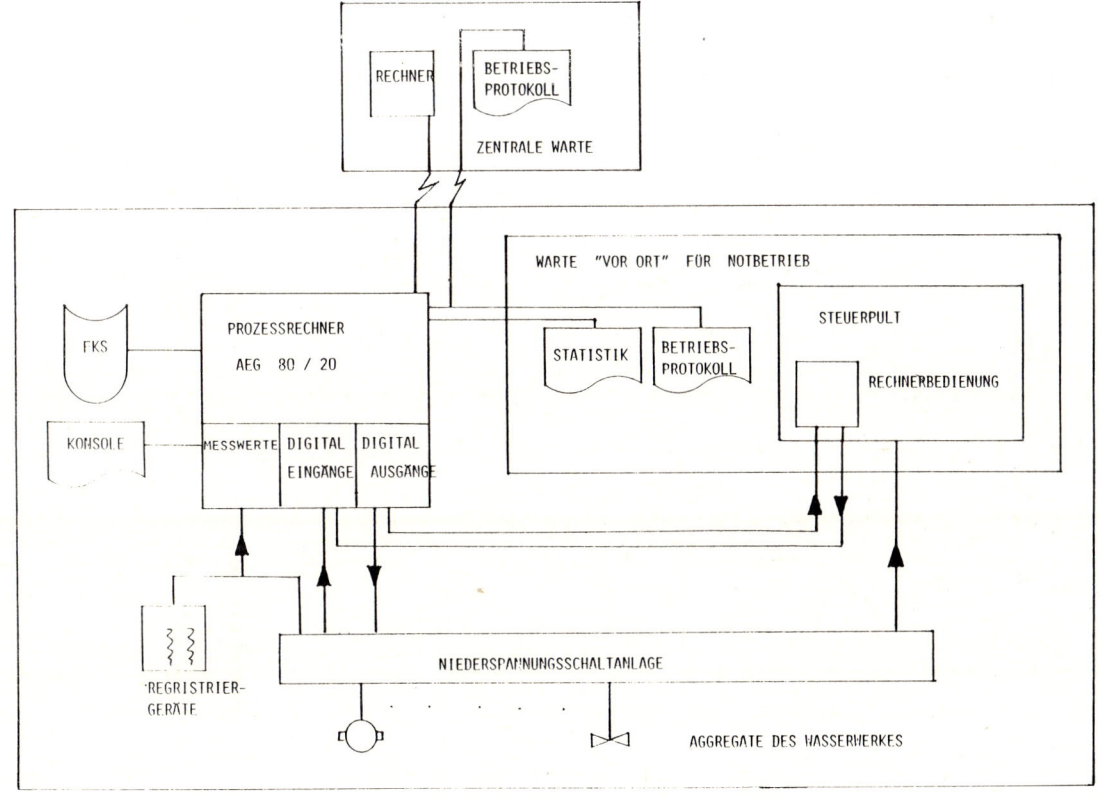

Bild 3 zeigt die Softwarestruktur der Prozeßsteuerung, mit deren Hilfe die Bereiche

Datenerfassung mit Ereignisprotokoll
Statistik und Protokolle
Werksoptimierung mit Prognose
Stromoptimierung: Einsatz des Notstromdiesels zur Abdeckung von Lastspitzen
(Viertelstundenleistung)
Steuerung für
   Brunnengruppenauswahl und Brunnen
   Entsäurer und Filter mit Filterspülprogrammen
   Reinwasserpumpwerk
   Chloranlage
   Schlammwasseranlage
   10 KV Stromschaltanlage
Betriebsartenumschaltung
Rechnerkopplung zur Zentralwarte
Dialogprogramme für Auskünfte an den Benutzer, zur Angabe von Stammdaten, für Fehlerdiagnose und für Wartungszwecke

implementiert sind.

2.1 Datenerfassung mit Ereignisprotokoll

Digitaleingänge werden mit einstellbarer Abtastzeit erfaßt und in einem ungeprüften Prozeßabbild hinterlegt. Aus der Erkennung von Signalwechseln werden gepufferte Protokollmeldungen erzeugt.

Dieses sind typischerweise
   Störmeldungen
   Vor-Ort-Meldungen
   Ein-/Ausschaltmeldungen

Interrupt-Eingänge werden identifiziert und lösen direkte Beauftragung einer Steuerung oder Überwachung aus.
Bei Signalwechsel werden grundsätzlich die für diesen Bereich zuständigen Meldebilder neu erstellt.

Bild 3:   Softwarestruktur des Automationssystems

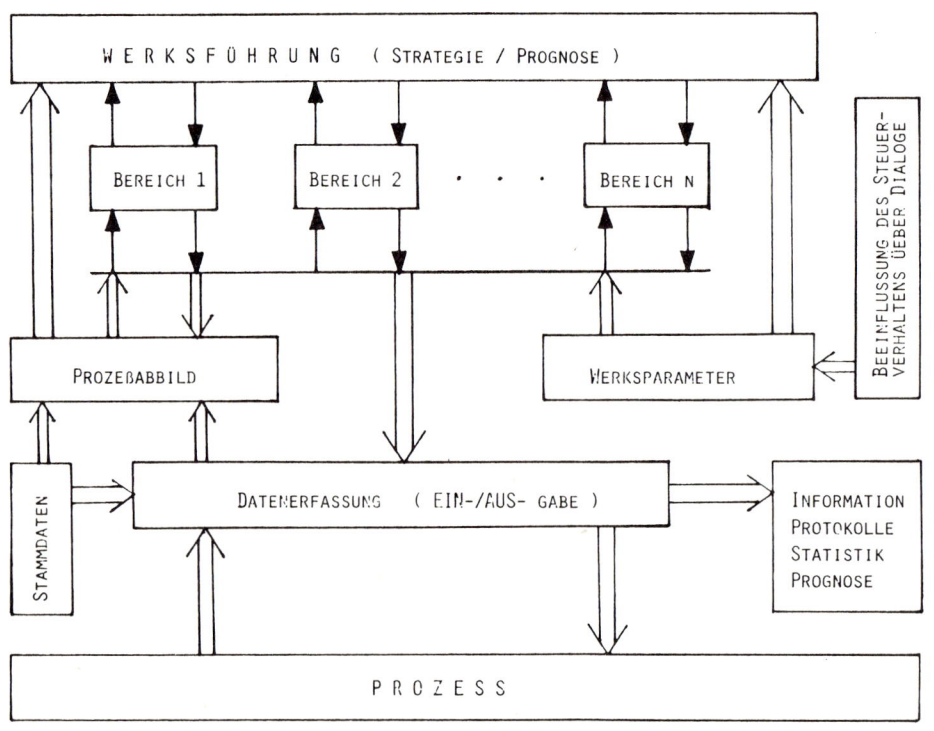

Die Erfassung der Analogwerte erfolgt zyklisch alle 15 Sekunden. Im Automatikbetrieb sind folgende Plausibilitätsprüfungen anwählbar:

  Grenzwertprüfung nach oben und unten
  Gradientenprüfung
  Filterung über Glättungskonstante
  Auswertung der Hardwarefunktionsprüfungen

Die Plausibilitätsprüfungen einzelner Werte können nach Schalthandlungen in den entsprechenden Werksbereichen für individuell definierte Zeiten abgeschaltet werden. (Z. B. nach Einschalten einer Brunnenpumpe schwappt das Wasser im Brunnen und die Höhenstandmessung des Grundwasserspiegels überschreitet die normalerweise zulässigen Gradienten, gleichzeitig schwappt das Wasser in der Entsäurereinlauftasse, die Positionen der Regelschieber erreichen Werte, die im Normalbetrieb nicht zulässig sind.)

Betriebsstunden und Mengen werden über elektromechanische Impulszähler alle 15 Minuten erfaßt. Diese Geräte sind keine Standardprozeßperipherie. Ein spezieller Monitor überwacht die Funktion dieser Zähler und gibt dem Betreiber detaillierte Informationen über die Betriebsbereitschaft der Geräte. Überwachungsfunktionen sind:
  Prüfung auf stetige Zunahme der Werte
  Kontrolle der Zählerausleseanwahl
  Mehrfachabfrage bei Zählersprüngen mit Zeitüberwachungen

Ausführliche Diagnosemöglichkeiten ermöglichen die Wartung und Fehlerdiagnose der elektromechanischen Zähler, ein Meßwertloggingprogramm dient der Wartung der Analogeingänge und der vorgeschalteten Meßgeräte, die aktuelle Beaufschaltung der Digitaleingänge kann kartenweise unabhängig von dem Datenerfassungssystem abgefragt und angezeigt werden. Digitalausgänge können unabhängig vom Steuerungssystem gesetzt und rückgesetzt werden, die Ausführung kann bis in die konventionellen Steuerungen hinein überwacht und geprüft werden.

Mit diesen Hilfsmitteln ist es dem Betreiber möglich, die Prozeßrechneranlage und die konventionelle Steuerung in einer für unbemannten Betrieb ausreichenden Qualität zu erhalten und die installierte Redundanz für die Störfallbehandlung dauerhaft zu nutzen.

2.2 Meldebilderstellung

Auf der Grundlage von
  geprüften Meßwerten
  ungeprüftem Prozeßabbild
  Alarmsituationen
  Systeminterner Verarbeitungszeichen
  Werksparameter
  Stammdaten

wird je nach Anforderung bereichsweise ein logisches Prozeßabbild generiert. Zu einem verfahrenstechnischen Aggregat gehören in der Regel mehrere Meßwerte und Digitaleingänge. Aus der Kombination und den Übergängen dieser Primärsignale wird der Schaltzustand des Aggregats ermittelt. Erst bei logischer Zustandsänderung mindestens eines der Aggregate des entsprechenden Bereiches wird die entsprechende Bereichsüberwachung beauftragt.

2.3 Statistik und Protokolle

Störstatistik

Sowohl die hardwareerzeugten Meldungen als auch die Betriebshinweise, die von den Steuerungen durch logische Betrachtungen ermittelt wurden, sind dem Betreiber nicht nur durch das fortlaufend geschriebene Ereignisprotokoll zugänglich, sondern können auch auf Tastendruck in einem Störungsprotokoll zusammenhängend abgerufen werden.

Die außerdem angezeigten Wartungshinweise geben dem Benutzer Auskunft über überschrittene Wartungsintervalle, die aufgrund der Betriebsstundenzählung überwacht werden. Die Anzeige dieser aktuellen Störungsliste ist sowohl im Wasserwerk vor Ort, als auch in der Zentralwarte möglich.

Die im Ereignisprotokoll angegebenen Störungen werden in einer Monatsstatistik nach Störhäufigkeit und Gesamtstördauer für jedes angetriebene Aggregat geführt. Diese Statistik wird bei jedem Monatswechsel automatisch auf dem Schnelldrucker im Anschluß an das normale Monatsprotokoll gedruckt. Da das Wasserwerk über viele identisch vorhandene Aggregate verfügt, sind Auffälligkeiten in mangelnder Zuverlässigkeit leicht auszumachen.

Protokolle

Stündlich fortgeschrieben werden die Stunden-, Tages-, Monats- und Jahresprotokolle.
Dazu wird ein rollierender Plattenspeicherbereich verwaltet, in dem die jeweils letzten Protokolle vorrätig gehalten werden:
  24 Stundenprotokolle
  31 Tagesprotokolle
  12 Monatsprotokolle
   1 Jahresprotokoll

Diese Protokolle können über einen Dialog zur Druckausgabe angewählt werden.

Alle Protokolle werden im Zeitpunkt ihres Entstehens außerdem automatisch gedruckt. Die Zentralwarte erhält über eine Rechnerkopplungsleitung Zugang zu der Datenbasis der Protokolle für weitergehende Nachbereitung für statistische Zwecke.

Die Protokolle beinhalten im wesentlichen die von den Mengen- und Betriebsstundenzählern erfaßten Differenzwerte für den Betrachtungszeitraum. Zusätzlich werden einige ausgewählte Analogwerte eingeblendet, die für die Beurteilung des allgemeinen Werkszustandes wesentlich sind.

Grundwassertopologie

Alle Brunnen sind mit Höhenstandssonden ausgerüstet. Die Werte werden verschieden nach dem Schaltzustand der Brunnen in Statistiken erfaßt. Die Differenz des Höhenstandes zwischen ausgeschaltetem und eingeschaltetem Brunnen ist ein Maß für die Versandung des Brunnens, der Wert bei ausgeschaltetem Brunnen ist der Grundwasserspiegel. Die langfristige Beobachtung dieser beiden Werte ist betrieblich interessant und wird vom Prozeßrechner unterstützt.

Werksleistung

Der Wirkungsgrad, die sog. 'Werksleistung', wird in den geläufigen Dimensionen berechnet und im Protokoll ausgedruckt. Eine unmotivierte Verschlechterung dieser Größe soll den Betreiber zu Nachforschungen veranlassen.

Prognosestatistiken

Die für die optimale Werkssteuerung notwendige Aufbereitung der Basisdaten und die Speicherung in relativer Form bezogen auf bestimmte Grundwerte, die unabhängig von saisonalen Einflüssen sind, wird ebenfalls vom Statistikpaket übernommen. Einzelheiten dazu sind im Abschnitt Prognose beschrieben.

Meß- und Zählwertstatistiken

Alle Meßwerte werden mit ihren Minima, Maxima und Mittelwerten in über Dialog abrufbaren Statistiken geführt. Die zuletzt gelesenen und die zur letzten vollen Stunde gelesenen Zählwerte und die Differenzen zur Vergangenheit werden zu Test- und Überwachungszwecken ebenfalls zugänglich gemacht.

2.4 Werksoptimierung und Prognose

Ziel der Werksoptimierung ist es, die Betriebsmittel des Werkes nach den Gesichtspunkten

Schaltspielminimierung
Energieoptimierung

einzusetzen, um den Wasserbedarf des Verbrauchers zu befriedigen. Darüber hinaus hat die Werksoptimierung einige globale operative Aufgaben zu erfüllen. Diese sind die

Spülzeitpunktbestimmung für die Filterstraßen
Schlammwasserpumpensteuerung
Überwachung der maximalen Stillstandszeiten für die Entsäurer und die Filter

Die Betriebsmittel des Werkes werden laufend global überwacht. Der Betreiber erhält von der Werksoptimierung möglichst früh Hinweise auf bevorstehende Betriebsmittelengpässe. Dazu wird entscheidend die Verbrauchsvorhersage (Prognose) benutzt. Die Datenerfassung und Statistik stellt in speziellen Dateien langfristige Erfahrungsdaten für das Vorhersageprogramm zusammen. Die relativen Stundenverbräuche der letzten 200 Tage je Saison (Sommer/Winter) werden vorrätig gehalten. Innerhalb dieses Datenbestandes sucht das Prognoseverfahren den relativen Verbrauchsverlauf eines vergleichbaren Tages in der Vergangenheit (Gesichtspunkte sind: Wochentag, Trockenperiode bestimmter Dauer, Naßperiode bestimmter Dauer, Feiertagseinfluß, Wochenendeinfluß, Sommer/Winter-Saison) und normiert diesen Verlauf nach einigen aktuellen Gesichtspunkten.

Das Prognoseprogramm liefert um Mitternacht Vorhersagewerte für den nächsten Tag. Bestimmt wird der zu erwartende Gesamtverbrauch des Tages sowie die Vorhersagewerte für jede Stunde. Mit fortschreitender Tageszeit wird die Vorhersage mit der tatsächlichen Wasserabgabe verglichen. Die Vorhersage wird bei großen Abweichungen stündlich nachberechnet.
Das Puffervolumen des Behälters wird für die oben genannten Ziele genutzt. Die Werksoptimierung sorgt dafür, daß der Reinwasserbehälter morgens gefüllt ist und zum Abend hin geleert wird. Dadurch wird preiswerter Nachtstrom zur Förderung und Aufbereitung von Rohwasser ausgenutzt. In der Tagzeit mit ihren starken Verbrauchsschwankungen wird versucht, im Bereich der Rohwasserförderung und Aufbereitung die Anzahl der Schaltspiele zu minimieren. Dazu wird alle zwei Minuten das Gleichgewicht zwischen Rohwasseraufbereitung und Wasserabgabe überprüft. Korrigiert wird der Zustand nur, wenn die im Reinwasserbehälter vorhandenen Mengen unter einen Sicherheitswert absinken oder wenn die aus der Prognose für die nächsten Stunden zu erwartende Abgabe dieses notwendig macht. Kurzzeitige Verbrauchsspitzen oder Einbrüche bleiben so ohne Einfluß auf die Rohwassergewinnung. Gerade im Bereich der Entsäurer ist eine gewisse Schaltruhe aus verfahrenstechnischen Gründen erwünscht. Filterspülungen werden von der Werksoptimierung nur in der Schwachlastzeit initiiert. Tagsüber als notwendig erkannte Spülungen werden für die Spülung vorgemerkt. Kriterien für notwendige Spülungen sind:

Überschreiten der definierten zulässigen Durchflußmenge
Messung eines zu hohen Wassertrübungsgrades
Herauslaufen aus dem Regelbereich der Ein- und Auslaufregelschieber
Zu hoher Wasseraufstau über dem Filter

Eintritt eines oder mehrerer Ereignisse führt zu unterschiedlichen Maßnahmen. Die Filter werden bei Notwendigkeit bis in die nächste Schwachlastzeit weiter betrieben, jedoch wird nach Möglichkeit auf ein sauberes Filter umgeschaltet.

Stromoptimierung

Die Stromoptimierung startet den Notstromdiesel, um ein Überschreiten des tarifvertraglichen Viertelstundenleistungsgrenzwertes zu vermeiden. Alternativ kann der Betreiber verfügen, daß eine Beschränkung in der Reinwasserabgabe eintritt.
Einmal monatlich wird der Notstromdiesel zum Probelauf gestartet.

2.5 Steuerungen

Die Steuerungen haben die Aufgabe, die Schaltung der Betriebsmittel des Wasserwerkes vorzunehmen, ihre Ausführung zu überwachen und auf Störungen adäquat zu reagieren, so daß die Wasserversorgung der Stadt möglichst lange aufrecht erhalten wird.

Da das Wasserwerk in seinen wesentlichen Teilen verfahrenstechnisch redundante Komponenten besitzt, kann in hohem Maße durch automatische Umschaltung auf Reservewege oder Reserveaggregate ein Ausfall der Wasserversorgung verhindert werden. Erst durch eine automatischen Störfallbehandlung ist ein unbemannter Werksbetrieb sinnvoll möglich.

Auch seltene Störfälle werden vom Rechner sicher beherrscht, wenn sie einmal programmiert sind. Komplizierte Anlagen, wie die 10 kV Umspannanlage können so optimal genutzt werden. Der Rechner ist - im Gegensatz zum bemannten Betrieb - in der Lage, innerhalb weniger Minuten nach einem Hochspannungsausfall die Wasserversorgung wieder aufzunehmen. Es wird nicht erst die Anwesenheit von Fachpersonal erforderlich.

Bild 4 zeigt den internen Aufbau einer einfachen Steuerung am Beispiel der Reinwassersteuerung.

Jede Steuerung besteht im Prinzip aus den drei Komponenten:
  Auftragsermittlung
  Bereichsoptimierung
  Schaltprogramm

Die Auftragsermittlung identifiziert aus der Vielzahl der Informationen den für die Steuerung relevanten Auftrag. (Z. B. Förderleistung erhöhen).
Die Bereichsoptimierung stellt fest, wie dieser Auftrag optimal unter Berücksichtigung der verfügbaren Betriebsmittel ausgeführt werden kann und welche Schaltungen erforderlich sind.
(Z. B. Pumpe 1C aus, 2D ein)
Das Schaltprogramm führt diesen Befehl aus und überwacht das richtige Schalten der Aggregate. Bei Störungen im Ablauf erfolgt eine Rückmeldung an die Bereichsoptimierung.

Bild 4: Interner Aufbau von Steuerungen am Beispiel Reinwassersteuerung

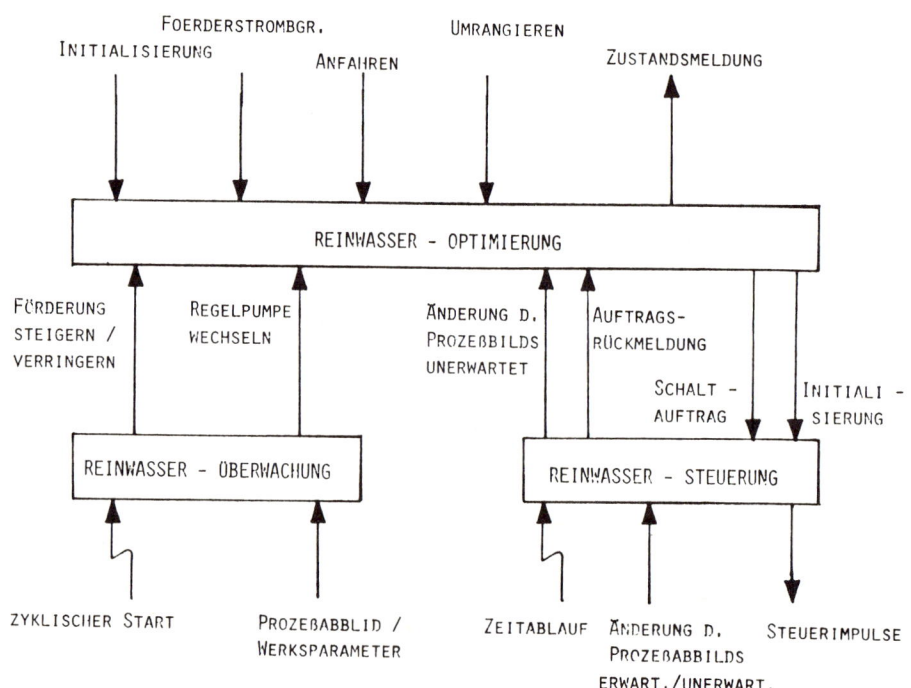

Hier der wesentliche Leistungsumfang einzelner Bereiche:

Brunnenauswahl und Steuerung

Definition von Brunnengruppen für optimale
Wassergemische durch den Betreiber
Reservegruppenschaltung
Reservepumpenschaltung
Auswahl nach Betriebsstunden der Pumpen für
gleichmäßigen Verschleiß
Stillstandsüberwachung mit Zwangseinschaltung
gegen Verkeimung
Vollautomatische Werksanfahrsteuerung nach
Spannungsausfällen

Entsäurer und Filter

Reservewegschaltung mit Ersatzentsäurer
Stillstandszeitüberwachung
Filterstraßenauswahl nach verfahrenstechnischen Kriterien
Umfangreiche Störfallbehandlung bei Filterstörungen
    Einbezug sämtlicher Einzelaggregate
    (Schieber, Klappen)
Filterspülung vollautomatisch mit Störfallbehandlung

Reinwasserpumpwerk

Überwachung von Druck/Durchfluß
Schaltung der starren Pumpen nach Förderklassen
Umschaltung auf Reservekombinationen bei Störungen an den Pumpen
Vollautomatische Werksanfahrsteuerung nach Spannungsausfällen

10 kV Stromschaltanlage

Umschaltungen bei folgenden Ereignissen
    Ausfall Einspeisung 1
    Ausfall Einspeisung 2
    Beidseitiger Ausfall
    Wiederkehr einer Einspeisung nach beidseitigem Ausfall
    Wiederkehr aller Einspeisungen

2.6 Dialogprogramme

Die Dialogprogramme dienen der Korrespondenz des Betreibers mit der Automatik. Über sie werden dem System Stammdaten bekanntgegeben. Informationen der Statistik werden angefragt und ggf. ausgedruckt, Schaltbefehle auf Wunsch ausgeführt wie z. B.:

Start des Notstromdiesels zum Probelauf
Werk stillsetzen
Werk anfahren
Protokoll eines bestimmten Monats ausdrucken
Grenzwerte für Analogwerte ändern
Neue Brunnengruppen definieren
Ausdrucken der Prognosedaten

3. Betriebserfahrung und Zuverlässigkeit

Das vorgestellte Automatisierungssystem hat sich in nunmehr 2-jähriger Betriebszeit bewährt. Es gab zwei nennenswerte Ausfälle des Systems, hervorgerufen durch Blitzeinschlag in unmittelbarer Nähe der Wassergewinnungsgebiete. Die Elektrik des Gesamtsystems hat sich für solche äußeren Störungen als zu durchlässig erwiesen.

In komplexen Automationssystemen spielt die Zuverlässigkeit von Rechnerhard- und -software nur eine gleichgewichtige Rolle zur Meßgeräteausrüstung, Steuerungstechnik und zur Auslegung der Verfahrenstechnik. Auch der zuverlässigste Rechner kann nicht eine fehlende Rohrleitung ersetzen. So sind die Notlaufeigenschaften eines geplanten Automatisierungssystems unter allen denkbaren Störfällen sorgfältig zu analysieren und durch adäquate interdisziplinäre Maßnahmen zu garantieren.

Der unbemannte Werksbetrieb über lange Zeiträume hinweg bedingt aber auch umfangreiche Diagnostik- und Testhilfsmittel, die die möglichen Störungen für die Reparaturmannschaft verständlich analysieren und rechtzeitig anzeigen.

EVALUATING SOFTWARE FAULT TOLERANCE

IN A REAL-TIME SYSTEM

T. Anderson
Computing Laboratory, University of Newcastle upon Tyne, England.

M. R. Moulding*
CAP Scientific Ltd., London, England.

ABSTRACT

Fault tolerance, in conjunction with the more traditional techniques of fault avoidance and removal, may have an important role to play in the construction of reliable software for real-time systems. A controlled experiment is being conducted to determine whether the use of software fault tolerance techniques can significantly improve the reliability of a practical system. By quantifying both the contribution to reliability and the costs due to the use of software fault tolerance, an assessment will be made of the cost-effectiveness of this approach to software reliability.

RÉSUMÉ

L'insensibilité aux défaillances, associée aux techniques plus traditionnelle permettant d'éviter ou de corriger les défaillances, peut avoir un rôle important à jouer dans l'élaboration de logiciel fiable pour les systèmes temps réel. On est en train de mener une expérience pour déterminer si l'utilisation de techniques d'insensibilité aux défaillances du logiciel peuvent améliorer de façon significative la fiabilité d'un système pratique. En évaluant, et la contribution à la fiabilité et les coûts résultant de l'utilisation de l'insensibilité aux défaillances, on pourra déterminer le rapport efficacité -coût de cette approche de la fiabilité du logiciel.

ZUSAMMENFASSUNG

Fehlertoleranz in Verbindung mit den mehr Konventionellen Verfahren der Fehlervermeidung und -Entdeckung Spielen eine wichtige Rolle beim Erstellen zuverlaessiger Software fuer Echtzeitsysteme. In einem Experiment wird untersucht ob die Verwendung fehler toleranter Software-Verfahren die Zuverlaessigkeit eines wirklichen Systems wesentlich verbessern Kann. Indem sowohl der Gewinn an Zuverlaessigkeit als auch die durch fehlertolerante Software verursachten Kosten quantifiziert werden, kann eine Aussage veber die Kosteneffizienz der Verfahren fewonnen werden.

* Current affiliation: Division of Electrical and Electronic Engineering,
　　　　　　　　　　　Hatfield Polytechnic,
　　　　　　　　　　　England.

INTRODUCTION

For many real-time applications the cost of a system failure is so high that stringent standards of reliability are rightly demanded of both hardware and software. Unfortunately, even for quite modest real-time systems, the software can be so complex that it inevitably contains residual design faults - despite very careful design and thorough testing. Just as fault tolerance may be used in hardware to cope with the consequences of physical deterioration of components, software fault tolerance techniques (1) have been proposed as a means of preventing program 'bugs' from causing system failure. An evaluation of the cost-effectiveness of software fault tolerance is in progress at the University of Newcastle upon Tyne.

RECOVERY BLOCKS

The recovery block construct (2) provides a convenient notation for incorporating tolerance to design faults within a program. The approach is essentially a stand-by sparing scheme applied to blocks of a program. It is able to cope with unpredictable design faults because it employs acceptance tests (rather than specific error checks), error recovery by prior state restoration (rather than selective amendments), and spare alternate blocks of independent design (rather than retry of identical code). In its basic form, the recovery block approach to software fault tolerance can only be applied in sequential programs; its adoption in concurrent systems entails the provision of a more complicated recovery mechanism. A mechanism has been devised (3) which enables recovery blocks to be utilised in MASCOT (4) systems of concurrent communicating processes.

OBJECTIVES

In order to investigate the use of software fault tolerance in a realistic application a demonstration system is being constructed. This system utilises recovery blocks, is supported by MASCOT, and models the functions of a naval shipborne command and control system. To ensure the realism of the demonstration system it is being implemented by a small team of software specialists to full commercial standards, and in consultation with naval personnel. Our aims in building the system are:

(i) To assess the effectiveness of software fault tolerance, by means of quantitative measures and (independent) qualitative judgements.
(ii) To estimate the costs of software fault tolerance, in terms of extra resources and performance penalties.

(iii) To confirm the applicability of software fault tolerance techniques to practical real-time systems (specifically command and control systems).

(iv) To promote the adoption of software fault tolerance as a means of enhancing software reliability.

EVALUATION

An experimental programme will be conducted to obtain a quantitative estimate of any improvement in reliability consequent to the use of recovery blocks. Because it is easy to disable the fault tolerance provided by recovery blocks, the reliability of the demonstration system will be measured both with and without tolerance to software faults. Hopefully, statistical tests will show a significant improvement in the mean time between failures for the fault tolerant system over the intolerant version. Should it be necessary to artificially increase the system failure rate (so as to obtain more data) this will be done under strict controls, using techniques such as fault injection or inclusion of a relatively untested block of code.

ACKNOWLEDGEMENT

This work is supported by the Science and Engineering Research Council and the Ministry of Defence (Great Britain).

REFERENCES

(1) Anderson, T. and Lee, P. A.,
Fault Tolerance: Principles and Practice.
(Prentice/Hall International, London, 1981)

(2) Horning, J. J. et al.,
A Program Structure for Error Detection and Recovery, Aspects Theoretiques et Pratiques des Systems d'Exploitation, 177-193.
(IRIA, Rocquencourt, 1974)

(3) Anderson, T. and Moulding, M. R.,
A Conversation Mechanism for MASCOT Systems, Report in Preparation.

(4) Jackson, K. and Moir, C. I.,
Parallel Processing in Software and Hardware - the MASCOT Approach. Proc. 1975 Sagamore Computer Conference on Parallel Processing, 71-78.

A MICROPROGRAM PRODUCTION ENVIRONMENT
TO SUPPORT COMPUTER SYSTEM RELIABILITY

P. WILK AND M, NORRIE
School of Information Sciences, The Hatfield Polytechnic
PO Box 109, College Lane, Hatfield, AL10 9AB

ABSTRACT

A microprogram production environment has been designed to encourage the production of computer systems with improved standards of reliability, performance, efficiency and maintainability.

This paper outlines how this microprogram production environment will lead to more reliable computer systems.

Whilst the paper concentrates on computer system correctness, it is envisaged that the tools and techniques used in the production life-cycle will encourage the inclusion of fault tolerant mechanisms in the computer architecture.

RÉSUMÉ

Un environnement de production de microprogrammes a été conçu pour encourager la production d'ordinateurs conformes à des standards élaborés de fiabilité, performance, efficacité et maintenabilité.

Ce papier souligne la façon dont cet environment conduira à des ordinateurs plus fiables.

Alors que cette présentation s'intéresse surtout à la qualité des ordinateurs, on espère que les outils et techniques utilisées au cours du cycle de production faciliteront l'introduction de mécanismes insensibles aux défaillances dans l'architecture des orinateurs.

ZUSAMMENFASSUNG

Es wurden Verfahren und Hilfsmittel entworten, die Erstellung von Rechnersystemen erleichtern, die nachweisbar zuverlaessiger, leistungsfaehiger, effizienter wartungsfreundlicher sind.

Die Abhandlung skizziert wie diese Mikroprogrammier-Verfahren zu zuverlaessigeren Rechnersystemen fuehrt.

Waehrend die Abhandlung sich im wesentlichen mit der Korrektheit von Rechnersystemen beschaeftigt, Kann erwartet werden, dass die Werkazeuge und Verfahren, die bei des Programmerstellung benutzt werden, auch den Einbau fehlertoleranter Mechanismen in die Rechnerarchitektur foerdern werden.

A MICROPROGRAM PRODUCTION ENVIRONMENT
TO SUPPORT COMPUTER SYSTEM RELIABILITY

P WILK AND M NORRIE
School of Information Sciences, The Hatfield Polytechnic
PO Box 109, College Lane, Hatfield. AL10 9AB

A microprogram production environment has been designed to encourage the production of computer systems with improved standards of reliability, performance, efficiency and maintainability.

This paper outlines how this microprogram production environment will lead to more reliable computer systems.

Whilst the paper concentrates on computer system correctness, it is envisaged that the tools and techniques used in the production life-cycle will encourage the inclusion of fault tolerant mechanisms in the computer architecture.

1. INTRODUCTION

The use of user microprogrammable computers to obtain improved computer performance is often achieved at the expense of reliability. However, we believe that when correctly managed and applied, user microprogramming can also improve the reliability of the computer system. Unfortunately, the complexity of extensive user microprogramming using a present day microprogram production environment is hindered by a phenomenon known as the semantic gap (11) - a measure of the difference between the concepts in high-level languages and the concepts in computer architecture. The resulting code violates every software premise. Clearly it was necessary to design a production environment to control this complexity.

In order that the requirements for a microprogram production environment could be identified (17, 18), an experiment in implementing a model processing system was carried out (16).

It was apparent that the underlying problem was to determine how to encourage the production of microprogrammed computer architectures directed towards the application requirements - a part of which contains the reliability requirements.

This paper describes how this microprogram production environment will lead to more reliable computer systems.

2. RATIONALE

A computer system would be deemed unreliable as a consequence of faults in the system. To provide a reliable system - one that will always conform to its specifications - there are two basic approaches:

1. Ensure the "correctness" of a system by a combination of techniques for fault prevention, fault avoidance and fault removal.

2. Incorporate measures and mechanisms for fault tolerance (a description of these may be found in (1)). Fault tolerance can be broken down into the four phases of error detection, damage confinement and assessment, error recovery and fault treatment.

Even under the naive assumption that a system could be guaranteed to be "correct", there are some faults which, at present, cannot be prevented or avoided e.g. hardware failure. To achieve a reliable system, there must be a combination of both approaches - a production method should be used that will increase the probability of "correctness" but at the same time will incorporate features to support fault tolerance.

Some previous research on the use of microprogramming has concentrated on the aspect of "correctness" - with the use of higher-level microprogramming languages and formal verification techniques being advocated.

Formal verification began with Ffloyd's inductive assertion method (4) and has since been elaborated on by Hoare (6), Dijkstra (3) and Manna (8). The method requires the addition of non-executable assertive text to the program. The assertions are made to describe the desired state of the variables of the program.

STRUM (13) uses structured programming, a higher-level language and a formal verification method to aid in the design of correct microprograms for the Burroughs D-machine. The language has PASCAL-like syntax and includes statements useful in proving programs, such as "assert". The verification system is modelled after London's Pascal verification system (5) which also uses Floyd's inductive assertion method. The STRUM compiler uses assertions written by the programmer to generate verification conditions. The verification conditions are simplified by one module of the system and then proven using user-supplied axioms. The failure to prove an assertion can mean either a mistake in the microprogram or a mistake in the assertion. Both of these possibilities must be checked by the verifier.

The logical flaw in the design of contemporary verification systems is that the verifiers themselves are unverified. Furthermore, showing that a microprogram is logically correct does not assure its correct running on the computer. For example, corrupt data might be supplied. All the verification proof techniques so far presented suffer from this flaw, by concentrating on logic errors and ignoring the main class of program errors: computation errors such as overflow, and incorrect input of programs or data.

A formal proof of the correctness of microprograms is the best means of guaranteeing that it meets the given specification. However, at present, formal verification is a technique not yet adopted by the majority of the computing fraternity. In addition, the formal techniques are non-rigorous, e.g. Floyd and Hoare do not prove the termination of a program.

Extensive simulation and testing of the microprogram is the usual alternative. For large systems, it is not cost-effective to use simulation techniques to test all the run-time alternatives. It is therefore a "weak" tool for determining the reliability of a system.

By correct, consistent and complimentary use of both types of support system the reliability of computer systems may be taken to a much higher standard. Their complimentary use prevents polarization of the meaning of the word "rigour" towards either one of these non-rigorous techniques.

The semantic gap is a significant contributor to software unreliability because a large set of programming error types that could be detected by the compiler go unchecked. For example, a common situation arises where a reference is made to an array element where one of the subscripts falls beyond the bounds of the corresponding dimension. Many compilers make the check for this error optional because of the substantial overhead it incurs, in both storage space and execution time, on many outdated computer architecture designs.

The aim of target system designers should be to close the semantic gap by producing application-directed computer architectures. One of the reasons that microprogramming has been misused as a system implementation technique, is the failure of computer system designers to recognise the need for a computer system architect as part of the design team. The function of this role involves making compromise decisions between the conflicting objectives of the system design in terms of the available environment resources. These decisions are important because they determine the framework within which the computer architect and other members of the production team work.

No mention is made, by any of the higher-level microprogramming language designers, of any attempt to determine the characteristics and requirements of the "user" before designing the language. Often, unsuitable microprogramming tools are made available to users, with the result that unreliable and inefficient microcode is produced. Furthermore, given the present "unreasonable" circumstances outlined by Rosin (14), the main reason for unreliable and inefficient microcode would seem to be a mis-directed research effort towards machine independence of language and portability of code. An effort towards the provision of reasonable hardware, based on functional building-blocks (19), would stem the proliferation of egotistical microprogrammable computers. This approach would encourage portability of code by the removal of the notion of host machine dependence. As a result, this would enable the derivation of a set of universally acceptable higher-level microprogramming language design rules.

The conclusion drawn from a survey of current work was that traditional microprogram production systems are not suited to the production of large amounts of microcode - from the consideration of reliability, performance and efficiency. The recommendation is made for a radical reconsideration of all the elements of a microprogram production system.

3. METHODOLOGY

It was evident that a formal method was required to guide and encourage a correct and rigorous production process. The software life-cycle methodology (2) is chosen as the basis for this production environment but with extensions to include firmware.

The methodology proposed embodies the use of generally accepted software engineering principles which help to identify the properties common to superficially different software. It complements this by making inessential information inaccessible to those who do not require it. Each stage in the methodology aims to encourage a consistent pattern of development to ensure that the system is complete thus enabling correctness checks to be carried out. This is aided by the use of appropriate analysis, design, development and implementation tools. A full description of a production life-cycle and methodology for producing target systems can be found in (19).

Figure 1. illustrates the microprogram production life-cycle. It should be noted that the stages which involve the production of microprograms are interchangeable with those in a conventional software life-cycle. One effect of this is to enable the use of Stockenberg's vertical migration technique (15) in target system implementation and tuning.

4. TECHNIQUES

The experiment in microprogramming led to the belief that the target machine design objective should be to achieve the composition of computer systems from a set of highly independent functions and data types encapsulated in target machines. This implies a bottom-up target system design method.

Myers' Composite Design technique is advocated for target system design (10).

Composite Design is based around the concept of a module. An important consideration in system design is having a set of objective measures of the design standard. Myers uses "module strength" and "module coupling". Module strength is a measure of the maintainability of an individual module. Module coupling is a measure of the interconnections and relationships among modules. The ideal module strength (functional strength) is where all the elements are related to the performance of a single function. The ideal module coupling is one which is loosely coupled, i.e. it is data coupled. Two modules are data coupled if one calls the other, and all input and output is passed as arguments. Furthermore, all of the arguments are data elements; not control elements or data structures.

With the use of the Composite Design technique, the qualities attributed to the microprograms can be summarised as follows.

1. "Correctness" of microprograms will be encouraged since decomposition into highly independent units implies that they will be less complex - this means that the design, coding, verification, debugging and testing of the microprograms will be easier.

2. The inherently local nature of errors contained in loosely coupled functions will benefit fault tolerance. At the very minimum, redundancy checks should be included within a module to ensure that any error will be detected before an exit from the module. Once an error has been detected, the resulting damage will be confined to the module, the error recovery will be simpler and the location and repair of the fault will be easier.

3. Any modifications to the system will normally only affect a minimal subset of the parts due to their highly independent nature.

Formal verification systems are necessary to ensure consistency of verification (although the verification software should itself be verified by hand). For this work, two formal verification techniques are advocated: Hoare's (6) and Floyd's (4).

Hoare's proof technique was chosen for the verification of the abstract data structures and their concrete representation. A full discussion of the suitability of Hoare's technique for this application may be found in (7). Floyd's technique is used for the verification of microprograms and the target system specification - although its limitations are well known.

Both "white" and "black" box test-case design methodologies are advocated for the support of testing (12).

White-box methodologies involve the derivation of test cases to meet criteria with respect to the logic of a microprogram. These criteria are:

1. Sufficient test cases to cause each branch operation in the program to be active at least once.

2. Sufficient test cases to cause the selection of each branching option of a branch operation to be executed at least once.

3. Sufficient test cases to cause every unique path within the program to be traversed at least once (a criterion that is likely to be achievable in this system).

Black-box methods should be used to complement the white-box methods. They involve the derivation of test cases based on an analysis of the external interfaces of the system.

The specific methodologies advocated are "Boundary-value analysis" and "Equivalence partitioning".

Boundary-value analysis is a methodology that recognises that the most useful test cases are those at, and just beyond, the boundaries of equivalence classes in both the input and output states of the program.

Equivalence partitioning is based on the effective measures of an effective test case being: it significantly reduces the number of other test cases that must be developed to achieve some predefined level of testing; it indicates the presence or absence of errors beyond the specific set of input values in the test case. The methodology involves partitioning the input state of the program into equivalence classes so that there is a high probability that one test case in an equivalence class is representative of all other test cases in an equivalence class.

5. TOOLS

A number of language tools are advocated for the use of the appropriate personnel involved at each stage of the microprogram production life-cycle. These are:

1. Command Language Grammar (9) - a tool for the system designer which encourages the adoption of the end-user's model of the system.

2. A Data Base Query Language - to aid the design of the target system architecture, by the system programmer, by accessing the target object library (stored in a data base).

3. Computer System Specification Language (CSSL) - to enable the system programmer to code the design of the target system.

4. Computer Architecture Data Abstraction Language (CADAL) - for use by the target machine designer (the computer architect) in the specification of abstract target machines and their concrete representation.

5. Computer Architecture Microprogramming Language (CAML) - for use by a specialist microprogrammer in coding target machine functions; and by the firmware engineer in control module specification (containing microcode functions).

6. CONCLUSIONS

The lack of an integrated microprogram production environment has resulted in the limited (and often incorrect) use of microprogramming at the expense of computer system reliability. However, the addition of suitable management skills, procedures and methods to the proposed microprogram production environment will encourage the design of features in the computer architecture to support fault tolerance - in particular, tagged storage, descriptors, executable assertions (for range, set and value checking, assertion and protection mechanisms. Furthermore, it will encourage the production of "correct" systems by bringing the level of complexity of large applications under control.

The microprogram production life-cycle proposed, allows the system production team to delay the implementation of the computer architecture until the application and its reliability requirements have been defined and specified. In terms of efficiency, a computer architecture designed in this way will probably contain some overhead that might not have been present if it had been designed in an unstructured way. However, it is more cost-effective to adopt this approach when marginally increased storage and decreased execution efficiency (in practice we have found these are not evident in large applications) is weighed against reduced development, maintenance and modification costs coupled with increased reliability.

7. REFERENCES

1. Anderson, T. and Lee, P. A., Fault Tolerance (Principles and Practice), (Prentice-Hall Intl., 1981).

2. Boehm, B. W., Software Engineering, IEEE Trans. on Computers Vol.C-25 No. 12 (1976) 1226-1241.

3. Dijkstra, E. W., A constructive approach to the problem of program correctness, BIT Vol. 8 (1968) 174-186.

4. Floyd, R. W., Assigning meanings to programs, Proc. of a Symposium in Applied Mathematics, American Mathematical Society, Vol. 19 (1967) 19-32.

5. Good, D. I., London, R. L. and Bledsoe, W. W., An Interactive program verification system, IEEE Trans. on Software Engineering, Vol. 1 No. 1 (1975) 59-67.

6. Hoare, C. A. R., An axiomatic approach to computer programming, Comm ACM Vol. 12 (1969) 576-580.

7. Hoare, C. A. R., Proof of correctness of data representations, Acta Informatica Vol. 1 (1972) 271-281.

8. Manna, Z., The correctness of programs, Journal of Computer and System Sciences, Vol. 3 No. 5 (1969) 119-127.

9. Moran, T. P., Introduction to the Command Language Grammar, Report SSL-78-4. XEROX Corp., Palo Alto, California, (1978).

10. Myers, G. J., Reliable Software through Composite Design (Litton Educational Publishing Inc., 1975).

11. Myers, G. J., Advances in Computer Architecture (Wiley, 1978).

12. Myers, G. J., The Art of Software Testing (Wiley, 1978).

13. Patterson, D. A., STRUM: structured microprogram development system for correct firmware, IEEE Trans. on Computers Vol. C-25 No. 10 (1976) 974-985.

14. Rosin, R. F., The significance of microprogramming, in ICS 73 237-242 (North-Holland, 1974).

15. Stockenberg, J. E., Optimization through Migration of Functions in a Layered Firmware-Software System, Ph.D. Thesis, Brown University (1977).

16. Wilk, P. F., An investigation into user microprogrammed software, TN-51, School of Information Sciences, The Hatfield Polytechnic, Herts., U.K. (1981).

17. Wilk, P. F. and Bull, G. M., A strategy, method and set of tools for a user, dynamic microprogramming life-cycle, in ICS 81 54-61 (Westbury House, 1981).

18. Wilk, P. F. and Norrie, M., A rationale for a microprogram production environment, TR-31, School of Information Sciences, The Hatfield Polytechnic, Herts., U.K. (1982).

19. Wilk, P. F., A Study of Dynamically Variable Computer Architecture, Ph.D. Thesis, The Hatfield Polytechnic, Herts., U.K. (1982).

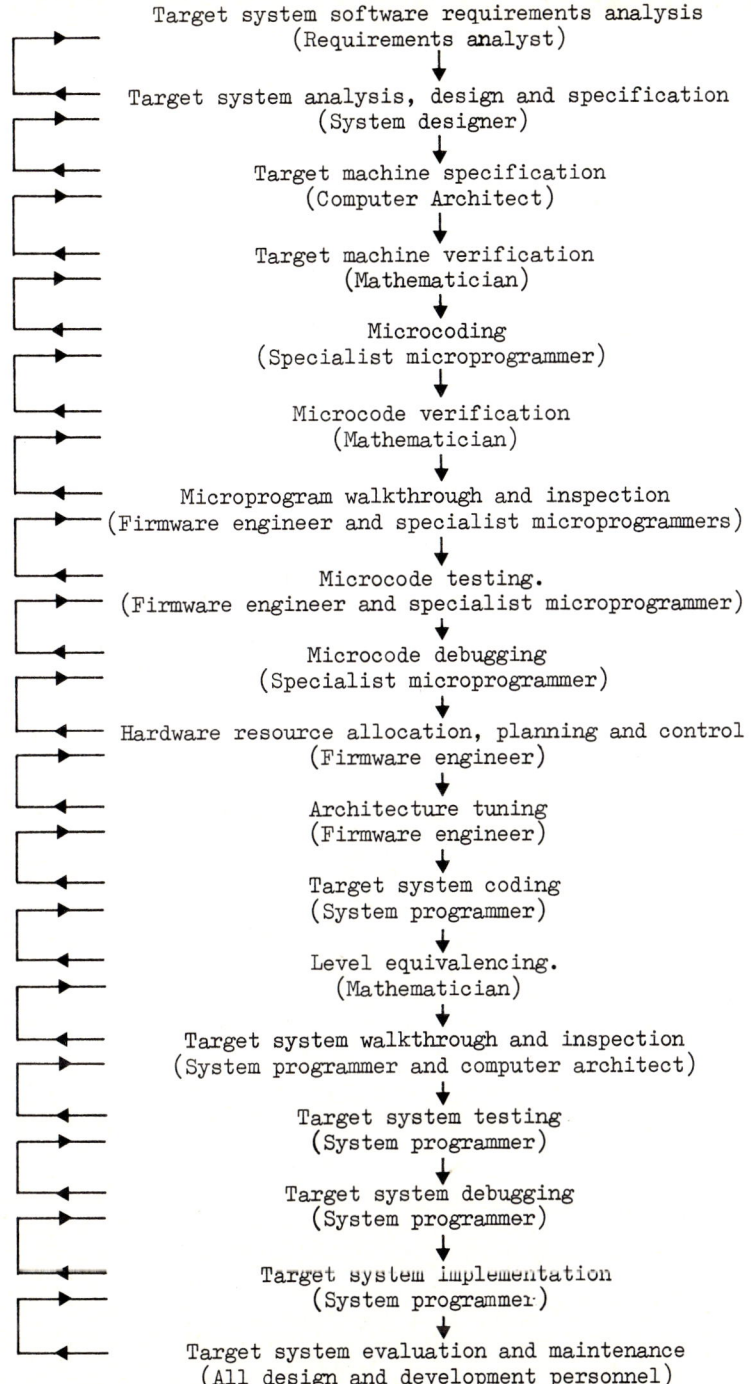

Figure 1. The Microprogram Production Life-cycle.

Session 10

"POSTERS"

RADAR LINE WATCH SYSTEM

Jörg Beyer, EUROCONTROL Brussels, Belgium
Joachim Heller, EUROCONTROL UAC Karlsruhe, Federal Republic of Germany

ABSTRACT

In the frame of the Technical Monitoring and Control System of the EUROCONTROL Upper Area Control Centre (UAC) at Karlsruhe, a system to monitor the radar data transmission has been implemented. The system consists of 6 radar input (RAD-IN) units and a concentrator. The RAD-INs as well as the concentrator are microprocessor-based systems. Radar data is received, processed by the RAD-INs and transmitted by them to the concentrator. The concentrator outputs the assembled data to a CRT display. The Radar Line Watch System monitors the line quality as well as radar station and extractor status. Given this information, the Technical Supervisor can take appropriate remedial actions.

RESUME

L'ensemble du système de supervision et de contrôle technique implanté au centre EUROCONTROL de contrôle de l'espace supérieur de Karlsruhe comporte un système de supervision de la transmission des données radar. Les données d'un radar sont reçues par une unité de réception. Elles y sont traitées, adaptées et transmises à une unité commune qui rassemble les données de tous les radars et les transmet vers l'écran de visualisation. Toutes ces unités sont de type "microprocessor". Le système de supervision contrôle la qualité de la ligne, la disponibilité et l'état des chaînes radar-extracteur. Par ces informations, le contrôleur technique responsable peut provoquer les actions correctives appropriées.

ZUSAMMENFASSUNG

Im Rahmen des Technischen Überwachungs- und Kontrollsystems von EUROCONTROLs Flugsicherungszentrale für den oberen Luftraum in Karlsruhe wurde ein Überwachungssystem für Radardatenleitungen in Betrieb genommen. Das System besteht aus 6 Radareingangseinheiten (RAD-INs) und einer Auswerteeinheit. Eingangseinheiten wie auch die Auswerteeinheit verwenden Mikroprozessoren. Radardaten werden empfangen, bearbeitet und zur Auswerteeinheit übertragen, die die Daten einem Sichtgerät übergibt. Das Überwachungssystem für Radardatenleitungen überwacht sowohl die Qualität der Übertragungsleitungen als auch den Zustand der Radars und des Extraktors. Mit diesen Informationen kann der technische Wachleiter die Instandsetzung fehlerhafter Geräte veranlassen.

1. INTRODUCTION

In the Upper Area Control Centres (UAC) at Karlsruhe and Maastricht, EUROCONTROL has put into operation centralized Technical Monitoring and Control Systems (TMCS) (1) (2). Those systems monitor and control from a central point the real time computers for radar and flight plan data handling as well as the supporting systems. In the frame of the TMCS, equipment to monitor the radar data transmission has been implemented. Whereas at Maastricht UAC conventional techniques with integrated circuits (ICs) have been used, the design of the Radar Line Watch System at Karlsruhe UAC has been based upon the 6800 microprocessor and Large Scale Integrated (LSI) peripheral interface adapters. This system provides the Technical Supervisor at the TMCS with the status of the complete radar data processing and transmission system from the radar heads up to the UAC and will be described in this paper.

2. SYSTEM DESCRIPTION (Karlsruhe)

2.1. Functional Description

Radar is a very important source of information for air traffic control. Thus, extracted and digitized radar data of 6 radars located at different places in the Federal Republic of Germany is transmitted via fixed data links (telephone lines) to the multi-radar tracking system at Karlsruhe UAC (3). Each radar station is connected to the Karlsruhe Centre by up to three data lines. In order to cope with line interruptions, these lines use different routings. The data is transmitted via modems with a speed of 4800 bits per second in a serial synchronous procedure (V 24 standard). The transmission is carried out in fields of 18 bits including 2 parity bits. A unique synchronisation message consists of one field. The first bit of this message is always "0", whereas all other messages start with a "1". This synchronisation message is also used as the "idle field", which is transmitted if all other type of information is absent.

The radar messages are composed of 3, 4, 5 or 6 fields. The first byte indicates the type of the message and the number of fields it contains. Radar station, extractor and data transmission system status are indicated in the so-called Extractor Monitor Message (EMM), which has a length of 3 fields. This information is used in the Radar Line Watch System together with information on line quality gained by counting parity errors and message format errors. When monitoring the radar lines, the radar messages are only used for statistics.

2.2. Hardware Description

2.2.1. RAD-INs

The Radar Line Watch System, which is connected to the radar data transmission system - the layout and design of the latter is not furthermore dealt with in this paper -, consists of 6 Radar Input units (RAD-INs). For each of the 6 radar sources one of those units is provided. These units are based on a standardized design used within EUROCONTROL and also by other Air Traffic Control Authorities for different applications.

Each unit is housed in a 19 inch rack and consists of a data, address and control bus, slots for up to 13 boards of "Eurocard" format and a power supply. For radar line watch purposes this rack is equipped with the following standard boards :

- 1 CPU card using a Motorola 6800 microprocessor with 4 K byte ROM (for program storage);
- 1 memory card with 2 K byte RAM (for data storage and stack);
- 3 serial synchronous receiver cards each with one 6852 SSDA (for connection to the radar lines);
- 1 serial synchronous transmitter card with one 6852 SSDA for test purposes (test loop to receivers);
- 1 serial asynchronous transmitter card with one 6850 ACIA (for transmission of sampled and processed data to the status display via a concentrator);
- 1 peripheral interface adapter with one 6821 PIA (for test and internal system status indication).

The receiver and transmitter cards for synchronous and asynchronous transmission have identical board lay-outs and are adapted to the different applications by the use of different components and placing of straps.

2.2.2. Concentrator

The concentrator receives the sampled and processed data from the 6 RAD-IN units and provides a composite status indication on a CRT display. Basically the concentrator utilizes the same hardware components as the RAD-INs (see block-diagram).

2.3. Software Description

For applications such as the Radar Line Watch System one can either use a straight-forward polling sequence for interrogating receivers or control receivers by interrupts. Since the events (i.e. reception of data fields) occur in any order on the different lines, control by interrupts has been preferred for this application. Updating of the internal system status and interrogations of a process table are handled during the idle loop (the program goes into the idle loop if all pending interrupts are processed).

The application program can be divided into three main tasks :

- Extractor Monitor Message (EMM) processing;
- provision of message statistics;
- line quality indication.

The EMM will be transmitted by the extractor during the North crossing of the radar head (i.e. the centre of the radar beam is pointing to the North) and contains amongst other information :

- radar station code;
- radar station status;
- transmission status;
- time between two consecutive north crossings.

This data will be decoded, processed and together with the message statistics and quality indications transmitted to the concentrator.

The following message statistics are processed :
- the number of data fields received;
- the number of idle fields received;
- the number of radar targets received.

The quality indication is based on the number of parity errors and format errors detected during one antenna revolution.

The processed data is written into an input table. On receipt of the next EMM this input table is closed and a second one opened, into which more processed data is written until a further EMM arrives and triggers the switch-back to the first table. The data of the closed table is converted to ASCII-code and put into an output table in order to be transmitted to the concentrator.

3. CONCLUSION

The summary of all this information provides the Technical Supervisor with sufficient data to determine possible sources of failures and to take appropriate remedial actions in the case of radar failures, broken lines or lines with bad quality.

Experience has shown the advantages of a solution using microprocessors, i.e. its flexibility in the case of a change of procedure or message structures. The required effort in such a case is reduced to software changes without significant interruptions of system operation.

As already mentioned above, the standardized design of the units have enabled similar boards to be used for different applications. Examples are :
- buffer multiplexer 8 to 1 line; and
- radar data concentrator (to combine data from two radar stations into one channel).

Further applications are already envisaged, for example :
- message filter (to sort out certain types of messages) and
- message switching system (to send different messages to different destinations).

4. REFERENCE

(1) J. Beyer, 'A Description of the Technical Monitoring and Control System installed at Maastricht UAC', EUROCONTROL 5-III (August 1975), p. 3.
(2) J. Beyer, J. Heller, J. Uhl, 'Technische Monitor- und Kontrollsysteme für Flugsicherungskontrollzentralen', Real Time Data Handling and Process Control, Proceedings of the First European Symposium (Berlin, Oct. 1979), p. 111.
(3) J.F. Pieri, G. Fairfax-Jones, 'Radar Tracking', EUROCONTROL Institute Press (Luxembourg, Oct. 1981).

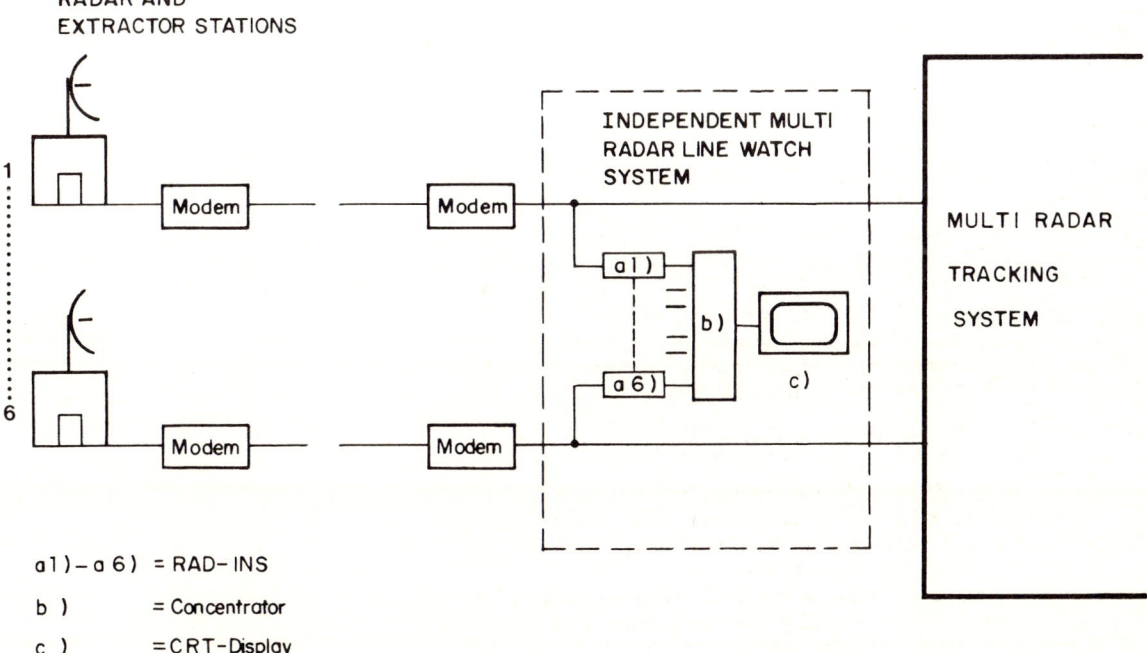

Fig. 1 : RADAR LINE WATCH SYSTEM

ANFORDERUNGEN UND ENTSPRECHENDE LÖSUNGSMETHODEN FÜR BUSZUGRIFFSSTEUERUNGEN

Herbert Kern
Institut für elektrische Meßtechnik
Technische Universität - Wien, Österreich

ZUSAMMENFASSUNG

Entsprechend der Bedeutung, die der Buszugriffssteuerung (Arbitration) in Multi-Mikroprozessorsystemen zukommt, wurden die Anforderungen an eine solche untersucht, die theoretischen Möglichkeiten verglichen und ein System entwickelt, das als günstigster Kompromiß von Geschwindigkeit und Leitungszahl 256 Prioritätsebenen über 3 Leitungen in 4 Signallaufzeiten unterscheiden und mit verteilter Logik den Bus für jede Busoperation neu dem höchstprioren Modul zuteilen kann.

ABSTRACT

According to the importance of the arbitration in multi-microprocessor-systems, its requirements have been analysed, the theoretical possibilities have been compared and a system has been developed as best compromise of speed and number of transmission lines that is able to define 256 priority-levels, using 3 transmission lines needing four times the transmission-delay. For each bus-operation the bus is assigned to the modul with highest priority with distributed logic.

RÉSUMÉ

Vue l'importance de l'arbitrage dans les systèmes à microprocesseurs multiples, ses impératifs ont été analysés, les possibilités théoriques comparées et un système a été développé pour aboutir au meilleur compromis possible de vitesse et de nombre de signes de transmission qui permette de définir 256 niveaux de priorité: 3 lignes de transmission ont été utilisées nécessitant 4 retards de transmission. A chaque utilisation du bus, celui-ci est affecté, grâce à une logique distribuée, au module de niveau de priorité le plus élevé.

1. ZUSAMMENSTELLUNG DER ANFORDERUNGEN AN EIN LEISTUNGSFÄHIGES ARBITRATIONSSYSTEM

Jede Buszugriffssteuerung hat im Wesentlichen zwei Aufgaben zu erfüllen: Sie muß den Modul mit der höchsten Priorität finden und diesem den Bus zuteilen können und sie muß für einen ausreichenden Kollisionsschutz sorgen können.

In einfachen Systemen erfolgte die Buszuteilung auf längere Zeit, z.B. für die Erfüllung einer bestimmten Teilaufgabe, für die Auswertung der Anforderungen stand also relativ viel Zeit zur Verfügung.

Da die Moduln meist bestimmte, gleichbleibende, bekannte Aufgaben zu erfüllen hatten, genügte eine Prioritätsebene pro Modul. Jede Prioritätsebene konnte nur einmal vorkommen, der Kollisionsschutz war dadurch gegeben.

In solchen Systemen müssen alle im Betrieb möglichen Zustände bekannt sein und berücksichtigt werden, da ein flexibles Reagieren in bestimmten Sonderfällen nicht möglich ist. Durch die Zuteilung des Busses auf längere Zeit sind Aufgaben mit Echtzeitanforderungen nur schlecht zu bewältigen.

Komplexe Multi-Mikroprozessor-Systeme bieten nun die Möglichkeit, Aufgaben flexibel im System zu verteilen, Teilaufgaben eines Moduls können für das Gesamtsystem verschieden wichtig sein. Ein Arbitrationssystem mit nur einer Priorität pro Modul kann hier nicht mehr ausreichen, der Modul muß die Möglichkeit bekommen, die Priorität seiner Anforderung wählen zu können.

Da dies zu Kollisionen führen könnte, wenn 2 Moduln die gleiche Priorität wählen, muß der "Betriebspriorität" eine "Kollisionsschutzpriorität" überlagert werden. Durch diese Anforderungen steigt die Zahl der benötigten Prioritätsebenen. Für 16 Moduln scheint z.B. ein System mit 16 Ebenen für den Kollisionsschutz (etwa vergleichbar den bisherigen Prioritäten) und 16 variablen Prioritäten für die Anpassung an die Systemerfordernisse wünschenswert. Die dazu notwendigen 16 x 16 = 256 Ebenen kann jedoch kaum ein gebräuchliches Arbitrationssystem genügend schnell unterscheiden.

Für das Echtzeitverhalten wäre es wichtig, daß jeder Modul nahezu sofort für kurze Zeit den Bus zugeteilt erhalten kann, um in einem Mindestausmaß sofort reagieren zu können. Dazu ist jedoch ein Mechanismus notwendig, der dafür sorgt, daß jeder Buszyklus einem anderen Modul zugeteilt werden kann. Das bedeutet, daß die Arbitration und die Buszuteilung innerhalb einer Buszykluszeit erfolgen müssen.

Dadurch und durch eine geeignete Strategie beim Ändern der flexiblen Priorität - Senken nach Erhalt des Buszugriffsrechtes, stetiges Erhöhen bei erfolgloser Arbitration - wird die Wartezeit auf eine Busoperation auch bei unvollständiger Kenntnis des Systemzustandes abschätzbar.

Um schließlich ein Gesamtsystem aus Moduln leichter aufbauen zu können, sollte die Hardware für die Arbitration auf den Moduln verteilt sein, damit ein Umprogrammieren oder eine Änderung der Verdrahtung beim Umstecken bzw. Ergänzen von Moduln nicht notwendig ist.

2. GRUNDLAGEN UND MÖGLICHKEITEN FÜR ARBITRATIONSSYSTEME

2.1 Minimale Dauer einer Informationsübertragung:

Die minimale Zeit, in der ein Signal vom Sender zum Empfänger laufen kann besteht aus 3 Komponenten:

Schaltzeit der Bustreiber
Buslaufzeit
Dekodierzeit im Empfänger

Mindestdauer eines einfachsten Übertragungszyklusses

Sie ist mindestens notwendig, um einfachste Information in Form eines Bits zu übermitteln. Üblicherweise sind mindestens 2 einfachste Übertragungszyklen für eine Busoperation notwendig, da der Commander den Responder zu einer Aktivität auffordert und dieser dann entsprechend reagieren muß.

2.2 Möglichkeiten zur Feststellung der höchsten Priorität

2.2.1 Leitungsführung

Spezialverdrahtung: Stichleitungen, Daisy-Chain
Sie ist für eine größere Zahl von Prioritätsebenen kaum geeignet, da sie bei großer Schnelligkeit zu viele Leitungen (Stichleitungen) benötigt oder bei einfacher Verdrahtung (Daisy-Chain) zu langsam ist. Außerdem ist eine Überwachung des Systems wenn überhaupt, so nicht von allen Steckplätzen aus möglich.

Parallelverdrahtung:
Bei Verwendung von Busleitungen können von jedem Steckplatz aus die Vorgänge beim Arbitrieren überwacht werden.
Die Auswahl der höchsten Prioritätsebene erfolgt dadurch, daß entsprechend einem Entscheidungsbaum in mehreren Stufen durch Auswahl der ranghöchsten Gruppe zuletzt, der ranghöchste Modul gefunden wird. Jede Entscheidung im Entscheidungsbaum entspricht einer Informationsübertragung.

Unterscheidbare Ebenen =
= (gleichz. unterscheidb. Zustände)^{Entscheidungen}

Da meistens mehrere Moduln an der Arbitration beteiligt sind, können zur Auswahl nur eindeutige Zustände herangezogen werden, die nicht durch Überdeckung niederpriorer Anforderungen entstanden sind.
Bei 1-aus-n-Codierung auf den Auswahlleitungen ist dies sicher gegeben, bei binärer Codierung müssen in zusätzlichen Entscheidungsschritten erst die Überdeckungen eliminiert werden, die durch die Wired-Or-Verknüpfung der Anforderungen entstanden sind.

2.2.2 Codierung

1-aus-n Codierung:

Die möglichen Zustände werden dadurch unterschieden, daß in jeder Auswahlstufe jeder Modul das ihm zugeordnete Bit setzt.
Das höchste gesetzte Bit kennzeichnet die Gruppe von Moduln, die in der nächsten Entscheidungsstufe weiterarbitrieren darf.
Kein gesetztes Bit kennzeichnet die Tatsache, daß kein Modul den Bus benötigt.

Zahl der Ebenen = (Leitungszahl)^{Übertragungszyklen}

Für 256 Ebenen:
$256 = 256^1$ oder 16^2 oder 4^4 oder 2^8

1-aus-n-Codierung mit Einbeziehung von 0 als eindeutigem Zustand

Verzichtet man auf Kennzeichnung des Zustandes in dem kein Modul den Bus zugeteilt bekommen hat, so kann man die niedrigstpriore Gruppe oder innerhalb dieser den niedrigstprioren Modul dadurch kennzeichnen, daß kein Bit gesetzt ist.
Dieser kann nicht eigentlich den Bus anfordern sondern erhält ihn zugeteilt, wenn kein wichtigerer Modul den Bus angefordert hat.
Da diese Tatsache jedoch nicht bedeutet, daß er eine Busoperation durchführen muß, bringt dies für den Betrieb keine Nachteile mit sich.
Diese Kennzeichnung ist auch eindeutig überdeckungsfrei und es sind keine unnötigen Entscheidungen zu treffen.

Ebenen = (Leitungszahl + 1)^{Übertragungszyklen}

Für 256 Ebenen:
$256 = (255 + 1)^1$ od. $(15 + 1)^2$ od. $(3 + 1)^4$ od. $(1 + 1)^8$

Serielle Abfrage:

Eine serielle Abfrage über eine einzige Leitung ist ein etwas abweichender Sonderfall eines Entscheidungsbaumes.
Sie entspricht einer binären Codierung, die seriell übertragen wird.
Für eine große Zahl von Prioritätsebenen ist dieses Verfahren in synchronen Bussen kaum geeignet, da pro Ebene ein Übertragungszyklus notwendig ist. Die Information über die Adresse des anfordernden Moduls ist im Zeitpunkt (in der Nummer der Entscheidungsstufe) der Antwort enthalten. Da das Verfahren bei einer Abfrage in Richtung fallender Priorität nach Auftreten der 1. Anforderung abgebrochen werden kann, könnte es in asynchronen Systemen sinnvoll anwendbar sein.

Binäre Codierung:
Die binäre Codierung, in der jeder Modul die Nummer seiner Prioritätsebene, binär codiert an den OC-Bus legt, ist eine Sonderform der 1-aus-2 Codierung bei der weitere Leitungen zur Speicherung der bereits ermittelten Entscheidungen dienen.
Das Verfahren beruht darauf, daß jeder Modul die niederwertigen Bits seiner Anforderung wieder wegnimmt, wenn er erkennt, daß auf einer Leitung ein Bit auf 1 gesetzt ist, in der er selbst eine 0 anlegt.
Die schlechtere Ausnützung kommt dadurch zustande, daß jeweils nur eine Leitung zur Entscheidung beiträgt und die übrigen anfangs unbeachtet bleiben und dann die Information speichern.

Für 256 Ebenen:
8 Leitungen 8 Übertragungszyklen

Die Tatsache, daß manche Kombinationen von Anforderungen schneller die höchste Priorität auf dem Bus anzeigen, kann kaum ausgenützt werden, da sich kein Signal ableiten läßt, das die Beendigung der Arbitration anzeigt.
Es müssen deshalb immer die maximal möglichen Zeiten abgewartet werden, wodurch die mögliche Geschwindigkeit genauso groß ist wie bei 1-aus-n Codierung in getakteten Bussystemen.

3. FOLGERUNGEN FÜR DIE OPTIMIERUNG EINES SYSTEMS DER BUSZUGRIFFSSTEUERUNG FÜR 16 MODULN

Innerhalb der oben erwähnten Möglichkeiten galt es nun, ein System zu erstellen, das zwischen den Extremen der möglichst schnellen (n-Leitungen - eine Übertragungs-einheit) und der möglichst leitungssparenden (1 Leitung - n Übertragungseinheiten) Methode die meisten Vorteile bietet.

Für die Codierung wurde die 1-aus-n Codierung gewählt, da nur in dieser Form reine Prioritätsebenenentscheidungen über den Bus übertragen werden müssen.
Binäre Codierung erfordert einen zu großen Zeit und Leitungsaufwand.

Für 16 Moduln ist 1-aus-4 Codierung günstig, da eine Erweiterung durch Hinzufügen von weiteren Übertragungseinheiten erreicht werden kann, ohne zusätzliche Lei-tungen verwenden zu müssen.

Für die Übertragung der 4 möglichen Zustände wurde die Form mit Einbeziehung der 0 gewählt, da damit eine Leitung gespart werden kann.

Aus den Möglichkeiten für 256 Prioritätsebenen wurden als günstiger Kompromiß zwischen Geschwindigkeit und Leitungszahl 3 Leitungen mit 4 Übertragungseinheiten gewählt.

Können diese 4 Übertragungseinheiten nicht in einer Busoperation untergebracht werden, so kann durch Hinzufügen von weiteren 3 Leitungen und Verschachtelung der Zeiten eine Anpassung erreicht werden. Die Datenrate bleibt dadurch gleich, die Einzelzyklusdauer wird jedoch länger.

4. REALISIERTES SYSTEM FÜR 16 MODULN

Die Arbitration wird auf die Karten verteilt über ein paralleles Bussystem mit 3 (6) Leitungen durchgeführt.

Für die reine Buszugriffssteuerung (Kollisionsschutz) werden 16 Ebenen unterschieden. Dazu werden 3 Leitungen und 2 Übertragungseinheiten benötigt.
Für eine Erweiterung in Form einer flexiblen Prioritätsänderung sind weitere 2 Übertragungseinheiten vorgesehen, die für sehr große Busgeschwindigkeiten auf 3 weiteren Leitungen zeitlich versetzt durchgeführt werden.

Die Hardware ist abgesehen von Leitungstreibern und Puffern sehr eng mit der Ablaufsteuerung verknüpft und erfordert deshalb kaum einen zusätzlichen Aufwand.

Die Arbeitsweise ist die, daß ein höherwertiges gesetztes Bit den Ablauf einer Busoperation in der Arbitrationsphase abbricht, sodaß Steuer- und Dateninformation nur von einem einzigen Modul, der nicht unterbrochen wurde, an den Bus gelegt werden. Erhält ein Modul nicht den Bus, so bleibt ein internes Anforderungsbit gesetzt, der Modul nimmt am nächsten Arbitrationszyklus (eventuell mit erhöhter flexibler Priorität) wieder teil.

The Equipment Alarm System in the Experimental
Areas of the CERN SPS particle accelerator

A. Cojan, M. Rabany, R.I. Saban

CERN, Geneva, Switzerland

ABSTRACT

The Experimental Areas of the Super Proton Synchrotron at CERN, contain various types of equipment for monitoring and control of the secondary beam lines. Several hundreds of critical parameters such as power supplies of different types, gas flow and others need to be surveyed to provide a means of debugging. This paper describes an Alarm System built around a set of Line Surveyor modules, a serial CAMAC highway, a highway surveillance system and a dedicated micro-processor.

RÉSUMÉ

Les zones expérimentales du Super Proton Synchrotron du CERN contiennent differents types d'équipement de surveillance et de contrôle des rayons secondaires. Plusieurs centaines de paramètres critiques, tels des alimentations électriques de plusieurs sortes, le débit du gaz et d'autres nécessitent une surveillance qui facilitera le dépannage. Ce papier décrit un système d'alarme construit autour d'un ensemble de modules de surveillance de lignes, d'un bus CAMAC série, d'un système de surveillance de bus et d'un microprocesseur spécialisé.

ZUSAMMENFASSUNG

Inden Experimentalzonen des Super Proton Synchroton des CERN stehen zahlreiche verschiedene Geraet fuer die Ueberwachung und Steverung der sekundoeren Strahlstrecken. Mehrere hundert kritische Messwerte, wie z.B. Stromversorgungen, Gasfluesse, muessen ueberwacht werden. Die Abhandlung beschreibt ein Alarm-System, das besteht aus einem Satz von Leitungsueberwachungs-Moduln, einer seriellen CAMAC-Schleife, einer Ueberwachungseinheit fuer die Datenvebertragungsstrecke und einem speziellen Milrorechner.

1. INTRODUCTION

The Experimental Areas of the Super Proton Synchrotron at CERN contain a number of detectors, collimators, vacuum pumps, magnets and other special equipment. In many cases, the hardware is organised to deliver digital information reflecting the working state of the electronics assemblies which are fundamental for the operation of an equipment. This is typically the case for power supplies. In order to be informed of failures we are

* providing, as much as we can each subsystem with a self check facility

* collecting individual alarm signals onto a Line Surveyor {1} CAMAC module

* concentrating the information from these modules to the Alarm Processor which filters, performs consequential analysis and finally presents the information to the operators

2. SYSTEM DESCRIPTION

The Experimental Areas are subdivided into regions over which the computer control is achieved by means of a Serial CAMAC Highway. Along the highway, entities called 'stations' serve the equipment in their vicinity. They provide

* a standard interface to the equipment

* an easy connection to the electronics of the experimental physicists

* a means of very easy installation of additional equipment

The Serial CAMAC Highways cover a total area of more than 28,000 m^2 and the longest of them stretches over a length of 2 km. They also provide the means of conveying the alarms from the equipment to the dedicated processor.

2.1 THE ALARMS

The stations contain one or more Serial CAMAC crates, NIM crates, Power Supplies and other specialised electronic assemblies. The power supplies of all these and other critical stati are surveyed by one or more Line Surveyor modules plugged into the CAMAC crate. A standard status card {2} which may be added to any DC power supply (when not provided by the manufacturer) has been designed. A change of state from logic 1 to logic 0 implies that the sub-system has ceased normal functioning. If the particular input on the Line Surveyor is enabled and the previous transition to a logic 1 had been memorized, this transition will cause a Look-At-Me (LAM) signalling that the module needs assistance.

2.2 THE COMMUNICATION MEDIUM AND ITS SURVEY

A LAM from a Line Surveyor is conveyed to the Multiprocessor Network {3} via the Serial CAMAC Highway. Since the faults of the elements belonging to the communication medium cannot, obviously, be transmitted through the medium itself, they have to be conveyed via a separate 'surveillance bus' which is directly connected to the Alarm Processor.

A set of modules implement the surveillance function :

* a module in each station called a 'local module' which surveys the master equipment, ie. those whose failure would jeopardize the integrity of the Serial CAMAC Highway.

* a module directly connected to the Alarm Processor, called 'master module' which collects the alarm information of all the Serial CAMAC Higways and presents it to the processor.

* the bus itself, consisting of the following lines

   a) a line to survey the master equipment. A logic 0 on this line will indicate that the proper functioning of the Serial Highway is no more guaranteed.

   b) a rearming line. Some units, such as CAMAC and NIM power supplies, are equipped with a rearming facility. When the processor detects that a line connected to such a unit has become bad (logic

0), it requests the master module to send a rearming pulse.

c) a line to request the interruption of the surveillance for that Serial Highway. This is used when an intervention in a station would cause known but undesirable alarms. The processor acknowledges this request by asserting, via the master module, that it has stopped the surveillance. A watch-dog mechanism will first announce via a buzzer that it will in a minute request that normal operation is resumed, and later, unless it is rearmed, request it to the Alarm Processor.

d) a line to indicate the status of surveillance for that Serial Highway

e) a line to inhibit the interruption request

2.3 THE PROCESSOR

The Alarm Processor is housed in an ACC {4} module. It performs the following tasks :

* execute user commands issued from NODAL {5} programs in order to read the status of a line, mask a line or disable the survey on one of the Serial Highways

* process alarm packets and present the alarm on an output device. This involves

  a) delaying the output in order to ensure that it was not caused by a spike

  b) calculation of the alarm frequency on each line to detect eventual oscillations

  c) consequential analysis in order to prevent dependent alarms from flooding the output device.

* respond to manual commands via the local and master module front panels. This involves the enabling and disabling of the survey on a particular Serial Highway

* detect the failure of a Serial Highway and demand the execution of a recovery program in one of the computers.

The Alarm Process is a set of programs running under a mini operating system which allows multi-tasking, real-time and task synchronisation {6}. The process communicates with the Multi-Processor Network via a set of interrupt-driven routines and uses all the facilities provided by the network. In particular, it communicates with the remote tasks located in control computers and network processors.

3. CONCLUSIONS

The system described constitutes a simple and complete means of informing the operators about the malfunctioning of a particular equipment. Furthermore, it does not flood the console with redundant alarms which would reduce the credibility of the system as a whole.

It is important to remember that the system is not designed to report alarms 'immediately' but that it can delay them for a reasonable amount of time (1 to 5 minutes) for filtering and special processing.

Also it was easily integrated into the Multiprocessor Network with which it now contitutes a whole.

REFERENCES

{1} 128 Line Surveyor type 128 LS 2079 manufactured by SEN Electronique.

{2} Power Supply Surveyor 8086-9070-B by G.Vismara, CERN-SPS-EBP/EL Note 78-2

{3} A Multiprocessor Network in a Multiuser CAMAC Environment by U. Beyschlag, M.J. Clayton, S. Patel, M. Rabany, R.I. Saban, A. Thys. Real Time Data 1982.

{4} A versatile CAMAC crate controller and computer, C. Guillaume and W. Heinze Nuclear Intruments and Methods, 177 (1980), 327-331.

{5} The NODAL System for the SPS, M.C. Crowley-Milling, G.C. Shering Yellow Report, CERN 78-07

{6} A Mini Multi-Tasking RT System for the ACC Module, R.I. Saban CERN-SPS/ELE/Note 82-04

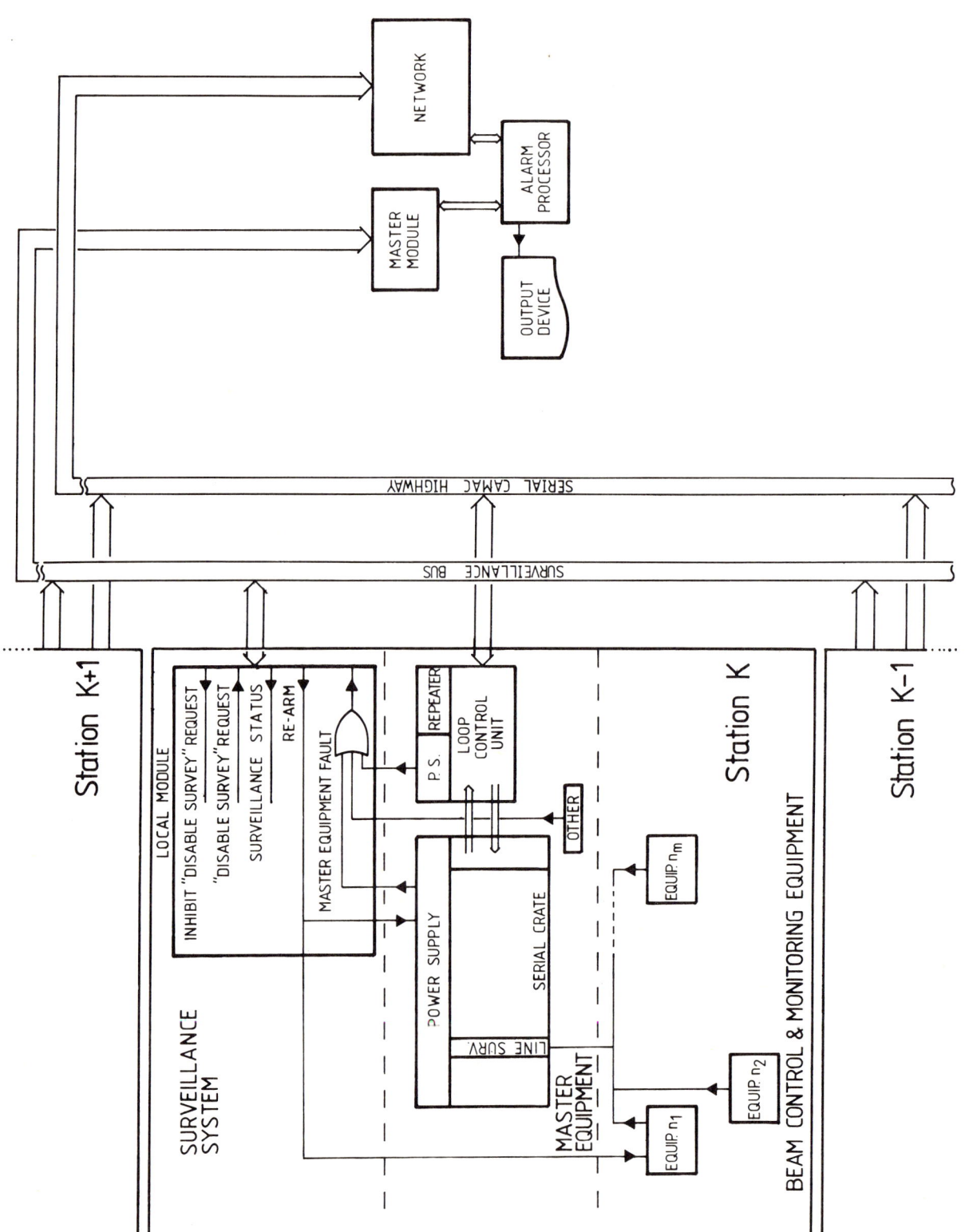

The Serial CAMAC Highway and the Surveillance Bus

A PASCAL-High-Level-Language Interface for an IEC-Bus Multiprocessor Environment

Ein PASCAL-Sprachinterface für Multiprozessor Betrieb an einem IEC-Bus

Erwin SCHOITSCH

Österreichisches Forschungszentrum Seibersdorf
(Austrian Research Center Seibersdorf)
A-2444 Seibersdorf, Austria

Abstract:
A software concept has been developed for studies of an IEC-bus multiprocessor environment based on the following principles:

- a PASCAL-Cross-development system producing code for a "virtual machine"

- handling of so called IEC-bus configurations (a group of several listeners and one talker) within the scope of the PASCAL file concept.

- Extension of the PASCAL-Run-Time System by simple external procedures for real-time synchronization (Conditional Critical Regions).

The concept described allows to produce loadable code for the IEC-bus controllers in a simple and comfortable manner on a host-machine using a high-level language. The concept of the "virtual machine" makes it easy to adapt the system to different processors. Set up, read or write -operations on any combination of devices connected to the IEC-bus may be done in a manner similar to that of standard file operations, and, additionally, the cooperation of controllers on process- (task-) level may be programmed on a high level by introducing a clear interface between handlers and programs on user level.

Zusammenfassung:
Für die Untersuchung des Multiprozessorbetriebes an einem IEC-Bus wird ein Softwaresystemkonzept vorgestellt, welches auf den folgenden Prinzipien beruht:

- PASCAL-Cross-System auf Basis einer "virtuellen Maschine"

- Behandlung der IEC-Bus Konfigurationen (zusammengehörende Listeners und der Talker) im Rahmen des PASCAL-Filekonzeptes

- Erweiterung des Pascal-Systems durch einfache externe Prozeduren zur Echtzeit-Synchronisierung (Bedingte Kritische Regionen).

Dieses Konzept erlaubt mit einfachen Mitteln die rasche und komfortable Erstellung von Versuchsprogrammen für IEC-Bus Controller in der höheren Programmiersprache am Host-System und ermöglicht die einfache Adaption des Laufzeitsystems für die "virtuelle Maschine" auf verschiedenen Prozessoren. Es gestattet die formale Gleichbehandlung von Fileoperationen und die Bedienung beliebiger Kombinationen angeschlossener Geräte sowie die Kooperation der Controller auf Prozeßebene durch saubere Trennung von Handler- und Sprach- (Benutzer-) ebene.

Résumé
Le concept logiciel développé a permis d'étudier un environnement multiprocesseur du bus CEI.

Les principes en sont les suivants :

- système de développement de compilateurs croisés PASCAL produisant le code destiné à une "machine virtuelle".

- gestion des configurations bus-CEI (groupe constitué de plusieurs appelés pour un appelant unique) dans le cadre du concept des fichiers PASCAL.

- extension du système exécutif PASCAL à l'aide de procédures externes simples pour une synchronisation en temps réel (zones conditionnelles critiques).

Le concept décrit permet de produire du code chargeable pour des contrôleurs de bus - CEI de façon simple sur un ordinateur central utilisant un langage de haut niveau. Le concept de "machine virtuelle" facilite l'adaptation due système aux différents processeurs. Les opérations de mise en forme, lecture ou écriture sur n'importe quelle combinaison d'appareils connectés au bus CEI peuvent se faire de façon similaire à celles effectuées sur des fichiers standard; de plus, on peut programmer à un haut niveau la coopération des contrôleurs au niveau des process ou des tâches en introduisant clairement un interface entre les systèmes de manipulation et les programmes au niveau utilisateur.

Ein PASCAL-Sprachinterface für Multiprozessorbetrieb an einem IEC-Bus

Für die Untersuchung des Multiprozessorbetriebes an einem IEC-Bus (Diplomarbeit am Elektronikinstitut, gemeinsam mit der Technischen Universität Wien, Institut für Elektrische Meßtechnik) wird ein Softwaresystem, basierend auf einem PASCAL-Cross-System für eine "virtuelle Maschine" mit dem zugehörigen Laufzeitinterpreter für den jeweiligen "realen" Prozessor, verwendet, welches in der Prozeßdatenverarbeitungsabteilung des Institutes für Physik entwickelt wurde.

Das Konzept der virtuellen Maschine, gekoppelt mit dem Einsatz der höheren Programmiersprache PASCAL in einem Cross-Entwicklungssystem /4/, ermöglicht einerseits die rasche und komfortable Erstellung von Versuchsprogrammen und die einfache Bedienung der minimal ausgestatteten IEC-Bus Controller, andererseits ist erhöhte Flexibilität und Wiederverwendbarkeit von Software bei Einsatz verschiedener Prozessortypen gewährleistet.

Der Crosscompiler erzeugt stets den gleichen Zwischencode, der jeweils von einem kleinen Laufzeitinterpreter der Zielmaschine sehr effizient (nach der Methode des threaded code) interpretiert wird.

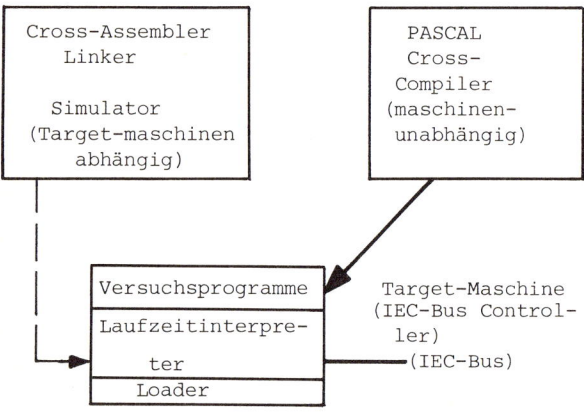

Abb. 1: PASCAL-Cross System

Adaptionen des vorhandenen Systems sind auf zwei Ebenen notwendig, wobei die Unveränderlichkeit des PASCAL-Sprachkonzeptes (und damit des Cross-Compilers) als Prämisse vorgegeben war:

- Behandlung der IEC-Bus Konfigurationen (welche aus den logisch zusammengehörenden, derzeit vom Controller belegten Listeners und dem Talker bestehen) und des damit gegebenen Datentransfers im Rahmen des PASCAL-Filekonzeptes.

- Ergänzung durch einfache externe Prozeduren zur Behandlung von Echtzeit-Synchronisierungsaufgaben gemäß dem Konzept der "Bedingten kritischen Regionen" (Conditional Critical Regions, CCR) /1/, /5/.

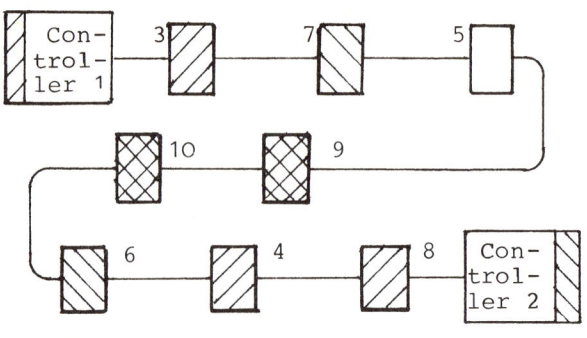

Abb. 2: IEC-Bus Konfiguration (Multiprozessorbetrieb)

Der Begriff der Konfiguration (siehe Abb.2) ist vor allem im Falle des Multiprozessorbetriebes von Bedeutung: Da zu einem Zeitpunkt nur eine CPU die Controllership ausüben kann, muß es möglich sein, nach jedem Transfer einer Message zwischen Stationen die Controllership zu wechseln. Um dennoch einen geregelten Ablauf zu gewährleisten, muß sich jeder Controller die von ihm länger benötigten Stationen logisch reservieren können. Dabei kann sich jeder Controller auch mehrere Konfigurationen zuordnen. Eine Konfiguration muß aus einem Talker und mindestens einem Listener bestehen, wobei die jeweilige CPU nicht Partner sein muß. Ein Lese- oder Schreibbefehl betrifft dann jeweils eine ganze Konfiguration. Deshalb ist das PASCAL-Filekonzept nicht direkt anwendbar.

Bei der Übergabe der Controllership müssen auch die Konfigurationsinformationen ausgetauscht werden, um Doppelbelegungen (Laufzeitfehler!) zu vermeiden.

Es wird eine Lösung vorgestellt, die die formale Gleichbehandlung von IEC-Bus-Konfigurationen und fileorientierter Peripherie in PASCAL erlaubt und daher nur geringfügige Modifikationen im Laufzeitsystem erfordert. Diese besteht in:

- der Einführung des "Configurationpointers"
- der Eintragung der vorhandenen Stationen per symbolischem Namen in ein (Pseudo-File-) Verzeichnis
- formaler Gleichsetzung von "Textzeile" und Datenmessage am IEC-Bus.

Der Configurationpointer entspricht dem Textfilepointer, es müssen aber mehrere Listeners und auf jeden Fall ein Talker eröffnet werden können, wobei auch ein Ansprechen der Geräte per Namen erlaubt sein soll. Für das Benutzerprogramm ergeben sich dabei logisch zwei Fälle:

a) Der Talker ist am Bus:
   Dann kann der Controller höchstens Listener sein, die erlaubte IO-Operation ist readln (configuration), d.h. der Talker muß mit Reset(config., IEC, 'Talkername') eröffnet werden.

b) Der Controller ist Talker:
Die erlaubte IO-Operation ist writeln (configuration), d.h. der (die) Empfänger (Listener) müssen mit Rewrite (config., IEC, 'Listenername') eröffnet werden. Der Controller muß über Reset (config., IEC, 'Controllername') zum Talker gemacht werden.

Diese Logik wird insofern verallgemeinert, als stets alle Listener mit Rewrite eröffnet werden, der Talker mit Reset. Mit Close (configuration) wird die Konfiguration wieder freigegeben (Abb. 3). Die logische Überprüfung der Konfiguration kann nur zur Laufzeit erfolgen. Die tatsächlichen Transfers am IEC-Bus werden mit Readln (conf.) bzw. Writeln (config) durchgeführt und betreffen stets die gesamte Konfiguration, die Aufbereitung erfolgt mit Hilfe des Handlers über die Configuration Pointers und den Configuration Header, der die zugeordneten Adressen (max. 31) enthält. Die Zuordnung der Namen der Geräte erfolgt im "Configuration Directory", welches alle Daten der Geräte enthält (ähnlich wie bei fileorientierten Speichermedien) (Abb. 4). Eine Variablenliste in Read/Writestatements ist nur sinnvoll, wenn der Controller selbst Listener oder Talker ist, dann gehen die Daten auch über den Message-Puffer. Die im Prozessor vorhandenen Konfigurationen sind wieder verkettet, um die notwendige Controller-Controller Kommunikation bei Übergabe den Controllership zu ermöglichen.

```
const   IEC = 3; (* interne logische Einheit *)
var   c1, c2: configuration;
              (* 1. Konfiguration *)
        begin
        rewrite (c1, IEC, ´drucker´);
              (* wird Listener definieren *)
        rewrite (c1, IEC, ´mayself´);
              (* Rechner hört mit *)
        reset   (c1, IEC, ´voltmeter´);
              (* zum Schluß Talker definieren *)
        rewrite (c2, IEC, ´Anzeige´);
              (* zweite Konfiguration *)
        reset   (c2, IEC, ´Temperatur`);
              (* Abschluß 2. Konfiguration *)
        readln (c1, message);   (* Meldung aktivieren, Rechner hört mit. *)
        :
        readln (c2);   (* Meldung aktivieren, Rechner nicht dabei *)
        :
        close   (c1);   (* Konfiguration freigeben *)
        close   (c2);
end.
```

Abb. 3: Aufbau und Behandlung einer IEC-Bus Konfiguration in PASCAL

Fast alle Interrupts und Fehlerbehandlungen können auf Handlerebene für sequentielle Benutzerprogramme aufbereitet werden. Dennoch sollen einige Probleme (Controllership-Übergabe, Service-Request u.a.) in höherer Programmiersprache formulierbar sein, wozu einige Echtzeiterweiterungen eingeführt werden müssen:

- einfaches Tasking (Prozesse)

- Prozeßsynchronisierung

Dafür gibt es drei externe Routinen ´Protect (region.condition)´, ´Release (region.new_condition)´ und ´Control (region.condition, region.new_condition)´, die Beginn und Ende sowie Wartebedingung und geänderte Bedingung der kritischen Region festlegen.

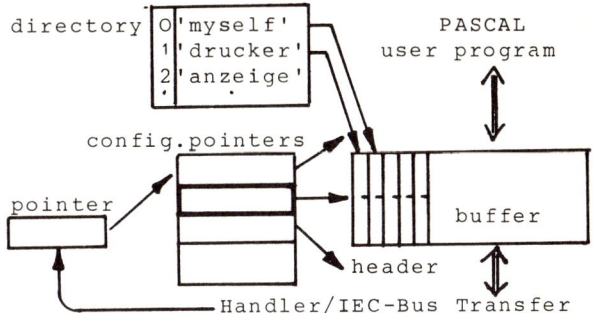

Abb. 4: Aufbau der IEC-Bus Configuration Datenstruktur

Der Schutz kritischer Programmabschnitte und das Warten auf Bedingungen erfolgt durch den protect-release Block. Das Control-Statement kombiniert beides und dient vor allem der effizienten Interruptbehandlung (Bedingung "Interrupt eingetreten" wird extern gesetzt). (/1/, /3/, /5/, Abb. 5)

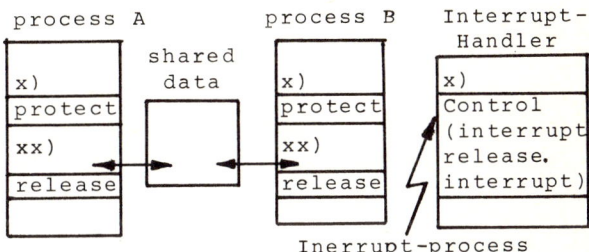

x) Wartepunkt, wenn anderer Prozeß zugr.

xx) Gesschützt v. Unterbrechung durch and. Prozeß m. selb. Zugriffsrechten

Abb. 5: Bedingte kritische Regionen zur Prozeßkontrolle

LITERATUR:

/1/ Brinch Hansen, P., Concurrent Programming Concepts. Summer School on Structured Programming and Programmed Structures, Munich 1973.

/2/ Jensen, K. and Wirth, N., PASCAL-User Manual and Report, Springer 1978.

/3/ Schoitsch, E., Implementation of Conditional Critical Regions in Systems Using Different Prozess Communication Techniques. IFAC-Real Time Programming 1980, pp. 29-38, Ed. Haase, V.M., (Pergamon Press, 1981).

/4/ Schoitsch, E. and Mayer, K., Portable Softwaresysteme für Cross-Entwicklung, Informationstagung Mikroelektronik (IE81), p. 172-176, Wien 1981.

/5/ Schoitsch, E., Bedingte kritische Regionen als Werkzeug zur Prozeßsynchronisation, Real-Time Data ´79, P. 553-561 (North Holland 1980).

TYPENUNABHÄNGIGES CROSS-ENTWICKLUNGSSYSTEM MIT MINIMALER HARDWARE-ANPASSUNG

Herbert Schweinzer
Institut für Elektrische Meßtechnik
Technische Universität Wien, Österreich

ZUSAMMENFASSUNG

Für die effektive Entwicklung von Mikroprozessorprogrammen, speziell solchen, die unter vorgegebenen Echtzeitbedingungen ablaufen, ist ein leistungsfähiges Testsystem von großer Bedeutung. Handelsübliche Entwicklungssysteme werden zwar mit Hilfe ihrer Emulations- und Testeinrichtungen den meisten Anforderungen voll gerecht, sind jedoch weder universell, noch kostengünstig, da sie eine relativ aufwendige, prozessorspezifische Spezialhardware darstellen.
In diesem Beitrag wird als Alternative eine Kombination von ROM-Simulator und Logikrekorder betrachtet, eine universelle Testhardware, die eine Anpassung weitgehend beliebiger Prozessoren mit geringem Aufwand ermöglicht. Dabei auftretende Einschränkungen werden in ihren entscheidensten Punkten dargestellt. Schließlich wird als konkrete Realisierung ein komplexes Universal-Entwicklungssystem beschrieben, das mit universeller Cross-Software betrieben wird.

ABSTRACT

For the effective development of microprocessor-programs, especially when working under defined real-time conditions, an efficient test-system is of great importance. Development-systems as available mostly fit in all demands with help of emulation- and test-equipment, but they are neither universal nor cheap, because they represent a relatively complicated, processor-specific hardware.
In this article a combination of ROM-simulator and logic-recorder is proposed as alternative, that is an universal test-hardware that brings about an accomodation of most processors with only little effort. Limitations of this technique are shown in their essential items. At last follows the description of a complex universal development-system as concrete realisation that works with universal cross-software.

RESUME

Pour le développement effectif des programmes de microprocesseur, surtout ceux qui se déroulent dans des conditions de temps réel, un système de test efficace est d'importance majeure. En effet les systèmes de développement commerciales satisfont tout à fait aux demandes, mais ils ne sont ni universels ni bon marché, parce qu'ils représentent un "hardware" assez compliqué et spécifique au processeur.
Dans cet article une combinaison de ROM-simulateur avec un "logic-recorder" est proposée comme alternative. Il s'agit d'un "hardware" universel de test qui rend possible une adaption de presque n'importe quel processeur avec une petite dépense de force. Les restrictions essentielles de cette méthode sont démontrées au cours de la communication. Finalement un système de développement universel est présenté comme réalisation concrète fonctionnant à l'aide d'un "cross-software" universel.

1. ALLGEMEINES ZUR PROGRAMMENTWICKLUNG

Die meisten Phasen der Programmentwicklung sind an beliebigen Rechnern durchführbar, deren Prozessortyp sich von dem des Zielsystems unterscheiden kann (Cross-System). Nur beim Programmtest ist besonders für echtzeitbezogene Programmteile die Miteinbeziehung des Ziel-bzw. Applikationssystems notwendig. Dazu müssen dem Zielsystem das neu entwickelte Programm zur Verfügung gestellt, seine Funktion beobachtet und beurteilt und der Programmablauf beeinflußt werden. Diese Aufgaben des Programmtests erfordern hard-und softwaremäßige Kopplungen zwischen Entwicklungsrechner und Zielsystem, die vom Prozessortyp abhängen. Es ist daher sinnvoll, für die schnelle Anwendbarkeit neuer, moderner Prozessoren diese Anpassung minimal zu machen.

2. KOPPLUNGSARTEN ZWISCHEN PROGRAMMENTWICKLUNGSRECHNER UND ZIELSYSTEM

2.1 Zielsystem mit Testmöglichkeiten

Der einfachste Fall liegt vor, wenn das Zielsystem selbst über grundlegende Testmöglichkeiten verfügt: Das kann durch einen einfachen Testmonitor gegeben sein. Eine modulares Mikrocomputersystem ist durch Zusatzplatinen erweiterbar, die für den Test notwendig sind. Jederzeit kann auch ein Logikrekorders angeschlossen werden. Das Programmladen erfolgt mit einer seriellen Schnittstelle und dazugehörigem Ladeprogramm. Damit ist Programmladen und -testen jedoch nur in einfacher Form möglich. Insbesonders müssen ROM-Bereiche für den Test mit RAMs bestückbar sein.

2.2 ROM-Simulator

Eine schnelle und flexible Programmlademöglichkeit bietet ein "ROM-Simulator", das ist ein Dual-Port-Memory, das vom Entwicklungssystem her geladen, vom Zielsystem aus als ROM oder auch RAM wirksam wird /6/. Die Verbindung mit dem Prozessorbus im Zielsystem erfolgt parallel, für den Programmtest ist jedoch wieder ein Testmonitor wie unter 2.1 nötig.

2.3 ROM-Simulator mit Hardware-Breakpoint

Durch die Verbindung mit dem ROM-Simulator ist der Bus des Zielsystems bereits herausgeführt. Es ist somit einfach möglich, eine universelle Hardware-Breakpoint-Einrichtung für Adresse und Daten mit dem ROM-Simulator zu kombinieren. Der Hardware-Breakpoint wird als Interrupt (z.B. NMI) wirksam und muß dementsprechend in der prozessortyp-abhängigen Software des Testmonitors implementiert werden.

2.4 ROM-Simulator mit integriertem Logikrekorder

Die Kombination eines ROM-Simulators mit einem Logikrekorder erscheint in dreifacher Hinsicht vorteilhaft: erstens ist durch die Busverbindung zum ROM-Simulator bereits im Hauptteil an Verbindungen gegeben, die auch für den Logikrekorder Verwendung finden und nur durch spezielle Steuerleitungen ergänzt werden. Weiters ist der Logikrekorder ("Trace Modul") durch den Programm-Entwicklungsrechner auslesbar, sodaß durch ein universelles Testprogramm des Entwicklungsrechners im ROM-Simulator angebotene Programme mit den im Logikrekorder aufgezeichneten Abläufen in Verbindung gebracht werden können.

Schließlich kann man sich bei Aufzeichnung des Programmablaufs auf die Interpretation der Aufzeichnung beschränken, womit die Notwendigkeit einer prozessorspezifischen Testsoftware im Zielsystem wegfällt. Ein Zugriff auf Hardware im Zielsystem oder auf die Prozessorregister ist damit jedoch nicht gegeben und muß gegebenenfalls durch prozessorabhängige Testsoftware, die im ROM-Simulator gespeichert sein kann, realisiert werden.

Die Steuerung des Zielsystems kann vom Entwicklungsrechner her erfolgen. Für den Programmstart ist primär RESET vorgesehen; ein Programmstop erfolgt mittels DMA-Request, der aufrecht bleibt, bis ein Weiterlauf erwünscht ist. Auch die Anwendung eines Interrupts (NMI) ist möglich.

2.5 Emulations- und Testeinrichtung (ICE)

Die handelsüblichen Entwicklungssysteme bieten für den Programmtest im Zielsystem Emulations- und Testeinrichtungen an. Die Anschlüsse des Prozessors im Zielsystem werden dabei parallel herausgeführt; der Prozessor selbst residiert im Emulator, womit die Verbindungen zum Zielsystem auch abschaltbar sind. Der Prozessor ist hard- und softwaremäßig voll vom Testsystem her kontrollierbar; das zu testende Programm wird vom Entwicklungssystem her angeboten; ein Teil des Programmablaufs kann ebenfalls in einem Trace-Modul gespeichert werden.

3. VERGLEICH EMULATIONS- UND TESTEINRICHTUNG MIT ROM-SIMULATOR/LOGIKREKORDER

Weitgehend unbeschränkte Möglichkeiten bietet die Emulations- und Testeinrichtung, da die Abschaltbarkeit der Verbindungen zum Zielsystem sicherstellt, daß die zu testenden Programme ablaufen können, ohne das Zielsystem unerwünscht zu aktivieren (Trennung bzw. Vervielfachung des Adreßraumes Zielsystem/Testsystem). Die Steuersignale des Prozessors sind ferner voll unter Kontrolle des Testsystems; ebenso sind die Statusmeldungen beliebig auswertbar.

Das Emulations- und Testsystem stellt eine relativ umfangreiche, prozessororientierte Spezialhardware dar, die überdies auch prozessorspezifische Testsoftware umfaßt. Es handelt sich somit um eine relativ teure Lösung, die sich außerdem bei jedem Entwicklungssystem nur für eine beschränkte Zahl von Prozessoren anbietet. Der Übergang auf einen anderen Prozessor kann somit durchaus auch den Übergang auf ein anderes Entwicklungssystem bedingen.

Die Verwendung eines ROM-Simulators mit integriertem Logikrekorder als universelle Testhardware ermöglicht die Unterstützung der Programmtestphase für einen neuen Prozessor in billiger, schnell realisierbarer Form. Der nötige Aufwand besteht vorwiegend in der Herstellung einer möglichst kurzen Kabelverbindung zwischen Prozessorbus und ROM-Simulator. Es ist auch hier vorteilhaft, den Prozessor z.B. durch eine in die Prozessorfassung gesteckte Zwischenplatine zu ersetzen, die neben dem Prozessor auch Treiber für das Verbindungskabel sowie Anpassungsschaltungen für Timing- und Kontrollsignale aufweist. In manchen Fällen ist allerdings ein Anschluß an anderer Stelle notwendig. Die Anpassung an ein bestimmtes Prozessortiming ist meist einfach mit wenigen Gattern möglich; als Steuersignale sind RESET und DMA-Request bzw. 'Single Step' (evtl. auch NMI) dem Prozessor zusätzlich zur anwendungsorientierten Verwendung im Applikationssystem anzubieten. Adresse und Daten müssen auf getrennten Leitungen übertragen werden. Sollen im Rekorder bestimmte Prozessorzustände aufgezeichnet werden (z.B. "Fetch of instruction" für einen Reassembler), so müssen diese Zustände gegebenenfalls durch Dekodierung gewonnen werden.

Die beschriebenen, recht einfachen Hardware-Anpassungsarbeiten ermöglichen bereits einen durchaus befriedigenden Testbetrieb. Für Operationen mit Prozessorregistern ist darüber hinaus eine minimale prozessorspezifische Testsoftware notwendig, die, solange sich die getesteten Programmbefehle im ROM-Simulator befinden, ohne Reservierung eines Interrupts (NMI) auskommt. Durch Auslagern und Überschreiben eines Teils des ROM-Simulator-Inhalts durch das Testprogramm ist eine völlige Transparenz des Adreßraumes für den Anwender gegeben. Die Tatsache, daß sich der Applikationsprozessor unmittelbar beim Zielsystem befindet und im direkten Datenaustausch mit dessen Komponenten stehen kann, ermöglicht bei Verwendung schneller RAMs für den ROM-Simulator durchaus auch schnellere Bustransfers als bei vielen Emulatoren.

Die meisten Nachteile dieser Lösung resultieren aus der Nichtabschaltbarkeit der Verbindungen Prozessor/Zielsystem: So ist der Adreßraum des Prozessors für ROM-Simulator und Zielsystem gemeinsam. Das bedeutet, daß nicht nur sichergestellt sein muß, daß keine überlappenden ROM-Bereiche im ROM-Simulator und Zielsystem auftreten, sondern es müssen auch Adreßdekoder und Treibersteuerung verhindern, daß bei gleichzeitigem Aussenden von Adressen in Richtung ROM-Simulator und Zielsystem Treiberbausteine beim Lesen nicht gegeneinander arbeiten, ohne daß im Zielsystem Speicher mit diesen Adressen vorliegen. In solchen Fällen muß der Datenbusanschluß nahe genug beim Speicher erfolgen, falls nicht abschaltende Treiberbausteine nicht entfernt werden können. Ferner ist für Breakpoints im Bereich von im Applikationssystem bestückten Speichern ein Interrupt (vor allem NMI) notwendig.

Es gibt also Einschränkungen in der Anwendbarkeit von ROM-Simulator/Logikrekorder gegenüber den Emulatoren. Dennoch ist die Kombination ROM-Simulator/Logikrekorder gut als Testsystem geeignet, wenn man mit etwas Überlegung vorgeht: eine Gliederung des zu entwickelnden Programmes in abgeschlossene Moduln, die beim Test vollständig in den ROM-Simulator passen, sichert den vollen Testkomfort. Strikte Beachtung der Treiberaktivierungen ist speziell beim Umkonfigurieren des Zielsystems wichtig, wenn fertige Programm-Moduln in ROM-Speicher des Zielsystems übertragen werden. Verzichtet man schließlich auf die Anwendung grundlegender Testprogramme (Registermodifikationen etc.) für bereits in ROM-Speichern installierte Programm-Moduln und beschränkt sich auf die Interpretation des aufgezeichneten Programmablaufes, so bleibt die volle Transparenz des Systems für den Anwender gewahrt, da kein Interrupt für Testzwecke reserviert werden muß.

4. DAS REALISIERTE CROSS-ENTWICKLUNGSSYSTEM

Bei dem Cross-Entwicklungssystem des Institutes für Elektrische Meßtechnik der TU Wien erfolgt die Programmentwicklung an einer VAX-11, in manchen Fällen auch auf einer institutseigenen PDP 11/10. Für die große Zahl paralleler Arbeitsplätze (derzeit 8) sind Standard-Mikroprozessor-Applikationssysteme mit Testmonitor und Ladeprogramm vorgesehen. Als leistungsfähiger Universal-Entwicklungsplatz sind ein ROM-Simulator und Logikrekorder an die PDP 11/10 angeschlossen.

Sowohl ROM-Simulator als auch Logikrekorder sind als CAMAC-Moduln aufgebaut /1/. Der ROM-Simulator ist byteorientiert und umfaßt 8 K Byte RAM, die in 1 K Byte-Blöcken über den gesamten Adreßraum verteilt werden können /7/. Jeder Block kann als ROM, RAM oder als nicht verwendet definiert werden. Schreiboperationen auf ROM-Bereiche führen zu einer Meldung (LAM) an die PDP 11/10.

Der Logikrekorder wird über zwei Spezialkabel direkt mit dem ROM-Simulator verbunden /7/. Er weist 32 Eingangskanäle auf, die mit einer externen Taktflanke übernommen werden (3 davon haben Taktflankenunabhängige Zwischenspeicher), und hat eine Speichertiefe von 1 K. Das Triggerwort umfaßt die vollen 32 bit; jedes einzelne bit kann auch in einen "Don't-Care"-Zustand gebracht werden. Nach dem Auftreten des Triggerereignisses erfolgt ein definierbarer Nachlauf, nach dem das eigentliche Ende der Aufzeichnung erfolgt. Dieses wird an die PDP 11/10 gemeldet (LAM); der getestete Prozessor kann gleichzeitig mit DMA-Request gestoppt werden, sodaß beliebig viel Zeit für die Auswertung der aufgezeichneten Daten vergehen kann und trotzdem der Weiterlauf des Prozessors beim folgenden Befehl möglich ist.

5. DIE SOFTWARE DES CROSS-ENTWICKLUNGSSYSTEMS

Vor allem findet prozessortypunabhängige Cross-Software Anwendung: Es existiert ein in PASCAL geschriebener Meta-Assembler, der ein Instruktionsfile zur Beschreibung des Zielprozessors benützt /5/. Der Assemblerrumpf weist nur eine geringe Zahl von Festlegungen auf, die die Formulierung von Macros, bedingten Assembleranweisungen, Pseudooperationen und Konstanten betreffen. Die eigentliche Syntax der Assembleranweisungen, Prozessorwortbreite und Op-Code-Generierung wird voll durch das Instruktionsfile gesteuert.

Das Gegenstück zum Assembler stellt ein Universaldebugger dar, dem man die Eigenschaften des Testsystems angeben kann /2/. Die Möglichkeiten des Logikrekorders und ROM-Simulators werden schließlich bei einem Spezialdebugger /4/ voll genutzt.

6. LITERATUR

/1/ Commission of the European Communities, CAMAC - A Modular Instrumentation-System for Data Handling, EUR 4100e (1972)

/2/ Fleck, H., Debugger mit Überprüfung des Realzeitablaufes und symbolischen Eingabemöglichkeiten, Diplomarbeit, Institut für Elektrische Meßtechnik, TU Wien (1981)

/3/ Hagmeister, Mikroprozessor-Entwicklungshilfen, elektronik industrie 7/8/9 (1979)

/4/ Hofstätter, J., Dialogprogramm zum Austesten eines rechnergesteuerten Mikroprozessor-Entwicklungssystems, Diplomarbeit, Institut für Elektrische Meßtechnik, TU Wien (1982)

/5/ Svoboda, E., Problem- und typenabhängig anpaßbarer Assembler, Diplomarbeit, Institut für Elektrische Meßtechnik, TU Wien (1981)

/6/ Schweinzer, H., Mikroprozessor-Cross-Entwicklungssysteme, in: K.P.Judmann u.a., Mikroprozessoren. Grundlagen und Anwendungen (Technischer Verlag Erb, Wien, 1978)

/7/ Troschl, M., ROM-Simulator und Logik-Recorder für Mikroprozessor-Entwicklungssysteme, Diplomarbeit, Institut für Elektrische Meßtechnik, TU Wien (1979)

REAL TIME DIAGNOSIS SOFTWARE WITH APPLICATION TO THE DIAGNOSIS AND THE TREATMENT OF TRADITIONAL CHINESE MEDICINE

Wang Zhi-Bao, Lu Gui-Zhang and Wang Xiu-Feng
Control Theory Section, Department of Mathematics,
Nankai University, Tianjin, China

Wang Zhi-Xiang and Din Xiu-Wen
Tianjin College of Medicine, Tianjin, China

ABSTRACT

This paper proposes a real time data handling software. The philosophy of this software is, to simulate the thinking of the diagnosis process. In this software, the symptoms and the herbal medicines are regarded as data. By real time handling these data it will carry out the diagnosis and the treatment. This is an intelligent simulation, so it can sum up experience nicely.

In this software we have defind a data structural form which cosisting of five files. The program of the real time diagnosis software contains five program blocks. We have applied the software to the diagnosis and the treatment of cough and asthma of children. More than 100 patients have been examined by this software and the results all are correct. This software is generalized, therefore it can use to any similar data handling process.

RÉSUMÉ

Ce papier propose un logiciel de traitement de données en temps réel. La philosophie en est la simulation du processus de diagnostic. Pour ce logiciel, les symptômes et les plantes médicinales sont considérées comme des données. En traitant en temps réel ces données, on aboutit au diagnostic et au traitement. C'est une simulation intelligente qui permet de tenir compte facilement de l'expérience.

Nous avons défini une structure de données constituée de cinq fichiers. Le logiciel de diagnostic en temps réel contient cinq blocs de programme. Nous avons appliqué ce logiciel au diagnostic et au traitement de la toux et de l'asthme chez des enfants. Ce logiciel a permis d'examiner plus de 100 patients et les résultats se sont, à chaque fois, révélés corrects. On peut généraliser ce logiciel pour l'appliquer à tout processus similaire de traitement de données.

ZUSAMMENFASSUNG

Diese Arbeit enthält den Vorschlag für eine Echtzeit-Datenverarbeitungs-software. Im Prinzip beruht diese Software auf der Simulierung der bei der Diagnose geübten Vorgehensweise. In dieser Software werden die Symptome und die Heilkräuter als Daten betrachtet. Durch Echtzeitverarbeitung dieser Daten wird vom Programm Diagnose sowie Behandlung ausgeführt. Als intelligentes Simulationsprogramm kann es zur Ergebnisauswertung eingesetzt werden.

In dieser Software haben wir eine, fünf Dateien umfassende Datenstruktur definiert. Das Programm der Echtzeit-Diagnosesoftware enthält fünf Programmblöcke. Die Software ist für die Diagnose und Behandlung von Husten und Asthma bei Kindern eingesetzt worden. Mehr als 100 Patienten wurden mit dieser Software untersucht und alle Ergebnisse waren korrekt. Als generalisierte Software kann sie auf jeden gleichartigen Datenverarbeitungsprozess angewandt werden.

REAL TIME DIAGNOSIS SOFTWARE WITH APPLICATIONS TO
THE DIAGNOSIS AND THE TREATMENT OF TRADITIONAL
CHINESE MEDICINE

Wang Zhi-Bao, Lu Gui-Zhang and Wang Xiu-Feng
Control Theory Section, Department of Mathematics,
Nankai University, Tianjin, China

Wang Shi-Xiang and Din Xiu-Wen
Tianjin College of Medicine, Tianjin, China

1. INTRODUCTION

Traditional Chinese medicine, one of the excellent cultural heritage of the Chinese nation, has played a great role in improving the health of the Chinese people over thousands of years.

However, to diagnose the diseases, the doctor of the traditional Chinese medicine uses "four methods of diagnosis", namely to see, to hear, to ask and to feel the pulse of the patient, which, to a great extent, be subjective and too much dependent on his experience.

It has been realized in our practice that evenmore dramatic development in the traditional Chinese medicine can be anticipated if the following two problems can be solved: systematization of diagnosis and standardization of the symptoms of the patients. Most logical and desirable solution to them is the combination of the doctor's sophisticated experience with computer system simulation so that the diagnosing process can be computerized. There are in the main three advantages:

(1) The experience of traditional Chinese medicine can be sumarized ratherly easily with the help of computers having better logical function.

(2) Computers memorize the information and permanently, which is beyond human possibility.

(3) Computers can accumulate a wealth of ever-increasing experience and bring about steady improvement in diagnosis and treatment.

So far, immease efforts have been made by many authods in an attempt to sum up the experience, using statistical methods, such as discriminant analysis, but no satisfastory results have been achieved.

This paper proposes a real time data handling software. The philosophy of this software is to simulate the thinking of the diagnosis process. In this software, the symptoms and the herbal medicines are regarded as data, by real time handling these data it will carry out the diagnosis and the treatment. This is an intelligent simulation, so it can sum up experience nicely. On the other hand, this software is generalized, therefore it can use to any similar data handing process.

2. DATA BASE

Basied on characteristics of diagnosis and treatment of traditional Chinese medicine, the symptoms of the patients and all herbal madicines in a prescription all are regarded as data. Therefore diagnosis and treatment can be changed into a process of data handling. For covenience of data handling, we must define a suitable data structural form in the real time diagnosis software. For this purpose, we form five files as following.

(1) Symptom Base File
The symptom base file contains all known symptoms of some kinds of patients. Each stmptom should be presented by a record. The form of the record is
/order of symptom/ dummy / name of
    symptom/

(2) Medicine Base File
The medicine base file contains all herbal madicines which we may use for some kinds of the diseases. Each herbal medicine should be presented by a record. The form of the record is
/order of medicine/ dummy /name of
    herbal medicine/

(3) Alternate Prescription File
In process of diagnosis, the different symptoms supply a doctor different informations. For convience's sake, the symptoms, based on which we can determine what the disease the patients have taken, are refered to as main symptoms. The prescription which determined by main symptoms is refered to as basic presciption. According to auxiliary symptoms we can increase or decrease some herbal medicines in basic prescription. Basic presciption together with all possible increasing herbal medicines is refered to as alternate prescription. The form of the add record is
/order of the record of the medicine

base file/
and the form of the even records is
/ dose /

(4) Symptiom File
This file is aimed at a patient. It begins with a table head and follows by all symptoms of the patient. Each symptom should be present by a record. The form of the record is same as that of the symptom base file. It ends with a table end.

(5) Presciption File
This file is also begins with a table head and follows by a prescription. The prescription is formed by several records. The form of record is
 /the record of the medicine base
     file/ dose /
And ends with a table end.

3. PROGRAM OF REAL TIME DIAGNOSIS SOFTWARE

Real time diagnosis software cosist of five program blocks. It carry out, by these subroutine, all tasks of the data handling.

(1) Main Program Block (MPB)
MPB accomplish a part of the general interative process and dispatch each subroutine to complete all tasks of data handling.

(2) Diagnosis Subroutine Block (DSB)
The function of this block is to ask, by data of the symptom base, what symptoms did the patients have. According to the results, DSB make a decision that the patients have catched a disease or not. If the patient falls ill DSB willdetermine what kind of disease has got. Then DSB asks whether the patient have any auxiliary symptoms or not.

(3) Print Symptom Subroutine Block (PSSB)
According to the results of DSB, the symptom file can be formed and printed by PSSB.

(4) Presciption Subroutine Block (PSB)
According to the results of DSB, alternate prescription file and medicine base, a condense form of the prescription is formed by PSB. This condense form is called the logical prescription. The logical prescription can be stored in the inner-storage.

(5) Print Presciption Subroutine Block (PPSB)
The function of this subroutine is to form, according to the logical presciption file and to print it.

4. APPLICATION TO DIAGNOSIS AND TREATMENT OF THE TRADITIONAL CHINESE MEDICINE

As a example, we have applied the software to diagnosis and treatment of the cough and asthma of children. Using this software we have examined over 100 patients and the results of diagnosis and treatment all are corret.

5. CONCLUSION

To a great extent, diagnosis and treatment of the traditional medicine are dependent on experience of the doctors. Therefore systematization of diagnosis and treatment of the traditional Chinese medicine is rather difficult. The software used here try to simulate the mode of thinking of the doctors.

The key point is that the herbal medicine and the symptoms are regarded as real time data. And diagnosis and treatment also become a process of data handling. Therefore, to bring the logical function of the computer into fall play, the systematization of the diagnosis and the treatment of the traditional Chinese medicine may be realized.

The philosophy of this software is to simulate the general diagnosis process, so it can also be used for any diagnosis process.

REFERENCE

1 Chou Shu-Yuan et al, A mathematical modal for traditional Chinese medicine computer diagnosis and treatment, <u>Information and Control,</u> vol.11,1982,24-25

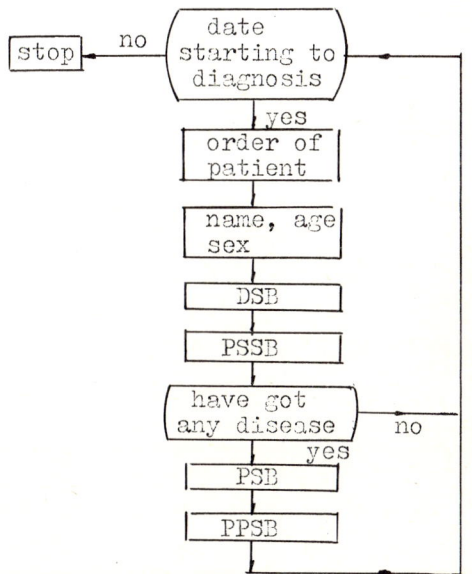

Fig. 1 Block diagram of main program

ORDERED BUS ACCESS BY LOW LEVEL TOKEN PASSING
FOR THE LOCAL NETWORK TOPAS

W. Wawer, K. Emmelmann, V. Tschammer

Hahn-Meitner-Institut für Kernforschung Berlin GmbH, Berlin (West)

ABSTRACT

In Ethernet-type local area networks, messages are transferred between stations via a single commonly used bus line. When several stations are going to access the communication channel simultaneously, the access is granted to the stations according to a non-deterministic algorithm. In particular cases message transmissions may be delayed to an extent which seems to be untolerable in real-time applications. Therefore, a control unit, which permits access to the common bus according to a deterministic algorithm, has been inserted between a commercially available Ethernet-type data link control board and the transceiver. The paper describes the access algorithm of the network TOPAS (fair ordered access by means of decentrally generated token messages, which start a time slice period) and the interaction of the data link and the access control unit.

ZUSAMMENFASSUNG

In lokalen Netzen, die nach dem Ethernet-Prinzip arbeiten, werden Nachrichten zwischen Stationen über ein gemeinsam benutztes Kabel übertragen. Wollen mehrere Stationen den Übertragungsweg gleichzeitig belegen, erfolgt die Gewährung des Zugriffsrechtes nach einem nicht-deterministischen Algorithmus. Dabei können im Einzelfall für eine Station hohe Wartezeiten auftreten, was bei Realzeitanwendungen nicht tragbar erscheint. Deshalb wird einer kommerziell verfügbaren Ethernet-Data Link-Steuerung eine Kontrolleinheit unterlegt, die das Zugriffsrecht zum Bus nach einem deterministischen Verfahren vergibt. Im folgenden wird der Buszugriffsalgorithmus des Netzwerks TOPAS (faire Gewährung des Zugriffsrechtes mit Hilfe dezentral erzeugter Token-Nachrichten, die eine Zeitscheibenfolge einleiten) und die Zusammenarbeit der Data Link-Steuerung mit der Buszugriffs-Steuerung beschrieben.

RÉSUMÉ

Dans les réseaux locaux de type Ethernet, les messages sont transférés entre les stations par l'intermédiaire d'un bus unique commun. Lorsque plusieurs stations doivent accéder simultanément au canal de communication, l'accès leur est garanti selon un algorithme non déterministe. Dans certains cas particuliers, les transmissions de messages peuvent être retardées à un point difficilement supportable pour des applications en temps réel. En conséquence, une unité de commande, qui permet l'accès au bus commun selon un algorithme déterministe, a été insérée entre le transmetteur et un système de commande de transfert de données du type Ethernet disponible dans le commerce.

Cette communication décrit l'algorithme d'accès du réseau TOPAS (accès ordonnancés au moyen de jetons générés à distance qui marquent le début de la période de temps partagé) et l'interaction entre le système de transfert de données et l'unité de contrôle des accès.

1. INTRODUCTION

In Ethernet-type local area networks /1/, messages are transferred between stations via a single commonly used bus line. CSMA/CD access control (Carrier Sense Multiple Access/Collision Detection) arbitrates contention of messages in a non-deterministic way: When a message collision has been detected by the transmitting station, the current message is aborted and a retransmission of this message is started after a randomly selected time /2, 3/. At heavy bus loads this time may increase drastically, and a worst case access time cannot be guaranteed. This may be intolerable for distributed process control systems, and an ordered access protocol with token control may be used as an alternative.

2. ORDERED ACCESS PROTOCOLS WITH TOKEN PASSING

2.1 High Level Token Passing

Ordered access protocols arbitrate contentions in a deterministic way by offering the opportunity to access the bus line explicitly to each individual station of the system. For this purpose, a specially marked message, called token, may be passed from station to station. A station, which is ready to transmit an information message, waits for a token message addressed to it, transmits its information message and then passes the token message to the next station according to an address list /4/. If both information and token messages are assembled and deassembled at the Data Link Layer of a station, the service of media access control (decision on receipt of a token message) will be performed at a relatively high level.

2.2 Low Level Token Passing

It seems to be advantageous, if the Data Link Layer can be relieved of the media access control service, because receipt, analysis and retransmission of a token message is an additional burden to the processing capacity at the Data Link Layer. Therefore, the access control should be performed at a lower layer, i.e. at a sublayer of the Physical Layer. At this layer there is typically no or only low computing power available. Token passing protocols must take this into account, e.g. handling of large address lists should be avoided.

2.3 TOPAS - Ordered Access Protocol with Individual Time Slice Assignment

The contention which of several requesting stations is permitted to access the transmission medium may be arbitrated in a deterministic manner by assigning individual time slices to each station. A station may start a transmission at its individually assigned time slice when the medium is free (see figure 1). When the communication channel is already busy, the station defers the transmission until the channel has become free and another time slice has been assigned to the station (listen before talk at a time slice). If the time slices are larger than two times the propagation delay of the system, collisions between messages are precluded /5/.

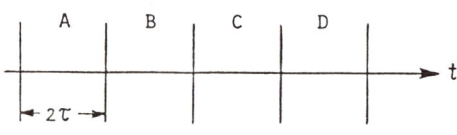

τ = Propagation delay of the system

a) Time Slice Allocation to Stations A,B,C,D

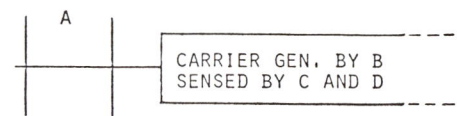

b) Start of a Transmission by Station B

Fig. 1 Access Arbitration by Time Slices

The time slices for all stations of the system are grouped together, thus representing a scheduling period (see figure 2a, b). The scheduling period is started by a specially encoded message (scheduling starter or token) which carries an address and which assigns the succession of time slices to successive station addresses. If the scheduling starter carries the address i, the first time slice may be allocated to the station address i+1 and the last one to the station address i, if for a total of K stations a mod K-count is assumed.

The scheduling period is stopped for all stations as soon as one of the K stations starts a message transmission at its time slice. When no station transmits a message, the scheduling starter is reissued after the scheduling period has elapsed.

A station with an address allocated to the first/last time slice has the highest/lowest priority for the channel access. If the address of the scheduling starter is changed on each scheduling period according to an appropriate algorithm, the access protocol will become fair to each station. A fair algorithm may assign the lowest priority to that station which has just transmitted a message (see figure 2b). In this case, the address of the scheduling starter is identical with the address of the last transmitting station. This is an ideal solution for decentralized access control: A station which relinquishes channel control generates a scheduling starter which carries the station address (source addressed scheduling starter, i.e. token).

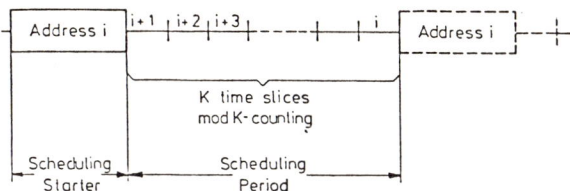

Fig. 2a Idle System – No Start of Transmission

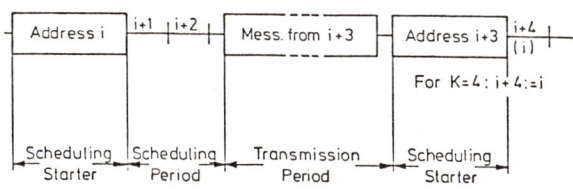

Fig. 2b Transmission of a Message

3. IMPLEMENTATION OF THE TOPAS CONTROL UNIT

The current implementation of the access method described above makes use of the DBC-100 data link control board of the Ethernet-type network DANUBE (developed by the project KAYAK at INRIA, France) in order to allow comparative investigations. The token passing control unit is inserted between the data link control board and the transceiver with a tap to the coaxial bus cable (see fig. 3).

Both information and token messages are transferred via the coaxial cable in a Manchester encoded form, but the messages have different preambles (by Manchester code violation) and can easily be distinguished by the token control unit. Information messages received are passed to the data link control board at any time without interference by the token control unit. Token messages received start the scheduling period and the calculation of the assigned time slice.

As long as no token has been received, the generation of information messages at the data link control board is inhibited by the simulation of a Bus Busy signal from the token control unit (using the carrier sense facility at the data link control board).

When a time slice allocated to this station is present, the Bus Busy simulation pauses, and the data link control board may transmit an information message which will be monitored by the token control unit. At the end of the information message, Bus Busy is simulated again and a token message is generated, which starts the next scheduling period.

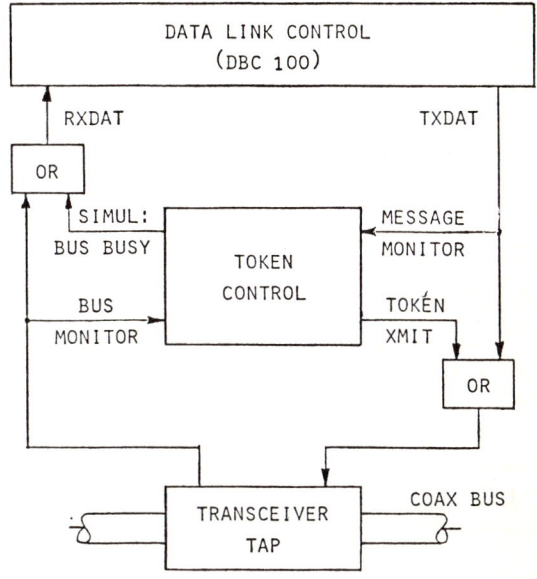

Fig. 3 Token Control at the Local Network TOPAS

The token control unit is an addition to the data link control unit which affects only the performance of the system. It remains invisible for the data link and higher protocols; thus these protocols need not to be modified.

4. REFERENCES

/1/ Metcalf, R.M., Boggs, D.R., Ethernet: Distributed Packet Switching for Local Computer Networks, Communications of the ACM, Vol. 19, No. 7 (July 1976).

/2/ DIGITAL, INTEL, XEROX, The Ethernet, A Local Area Network, Data Link Layer and Physical Layer Specifications, Version 1.0 (September 30, 1980).

/3/ IEEE 802 DLMAC Working Group on CSMA/CD, IEEE 802 CSMA/CD Media Access Standard (January 1981).

/4/ Token Ad Hoc Working Group, IEEE Project 802, DLMAC Subcommittee, "Token Structure", An Architecture, Revision 1 (January 21, 1981)

/5/ Chlamtac, I., Franta, W.R., Levin, K.D., BRAM: The Broadcast Recognizing Access Method, IEEE Transactions on Communications, Vol. COM-27, No. 8 (August 1979)

CLOSING ADDRESS

REAL-TIME DATA '82
CLOSING SPEECH

Palle Christensen
Risø National Laboratory, Denmark

The Symposium is coming to an end now, but before we leave for the cocktails I should like to address you shortly on behalf of the three organisations which took the initiative in arranging the Symposium. They are EWICS, ESONE, and ECA. Since we do not all of us want to speak, we made a small token passing, and the token eventually came to me being the present chairman of the ECA.

We have been together here for three days - people from informatics research and people from the real time instrumentation field, both interested in standards and common practices. I believe that we have reached our main goal, the exchange of experience. There have been many challenging topics and we have had excellent conditions of discussions, both during the Symposium and at the lunch table.

In connection with the talks of the opening session it is important to remember the words of Ken Thompson of the CEC. In my interpretation the message was as follows: "If you know what you want to do in informatics, the CEC can help you". It also became clear that the three organisations should collaborate more closely with the ISO in the future.

We have been led through a very interesting programme. We have heard about advanced architecture and distributed systems; time multiprocessing was touched, but it is clear that some software stantardization is missing here. About programming languages many ideas were presented, but do not forget that most RT programmes are written in DEC-FORTRAN.

From the debate about the software crisis I shall cite Rupert Patzelt: "The expectations have been too high". I agree and can add that I feel that this lack of software experts is not so bad for society - it leaves time for adapting.

We have heard about methods for software development, where a total environment concept is appearing now. For me the papers about PILS and EPOS were of special interest. Applications in control and automation were presented in a veritable forum for exchange of information. The conformity testing session was a kind of demystification for me; it seems to be an area where great and rapid progress is being made; certification of software seems to be the final goal. Networking is an area with great expectations; much work will have to be carried out in the upper levels of the ISO-model, also on gateways between networks. In the area of reliability, new techniques are giving us means for much safer software controlled steering.

This was the second symposium; a third one will be needed in some years' time to give us the chance to identify the status of all these challenging topics again. Please fill in the Symposium questionnaire - it will help the organisers in future planning.

Let me conclude by thanking all who contributed to the success of this symposium. First of all the CEC and the French Ministry of Research and Industry working through INRIA and BNI for sponsoring the Symposium, together with the international standardization organisations IEC and ISO. Thanks also to the International Organising Committee, the Local Organising Committee, and the Programme Committee headed by Ted Kingham, Nicolas Malagardis, and Gerard Le Lann, respectively. The interpreters did a great job and must be quite exhausted by all these EDP details. We also thank the INRIA and BNI staff for their help and hospitality with a special thanks to Miss Therese Bricheteau for a perfect organisation. Let me finally thank all invited speakers, all authors, all chairmen and finally all of you, the participants of the Symposium.

LIST OF PARTICIPANTS

LISTE DES PARTICIPANTS

TEILNEHMERLISTE

Mr. ABBES Mohammed	College de France 11, Place Marcelin Berthelot 75231 PARIS Cedex 05	FRANCE
Mr. AICHOUN Jean Marc	T.R.T. 5, Avenue Reaumur 92350 LE PLESSIS ROBINSON	FRANCE
Mr. ALANKO Yrjo	Outokumpu Oy Engineering Division P.O. Box 27 SF-02210 ESPOO 20	FINLAND
Mr. ALBOT Andre	SEMS 1, rue de Provence 38130 ECHIROLLES	FRANCE
Mr. ALLEGRE Maurice	Ministere de la Recherche et de l'Industrie Directeur de la DESTI 1, rue Descartes 75005 PARIS	FRANCE
Mr. ANDERSON Thomas	University of Newcastle upon Tyne Claremont Tower, Claremont Road NEWCASTLE UPON TYNE NE1 7RU	GREAT BRITAIN
Mr. ANTOINE Pierre	CEN Saclay 91191 GIF FUR YVETTE Cedex	FRANCE
Mr. ANTONOV Lubomir	Lano Central d'Automatisation 72, bul Lenin 1184 SOFIA	BULGARIA
Dr. ATTWENGER Wolfgang	Osterreichische Forschungszentrum Lenaugasse 10 A-1082 WIEN	AUSTRIA
Mr. AVON Anthony	Royal Greenwich Observatory Hertsmonceux Castle HAILSHAM, E. Sussex	GREAT BRITAIN
Mr. BACONNET Jean Pierre	CERCI 56, rue R. Salengro 94120 FONTENAY SS BOIS	FRANCE
Mr. BASTIE Bernard	THOMSON CSF 33, rue de Vouille 75015 PARIS	FRANCE

Mr. BELKHATIR Noureddine	Commissariat Nal a l'Informatique RN 5 Cinq Maisons EL HARRACH	ALGERIA
MR. BENABEN Albert	SEMS 36, rue da la Princesse 78430 LOUVECIENNES	FRANCE
Mr. BENNETT Philip	York Systems Consultants Ltd. 2, Brigg Road, Broughton Brigg S HUMBERSIDE	GREAT BRITAIN
Mr. BERRY G.	ENSMP Sophia Antipolis 06560 VALBONNE	FRANCE
Mr. BEYER Jorg	EUROCONTROL Rue de la Loi 72 B-1040 BRUXELLES	BELGIUM
Mr. BLOCH Armand	CEA Orme des Merisiers 91191 GIF SUR YVETTE Cedex	FRANCE
Mr. BOCK Norbert	Osterreichisches Forschungszentrum Lenaugasse 10 A-1082 WIEN	AUSTRIA
Mr. BORRIUS W.C.	Vrije Universiteit Amsterdam De Boelelaan 1081 1081 HV AMSTERDAM	NETHERLANDS
Mr. BOYCE Desmond	UK Atomic Energy Authority Bldg. 418 - Aere Harwell DIDCOT, Oxon OX11 ORA	GREAT BRITAIN
Mr. BRUCKNER Gottwalt	Badenwerk AG Badenwerkstr. 2 7500 KARLSRUHE 1	FED. REP. GERMANY
Mme. BRUN Nicole	CEN Saclay 91191 GIF SUR YVETTE Cedex	FRANCE
Mr. BUFFET Jean	Credit Lyonnais Tour Credit Lyonnais Cedex 10 92081 PARIS LA DEFENSE	FRANCE
Mr. BULL Gordon	The Hatfield Polytechnic PO Box 109 College Lane HATFIELD, Herts.	GREAT BRITAIN
Mr. BURGER Josef	SIEMENS AG Oestl. Rheinbrueckenstr. 50 7500 KARLSRUHE 21	FED. REP. GERMANY

List of Participants

Mr. CAIN Gerald	Polytechnic of Central London 115 New Cavendish St. LONDON WIM 8JS	GREAT BRITAIN
Mr. CAMUS Christian	SFENA Aerodrome de Villacoublay BP 59 78141 VELIZY VILLACOUBLAY	FRANCE
Mr. CHAUVIN Jean-Jacques	SESA 30, Quai de Dion Bouton 92806 PUTEAUX Cedex	FRANCE
Mr. CHERNYKH E.	Joint Institute for Nuclear Res. Head PO Box 79 101000 MOUSCOU	USSR
Mr. CHRISTENSEN Palle	Riso National Laboratory DK-4000 ROSKILDE	DENMARK
Mr. CLAYTON Mike	CERN 1211 GENEVE 23	SWITZERLAND
Mr. CLUZEL	CIT ALCATEL 33, rue Emeriau 75725 PARIS Cedex 15	FRANCE
Mr. COHEN Andre	CEA DRFC SCP CENFAR BP 6 92260 FONTENAY AUX ROSES	FRANCE
Mr. COMES Renato	SYSDATA Team SPA Corso Marconi 13	ITALY
Mr. CORCIA Yvon	CNET 22301 LANNION	FRANCE
Mr. CROUILLERE Michel	IRSID 185, rue du Pt Roosevelt 78100 ST GERMAIN EN LAYE	FRANCE
Mme. DARONIAN Denise	CEA 29-33, rue de la Federation 75015 PARIS	FRANCE
Mme. DELGADO Jacqueline	CIT ALCATEL 10, rue Latecoere 78140 VELIZY	FRANCE
Dr. DEMETROVICS Janos	Inst. de Recherche de Hongries Kende u.13-17 1111 BUDAPEST	HUNGARY
Mr. DESCLAUD Patrice	CNET/LAA/SLC-PGL Route De Tregastel 22301 LANNION	FRANCE

Mr. DIRKER Lee	Technische Hogeschool Delft Mekelweg 15 DELFT	NETHERLANDS
Mr. DOUCET Rene	TELEMECANIQUE Chemin du Vieux Chene 38240 MEYLAN	FRANCE
Mr. DUCORPS Antoine	Labo de l'Accelerateur Lineaire Centre d'Orsay 91405 ORSAY Cedex	FRANCE
Dr. DURCANSKY Georg	KFA-IFF Leitung von Elektroniklabor D-5170 JUELICH	FED. REP. GERMANY
Mr. EBERT-KNUDSEN Per	ELKRAFT Power Company Ltd. 5, Lautruphoj 2750 BALLERUP	DENMARK
Mr. EHRENBERGER Wolfgang	Gesellschaft fur Reaktorsicherheit Forschungsgelaende D-8046 GARCHING	FED. REP. GERMANY
Pr. EICHER Lawrence D.	ISO 1, rue de Varembe 1211 GENEVE 20	SWITZERLAND
Mr. ELLOY Jean Pierre	ENSM 1, rue Noe 44072 NANTES Cedex	FRANCE
Mr. FANGEL-JENSEN Jorgen	ELKRAFT Power Company Ltd. 5, Lautruphoj 2750 BALLERUP	DENMARK
Mr. FANGMEYEV Hermann	Kommission der Europerischen Gemei Casella Postale 1 21020 ISPRA	ITALY
Mr. FEBVRE Jacques	SEMS 1, rue de Provence 38130 ECHIROLLES	FRANCE
Mr. FEDERIGI Ferdinando	Via S. Franceschino 17 55040 RIPA-SERAVEZZA (LU)	ITALY
Mr. FEST Raymond	Universite de Haute Alsace 4, rue des Freres Lumiere 68100 MULHOUSE	FRANCE
Mr. FOLIGUET Gerard	Thomson CSF 52, rue Guynemer 92130 ISSY LES MOULINEAUX	FRANCE
Mr. FORSTER Michael	Science & Eng. Research Council Chilton, DIDCOT, Oxon OXII OQX	GREAT BRITAIN

List of Participants

Mr. FOUQUE Gerard	Ecole Polytechnique Route de Saclay 91128 PALAISEAU Cedex	FRANCE
Mr. FRUMAU Carel F.A.	Netherlands Energy Research Foundation PO Box 1 1755 ZG PETTEN	NETHERLANDS
Mr. GAGEY Bernard	CIMSA 12, Avenue de l'Europe 78140 VELIZY	FRANCE
Mr. GAGLIARDI Fabrizio	CERN 1211 GENEVA 23	SWITZERLAND
Mr. GALLICE Pierre	CEN Saclay BP 2 91191 GIF SUR YVETTE	FRANCE
Mr. GANDER Jean Gabriel	Borer Electronics AG Postfach, 4501 SOLOTHURN	SWITZERLAND
Mr. GANDOLFI Giorgio	Centro Sperimentale Metallurgico Via di Castel Romano 100 ROMA	ITALY
Mr. GERNER C.P.	Vrije Universiteit Amsterdam De Boelelaan 1081 1081 HV AMSTERDAM	NETHERLANDS
Mr. GIEN Michel	INRIA Rocquencourt Domaine de Voluceau B.P. 105 78153 LE CHESNAY Cedex	FRANCE
Dr. GILOI Wolfgang K.	Technical University of Berlin Franklinstr. 28/29 1000 BERLIN 10	FED. REP. GERMANY
Mr. GOEDICKE Michael	Universitat Dortmund Postfach 500 500 D-4600 DORTMUND 50	FED. REP. GERMANY
Mr. GOMEZ-GONZALEZ Jose	Junta de Energia Nuclear Avda Complutense, 22 3 MADRID	SPAIN
Mr. Gorry Alfred	CEN Saclay 91191 GIF SUR YVETTE Cedex	FRANCE
Mr. GOURCY Georges	CEN de Saclay DPHN/AL 91191 GIF SUR YVETTE Cedex	FRANCE
Mr. GROF J.	Inst. for Electric Power Research VEIKI zrinyi u.1. 1051 BUDAPEST	HUNGARY

Mr. GUERIN Jean Pierre	CEA 91191 GIF SUR YVETTE Cedex	FRANCE
Mr. HAASE Volkmar	Techn. Universitaet Graz Schiesstattg 4A 8010 GRAZ	AUSTRIA
Mr. HAINICH Rolf	Hahn-Meitner Institut Glienicker Str. 100 D-1000 BERLIN 39	FED. REP. GERMANY
Dr. HALLING H.	K.F.A. Postfach 1913 5170 JULICH 1	FED. REP. GERMANY
Mr. HAMEL Jean Louis	CEN SACLAY 91191 GIF SUR YVETTE	FRANCE
Mr. HARROCH Henri	IPN BP 1 91406 ORSAY	FRANCE
Mr. HASKELL P.	Rutherford Appleton Laboratory Chilton DIDCOT, Oxon OX11 OQX	GREAT BRITAIN
Mr. HENNION Francois	CEA DREC/SCP BP 6 92260 FONTENAY AUX ROSES	FRANCE
Dr. HERRMANN Gerhard	Boehringer Ingelheim KG Binger Strasse 6507 INGELHEIM	FED. REP. GERMANY
Mr. HUET Jacques	CEA BP n.7 93270 SEVRAN	FRANCE
Mr. HULOT Jean Paul	CEA CEN Saclay BP 2 91191 GIF SUR YVETTE	FRANCE
Mr. ISAY Jean Paul	COGI-CGO 22, rue d'Arras 92000 NANTERRE	FRANCE
Mr. ISSALY Alain	SINTRA Alcatel 1, Avenue Aristide Briand 94110 ARCUEIL	FRANCE
Mr. JONES Derek	Science & Engineering Research Council Bldg. R 63 Chilton DIDCOT, Oxfordshire OX11 OQX	GREAT BRITAIN
Mr. JONES Lewis R.	Royal Greenwich Observatory Hertsmonceux Castle HAILSHAM, E. Sussex	GREAT BRITAIN

List of Participants

Mr. JONES Peter E.	ISO 1, rue de Varembe case postale 56 1211 GENEVE 20	SWITZERLAND
Mr. KAISER Josef	CEA 91191 GIF SUR YVETTE Cedex	FRANCE
Dr. KERN H.	Technische Universitaet Wien Gusshausstr. 25 1040 VIENNE	AUSTRIA
Mr. KINGHAM Edward G.	Central Electricity Research Lab Kelvin Avenue LEATHERLAND, Surrey KT22 7SE	GREAT BRITAIN
Mr. KLESSE Reinhard	ILL-Grenoble BP 156 38042 GRENOBLE Cedex	FRANCE
Mr. KOHLER Christian	Industrieanlagen-Befriebsgesellsch Einsteinstrasse Geb. 21 8012 OTTOBRUNN	FED. REP. GERMANY
Mr. KOPETZ Hermann	TU-Berlin/TU-Wien Gubhausstrasse 30 1040 WIEN	AUSTRIA
Mr. KRUITHOF Albert	Cie Generale de Const. Teleph. 251, rue de Vaugiarard 75015 PARIS	FRANCE
Mr. LACOSTE Jean	Les Anciens Combattants BP 246 75264 PARIS Cedex 06	FRANCE
Mr. LAUTENBAG Cornelis	Netherlands Energy Research Foundation PO Box 1 1755 ZG PETTEN	NETHERLANDS
Mr. LE BARAILLEC Francois	S.N. LOGABAX Quartier des Epinettes 91000 EVRY	FRANCE
Mr. LE LANN Gerard	INRIA Rocquencourt Domaine de Voluceau B.P. 105 78153 LE CHESNAY Cedex	FRANCE
Mr. LECORCHE Eric	GANIL BP 5027 14021 CAEN Cedex	FRANCE
Mr. LEJUSTE Jean Marie	CGEE ALSTHOM 51, rue des Trois Fontanot BP 202 92002 NANTERRE Cedex	FRANCE

Mr. LEWIS Alan	UK Atomic Energy Authority DIDCOT, Oxon OX11 ORA	GREAT BRITAIN
Pr. LIONS J.L.	INRIA Domaine de Voluceau B.P. 105 78153 LE CHESNAY Cedex	FRANCE
Mr. LOHNERT Frieder	Technische Universitat Berlin Fachbereich 20 Franklinstr. 28/29 1000 BERLIN 10	FED. REP. GERMANY
Mr. LOUVET Olivier	CNET 22301 LANNION	FRANCE
Mr. LUONG Thanh-Tam	GANIL BP 5027 14021 CAEN Cedex	FRANCE
Mr. LUPTON W.	Royal Greenwich Observatory Herstmonceux Castle HAILSHAM, Sussex BN27 IRP	GREAT BRITAIN
Mr. MAGERSKY Petko	State Committee for Science 8 Slavjanska str. 1040 SOFIA	BULGARIA
Mr. MALAGARDIS Nicolas	BNI Domaine de Voluceau BP 105 Rocquencourt 78153 LE CHESNAY Cedex	FRANCE
Mr. MAREL Philippe	Electronique Actualites 49, rue de l'Universite 75007 PARIS	FRANCE
Mr. MARTIN Ralph	Royal Greenwich Observatory Hertsmonceux Castle HAILSHAM, E.Sussex	GREAT BRITAIN
Mr. MATRICON Pierre	LPNHE 91128 PALAISEAU Cedex	FRANCE
Mr. MENARD Jean Pierre	CREDIT LYONNAIS Tour Credit Lyonnais 92081 PARIS LA DEFENSE Cedex 10	FRANCE
Mr. MENCIK Maurice	Labo de l'Accelerateur Lineaire Bat 200 91405 ORSAY	FRANCE
Mr. MENENDEZ Claudio	Institut National des Sciences Appliquees 20, Avenue des Buttes de Coesmes 35043 RENNES Cedex	FRANCE
Mr. MEYER Horst	CEC Joint Research Centre B-2440 GEEL	BELGIUM

Mr. MEZCHICHE Mohammed	Commissariat Nal a l'Informatique Route Nationale 5 Cinq Maisons EL HARRACH	ALGERIA
Mr. MOLINARO Pierre	ENSM 1, rue de la Noe 44072 NANTES Cedex	FRANCE
Mr. MORLING Richard	Polytechnic of Central London 115 New Cavendish St LONDON WIN 8JS	GREAT BRITAIN
Mr. MOSSETTI Michel	Lal Orsay 91405 ORSAY	FRANCE
Mr. MOULIN Dominique	SINTRA ALCATEL 1, Avenue Aristide Briand 94117 ARCUEIL Cedex	FRANCE
Mme. MULLER Eveline	THOMSON-CSF 40, rue Grange Damerose 92000 MEUDON	FRANCE
Mr. MULLER Hans	CERN 1211 GENEVA 23	SWITZERLAND
Mr. MULLER Klaus-Dieter	Kernforschungsanlage Julich GmbH Postfach 1913 5170 JULICH	FED. REP. GERMANY
Mr. MUNTEAN Traian	IMAG BP 53X 38041 GRENOBLE Cedex	FRANCE
Mr. NERI Giovanni	Bologna University Viale Risorgimento 2 BOLOGNA	ITALY
Mr. NGUYEN NGOC Chan	LAL/IN2P3 91405 ORSAY	FRANCE
Mr. NGUYEN VAN DUONG Francois	CEN Saclay 91191 GIF SUR YVETTE	FRANCE
Ms. NORRIE Moira	The Hatfield Polytechnic PO Box 109 College Lane HATFIELD, Herts.	GREAT BRITAIN
Mr. NUNNS Stuart Russell	Imperial Chemical Industries PLC Wilton Works Middlesbrough CLEVELAND TS6 8JA	GREAT BRITAIN
Mr. OKSMAN Jacques	Ecole Superieure d'Electricite Plateau du Moulon 91190 GIF SUR YVETTE	FRANCE

Mr. OSTERWEIL Leon	University of Colorado Campus Box 430 BOULDER, Colorado 80309	UNITED STATES
Mr. OUERGHI Mohamed Said	CRIN Campus Scientifique Nancy I BP 239 54506 VANDOEUVRE LES NANCY Cedex	FRANCE
Mme. PAIN Colette	EDF 1, Avenue du Gal de Gaulle 92141 CLAMART Cedex	FRANCE
Mr. PAIN Jacques	CEN Saclay 91191 GIF SUR YVETTE	FRANCE
Mr. PAPILLON Andre	SNCF Paris Sud Ouest 1, Place Valhubert 75013 PARIS	FRANCE
Pr. PATZELT R.	Technische Universitaet Wien Gusshausstrasse 25 1040 VIENNE	AUSTRIA
Mr. PAUTHNER Georg	Technische Universitat Berlin Pariser Str. 53 1000 BERLIN 15	FED. REP. GERMANY
Mr. PAUTON Michel	CEA BP n 7 93270 SEVRAN	FRANCE
Mr. PEFFER Jacques	Laboratoire Central SOLVAY 310, rue de Ransbeek 1120 BRUXELLES	BELGIUM
Mr. PENNANEC'H Jean Claude	CIT ALCATEL 10, rue Latecoere 78140 VELIZY	FRANCE
Mr. PERRIN Yves	CERN Recherche Nucleaire Division DD 1211 GENEVE 23	SWITZERLAND
Mr. PFEUTY Patrice	Institut de Recherche de la Siderurgie 185, Avenue du Pt Roosevelt 78100 ST GERMAIN EN LAYE	FRANCE
Mr. PIQUET Bruno	CEN Saclay 91191 GIF SUR YVETTE	FRANCE
Mr. POPESCU-ZELETIN R.	Hahn-Meitner Institut Berlin Glienickerstr 100 1 BERLIN 39	FED. REP. GERMANY
Mr. PROME Michel	GANIL BP 5027 14021 CAEN Cedex	FRANCE

Pr. PYLE Ian C.	University of York Heslington YORK YO1 5DD	GREAT BRITAIN
Mr. QUERASSER Edwin	Osterreichisches Forschungszentrum Lenaugasse 10 A-1082 WIEN	AUSTRIA
Mr. RABANY Michel	CERN 1211 GENEVE 23	SWITZERLAND
Mr. RAFFI Gianni	European Southern Observatory 8046 GARCHING bei MUNCHEN	FED. REP. GERMANY
Mr. RAMBOZ Herve	SLIGOS 91, rue Jean Jaures 92807 PUTEAUX	FRANCE
Mr. REASON Christopher	Science & Eng. Research Council Chilton DIDCOT, Oxon OX11 0QX	GREAT BRITAIN
Mr. REYNDERS Deon	Trasys Rue d'Arlon 88 1040 BRUXELLES	BELGIUM
Mr. REYNIER Patrick	CEA BP n 7 93270 SEVRAN	FRANCE
Mme. RIMMER Elsie Margaret	CERN 1211 GENEVE 23	SWITZERLAND
Mme. ROLLAND C.	Universite Paris I 12, Place Jussieu 75231 PARIS Cedex 05	FRANCE
Mr. ROUGER Michel	CEA 91191 GIF SUR YVETTE	FRANCE
Dr. RUDIGER Ulrich	Scientific Control Systems GmbH 3 Hagen 43 D-43 ESSEN 1	FED. REP. GERMANY
Dr. RZEHAK Helmut	Hochschule der Bundeswehr Munchen Werner-Heisenberg-Weg 39 8014 NEUBIBERG	FED. REP. GERMANY
Mr. SABAN Roberto	CERN 1211 GENEVE 23	SWITZERLAND
Mr. SAIKAWA Natsuki	KAWASAKI Steel Corp. Chiba Works 1, Kawasaki-cho 260 CHIBA	JAPAN
Mr. SAVOYSKY	Laboratoire des Ponts et Chaussees 58, bd Lefebvre 75732 PARIS Cedex 15	FRANCE

List of Participants

Mr. SCHEUB Volker	Universitaet Stuttgart Seidenstr 36 D-7000 STUTTGART 1	FED. REP. GERMANY
Mr. SCHMIDT Volker	JET JOINT UNDERTAKING ABINGDON OX14 3EA	GREAT BRITAIN
Mr. SCHOBERT Daniel	CEA - CEN 91191 GIF SUR YVETTE	FRANCE
Mr. SCHOITSCH Erwin	Osterreichisches Forschungszentrum Lenaugasse 10 A-1082 WIEN	AUSTRIA
Mr. SCHOUKROUN Claude	CEA Orme des Merisiers 91191 GIF SUR YVETTE	FRANCE
Mme. SCHWAB Helga	ILL-Grenoble BP 156 38042 GRENOBLE Cedex	FRANCE
Mr. SCHWEINZER H.	Technische Universitaet Wien Gusshausstr. 25 1040 VIENNE	AUSTRIA
Dr. SCHWEPPE Heinz	Inst. fur Theorische Prakt Informat Gauss Str. 11 D-3300 BRAUNSCHWEIG	FED. REP. GERMANY
Mr. SCOWEN Roger	National Physical Laboratory TEDDINGTON, Middx. TW11 OLW	GREAT BRITAIN
Mme. SEDILLOT Simone	INRIA Rocquencourt Domaine de Voluceau B.P. 105 78153 LE CHESNAY Cedex	FRANCE
Mr. SEIPP Gundolf	CARL ZEISS Postfach 1369/1381 D-7082 OBERKOCHEN	FED. REP. GERMANY
Mr. SELLEM Robert	IPN BP 1 91406 ORSAY	FRANCE
Mr. SERIEYS Christian	Association d'Audit Informatique 8, rue de la Michodiere 75002 PARIS	FRANCE
Mme. SIDI J.	INRIA Rocquencourt Domaine de Voluceau B.P. 105 78153 LE CHESNAY Cedex	FRANCE
Mme. SILLY Maryline	ENSM 1, rue de la Noe 44072 NANTES Cedex	FRANCE

List of Participants

Mr. SIMON Yves	CNET 22301 LANNION	FRANCE
Mr. SIRATTANA Houmphanh	Serieys Consult 8, rue de la Michodiere 75002 PARIS	FRANCE
Mr. SPILLNER Andreas	Technische Universitat Berlin Fachbereich 20 Franklinstr. 28/29 1000 BERLIN 10	FED. REP. GERMANY
Mr. STANZEL Bertram	Beschleunigerlabor der Univ. Munch am Coulombwall D-8046 GARCHING	FED. REP. GERMANY
Mr. TAGESEN Siegfried	CEA 91191 GIF SUR YVETTE Cedex	FRANCE
Mr. TAILLIBERT Patrick	Electronique Marcel DASSAULT l'Information 55, Quai Carnot 92214 SAINT CLOUD	FRANCE
Dr. TAYLOR J.R.	Riso National Laboratory 4000 ROSKILDE	DENMARK
Mr. TAYLOR Philip	Royal Greenwich Observatory Hertsmonceux Castle HAILSHAM, E. Sussex BN27 IRP	GREAT BRITAIN
Mr. THOMPSON Kenneth	CCE rue de la Loi, 200 1049 BRUXELLES	BELGIUM
Mme. TRECA Catherine	Ecole Polytechnique 91128 PALAISEAU	FRANCE
Mr. TRIOLAIRE Christian	SEMS 1, rue de Provence 38130 ECHIROLLES	FRANCE
Mr. TSCHAMMER Volker	Hahn-Meitner Institut Glienicker Str. 100 D-1000 BERLIN 39	FED. REP. GERMANY
Mr. VERROUST Gerard	IPN BP 1 91406 ORSAY	FRANCE
Mr. Vogt Alois	CARL ZEISS Postfach 1369/1381 D-7082 OBERKOCHEN	FED. REP. GERMANY
Mr. WANG ZHI-BAO	Nankai University TIANJIN	PEOPLE'S REP. CHINA
Dr. WARD Stuart	Central Electricity Generating Board 15, Newgate Street LONDON EC1A 7AX	GREAT BRITAIN

Mr. WAWER W.	Hahn-Meitner Institute fuer DTV und Electronik Glienicker strasse 100 1000 BERLIN 39	FED. REP. GERMANY
Mr. WAWER Walter	Hahn-Meitner Institut Glienicker Str. 100 D-1000 BERLIN 39	FED. REP. GERMANY
Mr. WEILL J.	CEN 92191 GIF SUR YVETTE	FRANCE
Mr. WILK P.	Edinburgh University Forest Hill EDINBURGH	GREAT BRITAIN
Mr. WILLIAMS David	CERN CH-1211 GENEVE 23	SWITZERLAND
Mr. WOOD Graeme	Foxboro Company REDHILL, Surrey RH1 2HL	GREAT BRITAIN
Mr. ZALEWSKI Janusz	Instytut Badan Jadrowych ul, Dorodna 16 03195 WARSZAWA 91	POLAND
Pr. ZANDER Karl	Hahn-Meitner Institut Glienicker Str. 100 D-100 BERLIN 39	FED. REP. GERMANY
Mr. ZIMMERMAN H.	CNET 38-40 rue du General Leclerc 92131 ISSY les MOULINEAUX	FRANCE

AUTHOR INDEX

Albin, P., 103
Anderson, T., 327
Ansaldi, W., 277

Banasik, Z., 159
Barrelet, E., 175, 189
Berry, G., 111
Beyer, J., 339
Beyschlag, U., 75
Böck, N., 39
Buschbeck, F., 265

Camerini, J., 111
Chernykh, E.V., 251
Christensen, P., 369
Clayton, M.J., 75
Cojan, A., 347
Corcia, Y., 31, 131

Dauphin, J.L., 31, 131
David, L., 179
De Kerday, K., 175
Din, X.-W., 359

Eicher, L.D., 5
Elloy, J.P., 121
Emmelmann, K., 273, 363
Ezure, H., 165

Faro, A., 277
Faulle, C., 57
Federigi, F., 225
Fouque, G., 175
Froger, G., 31

Gagey, B., 57
Gagliardi, F., 193
Giloi, W.K., 15
Grof, J., 71
Guerin, J.P., 197

Hainich, R., 293
Heller, J., 339
Huet, J., 197

Jeandel, M., 57

Kern, H., 25, 343
Kingham, E.G., v, 3
Köhler, C., 139
Kopetz, H., 49

Lamarche, G., 239
Lecorche, E., 179
Le Lann, G., v
Lindner, M., 265
Lohnert, F., 49
Louvet, O., 31
Lu, G.-Z., 359
Luong, T.T., 179

Malagardis, N.E., v
Marbot, R., 189
Marmorat, J.P., 111
Matricon, P., 189
Merker, W., 49
Mikuriya, T., 165
Minematsu, T., 165
Mirabella, O., 277
Molinaro, P., 121
Moulding, M.R., 327
Müller, H., 287

Nitta, J., 165
Nguyen Phuoc, B., 111
Norrie, M., 329

Olobardi, M., 277
Osterweil, L., 89

Patel, S., 75
Patzelt, R., 25
Pauthner, G., 49
Pauton, M., 197
Perrin, Y., 175
Piquet, B., 179
Popescu-Zeletin, R., 257
Preineder, H., 265
Prome, M., 179
Pyle, I.C., 81

Querasser, E., 265

Rabany, M., 75, 347
Révész, G., 71
Rigault, J.P., 111
Rolland, C., 103
Rüdiger, U., 317
Russell, R.D., 147

Saban, R.I., 75, 347
Saikawa, N., 165
Savoysky, S., 201
Scharff-Hansen, P., 175
Scheub, V., 151
Schmidt, V., 205
Schoitsch, E., 351
Schweinzer, H., 355
Schweppe, H., 43
Sciré, M., 193
Scowen, R.S., 211
Sidi, J., 235
Simon, Y., 31, 131
Spandafora, I., 225
Stiege, G., 43

Takechi, T., 165
Taillibert, P., 239
Tamiya, T., 165
Taylor, J.R., 305

Thompson, K., 9
Thys, A., 75
Tremblet, L., 175
Triolaire, Ch. 247
Tschammer, V., 273, 363

Ulrich, M., 179

Vascotto, A., 193
Viallevieille, A., 57

Wang, S.-X., 359
Wang, X.-F., 359
Wang, Z.-B., 359
Wawer, W., 273, 363
White, V., 193
Wilk, P., 329
Williams, D.O., 147
Wood, G.G., 67

Yoshie, K., 165

Zalewski, J., 159
Zarándy, E., 71

RAYMOND H. FOGLER LIBRARY

DATE DUE

BOOKS ARE SUBJECT TO
RECALL AFTER TWO WEEKS

DEC 2 1 1990